# Animal Signals and Communication

Volume 3

*Series editors*

Vincent M. Janik
School of Biology
University of St Andrews
Fife, UK

Peter McGregor
Centre for Applied Zoology
Cornwall College
Newquay, UK

For further volumes:
http://www.springer.com/series/8824

Reginald B. Cocroft · Matija Gogala
Peggy S. M. Hill · Andreas Wessel
Editors

# Studying Vibrational Communication

Springer

*Editors*
Reginald B. Cocroft
Division of Biological Sciences
University of Missouri-Columbia
Columbia, MO
USA

Matija Gogala
Slovenian Academy of Sciences and Arts
Ljubljana
Slovenia

Peggy S. M. Hill
Faculty of Biological Sciences
University of Tulsa
Tulsa, OK
USA

Andreas Wessel
Museum für Naturkunde, Leibniz-Institut für Evolutions- und Biodiversitätsforschung
Humboldt-Universität zu Berlin
Berlin
Germany

and

Zoologisches Museum
Universität Hamburg
Hamburg
Germany

ISSN 2197-7305 ISSN 2197-7313 (electronic)
ISBN 978-3-662-43606-6 ISBN 978-3-662-43607-3 (eBook)
DOI 10.1007/978-3-662-43607-3
Springer Heidelberg New York Dordrecht London

Library of Congress Control Number: 2014942315



Printed on acid-free paper

Springer is part of Springer Science+Business Media (www.springer.com)

*To*
*Hildegard Strübing*
(8 May 1922–18 May 2013)

# Foreword

## The Emerging Field of Tremology

Vibrations and chemicals are the oldest modes of communication and both probably evolved from the original cell–cell mechanical and chemical interactions within early metazoans. In spite of vibrational communication being so widespread (more than 230,000 arthropods species and many vertebrates too), it is probably the least commonly known and least well known of all the sensory modes. This is partially because it is often lumped with auditory communication, with which it is related. It may also be because it is not as familiar to humans as other sensory modes (we only use it in the final stages of mating), and most people may not even think to ask if an animal is using vibrational communication, especially if it is using another form of communication very conspicuously. The possible vibratory communication through perches of night roosting birds is a good example. Scientific social facilitation is another possible reason; the authors of this book represent a significant fraction of people working in the field; other sensory modes are being researched by 1–2 orders of magnitude more people. Perhaps one reason is that there is not a word describing what we do; I suggest the term "tremology" (the study of tremors, vibrations, etc.). This book aims to redress this absurd lack of attention to an absolutely fascinating subject.

Vibratory communication is distinguished from auditory communication in that it is transmitted through solids or the air–water interfaces rather than only through air or water. This may sound like a small and arbitrary difference but it has major consequences for signal design and content. The major difference between vibrations and sounds is that sounds travel long distances through homogeneous media, whereas vibrations generally do not or travel shorter distances before losing their detectability. The distance between any impedance changes in air or water takes place over scales of thousands of wavelengths and hundreds or thousands of meters for sound. In contrast, changes in impedance take place on very small scales in solids, even in less than a wavelength. This means that the effects of the communication environment on the efficacy and evolution of signals are potentially much greater for vibrations compared to sound. In fact, vibratory communication may be the best model system for examining and predicting the effects of the environment on the function and evolution of signals, precisely because environmental effects are so strong.

This book provides an encyclopedic introduction to our current (2014) knowledge of vibratory communication or tremology. It covers the amazing diversity of vibration generating mechanisms, signal forms, receptors, neuroscience of reception, and some signal processing, in a variety of arthropods and vertebrates. Mollusks and "minor" phyla are not covered but this may be more a matter of unasked questions than a lack of vibratory use. Plants are not covered either, but vibrations are known to trigger pollination mechanisms in many taxa, and could possibly be used by vines for climbing cues. The general questions are addressed in various ways with different taxa as examples: What signals are important to the individual of a particular species? How are they generated? How do they impedance match with the environment for efficient transmission? How do environmental properties and environmental heterogeneity affect transmission from sender to receiver and from sender to eavesdropper (such as a rival or predator)? Does this lead to predictions about the form of vibrational signals under specified conditions? What kinds of receptors are used for detecting and gaining information on identification, distance, and direction, and how do they and the brain extract information? Over what range does detection work and is this used to intentionally communicate at shorter distance to conspecifics than to predators or other eavesdroppers? What is the effect of communication networks on the evolution of signal forms and content? And what are the effects of the environment on the evolution of signals?

One of the most interesting but also puzzling aspects of vibratory communication is how small arthropods use vibratory communication to find conspecifics and avoid predators and parasitoids when living on plants. The difficulty of finding a vibration source is not just a matter of noise induced by wind, rain, and vibrations from adjacent roads, but also a matter of reverberation. Localization may not be too difficult within a single leaf (as with leaf miners and parasitoids) but it is a difficult and puzzling problem in plant stems. Owing to the complex geometry and small-scale material heterogeneity of branches, the major difference from large homogeneous substrates is that in plants there is no necessary monotonic reduction in amplitude or other signal properties with distance from the signaler. For example, there are significant impedance changes within stems at nodes (denser parts of stems where buds and new branches form), branches, branch tips, and roots and these can result in geometric patterns of both resonance and multiple reflections within the plant. This results in standing waves with wave nodes (zones of very low amplitude at certain frequencies), where the locations of the wave nodes depend on vibration frequency. The consequence is that amplitude may go up and down as an insect or spider walks across the plant toward a signaler. However, not much is known about how often wave reflections and vibration nodes occur; models are needed which include transmission geometry, impedance change geometry, vibration axes, damping and other transmission loss, and transmission efficiency.

The mechanisms of determining location, distance, and direction on plants are largely unknown. Some progress is now being made by modeling various kinds of search and signal reception mechanisms and these show that searches are not simply random trial and error. Given that wave node positions depend on frequency, and different plant species have characteristic architecture (branching geometry), they may provide, on average, a predictable spatial sequence of nodes at different frequencies that might give host specialists clues to distance and direction and, with FM sweeps, the temporal pattern of changing node patterns might also give useful cues. Surprisingly, detailed mapping of spatial distributions of amplitude nodes with frequency over plant architecture is rarely done, and if this were done over the frequency ranges actually used by arthropods, new mechanisms combining frequency and nodal pattern may emerge. We have no idea whether or not there are general geometric nodal patterns generated in plants and how different are different plant species and families, and even the simple (rod) transmission properties are known for very few plant species. All may affect the ability of insects to specialize or generalize on ranges of plant hosts, their ability to detect invertebrate or vertebrate predators or parasitoids, and the ability of predators and parasitoids to find plant-dwelling prey, and the ability to separate useful signals from noise. There is clearly a lot to be done about distance and direction estimation in vibratory signals on plants.

The varying joint use of simultaneous vibratory and auditory signaling mechanisms is discussed at length. One can supplement the other, they can both be used to transmit different signals or both can be used to transmit the same signal content (redundancy) in an effort to reduce noise effects. The conceptual patterns are typical of multimodal signaling in general. However, what is different from other multimodal communication systems is that the two modes can be closely coupled. In many cases the joint use (bimodal signaling) probably coevolved because some sounds produce substrate vibrations, so any species could evolve emphasis on one or the other or both modes. Moreover, in some cases the same mechanisms produce and/or receive both kinds of signals, if so that might provide a constraint to divergence of function and use of the two modes. The transmission properties are so different that both the function (signal design) and purpose (signal content) of simultaneous vibratory and sound signals are often very different. For example, they could be separately used for short and long distance communication, or, due to different ambient noise levels, some may signal yes/no information (as in species and sex recognition) and others provide more detailed signal content (as in mate or territorial assessment).

This book will be valuable to anyone interested in vibratory communication but will also be valuable to anyone interested in the evolution of communication because the environmental effects on communication are covered more thoroughly here than they are in other sensory modes. Historians of Science will also find this a landmark book in the development of a new science because it gives a complete early history of the subject and the field is now just starting to expand rapidly. Perhaps the most valuable part of the book is the host of interesting and important unanswered questions it raises.

John A. Endler
Centre for Integrative Ecology
Deakin University
Waurn Ponds, Australia

# Acknowledgments

We are grateful to Hannelore Hoch, the Deutsche Gesellschaft für allgemeine und angewandte Entomologie (German Society for General and Applied Entomology), and the Deutsche Forschungsgemeinschaft (German Research Foundation) for sponsoring the symposium on vibrational communication, organized by AW and MG, which was the impetus for this volume. We appreciate the patience and kind cooperation of Andrea Schlitzberger, our Project Coordinator at Springer. RC thanks the National Science Foundation (IOB 0820533) for support.

Reginald B. Cocroft
Matija Gogala
Peggy S. M. Hill
Andreas Wessel

# Contents

# Part I
# Studying Vibrational Communication: Ideas, Concepts and History

# Chapter 1
# Fostering Research Progress in a Rapidly Growing Field

**Reginald B. Cocroft, Matija Gogala, Peggy S. M. Hill and Andreas Wessel**

**Abstract** Vibrational communication holds the unique position of being one of the most ancient and widespread forms of animal communication and yet the most poorly known. The long evolutionary history of vibrational communication, the remarkable diversity of species in which it occurs, and its central role in biotic interactions provide unparalleled opportunities for addressing general questions. Vibrational communication has also proven to be a key to understanding the behavior of individual species, across much of the tree of life. The goal of *Studying Vibrational Communication* is to inspire research into this important and fascinating communication modality by providing state-of-the-field reviews, historical perspectives, and technical advice and by suggesting new directions for groundbreaking studies. We also hope to convince those new to the field that studying this communication modality is surprisingly accessible, even for those with no prior experience.

R. B. Cocroft (✉)
University of Missouri, Columbia, MO, USA
e-mail: cocroftr@missouri.edu

M. Gogala
Slovenian Academy of Sciences and Arts, Ljubljana, Slovenia
e-mail: matija.gogala@guest.arnes.si

P. S. M. Hill
University of Tulsa, Tulsa, OK, USA
e-mail: peggy-hill@utulsa.edu

A. Wessel
Museum für Naturkunde—Leibniz-Institut für Evolutions- und Biodiversitätsforschung an der Humboldt-Universität zu Berlin, Invalidenstraße 43, 10115 Berlin, Germany
e-mail: andreas.wessel@mfn-berlin.de

A. Wessel
Zoologisches Museum, Universität Hamburg, Martin-Luther-King-Platz 3, 20146 Hamburg, Germany

R. B. Cocroft et al. (eds.), *Studying Vibrational Communication*,
Animal Signals and Communication 3, DOI: 10.1007/978-3-662-43607-3_1,

## 1.1 Introduction

The study of vibrational communication is a rapidly growing field. Researchers in animal behavior are recognizing the importance of substrate vibrations for understanding the behavior of their study organisms—and, just as importantly, those new to the field are finding that vibrational communication is surprisingly easy to study. In light of the growth of the field in recent years (Fig. 1.1), we see the need for a multiauthored volume that reaches out to two audiences. One is the broader community of researchers whose work deals with the mechanisms and evolution of behavior. We hope to convince this audience that vibrational communication is among the most important of communication modalities for understanding behavior, second only to chemical communication in its presence across the tree of life (also see Drosopoulos and Claridge 2006; Hill 2008; O'Connell-Rodwell 2010). Because the vibrational modality is so ancient and widespread, it offers unsurpassed opportunities for addressing general questions in animal communication.

Most of this volume is meant for our second intended audience: researchers, especially students and those new to the field, who want to incorporate the study of vibrational communication into their work. To foster the progress of vibrations research, we include chapters that provide state-of-the-field reviews of central concepts and suggest promising new directions; chapters that illustrate the importance of vibrations in particular taxa; chapters that address the mechanistic basis of signal detection and transmission through the substrate; and chapters that discuss how to deal with the challenges of studying vibrational communication in laboratory and field. We alsodiscuss the work of some early pioneers in the field, whose work addressed issues that are still very relevant today.

Some readers may be surprised that we see a need to draw attention to one particular mode of communication, given the increasing recognition that communication is often multimodal (Partan and Marler 1999; Hebets and Papaj 2005). However, there is still a tendency to overlook the role of vibrational communication; for example, the most recent edition of a major animal communication textbook (Bradbury and Vehrencamp 2011) mentions vibrational communication in the context of signal production, but provides no overview of this modality as it does for acoustic, visual, chemical, tactile, and electrical signals. This book provides that overview.

In the chapters of this volume, we provide an overwhelming body of evidence for the common, widespread use of vibrational signals in communication across the Arthropoda. Contexts include various stages of mating interactions (from advertisement to location to courtship), communication among groups (including family groups and insect societies), predator/prey interactions, and more. In some arthropod groups, substrate-borne vibrations are the basis of a complex signal repertoire encompassing signals used in multiple contexts, and they may be one of a suite of options in a multimodal signaling strategy. We also argue that vibrational communication is ancient, predating use of airborne signals audible to humans, and the primary communication modality used by a vast number of arthropods.

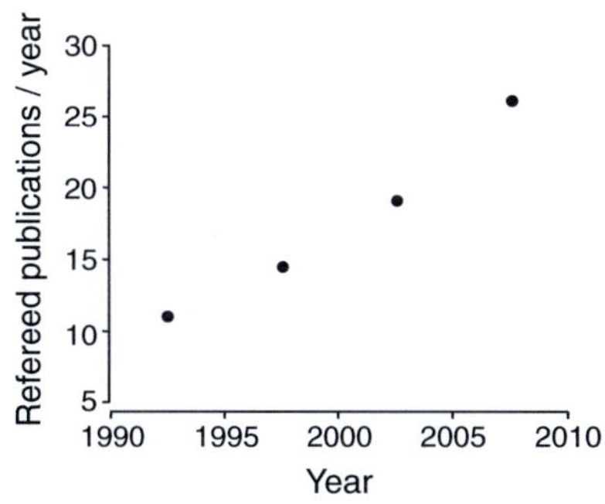


**Fig. 1.1** Average number of refereed publications per year in 5-year intervals, based on search in Google Scholar for the terms 'vibrational communication', 'vibratory communication', 'substrate-borne communication', 'substrate-borne signals', and 'seismic communication'. The results underestimate the number of publications (e.g., book chapters were not included) but reflect ongoing growth in the study of this communication modality

Vibrational communication is also widespread in vertebrates. In fish, it occurs in species in direct contact with the substrate and is usually considered separately from vibrations introduced to the watery environment by muscular manipulations of the swim/air bladder (which are considered to be sounds). We have much less evidence for vibrational communication in reptiles, but chameleons (Barnett et al. 1999) and snakes (Young 2003) both send vibrations through the substrate and are relatively less sensitive to airborne sounds (Hill 2008). Birds, with their complex visual communication and enhanced color vision, have received almost no consideration for potential use of vibrational communication. However, amphibians and mammals, in particular, often use the substrate-borne vibrational communication channel as a primary modality, especially in circumstances where vision or hearing is inefficient (Hill 2008; Caldwell et al. 2010). Vibrational communication is widespread in small mammals, especially rodents (Randall 2010). The importance of substrate-borne vibrations in communication among large mammals, most notably in elephants, is increasingly evident (O'Connell-Rodwell et al. 2006). Over and over again, in chapters of this volume, you will read that vibrational communication is not the minor player that it was considered to be three or more decades ago, but that it is found wherever we are forced to look for it when other interpretations fail. Indeed, one message of Caldwell (Chap. 6, this volume) is that whenever a 'singer' is in contact with a substrate—a leaf, a branch, and the soil—its airborne signal is inevitably accompanied by a substrate-borne counterpart. Once researchers recognize that most acoustic signals are inherently multimodal, this insight will undoubtedly enrich our understanding of many communication systems, as it has for katydids (Belwood and Morris 1987), red-eyed tree frogs (Caldwell et al. 2010), and elephants (O'Connell-Rodwell 2008). Birds that sing while airborne are one of the few exceptions, but even airborne birdsong contributes to the vibrational soundscape of nearby plants, in ways that may influence the singers' fitness (Lohrey et al. 2009). The vibrational communication modality allows for the same complexity and specificity of signals recognized for sound, vision, and chemical cues, if not more so, especially in arthropods. Yet, you will also learn that our study of vibrational communication is still in its infancy, evidenced by continuing discoveries, such as the use of vibrations in communication among model organisms studied in thousands of laboratories (Mazzoni et al. 2013).

Those of us with research interests in the area have formed collaborative groups and meet in professional societies to exchange ideas. It is no longer rare to find a number of papers in a behavior meeting devoted to vibrational communication, and students arrive at meetings aware of the ubiquity of vibrational communication even if they have not read all of the foundational works. Work on vibrations in the last two decades has answered questions that were previously inaccessible to those studying animal communication. We are excited by the work of our colleagues and compelled to dig deeper and increase the breadth of our searching for answers. Yet, the broader lay community, and even in many cases associates who study animal communication via other channels, is either blissfully unaware of the existence of vibrational communication or has not fully embraced the view that it overshadows other better studied communication modalities in importance to the animals who employ these signals. While some laboratory groups are relatively well funded, others still manage only to study vibrational communication as a side project to other funded works, and their funding agencies consider projects on vibrational communication to be risky, especially in the fact gathering stages of study of a taxon.

What can then be done to accelerate the perceived level of progress in the broader community for work that we know is solid and foundational? What can be done to increase access to the body of work that we share, to raise the awareness of others interested in communication that anomalies may be easily explained once vibrational communication is considered as an option? With this book, we hope to answer these questions and more.

## 1.2 Synopsis

The concept for this book is interwoven with that of a symposium series inaugurated at the *Entomologentagung* of the *Deutschen Gesellschaft für allgemeine und angewandte Entomologie* (*DGaaE*) in Berlin on March 23, 2011. Not all of our chapter authors spoke at that symposium, but many did and the meeting was the perfect vehicle to enthusiastically launch the developmental phase of this book. We have organized this volume into five sections of 3–5 chapters each. The chapters within a section have a common theme or focus or perspective on the issue of communication via substrate-borne vibrations. Just as our study of vibrational communication spans taxonomic boundaries and incorporates research lines from multiple, integrative disciplines, so the chapters of this book represent our current state of knowledge from across taxa and from multiple perspectives of investigation.

### 1.2.1 Studying Vibrational Communication—Ideas, Concepts, and History

The first section is composed of this chapter and four others that present a historical perspective but also consider the state of our current knowledge and directions for the future. Throughout the chapters of this book, we recognize the contributions of those early scientists who worked in isolation to document behaviors of animals that presented themselves as an anomaly. We thank them for leaving us detailed notes of their observations, even when their work was ignored and perhaps discouraged as a waste of time. Building on that work, we have developed a body of shared core concepts of vibrational communication that can be considered with respect to the dominant paradigm of animal communication and our future in that broader community (Chap. 2, this volume).

We now know that some mechanisms used by animals to produce substrate-borne vibrations also produce airborne vibrations, or sound, simultaneously. The debate of whether invertebrates use sound or vibration, or both, in communication is at least 100 years old (Chap. 3, this volume) but continues to this day. In a time when signaling via substrate-borne vibration is sometimes still doubted or ignored, it is refreshing to consider a time when invertebrate hearing was doubted, as well. Frej Ossiannilsson (1949) noted the gray area distinguishing sound from vibration as he argued that the Auchenorrhyncha were not silent, but it took the work of Strübing to definitively demonstrate that vibrational signals are required for mating in this group (Chap. 4, this volume). Her 1958 paper is included here in translation (Chap. 5, this volume). Likewise, it was Gogala's contributions in the early 1970s that gave us clear evidence that substrate-borne vibrations, alone, provide a sufficient and effective signal for cydnid bugs in mating interactions (Chap. 3, this volume).

### 1.2.2 The State of the Field: Concepts and Frontiers in Vibrational Communication

When we consider the major arguments or areas of concern from 50 to 100 years ago, we can only conclude that we have progressed a great deal in the depth of our understanding and the range of questions now asked about vibrational communication. One area of lingering concern, however, is the strict separation of study in many taxa of airborne and substrate-borne channels, a separation reflected in the literature. These distinctions may exist for some groups. Yet, we not only know that airborne and substrate-borne vibrations often are produced simultaneously by the same signal mechanism, but also know that energy traveling as waveforms in one medium will introduce vibrations into another at their boundary. The potential interactions between these two signaling modalities became more intriguing once we discovered that receivers are often sensitive to both airborne and substrate-borne vibrations (Chap. 6, this volume).

Likewise, 'noise' that we once worked to eliminate in controlled experiments, especially when studying sexual behavior, is now being seen as an important force of selection in the evolution of substrate-borne vibrational signals. Not only do insect groups interact in networks of individuals who send and receive signals that are available to all others. Along with the influence of abiotic 'noise', the biotic community includes related individuals, potential mates, parasites, and predators, all of which influence signal production in discrete ways (Chap. 7, this volume). Further, signaling takes place within an active space whose limits are defined by signal amplitude as well as background noise and the physical constraints of the medium (Chap. 8, this volume). One strategy employed by stinkbugs to maximize signal efficiency within the physical constraints imposed by the substrate and background noise is to limit the frequency range of the signals produced when communicating on plants (Chap. 10, this volume). One strategy employed by plant-dwelling pairs that use substrate-borne vibrational signals in duets is to increase the length of the female response, which is typically very short in airborne duetting. A strong case is made that these duetting pairs be used to model evolution of mate choice in the context of mutual male–female influence (Chap. 9, this volume).

### *1.2.3 Practical Issues in Studying Vibrational Communication*

Plant leaves and stems, spider webs and honeycombs, and all the kinds of substrates we find on the Earth's surface represent much more complex media for signal transmission than does the atmosphere. Almost none of these signaling environments, except for perhaps the surface of water, are ever homogeneous. Likewise, the receptors animals use to detect substrate-borne signals are different from the ones used exclusively as ears to detect airborne sound vibrations (although some organs detect both; Shaw 1994), and the medium can change the signal into something quite different on the receiving end than it was at the point of origin. For example, bending waves travel with dispersion, which is to say that the higher-frequency components of the signal travel faster than the lower-frequency components (Chap. 11, this volume).

Therefore, substrate-borne vibrational communication has evolved in environments with more than just biotic and abiotic 'noise'. Animals have adapted to environmental filtering by the substrate to produce signals that are the most efficient in transferring information in that specific environment, even if they are restricted to life on a single species of host plant. A close examination of both natural and artificial substrates, and the wave types produced as animals signal there, is essential when designing a controlled experiment (Chap. 12, this volume). Consideration of substrate characteristics and their potential for filtering a playback signal is crucial in the design of playback simulations for these very same reasons (Chap. 13, this volume).

### 1.2.4 Vibration Detection and Orientation

Up through the 1990s, advances were being made in our understanding of mechanisms for sending signals and in describing characteristics of the signals themselves. Yet, with a very few exceptions, little attention was being focused on signaling environment or on signal receivers. Here, the state of our knowledge of vibration receivers in insects is reviewed (Chap. 14, this volume), particularly the scolopidal organs, across the range from molecular mechanisms to systems analysis.

Likewise, because of the difficulty of locating information in the literature on substrate-borne vibration communication in the past several decades, theoretical models were formed for which the assumptions were left without broad empirical support. One such case is discussed here in testing an alternative hypothesis: do whirligig beetles actually use echolocation to locate immobile objects (including prey) on the water surface, or do they perceive a meniscus, or deformation of the water surface (Chap. 15, this volume)? Another look is also given to predator–prey interactions on sand, which was once considered a substrate incapable of conveying information that is biologically significant. In this case, the detection of prey by pit-building antlion larvae and orientation to and capture of the prey are presented (Chap. 16, this volume).

### 1.2.5 Biology and Evolution of Vibrational Communication in Some Well-Studied Taxa

Even in well-studied taxa, questions remain concerning how and what information is transferred via substrate-borne vibrations within the communication system. In honeybees, the function of hive-based substrate-borne vibrations is being studied by a number of research groups, but a case can be made that jet airflows created by dancing bees are transferring directional information to their nest mates (Chap. 17, this volume). Eusocial bees (stingless bees, honeybees, and bumblebees) that return from successful foraging trips produce pulsed thoracic vibrations that are introduced to the substrate upon entering the nest. At this time, evidence suggests a role for these thoracic vibrations in signaling profitability of a food source and generally coordinating foraging; however, whether or not the similarities in use of this mechanism reflect an origin in a shared ancestor has not been resolved. At least for stingless bees, experimental evidence has not found jet airflow to be an option, and navigation or distance information based on thoracic vibrations is not supported (Chap. 18, this volume).

Likewise, even though signaling in the Orthoptera has been considered a model for study of substrate-borne vibrational communication for some time, most of that research has been based on bimodal signals produced via stridulation. Other mechanisms for sending vibrations through the substrate are, in fact, known in the

Orthoptera, and signals produced by these mechanisms require an array of receiving and processing structures, as well. All these aspects are reviewed in this section, from perception to behavior, and particular stress is given to the understudied area of low-frequency vibrational communication in the Orthoptera (Chap. 19, this volume).

Lastly, so much of what we know about vibrational communication has come from our study of the insects that possess tymbals. This body of work has been produced over at least 65 years and represents contributions from research groups, large and small, across multiple continents. Aristotle knew that the cicada used a different way to communicate than grasshoppers, or flies and bees, but he did not know about the tymbal. The morphology of the tymbal was first described by the anatomist Casserius, who mostly worked with humans, in 1,600 and then lost, to be rediscovered a number of times independently beginning in the eighteenth century (Wessel 2013). Yet, many questions still remain about these 'silent singers'. Here, the body of knowledge on vibrational communication via a tymbal or tymbal-like apparatus is reviewed (Chap. 20, this volume), and a new taxonomic name, the Tymbalia, is proposed based on the autapomorphy of possession of this structure.

## 1.3 What Is Left to Be Learned?

After reviewing much of what we have learned about the function and evolution of vibrational communication systems in the past few decades, we are left with one overwhelming conclusion: What we have learned so far is a small island in a large sea of unanswered questions. The study of vibrational communication is indeed an exciting frontier in the study of animal behavior. Research with organisms that use a communication modality with such a long evolutionary history can address on the one hand questions specific to the modality—how can organisms locate a vibration source in complex environments, how are vibrations transmitted in a given medium, and for many taxa, we know relatively much about the behavior, but little about the structures used to produce vibrations (the 'hoppers' as one prime example). On the other hand, research with these organisms can address in a powerful way many general questions in animal behavior—questions about communication networks, about the influence of social environment on the development of behavior, and about the role of signals in speciation. The general theoretical questions that can be addressed with vibrationally signaling organisms are endless, so here we focus on some pressing questions that focus specifically on the vibrational modality.

When is vibration transmission dominated by reflected energy and standing waves, and when is it dominated by transient one-way wave propagation? Depending on the study system, researchers may observe that standing waves are ubiquitous, while other researchers find that they are nearly absent. We need a general framework, perhaps based on engineering tools such as finite element models, to help us understand the factors that determine when reflected waves will be important. The two kinds of vibration transmission environments will impose very different selection on vibration localization and mating strategies.

Are there conditions under which organisms can assess not only the direction of a propagating wave, but also the distance to the source? Suggestions that the dispersive propagation of bending waves in plants could allow distance estimation have not been tested. But under some conditions, amplitude gradients provide clear information about distance to the source, and anyone who has listened to the same insect signaling both near and far from the sensor can easily tell (even apart from amplitude) when the insect is close and when it is far. What are the signal features that make this possible, and do the animals use them?

How does impedance between the signaling animal and the substrate influence the evolution of traits such as signal frequency and substrate choice? This question has been long neglected, but could reveal an entirely new suite of adaptations for efficient communication.

How do animals couple vibrations to the substrate most efficiently? Across taxa, substrates include all the variations of plant structures, spider webs, honeycombs, water surfaces, as well as the varying levels of heterogeneity of soils, rocks, and litter, etc. on the ground. Bimodal signalers solve the challenge of efficient coupling of energy to both the substrate and the atmosphere. Some are narrow specialists and some are generalists, even when the signal production mode produces multiple types of vibrations, simultaneously. What can we learn about the interactions of these vibrations and their influence on effective energy coupling with the preferred substrate across taxa?

How is signaling strategy affected by abiotic changes across a season or an animal's lifetime? How is signaling strategy affected by biotic community and network changes in parameters such as operational sex ratio, density, or age structure across a season or a lifetime?

How can we more efficiently develop and utilize mathematical models and adapt cross-disciplinary methods from physics and engineering to make predictions that reflect the practical reality of a vibration-borne communication system? The simple answer is to develop collaborations. Often the 'big picture' requires a skill set rarely available to a single laboratory unit, and a brick wall for one scientist looks like a massive doorway to another. Still, recognizing issues such as filtering by the substrate and the accompanying attenuation beyond geometric spreading or variation in the suitability of artificial substrates for an experiment based on the question asked must be considered for many more taxa of signalers. There may be some hope arising, as well, from new technology such as high-resolution micro-CT and developing techniques such as digital 3D reconstruction for studying morphology.

These are just some of the questions posed in this book. When conventional wisdom collides with common sense, or you encounter a major obstacle in studying animal communication that has no solution, what do you do? Vibrational communication is a gold mine for continuing research and innovation, an exciting frontier in the study of animal behavior.

## References

Barnett KE, Cocroft RB, Fleishman LJ (1999) Possible communication by substrate vibration in a chameleon. Copeia 1999:225–228

Belwood JJ, Morris GK (1987) Bat predation and its influence on calling behavior in neotropical katydids. Science 238:64–67

Bradbury JW, Vehrencamp SL (2011) Principles of animal communication. Sinauer Associates Inc, Sunderland

Caldwell MS, Johnston GR, McDaniel JG, Warkentin KM (2010) Vibrational signaling in the agonistic interactions of red-eyed treefrogs. Curr Biol 20:1012–1017

Drosopoulos S, Claridge MF (2006) Insect sounds and communication: physiology, behaviour, ecology and evolution. Taylor & Francis Group, Boca Raton

Hebets EA, Papaj DR (2005) Complex signal function: developing a framework of testable hypotheses. Behav Ecol Sociobiol 57:197–214

Hill PSM (2008) Vibrational communication in animals. Harvard, Cambridge London

Lohrey AK, Clark DL, Gordon SD, Uetz GW (2009) Antipredator responses of wolf spiders (Araneae: Lycosidae) to sensory cues representing an avian predator. Anim Behav 77:813–821

Mazzoni V, Anfora G, Virant-Doberlet M (2013) Substrate vibrations during courtship in three *Drosophila* species. PLoS ONE 18:e80708

O'Connell-Rodwell CE (2008) The elephant's secret sense: the hidden life of the wild herds of Africa. University of Chicago Press, Chicago

O'Connell-Rodwell CE, Wood JD, Rodwell TC, Puria S, Partan SR, Keefe R, Shriver D, Arnason BT, Hart LA (2006) Wild elephant (*Loxodonta africana*) breeding herds respond to artificially transmitted seismic stimuli. Behav Ecol Sociobiol 59:842–850

O'Connell-Rodwell CE (ed) (2010) The use of vibrations in communication: properties, mechanisms and function across taxa. Transworld, Kerala

Ossiannilsson F (1949) Insect drummers. A study on the morphology and function of the sound-producing organ of Swedish Homoptera Auchenorrhyncha with notes on their sound production. Opuscula Entomol Suppl X:1–146

Partan S, Marler P (1999) Communication goes multimodal. Science 283:1272–1273

Randall JA (2010) Drummers and stompers: Vibrational communication in mammals. In: O'Connell-Rodwell CE (ed) The use of vibrations in communication: properties, mechanisms and function across taxa. Transworld, Kerala, pp 99-120

Shaw S (1994) Detection of airborne sound by a cockroach 'vibration detector': a possible missing link in insect auditory evolution. J Exper Biol 193:13–47

Wessel A (2013) Casserius and the secret of the cicada's voice. Deut Entomol Z 60:139–150

Young BA (2003) Snake bioacoustics: toward a richer understanding of the behavioral ecology of snakes. Quart Rev Biol 78:303–325

# Chapter 2
# Stretching the Paradigm or Building a New? Development of a Cohesive Language for Vibrational Communication

**Peggy S. M. Hill**

**Abstract** Before we can recruit the broader community to share our conviction of substrate-borne communication in animals as ancient, important, widely employed in vertebrates, and perhaps exclusively employed in a broad range of arthropod taxa, we first must assess our current status within the animal communication paradigm and plot a course with that focused goal in sight. We must agree on the words we use to unambiguously communicate research findings among ourselves. We can do this rapidly through consensus, or allow terminology and protocols to slowly evolve to cohesion over an extended period of time through inaction. This chapter briefly explores the current position of shared core concepts on vibrational communication within the framework of Thomas Kuhn's *Structure of Scientific Revolutions* and suggests that the study of substrate-borne vibrational communication really can be accommodated within the dominant paradigm of animal communication. We require a reinterpretation of what 'everyone knows to be true' in some cases where empirical studies now have falsified previous widely held assumptions. A first step might be to develop a concerted, coordinated strategy that is widely employed by those currently studying vibrational communication. The paradigm can be stretched without being replaced, or we can forge a separate paradigm for vibrational communication. It is simply time to collectively decide on a course of action.

## 2.1 A Revolution in Progress?

In his theoretical framework to explain the history of scientific revolutions, Kuhn (1996) described a series of events that typically define the stages in establishing a new scientific discipline. At first, there is a period of observing and gathering facts,

P. S. M. Hill (✉)
The University of Tulsa, Tulsa, OK, USA
e-mail: peggy-hill@utulsa.edu

R. B. Cocroft et al. (eds.), *Studying Vibrational Communication*,
Animal Signals and Communication 3, DOI: 10.1007/978-3-662-43607-3_2,

using existing tools and vocabulary from other established disciplines. Then, schools of thought are formed by like-minded individuals who seek to establish a paradigm, or shared methodology, that helps to answer the greatest number of questions of interest to the group. Competing schools of thought gradually disappear as evidence accumulates to support the paradigm of one group, and scholarly discourse moves away from general texts that are broadly read to journal articles that are accessible primarily to those who share the emerging paradigm. The school then may become an established discipline with its own societies, journals, and academic departments. Ultimately, the new discipline develops its own research technology and vocabulary, and finally its own textbooks.

### 2.1.1 Fact Gathering

The period of observations of vibrational communication among arthropods and other animals is centuries old (see Anderson 1973; Tributsch 1982; Snarr 2005; Hill 2008). Before there was a sense of the channel being employed, interested observers described what we now know to be substrate-borne vibrational signaling (e.g., book lice: Pearman 1928; termites: Emerson and Simpson 1929; Howse 1964). Even when the interactions among individuals were attributed to some other communication modality, sufficient evidence was included in the descriptions for us to now recognize them as classic vibrational communication systems.

Ossiannilsson (1949) produced an extensive treatise on leafhoppers where he described male alternation of calls, chorusing, male–female duets, and pitch variation with temperature. He suggested that vibrations produced by the tymbals of hoppers were conducted to receivers through the substrate. Working with bugs in the Cydnidae in the early 1970s, Gogala and his colleagues confirmed that the silent substrate-borne vibrations produced in signaling, rather than the airborne component audible to humans, was the adequate and essential stimulus used in mating interactions. They further described differences in signal frequency before there was much interest in more than the temporal component of vibrational signals (Gogala et al. 1974).

This period of fact gathering continues on at least two fronts. Those who have studied vibrational communication in one taxon are continuing to provide empirical data to fill knowledge gaps and are then often branching out to explore its use in unstudied groups. For example, Wignall and Herberstein (2013) described for the first time the repertoire of web-borne courtship vibrations of the well-studied model species *Argiope keyserlingi*. Mello and Dos Reis (1994) described a new species of phalangopsid cricket, but also included observations of foreleg drumming behavior and the substrate-borne vibrations produced that were required for successful mating. At the same time, those who have not yet been enlightened are still describing signaling that is highly likely due to substrate-borne vibrations, but ascribing the signals to visual or sound displays. Bringing this second group into the fold is a challenge that must be addressed.

## 2.1.2 Taking Control of the Terminology

While much of the work that has advanced our understanding of vibrational communication in arthropods has been carried out in relative isolation, the theoretical framework for this study composes a set of shared core concepts that have been elaborated in chapters of this volume and in earlier reviews for both arthropods and vertebrates (e.g., Kalmring and Elsner 1985; Devetak 1998; Virant-Doberlet and Čokl 2004; Cocroft and Rodríguez 2005; Drosopoulos and Claridge 2006; Hill 2008; O'Connell-Rodwell 2010). For example, we recognize that the term vibration as defined in engineering and physics encompasses all particle motion in a fluid medium, whether that medium is air, water, or a substrate such as the Earth or a plant body. Thus, our concept of sound is simply a subset of the larger category of vibrations that specifies a hearing organ as the receiver mechanism. We also recognize that at the boundary between two media, such as air/soil, water/soil, air/spider web, etc., energy being propagated as a vibration in one medium is always transferred to energy propagated in the other. So, even as one event can produce both airborne and substrate-borne vibrations, at the boundary between the air and substrate, we see exchange taking place between the two media that renders the entire distinction between sound and vibration a complex tangle. Ossiannilsson (1949) suggested long ago that whether we called something sound or vibration was simply a matter of taste.

Further, even though we recognize a range of vibrations that can be introduced to a substrate by a single event, not all of these appear to be used by animals as signals (see Markl 1983; Gogala 1985). In almost all cases, substrate-borne vibration signals travel as either Rayleigh waves (through the Earth) or bending waves (in plants). With both of these waveforms, particle motion is perpendicular to the direction of propagation or surface of the solid through which it is traveling (Gogala 1985), at least in part, and detection of the wave typically involves an inertial motion sensor that makes the animal's stance with respect to direction of wave propagation important (see Lewis 1984; Hill 2008). Animals, thus, may experience Rayleigh waves as a pushing up from the surface on which they stand, but those on a plant surface experience bending waves as a whirling motion that pushes both up and to the side (McNett et al. 2006, see Cocroft et al. Chap. 13, this volume). As we continue to use newly developed tools and protocols in our investigations, we will continue to reassess prior interpretations of complex waveform interactions within a substrate. For example, while we have held for some time that the substrate on which animals perch must be continuous for individuals to successfully signal to each other (see Hill 2008), we have recently learned that this does not hold true always for leafhoppers (Eriksson et al. 2011). We may also need to rethink conventional wisdom on substrate continuity for other taxa, as well as to consider the role of some of the waveforms created by an event that have previously been discarded as likely not important to vibrational communication (see Hill 2008) as we learn more about the dynamics of substrate-borne waveforms.

**Table 2.1** Examples of terms that could be assigned to a larger functional category

| Recommended Term | Synonym from Literature | Example of use by source |
|---|---|---|
| Drumming | Rapping | Clayton (2005) |
| – | Thumping | Salmon and Horch (1972) |
| – | Sounding | Broad and Quicke (2000) |
| – | Tapping | Pearman (1928), Zeigler and Stewart (1977) |
| Tremulation | Vibration | Fletcher et al. (2006) |
| – | Vibration signal | Lewis and Schneider (2000) |
| – | Dorsoventral abdominal vibration (DAV) | Rupprecht (1974) |
| – | Opisthosomal oscillation | Rovner and Barth (1981) |
| – | Body jerking | Henry (1979) |
| – | Thoracic muscle contraction | Kanmiya (2006a) |
| – | Tremble dance | Seeley (1992) |
| – | Begging signal | Esch (1961) |
| – | Stop signal | Nieh (1993) |
| – | Shaking signal | Seeley et al. (1998) |
| – | Spirit-tap | Milum (1955) |
| – | Queen piping: tooting, quacking | Michelsen et al. (1986) |

We also recognize common features among the mechanisms animals employ to generate substrate-borne vibrations in signal production. However, our historical literature is filled with terminology borrowed from physics and engineering and further refined to describe vibrational communication behaviors before we became aware of the size and scope of this communication modality. A range of jargon also was used to describe specific behaviors not reported before in the taxa being studied. This is part of the normal pattern during fact gathering in a scientific revolution before a more cohesive language has been developed. Still, some currently submitted manuscripts and grant proposals continue to use terms that have become ambiguous now that we have gathered enough data across taxa to see the commonalities in signaling mechanisms. This use of highly specific terms when describing behaviors that could be fit into a more generalized framework inhibits progress in pursuing both proximate and ultimate questions by limiting study to a novel behavior restricted to a single taxon.

In arthropods, we can distinguish at least four primary categories of signaling mechanisms that subdivide the general pool, based on shared signal parameters, into four rough functional groups: drumming, tremulation, stridulation, and tymbal buckling (Hill 2008, 2009, 2010). Yet, a range of terms, especially notable in describing drumming and tremulation events (Table 2.1), can be found in the past and current literature. Standardizing these terms is perhaps a very good point of focus for a concerted strategy.

#### 2.1.2.1 Drumming

Drumming is the term used more frequently than others in recent years to describe the production of substrate-borne vibrations using some body part to strike the substrate in a percussive event. The body parts could be the abdomen, one or more legs, one or two feet, tails, teeth, antennae, various parts of the head (including an elephant's trunk), or other specialized structures. Basically, any body part that can be used to strike the surface is likely being used by some animal group, had we only time and resources sufficient to survey every one. Sometimes, the body part is modified, but often there is little to hint at its role in vibrational signaling. Ossiannilsson (1949) used the term drumming to describe a very different phenomenon, and this has led to some confusion and proliferation of use of other terms.

Drumming on a substrate produces broadband, noisy signals that theoretically represent all frequencies produced by the event at equal intensity at the source of the signal. Spectral differences in the signal at the receiving end are due to filtering by the substrate through which the signal is propagated, and the temporal patterns are thus more important than spectral details to the individuals receiving drummed signals. Drummed signals often have both an airborne and a substrate-borne component and appear to be associated with habitats that are more heterogenous in substrate composition (Elias and Mason 2010). In the arthropoda, fiddler crabs (Aicher and Tautz 1990) and ghost crabs (Clayton 2005) signal via drumming, as do alderflies (Rupprecht 1975), ants (Kirchner 1997), Jerusalem crickets (Weismann 2001), wasps (Pratte and Jeanne 1984), gryllacrids (Field and Bailey 1997), heel-walkers (Eberhard and Picker 2008), and a vast number of spiders (Rovner 1975; Barth 1982; Uetz and Stratton 1982; Quirici and Costa 2005; Elias and Mason 2010), among others. Recently, Kojima et al. (2011) determined that pupae of a group-living beetle were able to signal via drumming to their larval family members that share the same soil space.

Vertebrate animals, especially mammals, are known to drum appendages as they signal (see Randall 2010). The first documented use of vibrational signaling in terrestrial mammals was for the Israeli mole rat (Heth et al. 1987; Rado et al. 1987), but drumming is known from at least 32 species in 11 families of mammals (Randall 2001). Drumming is the best known mechanism that mammals employ to produce substrate-borne vibrational signals, which are known almost exclusively as seismic signals that propagate through the Earth (Hill 2008).

#### 2.1.2.2 Tremulation

Morris (1980) first used the term tremulation to describe body motion without any percussive impact with the substrate, and since then, a broad range of behaviors have been recognized that would fit within the set of criteria that define tremulation. The term was coined to describe a trembling, shaking, sometimes jerking, body motion in katydids (bush crickets) that introduced substrate-borne vibrations into the plants on which the animals were perched. Substrate-borne tremulations are still known almost exclusively for animals that live and signal from plants. Typically, the vibrations pass to the substrate via appendages that couple with the surface (almost always legs/tarsi), but any body part (e.g., thorax) conceivably could be employed. Although animals across taxa produce non-percussive motions in communication that are distinctive, species specific and even context specific, thus arguably warranting a definitive descriptive term of their own, there are commonalities among the signals produced that allow us to group these into the single category of tremulation (Hill 2008).

Unlike drumming signals, tremulation produces substrate-borne vibrations in frequencies that are more pure tone, narrowbanded, and low in frequency. For example, in spiders, we find tremulation signals in the range of 1–300 Hz, and even very small oscillations of the body are sufficient to introduce vibrational signals into the substrate (Elias and Mason 2010). Arthropods that signal via tremulation have often been considered to be silent, due to the inability of humans to detect the low frequencies of their calls. Tremulation is a more effective strategy in environments with rather homogenous substrates, where low frequencies are not filtered or distorted as they are propagated (Elias and Mason 2010). Other than in katydids (Morris 1980; De Souza et al. 2011), tremulation is known, for example, from spiders (Rovner 1980; Dierkes and Barth 1995) and a variety of insects, among them planthoppers (Ichikawa 1976; Claridge 1985), whiteflies (Kanmiya 2006b), wetas (McVean and Field 1996), cave crickets (Stritih and Čokl 2012), groundhoppers (Kočárek 2010), lacewings (Devetak 1998), and bees (Sandeman et al. 1996; Kirchner 1997). There are surely others, including those with behaviors described in the literature that appear to be tremulations, but where the experimental design of a study did not include testing for possible use of vibrational signals.

We have also found production of tremulations in vertebrate animals that live and signal on plants (see Caldwell, Chap. 6, this volume). Caldwell et al. (2010) argue that arboreal vertebrates likely use substrate-borne vibrational signals to a much greater degree than is currently known, especially since almost all their movements excite the substrate. In the case of the red-eyed tree frog, all known signaling behavior produces stereotypical vibrational waveforms. Other modalities, especially visual displays, have been assumed in the past to be the mechanism of information transfer in tremulating vertebrates. However, in the red-eyed tree frog, which has been so well studied for its predator-induced rapid hatching response, experimental evidence confirms that substrate-borne tremulation vibrations provide the necessary and sufficient stimulus to elicit a tremulation response from other male competitors (Caldwell et al. 2010).

Tremulation by vertebrate animals that signal through the soil has not been described, but neither have we actually looked for this. It would take a very large animal to induce substrate-borne vibrations with a tremulation, but what about the rumbling vibrations that elephants produce with vocalizations (Hill 2010)? We may need to develop consensus after a dialog to consider a fifth category for vertebrates, vocalizations, but perhaps we can also manage to fit our examples into one of the four groups suggested here and consider whether these elephant vibrations are functionally close enough to plant-borne tremulations to warrant that label.

#### 2.1.2.3 Stridulation

Animals produce both airborne and substrate-borne vibrations as they rub one body part against another, generating friction. Stridulation is rather common in the Arthropoda, but in most cases, we do not yet know whether both the airborne and substrate-borne vibrations produced are actual signals used by the animals. Typically, a specialized file and scraper mechanism can be identified, and at least eight different types of these are known in spiders (Uetz and Stratton 1982). A variety of possibilities exist for the body components used (wing–wing, leg–wing, two body segments, etc.; see Wessel 2006), and signal characteristics vary with the body parts used (Hill 2008).

The substrate-borne vibrations produced by stridulation are generally higher in frequency than those produced by tremulation (Elias and Mason 2010), and they may include harmonics, unlike the broadband noisy signals produced by drumming (see Hill and Shadley 2001). Stridulation is known from ghost crabs (Clayton 2005), spiders (Hebets and Uetz 1999; Elias and Mason 2010), and a large number of insect groups (e.g., ants: Kirchner 1997; beetles: reviewed by Wessel 2006; dung beetles: Kasper and Hirschberger 2005; cerambycid beetles: Breidbach 1986; psyllid bugs: Tishechkin 2006; water bugs: Theiβ 1982, etc.).

An interesting interpretation has been suggested by Kenneth Stewart that a rubbing of a body part against the substrate is 'actually a body-substrate stridulation' (Stewart 2008, 4103). A 'scraping' with the mandibles has been described for larvae of hornets (Ishay and Landau 1972; Ishay and Schwartz 1973; Ishay et al. 1974), and both scraping and plucking are known in cherry leaf-roller caterpillars (Fletcher et al. 2006). In addition, caterpillars of the Lepidopteran genus *Drepana* exhibit mandibular and/or anal scraping behaviors (Yack et al. 2001). These substrate-borne vibrational signal-producing behaviors were classified under 'other mechanisms' in Hill (2008), but in keeping with the theme of this chapter, stretching the current definition to include rubbing a body part against the substrate seems logical, as well as provocative, in anticipation of the possibilities for additional uses of the label.

#### 2.1.2.4 Tymbal Buckling

Tymbals are specialized features of the exoskeleton of the lateral first one or two abdominal segments in cicadas, their non-cicada relatives in the Auchenorrhyncha, other bugs in the Hemiptera, and some tiger moths (Shaw et al. 1974; Shaw and Carlson 1979; Claridge 1985). Hoch et al. (2006) described substrate-borne vibration production via a putative simple tymbal by a member of the ancient Coleorrhyncha now found in Australia. Although the exact mechanism of signal generation has not been confirmed, physical characteristics of the first two abdominal segments and properties of the signal itself are consistent with tymbal-generated signals in cicadas and others. These observations support the argument that communication via tymbal vibrations, as well as drumming and stridulation, is at least 230 million years old in insects (Hoch et al. 2006).

Tymbal structures may be thinner walled or thicker regions of the exoskeleton, but vibrations are produced as muscles attached to the inner surface of the membrane contract to distort or buckle the tymbal in an in-and-out clicking motion (Hill 2008). In addition to his observations of behavior, Ossiannilsson (1949) described tymbal anatomy for 79 species and signals produced by tymbal buckling for 96 species of Auchenorrhyncha. Animals that possess tymbals may also produce substrate-borne vibrations through tremulation (Shaw and Carlson 1979) or stridulation (Gogala 1985); however, Ossiannilsson (1949) commented on the lack of information available on the tymbals and songs of the non-cicada relatives, especially when considered in light of the wealth of details available for cicadas. Work continues in a number of laboratories to document mechanisms of communication in species that possess tymbals.

### 2.1.3 Schools of Thought, Publications, and a New Sub-discipline: Increasing Accessibility and Broadening the Scope of Study

While some researchers who are new to the study of vibrational communication, or who are exploring the possible use of this modality in an unstudied group, are still gathering facts, for the most part, the community has moved beyond the school of thought to a functioning paradigm-based entity that communicates through scholarly articles and interactions at professional meetings. Yet, we have not become a discipline, even a sub-discipline, in the biological sciences with dedicated journals and academic departments or exclusive professional societies. I would suggest that individual working groups have not completely abandoned a degree of isolation in our scholarship, even though we are in clear agreement on basic core principles of animal communication via substrate-borne vibration. This is not to trivialize the practice. How does one make that leap? How do you define your peer group, and who is the audience of focus to whom you report research findings, when your questions are all interdisciplinary or cross-disciplinary in nature? How do you design experiments to test hypotheses of evolution of communication in your taxon of interest when no phylogeny exists, or when you observe a behavior but know nothing of the underlying morphology, or when the most closely related species and genera are just as poorly studied as the one that becomes your passion? How do you go back to the slogging efforts of descriptive studies after you have experienced the freedom to test models in your previous work? Fortunately, the synergistic effects of collaboration in interdisciplinary investigations can be just as rewarding. Seeking collaborations and forming alliances outside our own expertise helps to broaden the scope of study and may open unexpected doorways to new possibilities.

Even as we have studied signals and continue to document the array of sending mechanisms employed by the arthropoda, receiving mechanisms are still poorly known for many major groups, with a few exceptions. An array of invertebrate mechanoreceptors (reviewed in Hill 2008, Lakes-Harlan and Strauβ, Chap. 14, this volume) are known: scolopale joint receptor, campaniform sensilla, hair sensilla, Johnston's organ, subgenual organ, tarsal scolopidial organ, mid-coxal protuberance, trichobothria, slit sensilla (BCSS in scorpions and lyriform organ in spiders), and Barth's organ. However, receivers have been fully described only in the legs of two species of green lacewings (Neuroptera) (Devetak and Amon 1997; Devetak et al. 2004) and two species of heelwalkers (Mantophasmatodea) (Eberhard et al. 2010).

Studies of the complex tibial organ (a complex of the subgenual and intermediate organs and a crista acoustica homolog) in the Ensiferan Orthoptera are underway and yielding a wealth of data with far-ranging consequences for our collective progress (Strauβ and Lakes-Harlan 2008a, b, 2009, 2010, 2013, see also Chap. 14, this volume). The research initiative that has so far described the neuroanatomy and physiology of this complex in the legs of raspy crickets (Gryllacrididae), Jerusalem crickets (Stenopelmatidae), splay-footed crickets (Schizodactylidae), and stick insects (in the sister group of Orthoptera, the Phasmida) has yielded sufficient data to map this character onto existing phylogenetic trees to generate or support hypotheses on relatedness and evolution of communication systems in the Ensifera. In so many other groups where substrate-borne vibration signaling is well documented for a small taxon, or for one behavioral context in a larger group, insufficient comparative data are available to do more than speculate on evolution of the signal strategy. Therefore, being able to work across sub-discipline boundaries provides us with the opportunity to explore substantive questions once the low-hanging fruit has already been harvested from the safety of the mainstream, but at the same time, we face the challenges of working within methodologies and terminology that are less familiar, which can be daunting.

Taking up these challenges, Michelsen et al. (1982), in turn, challenged conventional wisdom on propagation of substrate-borne vibrations through plant tissue by actually testing a variety of insect species signaling on a variety of plant materials and producing a variety of context-specific signals. In the history of studies of animal communication via this channel, it still amazes that what everyone knew/knows to be true is sometimes based on little to no empirical evidence. Data from studies of other groups have been extrapolated to draw conclusions without confirmation. Actual experiments designed to falsify these established facts often have indeed falsified them. Michelsen's group thus set the bar for research on vibrational communication when the substrate is plant tissue. They were able to confirm, among other things, that plant-borne vibrations were filtered by the substrate in similar ways by both fresh and dry leaves and that vibrations could travel 1–2 m through a green stem without much loss of energy at the dominant frequency, even traveling up and down a stem several times at detectable levels. Since then, data continue to accumulate with the use of new technology to support our understanding of the signaling environment of a three-dimensional plant stem. McNett et al. (2006) used two transducers, rather than the single one used in prior studies, and placed them in position perpendicular to each other to more accurately model the world an animal encounters as it perches with all its legs in contact with (and gathering input from) a plant stem. Challenges continue to be addressed with creativity and effort.

One rich area for further study is to use newer technology or techniques from other sub-disciplines to examine substrate-borne vibration use by animals across contexts in groups for which signals in one context have already been confirmed. Virant-Doberlet et al. (2011) used molecular tools to examine predation of leafhoppers in the genus *Aphrodes* for which both male and female calling signals are known. The experimental evidence they obtained from PCR analysis of gut contents, along with playback and microcosm trials, was able to confirm for the first time that arthropod predators could not only use substrate-borne vibrations produced by prey to locate and capture them, but that they could use information in male mating signals to assess prey abundance, rather than randomly foraging on males, females, and nymphs.

My interest in vibrational communication sprang from necessity, as the male mole crickets I was studying in hopes of gathering data that could be used in making conservation decisions failed to respond to simulations of the loud airborne calling songs of other males. They did, however, respond with maddening consistency to the vibrations produced in the substrate by ceasing to call when one tried to approach them (Hill 2008). I was interested in revealing details of population ecology, life history, and mating behavior in order to construct quantitative models, but only two papers on the species existed at the beginning of my work. In the end, our research group needed expertise in mechanical engineering and bioacoustics, molecular genetics and phylogenetics, neurophysiology, soil science, biostatistics, and more. Yet, we have not really suffered as behavioral ecologists from the lack of an exclusive disciplinary home for vibrational communication studies as much as we have from the lack of field-worthy equipment that is not prohibitively expensive. This represents a related, but separate and ongoing challenge.

### 2.1.4 A New Paradigm, or a Challenge of the Common Ground?

The question then remains, do we really require a new and different paradigm that is separate from that employed by those who study communication via sound signals, or even from the general methodology important to studying communication in general? If not, then how do we best provide our 'silent singers' with a voice, at last? At this stage, and without a strong argument for increased isolation, making a place for ourselves within the dominant communication paradigm seems the best option. A concise vocabulary is essential for us to, at minimum, communicate efficiently with each other. We simply must agree upon the definitions of terms that we use. Most with research interests in substrate-borne vibration have already done so, but I suggest that we must rigorously require this of others as we share ideas in person and through peer reviews of manuscripts and funding proposals. For any 'new' ideas to be taken seriously by the extended community, they must first be defended seriously and consistently by those who require the dominant paradigm to be stretched to include them.

Once we examine our shared core concepts of substrate-borne vibrational communication, whether from vertebrates or invertebrates, none of them really is in conflict with traditional components of the dominant paradigm in animal communication. For example, the definitions for unidirectional and bidirectional airborne communication (Gerhardt and Huber 2002) could be slightly modified to accommodate a single individual of either sex searching for a silent sedentary mate, as well as the full range of mating strategies we know from substrate-borne vibrational dueting (Hill 2008). Does it really matter whether the female searches for a sedentary male, or the male searches for the female? Or, does it matter whether only one sex calls instead of both calling in an alternating sequence? Perhaps what is more important is how these communication systems might have evolved across taxa within the constraints of calling and mating investment strategies (de Groot et al. 2011). Our search for commonalities and standardization of terminology in communication simply requires that some established truisms be re-examined now that our technology is sufficiently advanced to make testing for substrate-borne vibrations accessible when anomalies persist in any body of work. We can be thankful to those whose curiosity and efforts provided adequate observational details upon which we can build evidence for vibrational communication, even as we appreciate the challenges of working in a climate that in some ways inhibited looking for the vibrational signals in the first place. When we first confirmed that leaf-cutter ants stridulate to produce signals that recruit help when a nest mate is buried (Masters et al. 1983), why would we have considered that the airborne component of this signal was not all that was produced, or that it would be in any way ineffective in recruitment? Yet, 10 years later, we learned that recruits respond to the substrate-borne component produced by stridulation when it is presented alone, but not to the airborne component alone (Roces et al. 1993). Likewise, wolf spiders exhibit clear visual mate-attraction displays as they wave their ornamented legs, but the substrate-borne vibrations produced by drumming are both sufficient and essential to elicit female response in a courting pair when the pair is not in visual contact (Gibson and Uetz 2008). Anyone who hears a calling cicada experiences a very clear signal that is impossible to ignore; however, Claridge et al. (1999) explained that this loud airborne vibration is derived from a substrate-borne vibration system still being used by their silent Auchenorrhyncha relatives. Adult cicadas have thousands of sensory cells that converge on only a few interneurons that carry hearing information to the brain for processing. The presence of so many receivers suggests that cicadas are capable of very fine-tuned discrimination in hearing, but our inability to find more interneurons suggests to Lakes-Harlan and Strauβ (2006) that these thousands of cells must have had a different function in the cicada's evolutionary history. They propose that larval cicadas in the soil have substrate-borne vibration-sensitive scolopidial organs that are merely retained in the adult, even if they have no function in adult hearing. Since this hypothesis was proposed, Chuche et al. (2011) have looked for, but been unable to confirm, use of vibrational communication in nymphs of the grapevine leafhopper. They found no vibrations in the substrate other than those that could be attributed to incidental movements, and nymphs did not respond to playback stimuli, nor did they respond to substrate-borne vibrations produced in the presence of actual predators. Still, this hypothesis opens a huge portal, not just a window of opportunity, for further research into the evolution of communication through substrate-borne vibration in the cicadas and their relatives, as well as other groups, with only a bit of stretching of the dominant paradigm.

When Philip Brownell was a PhD student working with Roger Farley to tease out the details of how sand scorpions were able to locate and capture their prey so quickly and efficiently, he worked within a paradigm where conventional wisdom held that vibrations in a natural substrate could not carry biologically meaningful information in a signal because of the inelastic medium and the propagation speed of the waveforms. The best that could be hoped for was some sort of alerting mechanism that a disturbance had occurred (see Schwartzkopff 1974). However, when Brownell falsified every hypothesis that did not involve vibrational signaling, he was forced to test and then challenge this conventional wisdom. He found that instead of damping low-frequency vibrations, the sandy substrate on which sand scorpions foraged conducted both Rayleigh waves and P-waves and conducted them slowly enough so that his scorpions could use information produced by prey in motion to locate and capture a meal (Brownell 1977, Brownell and Farley 1979a, b, c). Once the evidence was published, no dissenting camp actually emerged. These findings were then used by Briceno and Bonilla (2009) to explore vibrational communication by scorpions in other contexts and to identify a sort of tremulation ('judder') used in mate attraction and/or courtship.

The tick-tick-tick of the death-watch beetle was described in the 1600s, but Birch and Keenlyside (1991) were the first to test whether communication was through airborne or substrate-borne vibrations. Their very simple, but effective, experiment tested mating response by a female to a calling male. If they were on the same substrate, the female responded. If they were on adjacent pieces of wood, she did not. If the adjacent pieces were brought into contact, she then responded. No transducers or laser vibrometers were employed in this study, but the results strongly support the conclusion that the substrate-borne component provides the sufficient and essential stimulus to elicit a mating response. Both males and females drum in a reciprocal duet. She is stationary, and he searches for her, but he requires repeated responses from her in order to reach her location, perhaps because their body dimensions are small in comparison with the wavelengths in the drummed signals (Goulson et al. 1994). However, determining that the audible tick-tick-tick was only an artifact of the mechanism that produced substrate-borne vibrations that encoded the mating signals does not challenge the dominant paradigm. It simply requires a revised interpretation based on emerging evidence.

If we then continue to standardize our terminology, increase accessibility of our work, and broaden the scope of study, there are at least two areas that require additional efforts if we are to successfully challenge the common ground held with those who study airborne vibrations and hearing. As of yet, we do not have a consistently used metric to quantify substrate-borne vibration amplitude, and this impedes communication among researchers across disciplines. Airborne vibration amplitude is described in decibels, and even if others use different metrics at times, everyone knows what is meant by dB. We do not have a common amplitude metric for substrate-borne vibration studies, but perhaps a candidate is available in the wings, waiting for an advocate to step forward. Secondly, as a community, we must address the issue of waveforms and work beyond their mystery. Those who study airborne vibrations do not have to condition playback signals when designing manipulative experiments, because the medium of our atmosphere does not filter multi-frequency signals the way that substrates do. When researchers try to switch from sound to vibration studies, they sometimes try to play a signal recorded from a natural source through the substrate without any modifications, and the observations they make are compromised by this. Indeed, we might say the observations are without merit of any consideration because the experimental design is flawed. Even some editors do not understand why the signal conditioning is vital to substrate-borne vibrational studies and trivial to airborne vibrational studies. Both of these issues may be easily addressed, but they could represent serious barriers to integration of vibrational studies within the culture of animal communication.

## 2.2 Conclusion

The body of evidence documenting the importance of substrate-borne vibrational communication across arthropod and vertebrate taxa, across millions of years of the Earth's history, and throughout the Earth's habitats is growing in volume and richness. Using emerging technology, as well as basic empirical hypothesis testing, vibrational communication systems are being confirmed wherever questions are being asked. When we look for use of substrate-borne vibrational signaling, we find it. In making this wealth of information more accessible to the broader community, and in establishing vibrational communication as a component of the animal communication paradigm with the same standing as communication via sound, visual, or chemical signals, we must police our own use of descriptive terminology. Before we can speak authoritatively to the broader scientific and lay communities and command their attention, we must first define the limits of terminology we use to speak to each other. Our shared core concepts can be accommodated under the umbrella of the existing animal communication paradigm. We simply need to make sure we speak with a collective voice in our own very civil, and short-lived, revolution.

**Acknowledgments** I thank Hannelore Hoch and Andreas Wessel for the invitation to speak at the *Entomologentagung* of the *Deutschen Gesellschaft für allgemeine und angewandte Entomologie (DGaaE)* in 2011 and support to travel to Berlin. I also thank Rex Cocroft, Matija

Gogala, and Andreas for inviting me to contribute a chapter and to serve as co-editor of this volume. As always, I thank John Shadley for guiding my first forays into the study of vibrational communication and for his teaching, mentorship, and friendship over the years.

## References

Aicher B, Tautz J (1990) Vibrational communication in the fiddler crab, *Uca pugilator*. I. Signal transmission through the substratum. J Comp Physiol A 166:345–353

Anderson CJ (1973) Animals, earthquakes, and eruptions. Field Mus Nat Hist Bull 44:9–11

Barth FG (1982) Spiders and vibratory signals: sensory reception and behavioral significance. In: Witt PN, Rovner JS (eds) Spider communication. Princeton University Press, Princeton, pp 67–122

Birch MC, Keenlyside JJ (1991) Tapping behavior is a rhythmic communication in the death-watch beetle, *Xestobium rufovillosum* (Coleoptera: Anobiidae). J Insect Behav 4:257–263

Breidbach O (1986) Studies on the stridulation of *Hylotrupes bajulus* (L.) (Cerambycidae, Coleoptera): communication through support vibration—morphology and mechanics of the signal. Behav Process 12:169–186

Briceno RD, Bonilla F (2009) Substrate vibrations in the scorpion *Centruroides margaritatus* (Scorpiones: Buthidae) during courtship. Int J Trop Biol Cons 57(Suppl 1):267–274

Broad GR, Quicke DLJ (2000) The adaptive significance of host location by vibrational sounding in parasitoid wasps. P Roy Soc Lond B 267:2403–2409

Brownell P, Farley RD (1979a) Detection of vibrations in sand by tarsal sense organs of the nocturnal scorpion, *Paruroctonus mesaensis*. J Comp Physiol A 131:23–30

Brownell P, Farley RD (1979b) Orientation to vibrations in sand by the nocturnal scorpion *Paruroctonus mesaensis*: mechanism of target localization. J Comp Physiol A 131:31–38

Brownell P, Farley RD (1979c) Prey-localizing behaviour of the nocturnal desert scorpion, *Paruroctonus mesaensis*: orientation to substrate vibrations. Anim Behav 27:185–193

Brownell PH (1977) Compressional and surface waves in sand used by desert scorpions to locate prey. Science 197:479–482

Caldwell MB, Johnston GR, McDaniel JG, Warkentin KM (2010) Vibrational signaling in the agonistic interactions of red-eyed treefrogs. Curr Biol 20:1012–1017

Chuche J, Thiery D, Mazzoni V (2011) Do *Scapoideus titanus* (Hemiptera: Cicadellidae) nymphs use vibrational communication? Naturwissenschaften 98:639–642

Claridge MF (1985) Acoustic signals in the Homoptera: behavior, taxonomy, and evolution. Annu Rev Entomol 30:297–317

Claridge MF, Morgan JC, Moulds MS (1999) Substrate-transmitted acoustic signals of the primitive cicada, *Tettigarcta crinita* Distant (Hemiptera Cicadoidea, Tettigarctidae). J Nat Hist 33:1831–1834

Clayton D (2005) Substrate (acoustic/vibrational) communication and ecology of the ghost crab *Ocypode jousseaumei* (Brachyura: Ocypodidae). Mar Freshw Behav Phy 38:53–70

Cocroft RB, Rodríguez RL (2005) The behavioral ecology of insect vibrational communication. Bioscience 55:323–334

De Groot M, Čokl A, Virant-Doberlet M (2011) Search behavior of two hemipteran species using vibrational communication. Cent Eur J Biol 6:756–769

De Souza LR, Kasumovic MM, Judge KA (2011) Communicating male size by tremulatory vibration in a Columbian rainforest katydid, *Gnathoclita sodalist* (Orthoptera: Tettigoniidae). Behaviour 148:341–357

Devetak D (1998) Detection of substrate vibration in Neuropteroidea: a review. Acta Zool Fenn 209:87–94

Devetak D, Amon T (1997) Substrate vibration sensitivity of the leg scolopidial organs in the green lacewing, *Chrysoperla carnea*. J Insect Physiol 43:433–437

Devetak D, Pabst MA, Delakorda SL (2004) Leg chordotonal organs and campaniform sensilla in *Chrysoperla* Steinmann 1964 (Neuroptera): structure and function. Denisia 13:163–171
Dierkes S, Barth FG (1995) Mechanism of signal production in the vibratory communication of the wandering spider *Cupiennius getazi* (Arachnida, Araneae). J Comp Physiol A 176:31–44
Drosopoulos S, Claridge MF (eds) (2006) Insect sounds and communication: physiology, behaviour, ecology and evolution. Taylor & Francis, Boca Raton
Eberhard MJB, Picker MD (2008) Vibrational communication in two sympatric species of Mantophasmatodea (Heelwalkers). J Insect Behav 21:240–257
Eberhard MJB, Lang D, Metscher B, Pass G, Picker MD, Wolf H (2010) Structure and sensory physiology of the leg scolopidial organs in Mantophasmatodea and their role in vibrational communication. Arthropod Struct Dev 39:230–241
Elias DO, Mason AC (2010) Signaling in variable environments: substrate-borne signaling mechanisms and communication behavior in spiders. In: O'Connell-Rodwell CE (ed) The use of vibrations in communication: properties, mechanisms and function across taxa. Transworld, Kerala, pp 25–46
Emerson AE, Simpson RC (1929) Apparatus for the detection of substratum communication among termites. Science 69:648–649
Eriksson A, Anfora G, Lucchi A, Virant-Doberlet M, Mazzoni V (2011) Inter-plant vibrational communication in a leafhopper insect. PLoS ONE 6:e19692
Esch H (1961) Über die Schallerzeugung beim Webetanz der Honigbiene. Z Vergl Physiol 45:1–11
Field LH, Bailey WJ (1997) Sound production in primitive Orthoptera from western Australia: sounds used in defence and social communication in *Ametrus* sp. and *Hadrogryllacris* sp. (Gryllacrididae: Orthoptera). J Nat Hist 31:1127–1141
Fletcher LE, Yack JE, Fitzgerald TD, Hoy RR (2006) Vibrational communication in the cherry leaf roller caterpillar *Caloptilia serotinella* (Gracillariodea: Gracillariidae). J Insect Behav 19:1–18
Gerhardt HC, Huber F (2002) Acoustic communication in insects and anurans: common problems and diverse solutions. University of Chicago Press, Chicago
Gibson JS, Uetz GW (2008) Seismic communication and mate choice in wolf spiders: components of male seismic signals and mating success. Anim Behav 75:1253–1262
Gogala M (1985) Vibrational communication in insects (biophysical and behavioural aspects). In: Kalmring K, Elsner N (eds) Acoustic and vibrational communication in insects. Paul Parey, Berlin, pp 117–126
Gogala M, Čokl A, Drašlar K, Blaževič A (1974) Substrate-borne sound communication in Cydnidae (Heteroptera). J Comp Physiol 94:25–31
Goulson D, Birch MC, Wyatt TD (1994) Mate location in the deathwatch beetle, *Xestobium rufovillosum* De Geer (Anobiidae): orientation to substrate vibrations. Anim Behav 47:899–907
Hebets EA, Uetz GW (1999) Female responses to isolated signals from multimodal male courtship displays in the wolf spider genus *Schizocosa* (Araneae: Lycosidae). Anim Behav 57:865–872
Henry CS (1979) Acoustical communication during courtship and mating in the green lacewing *Chrysopa carnea* (Neuroptera: Chrysopidae). Ann Entomol Soc Am 72:68–79
Heth G, Frankenberg E, Raz A, Nevo E (1987) Vibrational communication in subterranean mole rats (*Spalax ehrenbergi*). Behav Ecol Sociobiol 21:31–33
Hill PSM (2008) Vibrational communication in animals. Harvard University Press, Cambridge
Hill PSM (2009) How do animals use substrate-borne vibrations as an information source? Naturwissenschaften 96:1355–1371
Hill PSM (2010) Introduction. In: O'Connell-Rodwell CE (ed) The use of vibrations in communication: properties, mechanisms and function across taxa. Transworld, Kerala, pp 1–8
Hill PSM, Shadley JR (2001) Talking back: sending soil vibration signals to lekking prairie mole cricket males. Amer Zool 41:1200–1214

Hoch H, Deckert J, Wessel A (2006) Vibrational signalling in a Gondwanan relict insect (Hemiptera: Coleorrhyncha: Peloridiidae). Biol Lett 2:222–224
Howse PE (1964) The significance of sound produced by the termite *Zootermopsis angusticollis* Hagen. Anim Behav 12:284–300
Ichikawa T (1976) Mutual communication by substrate vibrations in the mating behavior of planthoppers (Homoptera: Delphacidae). Appl Entomol Zool 11:8–21
Ishay J, Landau EM (1972) *Vespa* larvae send out rhythmic hunger signals. Nature 237:286–287
Ishay J, Schwartz A (1973) Acoustical communication between the members of the oriental hornet (*Vespa orientalis*) colony. J Acoust Soc Am 53:640–649
Ishay J, Motro A, Gitter S, Brown MB (1974) Rhythms in acoustical communication by the oriental hornet, *Vespa orientalis*. Anim Behav 22:741–744
Kalmring K, Elsner N (eds) (1985) Acoustic and vibrational communication in insects. Paul Parey, Berlin
Kanmiya K (2006a) Communication by vibratory signals in Diptera. In: Drosopoulos S, Claridge MF (eds) Insect sounds and communication: physiology, behaviour, ecology and evolution. Taylor & Francis, Boca Raton, pp 381–396
Kanmiya K (2006b) Mating behavior and vibratory signals in whiteflies (Hemiptera: Aleyrodidae). In: Drosopoulos S, Claridge MF (eds) Insect sounds and communication: physiology, behaviour, ecology and evolution. Taylor & Francis, Boca Raton, pp 365–379
Kasper J, Hirschberger P (2005) Stridulation in *Aphodius* dung beetles: songs and morphology of stridulatory organs in North American *Aphodius* species (Scarabaeidae). J Nat Hist 39:91–99
Kirchner WH (1997) Acoustical communication in social insects. In: Lehrer M (ed) Orientation and communication in arthropods. Birkhäuser Verlag, Basel, pp 273–300
Kočárek P (2010) Substrate-borne vibrations as a component of intraspecific communication in the groundhopper *Tetrix ceperoi*. J Insect Behav 23:348–363
Kojima W, Takanashi T, Ishikawa Y (2011) Vibratory communication in the soil: pupal signals deter larval intrusion in a group-living beetle *Trypoxylus dichotoma*. Behav Ecol Sociobiol 66:171–179
Kuhn TS (1996) The structure of scientific revolutions, 3rd edn. University of Chicago Press, Chicago
Lakes-Harlan R, Strauβ J (2006) Developmental constraint of insect audition. Front Zool 3:20. doi:10.1186/1742-9994-3-20
Lewis ER (1984) Inertial motion sensors. In: Bolis L, Keynes RD, Maddrell SHP (eds) Comparative physiology of sensory systems. Cambridge University Press, Cambridge, pp 587–610
Lewis LA, Schneider SS (2000) The modulation of worker behavior by the vibration signal during house hunting in swarms of the honeybee, *Apis mellifera*. Behav Ecol Sociobiol 48:154–164
Markl H (1983) Vibrational communication. In: Huber F, Markl H (eds) Neuroethology and behavioral physiology. Springer, Berlin, pp 332–353
Masters WM, Tautz J, Fletcher NH, Markl H (1983) Body vibration and sound production in an insect (*Atta sexdens*) without specialized radiating structures. J Comp Physiol A 150:239–249
McNett GD, Miles RN, Homentcovschi D, Cocroft RB (2006) A method for two-dimensional characterization of animal vibrational signals transmitted along plant stems. J Comp Physiol A 192:1245–1251
McVean A, Field LH (1996) Communication by substratum vibration in the New Zealand tree weta, *Hemideina femorata* (Stenopelmatidae: Orthoptera). J Zool 239:101–122
Mello FAG, Dos Reis JC (1994) Substrate drumming and wing stridulation performed during courtship by a new Brazilian cricket (Orthoptera: Grylloidea: Phalangopsidae). J Orthopt Res 2:21–24
Michelsen A, Fink F, Gogala M, Traue D (1982) Plants as transmission channels for insect vibrational songs. Behav Ecol Sociobiol 11:269–281
Michelsen A, Kirchner WH, Andersen BB, Lindauer M (1986) The tooting and quacking vibration signals of honeybee queens: a quantitative analysis. J Comp Physiol A 158:605–611

Milum VG (1955) Honey bee communication. Am Bee J 95:97–104
Morris GK (1980) Calling display and mating behaviour of *Copiphora rhinoceros* Pictet (Orthoptera: Tettigoniidae). Anim Behav 28:42–51
Nieh JC (1993) The stop signal of honey bees: reconsidering its message. Behav Ecol Sociobiol 33:51–56
O'Connell-Rodwell CE (ed) (2010) The use of vibrations in communication: properties, mechanisms and function across taxa. Transworld, Kerala
Ossiannilsson F (1949) Insect drummers. A study on the morphology and function of the sound-producing organ of Swedish Homoptera Auchenorrhyncha with notes on their sound production. Opuscula Entomol Suppl X:1–146
Pearman JV (1928) On sound production in the Psocoptera and on a presumed stridulatory organ. Entomol Monog Mag 64(Third series v14):179-186
Pratte M, Jeanne RL (1984) Antennal drumming behavior in *Polistes* wasps (Hymenoptera: Vespidae). Z Tierpsychol 66:177–188
Quirici V, Costa FG (2005) Seismic communication during courtship in two burrowing tarantula spiders: an experimental study on *Eupalaestrus weijenberghi* and *Acanthoscurria suina*. J Arachnol 33:159–166
Rado R, Levi N, Hauser H, Witscher J, Adler N, Intrator N, Wollberg A, Terkell J (1987) Seismic signalling as a means of communication in a subterranean mammal. Anim Behav 35:1249–1251
Randall JA (2001) Evolution and function of drumming as communication in mammals. Am Zool 41:1143–1156
Randall JA (2010) Drummers and stompers: vibrational communication in mammals. In: O'Connell-Rodwell CE (ed) The use of vibrations in communication: properties, mechanisms and function across taxa. Transworld, Kerala, pp 99–120
Roces F, Tautz J, Hölldobler B (1993) Stridulation in leaf-cutting ants: short-range recruitment through plant-borne vibrations. Naturwissenschaften 80:521–524
Rovner JS (1975) Sound production by Nearctic wolf spiders: a substratum-coupled stridulatory mechanism. Science 190:1309–1310
Rovner JS (1980) Vibration in *Heteropoda venatoria* (Sparassidae): a third method of sound production in spiders. J Arachnol 8:193–200
Rovner JS, Barth FG (1981) Vibratory communication through living plants by a tropical wandering spider. Science 214:464–466
Rupprecht R (1974) Vibrationssignale bei der Paarung von *Panorpa* (Mecoptera/Insecta). Experientia 30:340–341
Rupprecht R (1975) Die Kommunikation von *Sialis* (Megaloptera) durch Vibrationsignale. J Insect Physiol 21:305–320
Salmon M, Horch KW (1972) Acoustic signaling and detection by semiterrestrial crabs of the family Ocypodidae. In: Winn HE, Olla BL (eds) Behavior of marine animals. Plenum Press, New York, pp 60–96
Sandeman DC, Tautz J, Lindauer M (1996) Transmission of vibration across honeycombs and its detection by bee leg receptors. J Exp Biol 199:2585–2594
Schwartzkopff J (1974) Mechanoreception. In: Rockstein M (ed) The physiology of Insecta, vol 2. Academic Press, New York, pp 273–352
Seeley TD (1992) The tremble dance of the honey bee: message and meanings. Behav Ecol Sociobiol 31:375–383
Seeley TD, Weidenműller A, Kűhnholz S (1998) The shaking signal of the honey bee informs workers to prepare for greater activity. Ethology 104:10–26
Shaw KC, Carlson OV (1979) Morphology of the tymbal organ of the potato leafhopper *Empoasca fabae* Harris (Homoptera: Cicadellidae). J Kansas Entomol Soc 52:701–711
Shaw KC, Vargo A, Carlson OV (1974) Sounds and associated behavior of some species of *Empoasca* (Homoptera: Cicadellidae). J Kansas Entomol Soc 47:284–307
Snarr KA (2005) Seismic activity response as observed in mantled howlers (*Alouatta palliata*), Cuero y Salado Wildlife Refuge, Honduras. Primates 46:281–285

Stewart KW (2008) Vibrational communication. In: Capinera JL (ed) Encyclopedia of Entomology, 2nd edn. Springer, London, pp 4103–4105

Strauβ J, Lakes-Harlan R (2008a) Neuroanatomy and physiology of the complex tibial organ of an atympanate Ensiferan, *Ametrus tibialis* (Brunner von Wattenwyl, 1888) (Gryllacrididae, Orthoptera) and evolutionary implications. Brain Behav Evolut 71:167–180

Strauβ J, Lakes-Harlan R (2008b) Neuroanatomy of the complex tibial organ of *Stenopelmatus* (Orthoptera: Ensifera: Stenopelmatidae). J Comp Neurol 511:81–91

Strauβ J, Lakes-Harlan R (2009) The evolutionary origin of auditory receptors in Tettigonioidea: the complex tibial organ of Schizodactylidae. Naturwissenschaften 96:143–146

Strauβ J, Lakes-Harlan R (2010) Neuroanatomy of the complex tibial organ in the splay-footed cricket *Comicus calcaris* Irish 1986 (Orthoptera: Ensifera: Schizodactylidae). J Comp Neurol 518:4567–4580

Strauβ J, Lakes-Harlan R (2013) Sensory neuroanatomy of stick insects highlights the evolutionary diversity of the orthopteroid subgenual organ complex. J Comp Neurol doi: 10.1002/cne.23378. [Epub ahead of print]

Stritih N, Čokl A (2012) Mating behavior and vibratory signalling in non-hearing cave crickets reflect primitive communication of Ensifera. PLoS ONE 7:e47646

Tishechkin DY (2006) Vibratory communication in Psylloidea (Hemiptera). In: Drosopoulos S, Claridge MF (eds) Insect sounds and communication: physiology, behaviour, ecology and evolution. Taylor & Francis, Boca Raton, pp 357–363

Theiβ J (1982) Generation and radiation of sound by stridulating water insects as exemplified by the Corixids. Behav Ecol Sociobiol 10:225–235

Tributsch H (1982) When the snakes awake: animals and earthquake prediction. MIT, Cambridge

Uetz GW, Stratton GE (1982) Acoustic communication and reproductive isolation in spiders. In: Witt PN, Rovner JS (eds) Spider communication. Princeton University Press, Princeton, pp 123–159

Virant-Doberlet M, Čokl A (2004) Vibrational communication in insects. Neotrop Entomol 33:121–134

Virant-Doberlet M, King RA, Polajnar J, Symondson WOC (2011) Molecular diagnostics reveal spiders that exploit prey vibrational signals used in sexual communication. Mol Ecol 20:2204–2216

Weismann DB (2001) Communication and reproductive behavior in North American Jerusalem crickets (*Stenopelmatus*) (Orthoptera: Stenopelmatidae). In: Field LH (ed) The biology of wetas, king crickets and their allies. CAB International, Wallingford, pp 351–375

Wessel A (2006) Stridulation in the Coleoptera—an overview. In: Drosopoulos S, Claridge MF (eds) Insect sounds and communication: physiology, behaviour, ecology and evolution. Taylor & Francis, Boca Raton, pp 397–403

Wignall AE, Herberstein ME (2013) The influence of vibratory courtship on female mating behavior in orb-web spiders (*Argiope keyserlingi*, Karsch 1878). PLoS ONE 8:e53057

Yack JE, Smith ML, Weatherhead PJ (2001) Caterpillar talk: acoustically mediated territoriality in larval Lepidoptera. P Nat Acad Sci USA 98:11371–11375

Zeigler DD, Stewart KW (1977) Drumming behavior of eleven Nearctic stonefly (Plecoptera) species. Ann Entomol Soc Am 70:495–505

# Chapter 3
# Sound or Vibration, an Old Question of Insect Communication

**Matija Gogala**

**Abstract** About one hundred years ago, one of the pioneers of bioacoustics, Johann (Ivan, Joannes) Regen, born in Slovenia and living later in Vienna, investigated acoustic communication in crickets and bushcrickets. Despite many convincing results, he had a difficult dispute with a physiologist Otto Ernst Mangold to prove his ideas about airborne sound communication in insects. Eventually, he succeeded to persuade him with a series of imaginative experiments. However, his findings are by far not valid for all groups of insects. When I started to investigate acoustic communication in Heteroptera with my students and coworkers about half a century later, the question of their communication channel was not clear. After some critical experiments, it became evident that they emit and receive substrate-borne vibrational signals. Similar experiments were performed with "small cicadas" by Ichikawa, Strübing and Traue, who also came to the conclusion that they use substrate vibration as a communication channel. Nowadays, we know that the majority of Hemiptera and also many other insects use the vibrational channel for acoustic communication, some others use true sound or near field airborne vibrations, but not to forget acoustic signalization in aquatic and semiaquatic insects. However, some insects apparently use both channels for acoustic communication or orientation.

## 3.1 Introduction

From old books and prints, we know that people have been aware of insect sounds for centuries and even millennia. They included singing insects in pictures and mentioned them also in texts and poems. In some cultures, especially in the Far

M. Gogala (✉)
Slovenian Academy of Sciences and Arts, Ljubljana, Slovenia
e-mail: matija.gogala@guest.arnes.si

R. B. Cocroft et al. (eds.), *Studying Vibrational Communication*,
Animal Signals and Communication 3, DOI: 10.1007/978-3-662-43607-3_3,

East, people have appreciated from the old days until now the songs of crickets and keep them in captivity to hear and enjoy their voices. The old Greeks knew such details, like that only male cicadas emit loud songs and females do not. The Greek poet Xenarchus (fourth century BC) mentioned this in one of his poems with appreciation and a reflection on human life. Xenarchus says in the *Sleep*: *Are then the male cicadas not happy, say you? When they have wives who cannot speak a word?* (Athenaeus 1854). However, they could not know much about the details of sound production or reception and did not care how insects communicated among themselves.

Only with the invention of the microscope and development of natural sciences in the seventeenth and eighteenth century were papers with first descriptions of sound producing organs published (e.g., tymbal: Casserius 1600, stridulatory structures in bugs: Ray 1710). There appeared during the same time also first descriptions of acoustic behavior in some singing insects (e.g., Rösel von Rosenhof 1746–1755; Poda 1761: p. 58, "*Cimex iracundus* sonum edit"). However, the question of how insects communicate, and if they can receive airborne sound or just substrate vibration, scientists began to discuss much later, at the end of the nineteenth century.

## 3.2 About 100 Years Ago…Ernst Mangold and Johannes Regen

Just about 100 years ago, Ernst Mangold wrote in a renowned German Handbook of Physiology (Mangold 1913) the chapter on hearing and static senses in vertebrates and invertebrates. Pages 885–898 and 905–906 are devoted to insects where the author critically presented and discussed the observations and experiments on this topic that had been published during past decades. He cited publications of V. Graber (1875, 1877, and 1882), who reported simple behavioral reactions of insects to vibrations and sound. However, he mentioned that there were still no exact proofs for biologically relevant reactions to acoustic cues, for instance for orientation of grasshopper females to the singing males.

He mentioned and cited also papers of W. Nagel (1892), L. Oyen (1901), E. Radl (1905), and J. Regen (1909). Mangold's conclusion in his book chapter was that there was no clear evidence for hearing ability of insects despite the opposite but convincing results of Regen's experiments (Figs. 3.3, 3.4 and 3.5). Nevertheless, he admitted (Mangold 1913, p. 887) that:

> Die interessanten Versuche von Regen (s. weiter unten!) an Orthopteren machen denn hier auch einen willkommenen Anfang
>
> The interesting experiments of Regen (see below!) on Orthoptera make for a welcome beginning here.

Mangold expressed his skepticism further on the p. 888, where he wrote:

> Ausserordentlich bemerkenswert erscheinen die Resultate der Beobachtungen von Regen (349) an Männchen von *Thamnotrizon apterus* Fab., da sich daraus ein nicht unwesentlicher Einfluss der tympanalen Sinnesorgane auf die eigene Stridulation der Tiere ergab. Nur lassen sich leider zunächst noch gar zu viele Einwände machen,

> The results of Regen's observations on males of *Thamnotrizon apterus* Fab. appear highly remarkable, showing a significant influence of the tympanic sensory organs on their own stridulation. Unfortunately, though, for the time being too many objections can still be raised.

After this skepticism, Mangold wrote in his chapter (p. 889):

> … die Frage offen bleibt, ob die Tympanalorgane imstande sind, das Stridulationsgeräusch durch Luftleitung als Reiz anzunehmen, oder ob die Übertragung nicht vielmehr nur durch den festen Untergrund, auf dem die Tiere sitzen, vermittelt wird. Letzteres scheint mir nach Regen's Versuchen zunächst das Wahrscheinlichere …

> … the question persists, whether the tympanal organs are capable to receive a stridulation sound as an airborne signal, or rather as vibrations transmitted via the solid substrate on which the animals are sitting. The latter seems to me according to Regen's experiments more probable …

At this point, I would like to introduce both persons, involved in this dispute, Mangold and Regen (Figs. 3.1 and 3.2).

Otto Ernst Mangold (Fig. 3.1) was born on February 5, 1879 in Berlin and died in Hahnenklee-Bockswiese (today a borough of Goslar) on July 10, 1961. He studied medicine and zoology in Jena, Germany. In the year 1905, he received habilitation in zoology and began to teach physiology at the universities in Jena, Greifswald, and finally Freiburg. In the year 1923, he returned to Berlin, where he worked as professor of animal physiology at the Friedrich-Wilhelms-University at the School of Agriculture. In 1933, in the Nazi time, he was eliminated from the University and reactivated only in 1945. In 1921, he was elected as a member of the German Academy Leopoldina. He was known for his strong criticism. More details about his life one can read at the website http://www.sammlungen.hu-berlin.de/dokumente/7679/.

Johann or Ioannes Regen (Fig. 3.2), in his homeland called Janez (Ioannes) or Ivan, was born on December 9, 1868 in a small village, Lajše in Poljanska valley, not far from Škofja Loka in the country that is nowadays Slovenia. At that time, it was a duchy, Krain, in the Austrian-Hungarian monarchy. Regen studied biology at the University of Vienna and defended his doctoral thesis in 1897 [Einige Beobachtungen über die Stridulationsorgane der saltatoren Orthopteren—Some observations on the stridulatory organs of Orthoptera (Saltatoria)]. He devoted his research mainly to questions of sound production, transmission, and perception of acoustic signals in insects and is known as one of the founders of the modern bioacoustics of insects. His main experimental animals were crickets, *Gryllus campestris* (in most of his papers referred to as *Liogryllus campestris*), and bushcrickets, *Pholidoptera aptera* (in Regen's works *Thamnotrizon apterus*).

**Fig. 3.1** Prof. Otto Ernst Mangold (5. 2. 1879–10. 7. 1961) (Archive of the Humboldt University, Berlin, with permission)

**Fig. 3.2** Prof. Ivan Regen (9. 12. 1868–27. 7. 1947) (Library of the Slovenian Academy of Sciences and Arts)

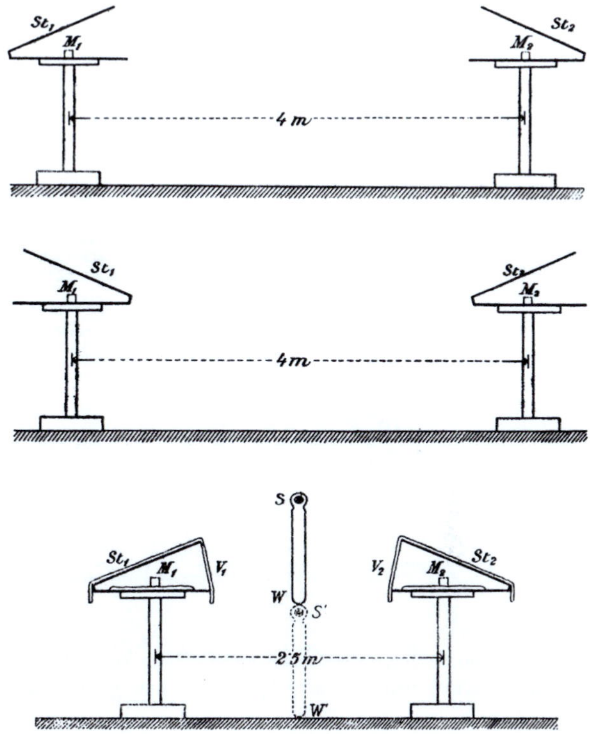


**Fig. 3.3** Regen's experiments with *Pholidoptera aptera* bushcrickets. *Above*, males in the cages $M_1$ and $M_2$ in the rectangular funnels $St_1$ and $St_2$ alternated regularly. If the funnels were rotated for 180° so that the transmission of sound was reduced, the alternation was interrupted, or better, did not occur. In the experiment shown *below*, the funnels were oriented to each other with the open end and insulated by a cotton wool material. When the sevenfold insulation curtain was raised to the upper position S W in complete darkness, the coordinated alternation between males was discontinued (adapted from Regen 1914)

He lived and worked in Vienna and organized his private laboratories. However, this would not have been possible without the financial support of his friend Willy Gutmann and partly also by the Austrian Academy of Sciences. He had vivid contacts with his homeland all the time, became after the establishment of the Academy of Sciences and Arts in Slovenia its corresponding member, and was also invited (1921) to become a professor at the newly founded University of Ljubljana. He was also one of the founders and an honorary member of the Natural History Society of Slovenia. For various reasons, also due to weak possibilities for research there, he decided to remain in Vienna, where he died on July 27, 1947.

He did not publish many papers (about 25), but among them are some that are very important or even crucial for understanding sound communication in insects. In his short paper, published in 1908 about the alternation behavior of *Thamnotrizon apterus*, he claimed that only males with intact tympanal organs were able to respond regularly to the chirps of another male. As mentioned before, Mangold (1913) did not accept his results as a proof for sound communication in insects, nor as a proof that tympanal organs are indeed true hearing organs.

In the following years, Regen published some new papers with a detailed explanation of his experiments on *Thamnotrizon apterus* and *Liogryllus campestris*. In the paper "Untersuchungen über die Stridulation und das Gehör von *Thamnotrizon apterus* Fab."(Regen 1914), he answered exactly all open questions put by

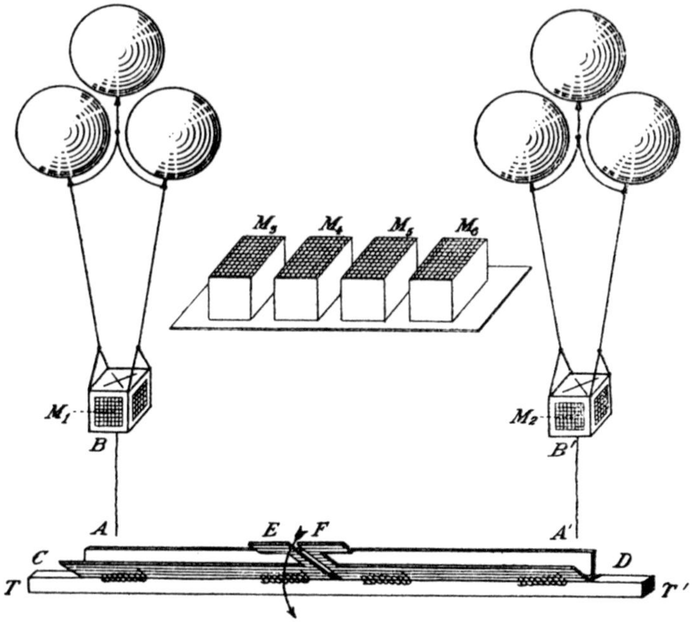


**Fig. 3.4** Another interesting experiment of Ivan Regen with *Pholidoptera aptera*. Some males in the cages on the shelf were singing and alternating with the males in the air (Regen 1914)

Mangold. In the series of experiments based on sound alternation in funnels, enabling sound propagation only in one direction, with and without sound insulation material in between, Regen succeeded in proving that these bushcrickets perceive and react to airborne sound (Fig. 3.3). For the final proof, he let the bushcrickets alternate in the air in small paper cages suspended below hydrogen-filled balloons without any contact with the substrate, where the conspecific males were singing (Fig. 3.4).

The biological function of the male song of (*Lio*)*gryllus campestris* he showed in another famous experiment using the telephone for transmission of a cricket male's song to attract a virgin female (Regen 1913, Fig. 3.5). Later, he organized a large-scale experiment on phonotaxis of female crickets toward singing males in a huge insectarium with a 576 $m^2$ surface area. He called it the "geobiological laboratory" (Fig. 3.6). He used 1600 female crickets in the peripheral part of the experimental field and some males in the central part. Around the singing males, he put traps with electric contacts in such a way that he recorded each capture of females approaching the singing male. He collected the animals, marked them, and released them again in their holes. One part (half) of the females had tympanal organs destroyed. He could show that only animals with intact tympanal organs in the legs were able to locate the singing male and showed efficient positive phonotaxis (Regen 1928).

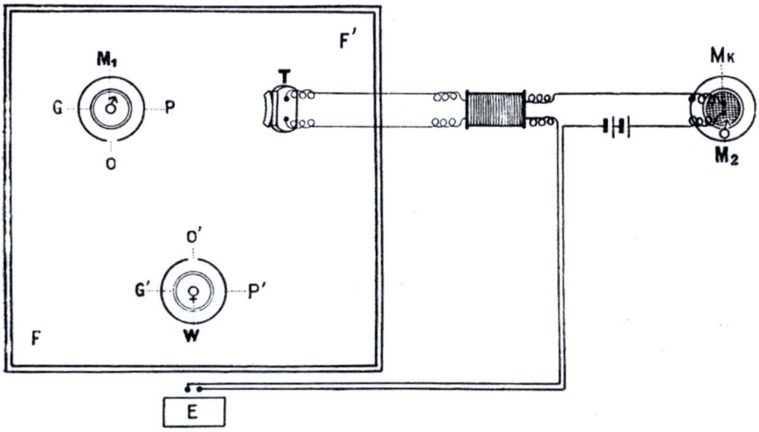


**Fig. 3.5** The schematic drawing of the famous Regen experiment with attraction of a *Gryllus campestris* female to the telephone speaker, which was transmitting the calling song of a male, $M_2$, from a distant room. The experimental female was not attracted to the other silent male sitting in the chamber, $M_1$ (Regen 1913)

**Fig. 3.6** Geobiological station built by I. Regen for the experiments with *Gryllus campestris*. In the experimental surface of 576 $m^2$, he used 1,600 animals for experiments on the phonotaxis of females to singing males (Library of the Slovenian Academy of Sciences and Arts)

The whole story of the dispute between O. E. Mangold and J. Regen ended with complete victory for the latter. In the year 1924, Mangold wrote a letter to Regen after receiving two recent papers from him (Regen 1922, 1923):

Ich habe sie mit grossem Interesse gelesen und mich besonders gefreut, dass es Ihnen nunmehr durchaus gelungen ist, sichere Beweise für das Hören von Wirbellosen zu erbringen. Sie haben dadurch meine, noch in der Handbucharbeit in Wintersteins Handbuch zum Ausdruck gebrachte Skepsis vollkommen besiegt… (Archive of the Slovenian Academy of Sciences and Arts Library).

In translation his words were: … I have read them with great interest and I am particularly pleased that you certainly have succeeded to provide firm evidence for hearing in invertebrates. So you succeeded to completely defeat my scepticism, that was still expressed in the chapter of the handbook of Winterstein… (see Mangold 1913).

Mangold later in another letter also supported Regen's application to the Austrian Academy to support financially the construction of the new "geobiological laboratory." Despite this, even in 1924, F. E. Lutz wrote in his publication, *Insect sounds*:

… I am not aware of a single experiment that has furnished indisputable evidence of communication between insects by sound… (p. 367).

… The suggested purpose of the well-developed insect sounds, a "sex call" is only imagined: it has not been proved and the chief evidence is that usually the females do not make a sound that we can hear (p. 371).

Also R. E. Snodgrass in his booklet, *Insect musicians, their music and their instruments* (1925), was not convinced about the true (airborne) hearing in insects. He wrote: …

… Experimental evidence of the hearing powers of insects is at present very meager, but it would be surprising if insects do not hear the sounds they themselves produce… (p. 451).

And about the function of tympanal organs in bushcrickets, he also expressed his doubts (p. 417):

…No one can state positively that any of these organs are ears, the principal reasoning in favor of their auditory nature being "if they are not ears, what are they?" …

Nowadays, there is, of course, no question whether some insects are able to receive airborne vibrations, whether the tympanal organs are true hearing organs, or if insects are able to communicate with sound signals.

## 3.3 About 50 Years Ago

Fifty years later, some insect physiologists and bioacousticians wondered about the very low level acoustic signals of many Hemiptera, especially Heteroptera, the missing of obvious sound receptors and the possible role in their intraspecific communication (Dumortier 1963; Haskell 1957; 1961; Jordan 1958; Leston 1954, 1957; Leston and Pringle 1963; Moore 1961).

Approximately 50 years ago, also, I got interested in the acoustic communication of Heteroptera, since I already had observed as a young entomologist in middle school the unusual behavior of the bugs from the family Cydnidae, with body vibrations during courtship and mating. At that time and even during my studies at the university in Ljubljana, I practically did not have any possibilities to record and investigate sounds or vibrations of these insects. On a few occasions, I got the opportunity to make some tape recordings in Radio Ljubljana, but since the time was limited in minutes I only succeeded in recording there some disturbance or alarm sounds. In the fifties, I used a stethoscope for listening in most cases in a similar way to that described by Leston (1954) and Jordan (1958). Later, it turned out that this was actually the best method to get an idea about their sound or vibration emissions. Only after 1965, when I received the basic equipment for bioacoustic investigations (tape recorder Revox A77, Oscilloscope Tektronix 502) from the Alexander von Humboldt foundation, was I able to carry out extensive investigations of the acoustic communication of bugs (Heteroptera).

From my field observations, I knew that many Heteroptera, like Cydnidae, perform courting and mating preferably in the early spring, when most of the other insects are still hidden in overwintering places. During this time period, it was not difficult to observe, listen to, or record complicated premating acoustic signals of various species. The important condition for such experiments and observations was, of course, that the males and females had not copulated before, and so, sexual motivation was at a high level.

One of the most important pioneers in investigations of acoustic signals, signal production and perception in the group of land bugs (Heteroptera, Geocorisae) was the German zoologist Prof. K. H. C. Jordan. In his publication (1958), he described the sound producing mechanisms and sounds of some species from the families Cydnidae, Pentatomidae, and Acanthosomatidae. However, the available technical devices for recording and analysis of sounds were not adequate. He was using, in addition to the condenser microphone and indirectly an oszillograph, a stethoscope in a similar way as it was described by Leston (1954). He did not tackle the question of airborne sound transmission or substrate vibrations in this group of insects. However, he discovered that bugs do not use only stridulatory mechanisms for sound production. His conclusion was that some Pentatomidae and Acanthosomatidae emit sounds by the movement (or deformation) of the first two abdominal terga and dorsoventral vibration of the abdomen.

In the sixties, I studied mainly bioacoustics of various bug species of the family Cydnidae (Gogala 1969, 1970). After describing song repertoires of males and females of single species (Gogala 1969) and showing the species specificity of different genera and species of Cydnidae (Gogala 1970, 1978, Gogala and Hočevar 1990), I began also to investigate with my team the question of communication medium. With limited equipment, we succeeded in gaining enough evidence for a conclusion that investigated species of Heteroptera use the substrate as a communication channel (Gogala et al. 1974, Fig. 3.7) and not the air, as supposed by some other authors mentioned above. We used the alternation in rivalry songs as the criterion for successful communication in a similar way as did Regen many years ago. Only the conclusions with our animals were different. Only vibrations transmitted through the substrate were sufficient in cydnid bugs to elicit alternation

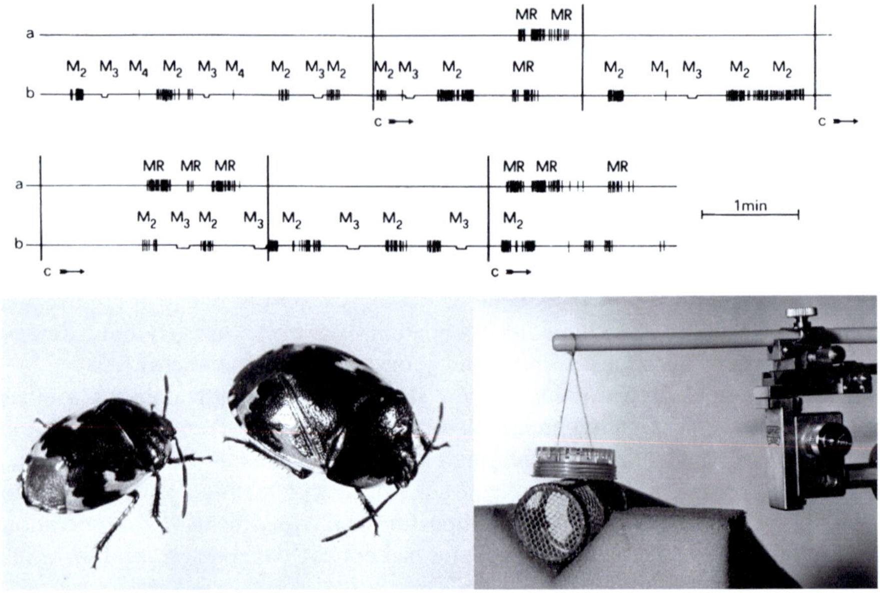


**Fig. 3.7** Simple graphic showing the alternation between a couple of the bug *Tritomegas bicolor* during courtship in one cage and another male in the second cage, just 1–2 mm away. When the cages were brought into contact by micropositioner (C, three times), the alternation with the rivalry song started (adapted from Gogala et al. 1974). *Below left* a couple of *Tritomegas bicolor*, *right* experimental setup with two cages, in the upper cage was a single male, and in the lower cage a couple of bugs (male and female)

between males. So, half a century after Regen's papers proved airborne communication in crickets and bushcrickets, we have proven the opposite for another group of insects–bugs communicating by substrate vibration.

Similar questions about the communication channel were asked around half a century ago by some researchers working on acoustic behavior of small Auchenorrhyncha. There, probably the first proof for vibrational communication in intraspecific behavior was published by Ichikawa (1976). Hildegard Strübing, another pioneer in bioacoustic investigations of Auchenorrhyncha (her first publication on this topic was published in 1958; see also Chap. 5, this volume), in a paper discussing the acoustic communication of *Dictyophara europaea* (Fulgoridae) (1977) came to the following conclusion: "…so sprechen doch alle Indizien für eine Verständigung über Substratvibration" (*…yet all the evidence points to an understanding* via *substrate vibration*). Traue worked in the laboratory of Strübing with *Euscelis incisus* (Cicadomorpha: Cicadellidae) and *Euides speciosa* (Fulgoromorpha: Delphacidae) and published two papers (Traue 1978a, b), where he showed evidence for the vibratory communication in premating behavior of these planthoppers and leafhoppers in a similar way as we did with the Heteroptera (Fig. 3.8).

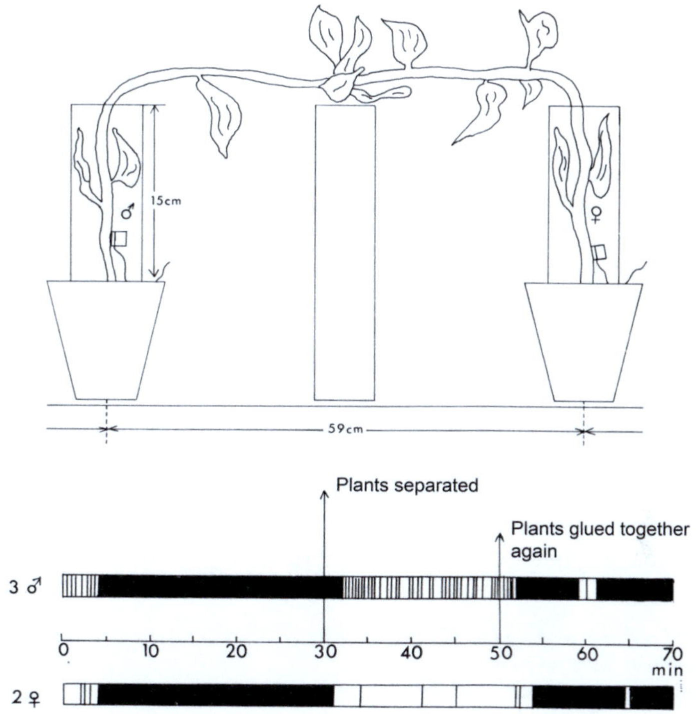


**Fig. 3.8** Experiment with substrate-borne communication of *Euscelis incisus*. Three males in one chamber (*left*) started to communicate by vibrational signals with the two females in the *right chamber* as long as the feeding plants were glued together. When the plants were separated, the acoustic (vibrational) activity fell to a very low level (Traue 1978a)

## 3.4 Recent Investigations and Open Questions

During newer investigations in the last decades, substrate-borne communication of many species of Hemiptera: Heteroptera, Cicadomorpha, and Fulgoromorpha have been studied in detail from ethological and physiological aspects (e.g., Drosopoulos and Claridge 2006; Cocroft and McNett 2006; Čokl and Virant-Doberlet 2003; Gogala 2006; Hill 2008; Michelsen et al. 1982). Many authors using modern techniques also studied in this group of insects signal production mechanisms, transmission of vibrational signals through various substrates, as well as sensory organs and structures. Many results of such investigations (studies) are presented also in this volume. Nevertheless, due to the extreme diversity among insects, and also within the Hemiptera, we have to be open for surprises.

Every sound emission in the air inevitably produces vibrations in the substrate, and vice versa. Therefore, one can expect that many animals are using acoustic signaling either in one, the other, or both media. The question is only if the animals possess suitable sensory structures sensitive for both associated acoustical communication channels and if the vibrations in both media are strong enough to be perceived by the target animal.

There is no doubt that all insects have some kind of mechanoreceptors capable of perceiving substrate vibrations. However, many insects also have true auditory sense organs, which are well known and investigated by many authors from various aspects. But did not we miss in many insect groups, believed to have only vibrational communication, true "ears" that were overlooked? Roeder's discovery of ultrasonic ears in Sphingid moths is a good example of how inconspicuous such sensory organs can be (Roeder et al. 1968). The other good example is the auditory organs of praying mantis (Yager and Hoy 1986, 1987; Yager 1999). The third example for a surprising site of sensory organs was found in parasitoid flies, which search for their prey, singing crickets, by a specialized auditory prothoracic organ. Such ears were first described by Lakes-Harlan and Heller (1992) in the tachinid fly, *Therobia leonidei*, and by Robert et al. in *Ormia ochracea* (Robert et al. 1994, 1996).

The discovery of ultrasonic ears in Sphingid moths by Roeder et al. (1968), but also similarly surprising findings by Miller (1970) on the wings of *Chrysopa*, and by others, who found such ears "on all unlikely places," provoked Pye to write a short poem that was published as a letter in *Nature* (Pye 1968). He pointed out that the ears of insects can be "…on all unlikely places."

In days of old and insects bold
  (Before bats were invented),
  No sonar cries disturbed the skies—
  Moths flew uninstrumented.

The Eocene brought mammals mean
  And bats began to sing;
  Their food they found by ultrasound
  And chased it on the wing.

Now deafness was unsafe because
  The loud high-pitched vibration
  Came in advance and gave a chance
  To beat echolocation.

Some found a place on wings of lace
  To make an ear in haste;
  Some thought it best upon the chest
  And some below the waist.

Then Roeder's key upon the breeze
  Made Sphingids show their paces.
  He found the ear by which they hear
  In palps upon their faces.
Of all unlikely places!

In the systematic group of Hemiptera, which I know best, there are in acoustic communication (in a broad sense, vibrational communication included) some interesting phenomena. In Auchenorrhyncha, or better in the suborder Cicadomorpha, we know a big group of insects with undisputable airborne communication—Cicadidae. And even the closest relatives, Tettigarctidae, apparently use only a substrate-borne vibrational communication (Claridge et al. 1999). On the other hand,

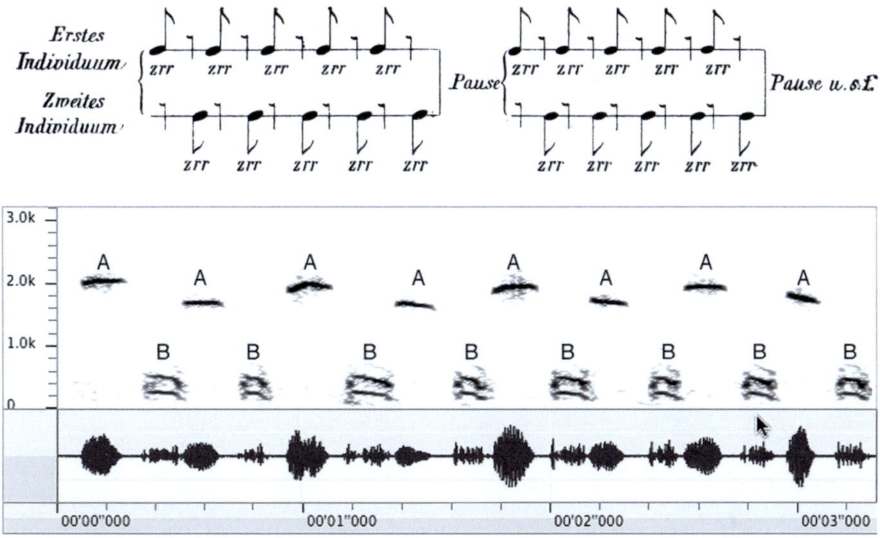


**Fig. 3.9** Alternation between two *Pholidoptera aptera* males as shown in Regen's paper (1914) (*above*) and alternation between a whistling person (*A*) and a bug (*Phymata crassipes*) answering with a non-stridulatory vibrational signal (*B*) (Gogala 2008)

there is an interesting case of cicadas of the genus *Pagiphora*, which emit songs with frequencies much too low compared to resonant frequencies of other cicada species of the same size (Gogala et al. 2005; Trilar and Gogala 2012; Bennet-Clark and Young 1994). Is this due to vibrational communication in these species?

In Heteroptera, the other big group of Hemiptera, we know quite a lot about their vibrational communication (see Čokl et al., Chap. 8, this volume, Gogala 2006). Nevertheless, there are some unsolved questions.

One question is why many species of Heteroptera, in addition to low-frequency body vibration by tremulation or the tymbal system, use stridulation or, according to some authors, "strigilation." Stridulatory signals are usually much higher in frequency and are even audible to the unaided human ear. Are these signals directed toward vertebrates? Why then are they used as an important part of the acoustic or vibratory mating behavior? In the order of Heteroptera, the stridulatory mechanisms evolved independently in many families, genera, and species.

Another question is connected with the interesting acoustic behavior of the Ambush bug, *Phymata crassipes*. It has been shown that this predatory insect responds to vibrational and airborne stimuli with low-frequency vibratory signals (Fig. 3.9) (Gogala and Čokl 1983, Gogala et al. 1984, see also Virant-Doberlet et al. Chap. 20, this volume). Apparently, the human voice produces substrate vibrations strong enough to be perceived by bugs. In these species, the stridulatory apparatus has been known for a long time, but we do not know yet how they produce low-frequency signals. Did we also miss auditory organs in *Phymata*?

Stölting et al. (2002) showed that airborne sounds of *Okanagana rimosa* can produce vibrations strong enough to be perceived by other insects through their vibrational sense organs. Is this also the answer to the questions in the case of *Phymata crassipes*?

Similar questions have been put forward also by Caldwell (Chap. 7, this volume). Anyway, if we look back in the history of bioacoustic research, we can see that some authors put forward true vibrational communication and others airborne acoustic communication. But we should be aware that animals are complex organisms with a variety of sensors in their body and can react to various stimuli in the environment in such a way that they do the best for them and their species based on the limits of their reaction norms and adaptations. And this is true also for acoustic or vibrational communication in the broadest sense.

## References


Athenaeus (1854) The deipnosophists—Book 13, "About Women", pp 555-571, Trans Yonge CD A few words (tettigia are cicadas, not grasshoppers) and spellings have been changed http://www.attalus.org/old/athenaeus13a.html. Cited 4 Aug 2013

Bennet-Clark HC, Young D (1994) The scaling of song frequency in cicadas. J Exper Biol 191:291–294

Casserius J (1600) De vocis auditusque organis historia anatomica. Baldini, Ferrara, p 317

Claridge MF, Morgan JC, Moulds MS (1999) Substrate-transmitted acoustic signals of the primitive cicada, *Tettigarcta crinita* Distant (Hemiptera Cicadoidea, Tettigarctidae). J Nat Hist 33:1831–1834

Cocroft RB, McNett GD (2006) Vibrational communication in treehoppers (Hemiptera: Membracidae). In: Drosopoulos S, Claridge MF (eds) Insect sounds and communication: physiology, behaviour, ecology and evolution. Taylor & Francis, Boca Raton, pp 305–317

Čokl A, Virant-Doberlet M (2003) Communication with substrate-borne signals in small plant-dwelling insects. Annu Rev Entomol 48:29–50

Drosopoulos S, Claridge MF (2006) Insect sounds and communication: physiology, behaviour, ecology and evolution. Taylor & Francis, Boca Raton, p 532

Dumortier B (1963) Acoustical behaviour of Hemiptera. In: Busnell RG (ed) Acoustic behaviour of animals. Elsevier, Amsterdam, pp 391–411, 798–799

Gogala M (1969) Die akustische Kommunikation bei der Wanze *Tritomegas bicolor* (L.) (Heteroptera, Cydnidae). Z vergl Physiol 63:379–391

Gogala M (1970) Artspezifität der Lautäusserungen bei Erdwanzen (Heteroptera, Cydnidae). Z vergl Physiol 70:20–28

Gogala M (1978) Acoustic signals of four bug species of the fam. Cydnidae (Heteroptera) (*in Slovenian*). Biol vestn (Ljubljana) 26:153–168

Gogala M (2006) Vibratory signals produced by Heteroptera—Pentatomorpha and Cimicomorpha. In: Drosopoulos S, Claridge MF (eds) Insect sounds and communication: physiology, behaviour, ecology and evolution. Taylor & Francis, Boca Raton, pp 275–295

English edition: Gogala M (2008) Pioneer of Bioacoustics Ivan Regen and his Legacy (trans: Gogala M). In: Fabjancic M, Merhar D, Samec D, Koman D (eds) Library seventy years of the Slovenian Academy of Sciences and Arts, pp 237-261. Slovenian Academy of Sciences and Arts, Ljubljana

Gogala M, Čokl A (1983) The acoustic behaviour of the bug *Phymata crassipes* (F.) (Heteroptera). Rev Can Biol Exptl 42:249–256

Gogala M, Hočevar I (1990) Vibrational songs in three sympatric species of *Tritomegas*. Scopolia 1:117–123
Gogala M, Čokl A, Drašlar K, Blaževič A (1974) Substrate-borne sound communication in Cydnidae. J Comp Physiol 94:25–31
Gogala M, Virant M, Blejec A (1984) Mocking bug *Phymata crassipes* (Heteroptera). Acoust Lett 8:44–51
Gogala M, Trilar T, Krpach VT (2005) Fauna of singing cicadas (Auchenorrhyncha: Cicadoidea) of Macedonia—a bioacoustic survey. Acta Entomol Slovenica 13(2):103–126
Haskell PT (1957) Stridulation and its analysis in certain Geocorisae (Hemiptera, Heteroptera). P Zool Soc Lond 129:351–358
Haskell PT (1961) Insect sounds. Witherby, London
Hill PSM (2008) Vibrational communication in animals. Harvard University Press, Cambridge
Ichikawa T (1976) Mutual communication by substrate vibrations in the mating behavior of planthoppers (Homoptera: Delphacidae). Appl Ent Zool 11:8–21
Jordan KHC (1958) Lautäusserungen bei den Hemipteren-Familien der Cydnidae, Pentatomidae und Acanthosomidae. Zool Anz 161(5/6):130–144
Lakes-Harlan R, Heller K-G (1992) Ultrasound-sensitive ears in a parasitoid fly. Naturwissenschaften 79:224–226
Leston D (1954) Strigils and stridulation in Pentatomoidea (Hem.): some new data and a review. Ent Month Mag 90:49–56
Leston D (1957) The stridulatory mechanisms in terrestrial species of Hemiptera Heteroptera. P Zool Soc Lond 128:369–386
Leston D, Pringle WS (1963) Acoustical behaviour of Hemiptera. In: Busnell RG (ed) Acoustic behaviour of animals. Elsevier, Amsterdam, pp 391–411, 798–799
Lutz FE (1924) Insect sounds. Bull Amer Mus Nat Hist 50:333–372
Mangold E (1913) Gehörsinn und statischer Sinn. A, VII Insekten und Spinnen; VIII Zusammenfassung der Ergebnisse bei Wirbellosen. In: Winterstein H (ed) Handbuch der vergleichenden Physiologie IV, pp 885–898, 905–906. http://www.archive.org/stream/handbuchdervergl04wint#page/n5/mode/2up. Cited 4 Aug 2013
Michelsen A, Fink F, Gogala M, Traue D (1982) Plants as transmission channels for insect vibrational songs. Behav Ecol Sociobiol 11:269–281
Miller LA (1970) Structure of the green lacewing tympanal organ (*Chrysopa carnea*, Neuroptera). J Morphol 131:359–382
Moore TE (1961) Audiospectrographic analysis of sounds of Hemiptera and Homoptera. Ann Ent Soc Am 54:273–291
Poda F (1761) Insecta Musei Graecensis. Graz, p 168
Pye JD (1968) How insects hear. Nature 218:797
Ray J (1710) Historia insectorum, London
Regen I (1913) Über die Anlockung des Weibchens von *Gryllus campestris* L. durch telephonisch übertragene Stridulationslaute des Männchens. Ein Beitrag zur Frage der Orientierung bei den Insekten. Pflüg Arch ges Phys 155:193–200
Regen I (1914) Untersuchungen über die Stridulation und das Gehör von Thamnotrizon apterus Fab. Sitzber Akad Wien, Mathematisch-naturwissenschaftliche Klasse, Abt I 123:853–892
Regen I (1922) Der Kropf von Liogryllus campestris L. als Organ zur Aufnahme von Luft zur Zeit der Häutung. Sitzber Akad Wien, Mathematisch-naturwissenschaftliche Klasse, Abt III 131:21–23
Regen J (1923) Über die Orientierung des Weibchens von Liogryllus campestris L. nach dem Stridulationsschall des Männchens. Sitzber Akad Wien, Mathematisch-naturwissenschaftliche Klasse, Abt I 132(4–6):81–88
Regen J (1928) Besitzen die Insekten einen Gehörsinn? Forsch Fort (Berlin) 4(5):47–48
Robert D, Read MP, Hoy RR (1994) The tympanal hearing organ of the parasitoid fly *Ormia ochracea* Diptera, Tachinidae, Ormiini. Cell Tissue Res 275:63–78

Robert D, Edgecomb RS, Read MP, Hoy RR (1996) Tympanal hearing in tachinid flies (Diptera, Tachinidae, Ormiini): the comparative morphology of an innovation. Cell Tissue Res 284:435–448

Rösel von Rosenhof AJ (1746–1755) Insektenbelustigungen. Nürnberg

Roeder KD, Treat AE, Vandeberg JS (1968) Auditory sense in certain Sphingid moths. Science 159:331–333

Snodgrass RE (1925) Insect musicians, their music and their instruments. Smithsonian Institution, Washington

Stölting H, Moore TE, Lakes-Harlan R (2002) Substrate vibrations during acoustic signalling in the cicada *Okanagana rimosa*. J Insect Sci 2:1–7

Strübing H (1958) Lautäußerung—der entscheidende Faktor für das Zusammenfinden der Geschlechter bei Kleinzikaden (Homoptera—Auchenorrhyncha). Zool Beiträge NF 4:15–21

Strübing H (1977) Lauterzeugung oder Substratvibration als Kommunikationsmittel bei Kleinzikaden? (diskutiert am Beispiel von Dictyophora europaea – Homoptera-Cicadina: Fulgoroidea). [Sound production or substrate vibrations as a means of communication in small Cicadidae discussed using as an example Dictyophara europaea Homoptera Cicadina Fulgoroidea]. Zool Beiträge 23(2):323–332

Traue D (1978a) Zur Biophysik der Schallabstrahlung bei Kleinzikaden am Beispiel von *Euscelis incisus* Kb. (Homoptera-Cicadina: Jassidae). Zool Beiträge NF 24:155–164

Traue D (1978b) Vibrationskommunikation bei Euides speciosa Boh. (Homoptera-Cicadina: Delphacidae). Verh Dtsch Zool Ges 1978:167

Trilar T, Gogala M (2012) *Pagiphora aschei* Kartal (Hemiptera: Cicadidae) from Crete (Greece) – Distribution and description of its song. Acta Entomol Slovenica 20:17–30

Yager DD, Hoy RR (1986) The cyclopean ear: a new sense for the praying mantis. Science 231:727–729

Yager DD, Hoy RR (1987) The midline metathoracic ear of the praying mantis, *Mantis religiosa*. Cell Tissue Res 250:531–541

Yager DD (1999) Structure, development, and evolution of insect auditory systems. Microsc Res Techniq 47:380–400

# Chapter 4
# Hildegard Strübing: A Pioneer in Vibrational Communication Research

**Andreas Wessel**

**Abstract** This laudation honors the role of H. Strübing as one of the early researchers of vibrational communication. In 1958, Strübing was the first to show the significance of vibrational signals in mate finding of small Auchenorrhyncha. She then devoted her entire scientific life to the further study and deeper understanding of this particular behavior.

## 4.1 Laudation for Hildegard Strübing

(On the occasion of the bestowal of the "Insect Drummer Award" at the International Symposium "Vibrational Communication in Arthropods—A Comparative Approach" on March 23–24, 2011, in Berlin)

Today, on the occasion of this symposium, we award Professor Hildegard Strübing (Fig. 4.1) with the newly created "Insect Drummer Award" for her "lifetime achievement in vibratory communication research." Unfortunately, Dr. Strübing cannot be with us today, but we will shortly hand it over to her with the regards of her colleagues. Her role in the development of our field has already been illuminated by Matija Gogala in his keynote lecture (see also Gogala, Chap. 3, this volume), so I can limit myself to a few remarks on her life and work.

Maybe, you wonder why I—of all possible candidates—have the honor to give this laudatory speech. Well, apart from my strong personal appreciation for her work, Hildegard Strübing and I shared the same academic teacher: Professor

A. Wessel (✉)
Museum für Naturkunde—Leibniz-Institut für Evolutions- und Biodiversitätsforschung an der Humboldt-Universität zu Berlin, Invalidenstraße 43, 10115 Berlin, Germany
e-mail: andreas.wessel@mfn-berlin.de

A. Wessel
Zoologisches Museum, Universität Hamburg, Martin-Luther-King-Platz 3, 20146 Hamburg, Germany

R. B. Cocroft et al. (eds.), *Studying Vibrational Communication*, Animal Signals and Communication 3, DOI: 10.1007/978-3-662-43607-3_4,

**Fig. 4.1** Hildegard Strübing in 2002 (*photo* by Werner Holzinger, with permission)

Günter Tembrock (1918–2011). Tembrock, who died in January this year aged 92, was a pioneer of modern behavioral research and one of the founding fathers of bioacoustics (see Wessel 2011). Hildegard Strübing insisted on having been Tembrock's first student, even though she was just 4 years younger and I, being 55 years younger than Tembrock, was his last and youngest student and I am now the trustee of his scientific legacy.

Hildegard Strübing took up her studies of chemistry, biology, and geography in 1940 here at the Humboldt-Universität zu Berlin, then called the Friedrich-Wilhelms-Universität. She originally intended to become a teacher. The director of the Zoological Institute of the University, Konrad Herter (1891–1980), could persuade her to do a Ph. D. on temperature adaptation in amphibians, though, and she managed to take her last examinations at the end of March 1945, just a few days before the invasion by the Red Army (see Strübing 1954, 2008). After July 1945, she assisted Tembrock in the reconstruction of the Zoological Institute of the Humboldt University, and in 1948, she was employed as a scientific assistant there. Her tasks also comprised the identification of samples from student excursions, where the planthoppers proved to be particularly difficult due to the lack of suitable literature. She addressed the planthopper taxonomists Hans Joachim Müller (1911–2007) and Wilhelm Wagner (1895–1977) for help, who awoke an enthusiasm for the group in her. This prompted her, initially without a precise aim, to keep and observe plant- and leafhoppers. At the Zoological Institute, she still experienced the inception of the animal sound archive, which is today the third largest in the world (see Wessel 2013). In 1952, she followed Herter to the Free University in Berlin Dahlem (see Herter 1979), where she largely spent her time teaching classes for biologists as well as medical and veterinary students. In 1971, she was appointed full professor.

Starting from the early 1950s, Strübing published on various aspects of the biology, morphology, and phylogeny of planthoppers, e.g., on parasitizing pipunculids, the formation of spermatophores, peculiar oviductal glands, diapause, and photoperiodically caused polymorphisms. It was the discovery of acoustic communication in a delphacid—*Calligypona lugubrina*, today *Struebingianella lugubrina*—though, which proved seminal for directing her ensuing scientific work. In 1958, 9 years after the groundbreaking study by Frej Ossiannilsson (1908–1995) in 1949, she could show that the vibrational signals play a crucial role in mating.

In her own words:

> [Ossiannilsson's] experiments to prove the role of those signals were not conclusive, although he could discern duet-calling between males and females of *Doratura stylata* (Boh.), as well as search actions by the male up to attempted copulation. He did, however, prove an ability of producing acoustic signals in Auchenorrhyncha, which at that time was entirely unknown.
>
> The significance of acoustic communication between male and female for finding the proper partner for copulation could be proved for the Delphacid *Struebingianella (Calligypona) lugubrina* Boh. When a few virgin males and females that had been separated in the last larval stage were introduced into a test cylinder [...] put over a flowerpot planted with *Glyceria aquatica*, mutual responses could be observed after a very short time. [...] Thanks to my knowledge of Ossiannilsson's observations, I immediately recognised all this behaviour as acoustic communication. However, I did not hear anything during all these observations. When not just a few but 25 males were put into the glass cylinder closed with a moistened pig bladder instead of gauze, I could hear the calling clearly. [...] At that time our institute did not possess a steel tape recorder and the first recording was made by courtesy of my student, Johannes Kinzer and his friend Fred Methner, who possessed a gigantic tape recorder with a recording speed of 76 cm/s and a recording frequency of 40–16,000 Hz, an amplifier and a special underwater microphone, which was particularly sensitive (Strübing 2006: 352).

She published about 25 papers on species-specific differences in vibrational signals and experiments with hybrids, and she supervised Diploma and Ph. D. theses on the morphology of the tymbal organs, stimulus reception, and signal transmission through the substrate. Only her last three papers on this topic were published in English; consequently, her work is generally poorly known internationally. We have thus decided that we will not only publish a short overview of her work in the volume that will arise from this symposium, but also an English translation of her 1958 paper in order to honor this pioneering feat (see Chap. 5, this volume).

## 4.2 Addendum 2013

Shortly after the symposium, we could present Prof. Strübing with the "Insect Drummer Award" as well as with the greetings and wishes of numerous of her colleagues; especially, the plan to include a translation of her 1958 paper into the present volume found her appreciation.

On May 18, 2013, shortly after her 91st birthday, Hildegard Strübing died in the nursing home where she had lived for the last years. Her detailed work on the biology of the smaller relatives of the singing cicadas has inspired researchers working on many aspects of plant- and leafhopper evolution such as systematics, morphology, ecology, and behavior. While this is not the place to dig deeply into her biography, two obituaries of Prof. Strübing will be published shortly by H. Günthart and H. Hoch in the *Sitzungsberichte der Gesellschaft Naturforschender Freunde zu Berlin* and by R. Mühlethaler and A. Wessel in *Cicadina* (open access, http://public.bibliothek.uni-halle.de/index.php/cicadina). A complete bibliography of Mrs. Strübing will be provided by the latter authors.

## References

Herter K (1979) Begegnungen mit Menschen und Tieren. Erinnerungen eines Zoologen 1891–1978. Duncker & Humblot, Berlin

Ossiannilsson F (1949) Insect Drummers. A study on the morphology and function of the sound-producing organ of Swedish Homoptera Auchenorrhyncha with notes on their sound-production. Opusc Entomol Suppl 10:1–145

Strübing H (1954) Über Vorzugstemperaturen von Amphibien. Z Morph Ökol Tiere 43:357–386

Strübing H (2006) Vibratory communication and mating behaviour in the European lantern fly, *Dictyophara europea* (Dictyopharidae, Hemiptera). In: Drosopoulos S, Claridge MF (eds) Insect Sounds and Communication. CRC–Taylor & Francis, Boca Raton, pp 351–356

Strübing H (2008) Zur Erinnerung an Günter Tembrocks 85. Geburtstag. In: Wessel A (ed) Ohne Bekenntnis keine Erkenntnis. Günter Tembrock zu Ehren. (Berliner Studien zur Wissenschaftsphilosophie und Humanontogenetik, 25) Kleine Verlag, Bielefeld, pp 41–52

Wessel A (2011) Günter Tembrock—pioneer of behavioural biology who understood the language of animals. Ethology 117:469–471

Wessel A (2013) Ethology in space and time—Berlin in the light. Nova Acta Leopoldina N. F. 111:15–27

## Publications on Vibrational Communication by H. Strübing

Strübing H (1958a) Paarungsverhalten und Lautäusserung von Kleinzikaden, demonstriert an Beispielen aus der Familie der Delphacidae (Homoptera, Auchenorrhyncha). Zool Beiträge (NF) 4:8–14

Strübing H (1958b) Lautäusserungen—der entscheidende Faktor für das Zusammenfinden der Geschlechter bei Kleinzikaden (Homoptera, Auchenorrhyncha): Vorläufige Mitteilung. Zool Beiträge (NF) 4:15–21 [for translation see Chap. 5, this volume]

Strübing H (1959b) Lautgebung und Paarungsverhalten von Kleinzikaden. Verh Deut Zool Ges Münster/Westf. 1959, Zool Anzeiger 23, Suppl:118–120

Strübing H (1960) Paarungsverhalten und Lautäusserung von Kleinzikaden, demonstriert an Beispielen aus der Familie der Delphacidae (Homoptera, Auchenorrhyncha). Proc Int Congr Entomol Wien 11:12–14

Strübing H (1963) Lautäusserungen von *Euscelis*-Bastarden (Homoptera, Auchenorrhyncha. Verh Deut Zool Ges München 57:268–281

Strübing H (1964) Gesänge, das Verständigungsmittel auch der Zikaden. Umschau in Wissenschaft und Technik (Frankfurt/M.):360–364

Strübing H (1965) Das Lautverhalten von *Euscelis plebejus* Fallén und *Euscelis ohausi* Wagn. (Homoptera, Cicadina). Zool Beiträge (NF) 11:289–341

Strübing H (1966) Ein Vergleich von Lautäusserungen verschiedener *Euscelis*-Arten (Homoptera, Cicadina). Deut. Entomol. Z. (NF) 13:351–358

Strübing H (1967) Zur Untersuchungsmethodik der Lautäusserungen von Kleinzikaden (Homoptera, Cicadina). Zool Beiträge (NF) 13:265–284

Strübing H (1970) Zur Artberechtigung von *Euscelis alsius* Ribaut gegenüber *Euscelis plebejus* Fallén (Homoptera, Cicadina): Ein Beitrag zur Neuen Systematik. Zool Beiträge (NF) 16:441–478

Strübing H, Hasse A (1975) Ein Beitrag zur Neuen Systematik demonstriert am Beispiel zweier *Javesella*-Arten (Homoptera, Cicadina: Delphacidae). Zool Beiträge (NF) 21:517–543

Strübing H (1976) *Euscelis ormaderensis* Remane 1968: 1. Saisonformenbildung und akustische Signalgebung. Sber Ges Naturforsch Freunde Berlin (NF) 16:151–160

Strübing H (1977a) Lauterzeugung oder Substratvibration als Kommunikationsmittel bei Kleinzikaden? Diskutiert am Beispiel von *Dictyophara europaea* (Homoptera, Cicadina: Fulgoroidea). Zool Beiträge (NF) 23:323–332

Strübing H (1977b) Neue Ergebnisse zur Kommunikation bei Kleinzikaden. Verh Deut Zool Ges Stuttgart 70:336

Strübing H (1978a) *Euscelis lineolatus* Brullé 1832 und *Euscelis ononidis* Remane 1967: 1. Ein ökologischer, morphologischer und bioakustischer Vergleich. Zool Beiträge (NF) 24:123–154

Strübing H (1978b) *Euscelis ohausi* Wgn. 1939 and *Euselis singeri* Wgn. 1951: Separate species or not? Tymbal 1:15

Strübing H (1980) *Euscelis remanei*, eine neue *Euscelis*-Art aus Südspanien im Vergleich zu anderen *Euscelis*-Arten (Homoptera, Cicadina). Zool Beiträge (NF) 26:383–404

Strübing H (1981) *Euscelis remanei* Strübing, 1980, from southern Spain compared with other *Euscelis* species. Acta Entomol Fenn 38:44

Strübing H (1983) Die Bedeutung des Kommunikationssignals für die Diagnose von *Euscelis*-Arten (Homoptera, Cicadina). Zool Jb, Allg Zool Physiol Tiere 87:343–351

Strübing H, Schwarz-Mittelstädt G (1988) The vibratory membranes of *Euscelis incisus* and *Struebingianella lugubrina* (Homoptera, Cicadina). In: Vidano C, Arzone A (eds) 6th Auchenorrhyncha Meeting Turin, Italy. Cosiglio Nazionale delle Ricerche, pp 343

Strübing H, Rollenhagen T (1988) Ein neues Aufnehmersystem für Vibrationssignale und seine Anwendung auf Beispiele aus der Familie Delphacidae (Homoptera-Cicadina). Zool Jb, Allg Zool Physiol Tiere 92:245–268

Strübing H (1992) Vibrationskommunikation im Paarungsverhalten der Büffelzikade *Stictocephala bisonia* Kopp und Yonke, 1977 (Homoptera-Auchenorrhyncha, Membracidae). Mitt Deut Ges allg angew Entomol 8:60–62

Strübing H (1999) Comments on the vibratory signals of *Stictocephala bisonia* Kopp et Yonke and *Dictyophara europaea* (L.) (Hemiptera: Auchenorrhyncha: Cicadomorpha et Fulgoromorpha: Membracidae et Dictyopharidae). Reichenbachia. Z taxonom Entomol (Dresden) 33:61–69

Strübing H (2006) Vibratory communication and mating behaviour in the European lantern fly, *Dictyophara europea* (Dictyopharidae, Hemiptera). In: Drosopoulos S, Claridge MF (eds). Insect Sounds and Communication. CRC–Taylor & Francis, Boca Raton, pp 351–356

Strübing H, Drosopoulos S (2006). Photoperiodism, morphology and acoustic behaviour in the leafhopper genus, *Euscelis*. In: Drosopoulos S, Claridge MF (eds). Insect Sounds and Communication. CRC–Taylor & Francis, Boca Raton, pp 255–262

# Chapter 5
# Sound Production: The Crucial Factor for Mate Finding in Planthoppers (Homoptera: Auchenorrhyncha) (Preliminary Communication)

**Hildegard Strübing**

**Abstract** Following the discovery of acoustic or rather vibrational communication in small Auchenorrhyncha by Ossiannilsson (1949), the role of those signals in mate finding is studied. For the delphacid planthopper *Calligypona lugubrina*, we present conclusive evidence that optical as well as chemical stimuli play no role in mate finding, which is rather facilitated through acoustic/vibrational stimuli. A note on the current taxonomic status of the species studied is provided.

---

This paper was originally published in German: Strübing, H. 1958. Lautäußerung—der entscheidende Faktor für das Zusammenfinden der Geschlechter bei Kleinzikaden (Homoptera—Auchenorrhyncha) (Vorläufige Mitteilung). *Zoologische Beiträge, Neue Folge (Berlin)* **4**(1): 15–21. [Manuscript received on 27.6.1958] English translation by S. Wendland-Tembrock (Berlin), edited by P. S. M. Hill (Tulsa), T. von. Rintelen and A. Wessel (both MfN, Berlin). Translated and reprinted with permission of Duncker & Humblot, Berlin.

---

H. Strübing (✉)
Abteilung für Tierphysiologie und Tierpsychologie, Zoologisches Institut der Freien Universität Berlin, Berlin, Germany

R. B. Cocroft et al. (eds.), *Studying Vibrational Communication*, Animal Signals and Communication 3, DOI: 10.1007/978-3-662-43607-3_5,

Herrn Dr. G. Tembrock
mit herzlichem
Gruss
H. Strübing

Sonderdruck aus:

# Zoologische Beiträge

Herausgegeben von

Prof. Dr. Konrad Herter

Berlin

Neue Folge. Band 4. Erstes Heft

1958

Duncker & Humblot · Berlin

(Aus dem Zoologischen Institut der Freien Universität Berlin
Abteilung für Tierphysiologie und Tierpsychologie
Direktor: Prof. Dr. K. Herter)

# Lautäußerung — der entscheidende Faktor für das Zusammenfinden der Geschlechter bei Kleinzikaden (*Homoptera – Auchenorrhyncha*)

## (Vorläufige Mitteilung)

Von

HILDEGARD STRÜBING

Mit 1 Abbildung

Eingegangen am 27. 6. 1958

Larven der Delphacide *Calligypona lugubrina* Boh. wurden im fünften Larvenstadium nach Geschlechtern getrennt und weiterhin isoliert gehalten. In genau definiertem Alter wurden die verschieden geschlechtlichen Imagines auf ihrer Nährpflanze *(Glyceria aquatica)* in Glaszylindern von 10 cm Durchmesser und 18 bzw 36 cm Höhe zusammengesetzt und ihr Verhalten beobachtet. Dabei zeigte es sich, daß jungfräuliche Weibchen vom dritten Lebenstage an kopulationsbereit waren. Die Kopulationsbereitschaft wird ausgedrückt durch ein heftiges Zittern des gesamten weiblichen Abdomens. Die Behauptung, daß dieses Abdomenzittern des Weibchens seine Paarungsbereitschaft ausdrückt, wird durch folgende Beobachtungen erhärtet:

Schon kurz nach Zusammenbringen beider Geschlechter beginnen die Weibchen mit Abdomenzittern, wobei sie sich ganz offensichtlich gegenseitig stimulieren. Besonders bemerkenswert ist, daß die Männchen schlagartig auf dieses Verhalten der Weibchen reagieren. Sie beginnen augenblicklich aufgeregt umherzulaufen, sobald nur irgendwo in dem Behälter ein Weibchen zittert. Sie bewegen sich mehr oder weniger direkt auf das zitternde Weibchen hin, „stutzen", d. h., verharren einen Augenblick, sobald sie sich wieder weiter von dem Weibchen entfernen, schlagen eine neue Richtung ein und finden durch eiliges Halmauf- und -abwärtslaufen

## 5.1 Translation of the Original Text

Nymphs of the delphacid planthopper *C. lugubrina* Boh. were separated according to sex during the 5th larval stage and then kept isolated. At a precisely defined age, the imagines of both sexes were put together on their food plant *(Glyceria aquatica)* in glass cylinders of 10 cm in diameter and 18 or 36 cm in height, respectively, and their behavior was observed. It became apparent that virginal females were ready to copulate from their third day onwards. The readiness to copulate is expressed by vigorously vibrating the entire abdomen. The claim that the female's vibrating the abdomen expresses her receptiveness is substantiated by the following observations:

Shortly after both sexes were put together, the females began vibrating their abdomens, thus apparently stimulating each other. It is particularly noteworthy how promptly males react upon observing this behavior of the females. They start running around immediately in an excited fashion as soon as a female vibrates anywhere in the cylinder. They move more or less directly toward the vibrating female, "hesitate", i.e., pause for a moment as soon as they diverge from the female, take a new direction, and find the female of concern rather quickly by running up and down blades, sit behind her, and begin to woo her by excitedly beating their wings. Thereupon, copulation ensues quickly in most cases. The female often stops vibrating the moment the male beats his wings, and while the latter bends round his abdomen, the female lifts herself slightly from the blade to facilitate copulation.

Now, the decisive question is as follows: Which stimulus modalities were operating here enabling mate finding? Optical and chemical stimuli seem excluded: Females vibrated without having been able to see the males (females may vibrate when strongly motivated for copulation without the undoubtedly strong stimulating presence or proximity of males who in most cases function as releasers in the first place), whereas the males appeared to act spontaneously; it was certain that they could not have seen the females in many cases. When they happened to meet another male by chance during their search, triggered by female abdominal vibrations, they would court it, which argues against an accurate optical ability to differentiate as well as against any chemical perceptions. If a female had just stopped vibrating while a searching male was running along the same blade, he often failed to find her and only renewed female vibrations would get him on to the right track. Hence, it is highly improbable that any scents would emanate from the female that might facilitate the male's approach at a close distance. Furthermore, it might be conceivable that the female prompted tremors when vibrating to which the male reacted. However, males were always activated in the same manner regardless of whether they were sitting on the ground, the glass walls of the cylinder, the gauze lid covering it, or at the *Glyceria* blades.

Nevertheless, the female ready for copulation must have perceived the male's proximity somehow since he triggered her abdominal vibrations, and on the other hand, the males must possess a fine capability to detect this behavior of the females in order to be able to react so promptly. The "spontaneous", hasty running off of males who had until then been sitting silently was often enough a safe criterion to assume a female vibrating somewhere in the cylinder, which I would notice only then. According to a comprehensive behavioral analysis, one could assume from the very beginning that there was an acoustic trigger and perception operating. However, this needed to be proven.

The fact that not only the big singing cicadas from the family Cicadidae produce sound but also our indigenous leafhoppers has been known ever since the investigations of Ossiannilsson (1949). Accordingly, delphacid males produce a song that has been named "common song" by Ossiannilsson, since it is completely independent from certain types of behavior. In contrast, in some species, he could additionally distinguish a species-specific courtship song of the males, which they would only do in the vicinity of females. In one case (the jassid *Doratura stylata* Boh.), even a reply song from the female could be heard (see Ossiannilsson 1949, 1953). However, he could not bring those songs into accordance with certain types of behavior and writes (1953) that he could not detect any effect of the male's courtship song on the female and that it remains unresolved whether the virginal female would be ready to copulate without any preceding song. It appeared clear to me that the female of *C. lugubrina* produces sounds while vibrating the abdomen that the male can perceive.

In a first attempt to make any sounds discernible for the human ear, I stretched a humid pig's bladder instead of gauze over the glass cylinder in which the animals were located. Due to this simple trick, the males' lively sounds were partially quite loudly audible when holding a stethoscope or one's ear on the pig's bladder. It did not make a difference whether the males were alone in the cylinder or together with the females. However, the females' abdominal vibrations could not be made audible in this way. Nevertheless, recordings with a tape recorder brought clarification.

Mr. Fred Methner, Berlin, was so kind as to lend me his AEG K 8 magnetic tape recorder with the tape speed of 76 cm/sec and a recording frequency of 40–16,000 Hz, as well as an electrodynamic microphone, constructed as a special underwater microphone. With this equipment, it was possible not only to reproduce the song of *Calligypona* males loudly and in its characteristic manner but also to prove that the female's abdominal vibrations do, in fact, cause sounds that can indeed be perceived by the human ear. Thus, the males' song consists of a shorter or longer sequence of staccato, fast buzzing sounds, or clicks (gog-gog-gog-gog) ending sooner or later in a long-drawn-out and substantially louder Goooooooh… ——. The answer to this song by the females ready to copulate, which can be visually detected through vigorous abdominal vibrations, becomes acoustically discernible as a distinct drumming. The transcription of the magnetic tape recordings onto the electron-beam oscillograph (Philips GM 3156) yielded oscillograms (see Fig. 5.1).

The fact that a female ready to copulate can actually hear this song and will then herself be stimulated to react by drumming could be demonstrated by the following experiment: The microphone was designed so that it could also serve as a loudspeaker. While the tape recorder was in a different room, the pitch of the reproduction of the male's song was adjusted such that it was hardly discernible by

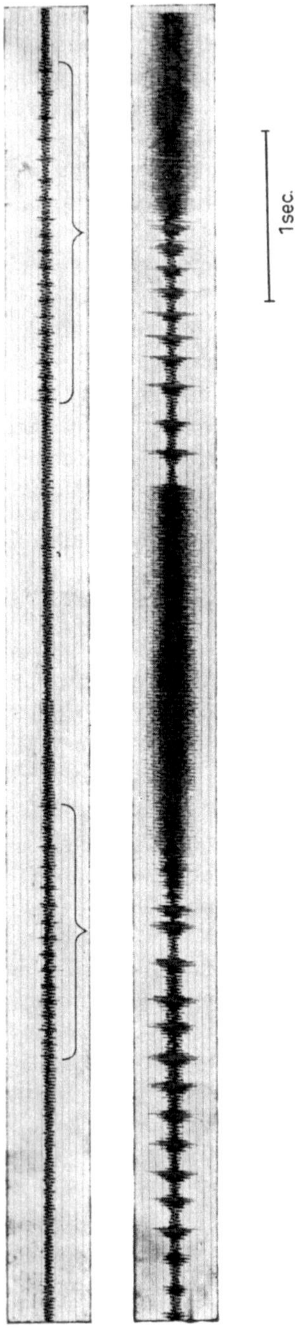


**Fig. 5.1** Oscillograms of the magnetic tape recordings. The 50-Hz disturbance of the baseline was caused by the amplifier of the magnetic tape recorder. *Above* the rhythmic drumming of the female (designated by parentheses). *Below* the male's song (in this case, particularly, stimulated by the female's drumming; bottom oscillogram is the continuation of the one above)

the human ear. Then, the microphone was placed into the upper opening of the glass cylinder containing completely motionless sitting females ready to copulate. In several experiments, normal abdominal vibrations could thus be elicited in the same manner as if one had added the males.

A specific courtship song of the males could not yet be proven by this method. However, it is assumed that it may be the males' intensive beating of the wings when courting a female. Possibly, sound production comes about without the drumming apparatus specific to cicadas in a kind of stridulation, which Pringle mentions as "wing-clacking" of some cicadas, perhaps generating sounds of quite different frequencies. Furthermore, no correlation could be established between the often distinct abdominal vibrations that males display when aroused by females' drumming and a sound that is characteristic for this behavior. Perhaps, this is just a crescendo of common song as shown by the tape recordings when the males ready to copulate ran toward the vibrating females (see Fig. 5.1). Likewise, the mechanics of the females' drumming is still uncertain. Again, it immediately suggests that it is produced by a particular kind of stridulation since the *Calligypona* females do not have a sound-producing apparatus corresponding to that of the males. A corresponding behavioral study (until now still without parallel sound analysis) already exists for eight other species of the family Delphacidae. In all cases, the female reacts to the male's proximity with abdominal vibrations and thus alerts the male that runs in a more or less directed fashion toward the female, courts it with species-specific behavior, and finally initializes copulation. In all species investigated, the male is ready to copulate again immediately after copulation, whereas the female refuses a renewed copulation by a particular defensive reaction. It will be discussed elsewhere at length that later on it may possibly come to further copulations.

Ossiannilsson's statement that males are not stimulated into song by the presence of females is only true for already mated females. Virginal females drumming in the way described do definitely stimulate the males' song (see above), while non-drumming females are probably not recognized as potential mates.

Neither Pringle nor Ossiannilsson could make any satisfying statement on the biological significance of cicada song based on their investigations. Pringle (1954) assumes that the song of cicadas only serves to bring the sexes together in a confined space and for separating the single species into certain groups, whereas the final approach is triggered by optical and chemical stimuli. Apparently, exact behavioral analyses for Cicadidae are not yet available; only they could produce evidence as to whether such a conclusion is justified. It seems more probable, though, that sound production is of more significance than previously assumed in cicadas, as well. How a virginal cicada female behaves concerning male song has not yet been investigated, as far as I know.

For planthoppers of the family Delphacidae, optical as well as chemical stimuli play no role in mate finding, which is rather facilitated through acoustic stimuli. Albeit the song of male planthoppers is not important as a territorial demarcation (which is not necessary in this case), it should announce, however, similarly to the behavior in birds: Here is a male ready to mate, thus triggering appropriate behavior in the females. The delphacids (and presumably all cicadas and planthoppers) thus belong together with locusts and crickets as those insects where mate finding is enabled by species-specific acoustic stimuli.

The fact that the song of *C. lugubrina* males is indeed species-specific resulted from a comparison to a magnetic tape recording of the male's song of the closely related *Calligypona adela*. Extensive and comparative behavioral and sound analyses are in preparation, for the time being, in further delphacids, and later on in other plant- and leafhoppers.

## 5.2 Taxonomic Note by the Editor (A.W.)

Wagner (1962) splits the genus *Calligypona* Sahlberg, 1871 in the course of a taxonomic revision of Central European delphacids and erected a new genus for five species including *C. lugubrina* Boheman, 1847. The new genus was named *Struebingianella* Wagner, 1963 (Wagner 1962) in recognition of Hildegard Strübing's ground-breaking work on the vibrational communication in planthoppers; thus, the valid name of the here studied species is *Struebingianella lugubrina* (Boheman, 1847).

The name for the "closely related *Calligypona adela*"—as mentioned in the last paragraph—is a little confusing, as this combination never constituted a valid species name. The species in question was originally described as *Delphax adela* Flor, 1861. In 1917, A. C. Jensen-Haarup established a new genus *Paraliburnia* Jensen-Haarup, 1917 with the new described type species *Paraliburnia jacobseni* Jensen-Haarup, 1917, what was then identified as a junior synonym of *D. adela* by Wagner (1962). The valid species name is therefore *Paraliburnia adela* (Flor, 1861), for further synonyms see Hoch (2004, Fauna Europaea, urn:lsid:faunaeur.org:taxname:152608).

Curiously, the invalid combination "*C. adela*" appears in exactly one other publication, a faunistic list by Moosbrugger (1946), who explicitly states W. Wagner as identifier of the problematic species as well as solely responsible for the nomenclature. The species is here listed as "*Calligypona adela* Flor. (*Liburnia*)" (pp. 68). Thus, H. Strübing probably got the name directly from W. Wagner, who was at that time her main source for taxonomic information. *S. lugubrina* and *P. adela* are not closely related (M. Asche, pers. comm.).

## References

### Original Publication (5.1)

Ossiannilsson F (1949) Insect drummers. A study on the morphology and function of the sound-producing organ of Swedish Homoptera Auchenorrhyncha with notes on their sound-production. Opusc Entomol Suppl 10:1–145

Ossiannilsson F (1953) On the music of some European leafhoppers (Homoptera-Auchenorrhyncha) and its relation to courtship. Trans IX. Int Congr Entomol 2:139–141
Pringle JWS (1954) A physiological analysis of cicada song. J Exp Biol 31:525–560

# Taxonomic Note by the Editor (5.2)

Boheman CH (1847) Nya Svenska Homoptera. Kongl. Vetenskapsakademiens Handlingar (Stockholm) 1847:23–67
Flor G (1861) Die Rhynchoten Livlands in systematischer Folge beschrieben. Zweiter Theil: Rhynchota gulaerostria Zett. (Homoptera Aut.). Cicadina und Psyllodea. Archiv für die Naturkunde Liv-, Ehst-und Kurlands, 2. Serie, Biologische Naturkunde (Dorpat) 4:1–567
Hoch H (2004) Fauna Europaea: Hemiptera: Cicadomorpha etc. Fauna Europaea version 1.0, http://www.faunaeur.org Date modified: 09 May 2004; release date: 27 Sept 2004
Jensen-Haarup AC (1917) Some new Delphacinae from Denmark. (Hem. Hom.). Entomologiske Meddelelser (Kjobenhavn) 11:1–5
Moosbrugger J (1946) Die Zikadenfauna von Vorarlberg. Zentralblatt für das Gesamtgebiet der Entomologie 1:65–75
Sahlberg JR (1871) Ofversigt af Finlands och den Skandinaviska hälfons Cicadariae. I. Notiser ur Sällskapets pro Fauna et Flora Fennica Förhandlingar (Helsingfors) 9:1–506
Wagner W (1962) Dynamische Taxionomie, angewandt auf die Delphaciden Mitteleuropas. Mitt. Hamburg. Zool. Mus. Inst. 60:111–180 (issued in April 1963)

# Part II
# The State of the Field: Concepts and Frontiers in Vibrational Communication

# Chapter 6
# Interactions Between Airborne Sound and Substrate Vibration in Animal Communication

**Michael S. Caldwell**

**Abstract** Airborne sound and substrate vibrations are communication channels inextricably linked through commonalities in signal production, propagation, and reception. Bimodal recordings of acoustic calls reveal that signal components in one modality often excite energy in the other and can thus be propagated to receivers via either channel. While studies of communication via airborne sound and substrate vibrations have proceeded largely independently of one another, the frequency at which these modalities co-occur and the number of receivers sensitive to both kinds of energy underscore the broad potential importance of interactions between the two communication channels. Nevertheless, only a handful of species are known to use bimodal acoustic signals. This chapter summarizes what is known about the interactions between airborne and substrate-borne signal components, discusses how interactions between modalities may shape the evolution of bimodal acoustic signals, and identifies outstanding issues in the field along with promising avenues for future study.

## 6.1 Introduction

Acoustic signals, such as insect and bird song, are used by a wide range of animal taxa and are recognized as key traits influencing important biological processes, such as mate choice and speciation (Bradbury and Vehrencamp 2011). Much of our understanding of the behavioral ecology of animals has come from the study of such signals. In many species, however, the function of "acoustic" calls and songs

M. S. Caldwell (✉)
Department of Ecology, Evolution, and Behavior, University of Minnesota, Saint Paul, MN, USA
e-mail: caldwell@umn.edu

M. S. Caldwell
Smithsonian Tropical Research Institute, Panama, Republica de Panama

R. B. Cocroft et al. (eds.), *Studying Vibrational Communication*, Animal Signals and Communication 3, DOI: 10.1007/978-3-662-43607-3_6, 

is likely more complex than generally assumed and may not be limited to a single communication channel.

The production of acoustic signals is a highly energetic behavior and unavoidably excites vibrations in both the fluid medium (air or water) surrounding the signaler and any substrate on which the signaler is resting (Caldwell et al. 2010; Fahy and Gardonio 2007; O'Connell-Rodwell et al. 2000; Stölting et al. 2002). No two other sensory modalities are so intertwined as audition and the perception of substrate vibrations. Airborne sound and substrate-borne vibrations are both generated by physical movement in the signaler's body, usually the vibration of specially adapted structures, both propagate as mechanical waves, and they are often transduced by the same sensory receptors. Only the paths and wave types through which they propagate differ.

The communication implications stemming from these differences and from what could be a particularly common form of bimodal communication, however, merit some special attention. While a handful of species have been shown to exploit both the airborne and substrate-borne call components (Hill and Shadley 2001; Latimer and Schatral 1983; Lewis et al. 2001; O'Connell-Rodwell and Wood 2010), the potential importance of bimodal transmission to the function and evolution of acoustic calls has generally been overlooked.

The purpose of this chapter is to make a case for further research into the interactions between airborne and substrate-borne components of acoustic signals, to provide a conceptual starting place for those studies, and to identify promising directions for this work.

Major questions regarding the interactions between airborne and substrate-borne components of acoustic signals:

- How commonly and in what contexts do airborne sounds and substrate vibrations interact during animal communication?
- Are airborne and substrate-borne components of bimodal acoustic signals largely redundant or do receivers extract different information from each component? What sorts of non-redundant information are encoded in each component?
- Do receivers of bimodal acoustic signals extract information from comparisons between the airborne and substrate-borne signal components?
- Do signalers facultatively alter the relative characteristics of airborne and substrate-borne signal components (e.g., the relative amplitude of each component) depending on behavioral context?
- What is the relative importance of acoustic energy that is coupled to either the air or a solid substrate directly via the signaler's body, and which propagates to receivers exclusively through this medium, as compared to energy which transfers from one medium to the other during signal propagation?
- What roles do the interactions between airborne and substrate-borne components play in the evolution of multimodal signals?

## 6.2 Examples of Bimodal Sound/Substrate Vibration Signals

Thus far, there have been very few studies on bimodal sound/substrate vibration communication, and only a few species have been experimentally confirmed to use such signals. These examples do, however, come from disparate taxa, and the signals they employ differ greatly in function, hinting at the hidden diversity of bimodal acoustic communication systems yet to be described.

### *6.2.1 Katydids*

Male katydids (*Tettigonia cantans*) stridulate from low vegetation to attract mates. In addition to airborne sound, these songs produce substrate vibrations in the plant, detectable for up to 2 m (Keuper and Kuhne 1983). Male and female conspecifics approach the airborne component of this signal, but in choice trials where the source of the airborne component was difficult to localize, both sexes were more likely to walk onto branches through which the substrate component of the song was being played (Latimer and Schatral 1983). The airborne component of katydid stridulatory songs likely serves to attract mates from afar, while the substrate-borne component aids mates and rivals in localizing the signaler once they reach the plant. Results from bimodal playback to another katydid species (*Ephippiger ephippiger*) are consistent with the pattern of receivers using the substrate vibrations excited by acoustic calls to localize signalers on a plant (Stiedl and Kalmring 1989), although the latter study was not specifically designed to test this hypothesis.

### *6.2.2 Mole Crickets*

Male prairie mole crickets (*Gryllotalpa major*) chirp from underground to attract flying females. These stimulatory songs produce a strong airborne component that is both amplified and coupled to the air with the help of an acoustically tuned burrow (Hill et al. 2006, but see also Bennet-Clark 1987). Stridulation also excites lower frequency seismic vibrations that carry to neighboring males. During playback, females responded to the airborne component of the advertisement call, but males did not (Hill and Shadley 2001, see also Walker and Figg 1990). When played a synthetic approximation of the call's substrate component, however, males responded by adjusting their call timing in a variety of ways (Hill and Shadley 2001). It appears that the two components of this stridulatory signal are used by different receivers. Females are attracted to the airborne sound, and males respond to substrate vibrations generated by their rivals.

### 6.2.3 White-lipped frogs

Male white-lipped frogs (*Leptodactylus albilabris*) call from moist ground to attract females. As in most frogs, these calls produce a conspicuous airborne component (1–3 kHz). When calling from a muddy substrate, however, males also produce impulsive, low-frequency seismic vibrations ('thumps', <150 Hz) as their vocal sacs impact the ground with each call (Lewis and Narins 1985). Bimodal playbacks reveal that in conditions of high-amplitude airborne noise, like those commonly encountered in frog choruses, males use these seismic thumps to adjust call timing, insuring that their calls do not overlap temporally with those of neighboring frogs (Lewis et al. 2001). Although it has been suggested that thump vibrations may influence spacing between signaling males, this has yet to be confirmed (Lewis and Narins 1985).

### 6.2.4 African Elephants

African elephants (*Loxodonta africana*) produce low-frequency alarm calls (fundamental frequency ~20 Hz, dominant frequency ~40 Hz) when predators are nearby. These calls excite intense airborne and substrate-borne components with similar spectral and temporal properties (O'Connell-Rodwell and Wood 2010). During playback trials, both components of the signal elicited defensive responses from other elephants. When played the airborne component, receivers became aggressive and fled the area (O'Connell-Rodwell et al. 2000). Responses to the substrate component were not as strong, but receivers did display defensive behaviors, such as grouping more tightly with troupe-mates and increased vigilance (O'Connell-Rodwell et al. 2000, 2007). For both call components, receivers appear to show a greater response to the recorded calls of familiar individuals than to the calls of elephants from other troupes, indicating that they are able to extract a considerable amount of information from both airborne and substrate-borne vibrations (O'Connell-Rodwell et al. 2000, 2007).

Both African and Asian elephants also produce intense, low-frequency "rumble" calls thought to coordinate group movements when separated by long distances (O'Connell-Rodwell and Wood 2010). Just as with alarm calls, rumbles excite strong airborne and substrate-borne components (Günther et al. 2004; O'Connell-Rodwell et al. 2000). Receivers respond to the airborne component when played alone (McComb et al. 2003), but it not known whether the seismic component functions as part of the signal.

## 6.3 A Comparison of the Airborne and Substrate-borne Acoustic Communication Channels

### *6.3.1 The Production of Airborne Sound and Substrate Vibrations*

Animals have evolved a great diversity of mechanisms for the production of acoustic signals. Most arthropod species use exoskeletal stridulatory structures, tymbal organs, drumming on the substrate, or whole body tremulation to produce acoustic calls (Drosopoulos and Claridge 2005; Taylor and Patek 2010; Virant-Doberlet and Cokl 2004). Fish call either by stridulating skeletal structures, by drumming parts of their bodies against the substrate, or by muscular vibration of the swim bladder (Collin and Marshall 2003; Whang and Janssen 1994). In most terrestrial vertebrates, calls are generated by modulating the pressure of air flowing between body cavities through the vibration of a resonant vocal valve organ (syrinx or larynx) (Bradbury and Vehrencamp 2011). Drumming is also found across 15 families of fossorial and other mammals (Randall 2010).

For these vibrations to be useful acoustic signals, they must be well coupled with the air (or water), the substrate, or both signaling media (Bradbury and Vehrencamp 2011; Fahy and Gardonio 2007).While there has been considerable research on the function and structure of acoustic radiators that couple body vibrations to the air (Bradbury and Vehrencamp 2011), far less is known about how signalers effectively couple vibrations to the substrate (Hill 2008). Even less well-understood is how animals simultaneously couple energy to both modalities during the generation of bimodal acoustic signals.

#### 6.3.1.1 Schemes of Bimodal Acoustic Signal Production

Bimodal acoustic signals can be produced through three general schemes:

1. Airborne and substrate-borne signal components can be produced by the same structures and mechanisms. The bimodal acoustic stridulation signals of some katydids (Keuper and Kuhne 1983), mole crickets (Hill and Shadley 2001), some spiders (Rovner 1975), and the vocal calls of elephants (O'Connell-Rodwell and Wood 2010) are examples of this scheme. Some vocal calls produced by frogs and birds (Fig. 6.1; Caldwell et al. 2010), insect tymbal songs (Stölting et al. 2002), and drumming signals used by insects (Morris 1980), crustaceans (Taylor and Patek 2010), and mammals (Randall 2010) also employ a single mechanism to generate intense airborne and substrate-borne components. Although in many cases, it is not known whether these signals function bimodally.

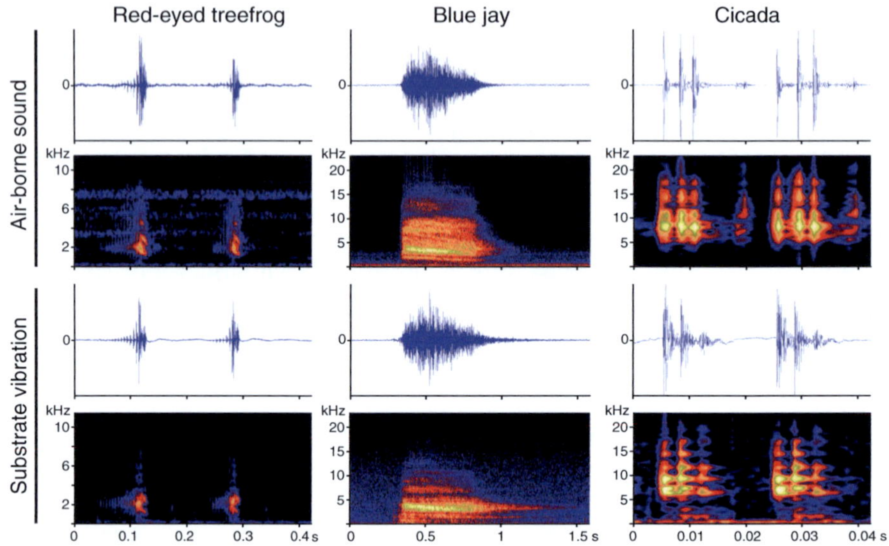


**Fig. 6.1** Airborne sound and substrate vibration recordings of animal vocalizations. Shown are waveforms and spectrograms for airborne and substrate-borne vibrations excited by (1) The "double chack" sexual advertisement call of a male *red-eyed treefrog* (*Agalychnis callidryas*) calling on a small sapling. The airborne component was recorded with a microphone at 1 m. The substrate component was recorded at 50 cm with an accelerometer attached to the plant. Noise at 4 and 8 kHz in the airborne sound recording is due to insects and other frogs calling in the same pond-side chorus, and is not present in the substrate recording. (2) The "scold" call of a *blue jay* (*Cyanocitta cristata*) perched on a thin wooden dowel. The airborne component was recorded with a microphone at 2 m. The substrate component was recorded at 30 cm with an accelerometer attached to the perch. Note the extended 4 kHz ring down in the substrate recording. (3) Excerpt of song from the *cicada* (*Okanagana rimosa*) resting on a twig elicited by electrical stimulation (adapted with permission from supplemental recordings published with Stölting et al. 2002). The airborne component was recorded with a microphone at 20 cm. A separate substrate component was recorded at 10 cm with a laser vibrometer focused on the twig

Tremulation may be the signal production mechanism least likely to excite an intense airborne component and thus the least common form of single mechanism bimodal acoustic signal production. Indeed, tremulatory signals are often described as producing no audible sound (Belwood and Morris 1987; Caldwell et al. 2010). Because tremulation does not benefit from the frequency multiplying structures of other signal production mechanisms, the upper limit of tremulation signal frequency is constrained by muscular physiology (Bradbury and Vehrencamp 2011; Hill 2008) and thus often limited to low-frequency ranges that for small animals are inefficiently coupled to the air (Bennet-Clark 1998).

2. Airborne and substrate-borne components can be produced by the same set of structures but by different mechanisms. The seismic thumps of calling white-lipped frogs are a good example of this. Inflation of the frog's vocal sac is

integral to the production of airborne call components, but this same structure also produces seismic thumps when it impacts the substrate (Lewis et al. 2001). Consequently, the two signal components have very different frequency and temporal properties.

3. Airborne and substrate-borne components could be produced by separate, independent structures. Multimodal signals involving components produced by separate structures are common in nature (e.g., Holldobler 1999; Narins et al. 2005; Partan and Marler 2005; Uetz and Roberts 2002). Males of the wolf spider *Schizocosa ocreata*, for example, include bimodal signals in their courtship displays, with visual components produced by the movement of decorated front legs, and acoustic components produced with stridulatory structures on the palps (Uetz et al. 2009). However, examples of this scheme for the production of purely acoustic bimodal signals are scarce (but see Stiedl and Kalmring 1989). It is possible that bimodal airborne and substrate-borne acoustic signals where both components are generated by the same structures are more likely to evolve than signals where each component is generated separately because both modalities are already inescapably excited by any single vibrating structure.

#### 6.3.1.2 Asymmetry in Bimodal Acoustic Signals

While signals with conspicuous airborne components are likely to also produce intense substrate vibrations, the converse is not always true. In some situations, signals with conspicuous substrate components may not produce airborne components that are detectable at any useful range. Red-eyed treefrogs (*Agalychnis callidryas*), for example, produce several airborne calls that also excite strong substrate vibrations. Tremulatory signals produced by the same frog, however, produce only a substrate component, with no audible sound (Caldwell et al. 2010). This asymmetry is also seen more broadly among insects. As mentioned above, approximately 20 % of insects communicate with both airborne and substrate-borne acoustic signals. However, while a further 70 % of insects are thought to acoustically communicate with substrate vibrations alone, only about 5 % use airborne sound alone (Cocroft and Rodriguez 2005). At the heart of this asymmetry is the greater impedance mismatch between a signaler's body and the surrounding air than between that body and the substrate, which is often itself composed of live, or once living, tissue. Thus, many body vibrations are likely to couple more strongly with the substrate than they are to the air. Whether unimodal substrate-borne signals are less common for benthic animals in water, where the impedance of a signaler's body is more closely matched to the surrounding fluid medium, has not been fully resolved.

## 6.3.2 The Propagation of Airborne Sound and Substrate Vibrations

While both airborne sound and substrate vibrations propagate as mechanical waves, solid media support a greater diversity of wave types. Airborne and waterborne far-field sound signals travel as compressional (pressure) waves (Bradbury and Vehrencamp 2011; Cremer et al. 2005). By contrast, vibrational signals in solid substrates usually travel as boundary waves, with Rayleigh waves accounting for most ground-borne signals and bending waves accounting for most plant-borne signals (Markl 1983; Michelsen et al. 1982). Surface waves that propagate at the interface between air and water are used for communication by some aquatic animals. The mechanics of surface waves in water shape these signals in unique ways, but they can generally be considered a form of substrate vibration (Markl 1983).

### 6.3.2.1 Propagation Characteristics of Each Modality

There are consistent differences in the global- and frequency-dependent attenuation characteristics of compressional waves in fluid media and those of boundary waves in solid substrates and liquids. In addition, solid substrates are often more structurally complex and spatially heterogeneous than air or water, which are each comparatively uniform in their effects on propagating signals (Markl 1983). Thus, the attenuation and degradation of substrate vibration components will be more variable than that of their airborne counterparts (Elias and Mason 2010). Each of these factors will strongly influence the function and evolution of bimodal acoustic signals (Table 6.1; Bradbury and Vehrencamp 2011; Cremer et al. 2005; Forrest 1994).

Mechanical waves traveling through solid substrates and at the air–water interface show more pronounced frequency-dependent filtering than compressional waves in fluid media (Markl 1983; Michelsen et al. 1982), with many substrates acting as low-pass filters (Cokl et al. 2007; Magal et al. 2000). The propagation of substrate vibrations can be further complicated by reflections, standing waves, changes in the axes of medium vibration, and spatial variation in the mechanical properties of the substrate. Thus, in some substrate types, most notably plants, amplitude does not always decrease monotonically with distance from the vibration source as it often does in air (Fahy and Gardonio 2007; Michelsen et al. 1982), and the transfer function of energy passing through the substrate between a signaler and receiver will depend, in complex ways, on the position of each animal.

While the speed of sound in air is fairly constant ($\sim 340\ \mathrm{ms}^{-1}$) and varies only slightly with temperature, pressure, and humidity, the propagation velocity of substrate-borne acoustic waves varies widely (phase velocities for acoustic communication signals in several substrates are reviewed in Virant-Doberlet et al. 2006) and is dependent on wave type, substrate density, substrate stiffness, and a host of

**Table 6.1** Comparison of the airborne and substrate-borne acoustic communication channels

| | Airborne sound | Substrate vibration |
|---|---|---|
| Wave types | Longitudinal pressure waves, with minimal particle movement (far field) | Rayleigh and bending waves: particle movement is transverse with displacements normal to the surface of the substrate, and in the case of Rayleigh waves, also longitudinal |
| Spatial variability of transmission medium | Relatively homogeneous | Highly variable |
| Propagation velocity | Relatively constant within and across habitats | Highly variable: substrate vibrations used for communication generally travel more slowly than pressure waves in air, but velocities can vary greatly within a substrate and between substrate types |
| Active space | Large and predictable | More variable: Often smaller, with well-defined limits corresponding to the edges of the vibrating substrate |
| Global attenuation | Moderate and predictable: Loss due to spreading scales with the square of distance. Medium absorption is relatively low. Amplitude decreases monotonically with distance | Variable: loss due to spreading is negligible or scales in direct proportion with distance. Medium absorption is high. Amplitude often does not decrease monotonically with distance |
| Frequency-dependent attenuation | Weak frequency-dependent filtering | Frequency-dependent filtering can be substantial |
| Frequency dispersion | All wavelengths propagate at the same speed. | Propagation velocity often depends on wavelength |
| Energetics of production | For small animals, coupling of low frequencies to the air is inefficient | High frequencies often do not propagate efficiently through the substrate. High-frequency tremulatory signals are difficult to produce |
| Environmental noise | Variable, often high | Variable can be lower than noise levels for airborne sound in the same environment due to the smaller active spaces of many substrate vibration signals |

other factors (Cremer et al. 2005). Unlike compressional waves, each wave type commonly used for substrate vibrational communication is dispersive, meaning that wave propagation velocity (phase velocity) varies with frequency. For Rayleigh waves, the exact nature of this frequency dispersion is dependent on how mechanical properties of the particular substrate vary with depth. For bending waves, dispersion is influenced by the thickness and mechanical properties of the vibrating structure (Fahy and Gardonio 2007). Water surface waves are also dispersive due to interactions between gravity, surface tension, and water depth, with energy at very low and very high frequencies generally propagating faster than that at intermediate frequencies (Markl 1983). It is not clear how relevant frequency dispersion is to biological signaling. It could, in theory, degrade frequency-modulated or broadband amplitude-modulated signals (Casas et al. 2007) or be used to determine range to a signal source (Elias et al. 2006; Michelsen et al. 1982). In many cases, however, dispersion of high-frequency substrate signals will be minimal (Brownell 1977; Casas et al. 2007).

The theoretical maximum effective range of substrate vibration signals is larger than that of compressional waves in air. This is because airborne sound is subject to spherical spreading, attenuating with the square of distance from the signal source, while Rayleigh waves in the earth exhibit cylindrical spreading, attenuating in direct proportion to distance, and bending waves in plants exhibit linear propagation, showing no spreading loss (Bennet-Clark 1998). In practice, damping due to material properties can rapidly attenuate substrate signals (Cocroft and Rodriguez 2005; Markl 1983). The maximum effective range for seismic vibrations excited by elephant rumbles, for example, is approximately 2 km under a best case scenario, similar to the range of the airborne component (Günther et al. 2004). The active space of substrate signals can be further limited by the physical boundaries of the vibrating medium. For instance, vibrational signals traveling through a plant will not generally propagate far beyond that plant (but see Eriksson et al. 2011; Michelsen et al. 1982). The pattern of substrate-borne vibrations having a more limited active space than their airborne counterparts is probably reversed in the case of burrowing animals such as fossorial mammals, however. Here, continuous lengths of airspace for the transmission of auditory signals are limited to burrows themselves, while a contiguous solid medium, the soil, connects even distant burrows. Indeed, several species of burrowing mammal use seismic signals for inter-burrow communication, and airborne signals appear to be limited to short-range interactions (Mason and Narins 2001; Narins et al. 1992).

Given the characteristics of wave propagation in each modality, what sorts of calls make the best bimodal acoustic signals? Generally, transmission of high frequencies is strongly attenuated in solid substrates (Markl 1983; Michelsen et al. 1982), while low frequencies are difficult for a signaler to couple to the air (Bennet-Clark 1998). Furthermore, global amplitude variation may not be informative for signals traveling in many solid media (Michelsen et al. 1982; Polajnar et al. 2012). We might, therefore, expect that effective bimodal acoustic signals have energy in intermediate frequency ranges efficiently propagated in both communication channels, or broad spectra, such that subsets of the energies generated by the signaler will propagate well in each channel. We might also expect that much of the information is encoded in the temporal, rather than frequency or global amplitude properties of the signal. Broadband, pulsatile or drumming signals, or temporally amplitude-modulated signals restricted to a narrow range of intermediate frequencies are, therefore, promising candidates for bimodal acoustic communication.

#### 6.3.2.2 Induction of Substrate Vibrations by Airborne Sound

Acoustic signals excite substrate-borne vibrations in one of two ways. Either vibrations are coupled directly from the signaler's body to the substrate or vibrations radiate from the signaler as airborne waves that then induce vibrations in the substrate (Fig. 6.2). This second mechanism is a singular issue in bimodal acoustic communication, and the function and biological importance of airborne-sound-induced substrate vibrations is poorly understood.

The mechanical response of substrate to airborne pressure waves will depend on the frequency of the pressure waves, their intensity, the spatial distribution of that intensity, as well as the acoustic impedance of the substrate (Cremer et al. 2005). Animal ears and vocal systems employ a variety of impedance matching structures adapted to more efficiently couple body vibrations to the surrounding fluid media. In the case of sound waves acting directly on solid substrate, these structures do not exist and coupling is not as efficient.

Attempts to describe the transfer of acoustic energy from airborne pressure waves to substrate vibration using layered fluid/solid models predict that vibrations in the two media will be maximally coupled at frequencies where the phase velocities of dispersive substrate waves are closest to the propagation velocity of pressure waves in air (Jardetzky and Press 1952; Press and Ewing 1951). For soil, where phase velocities decrease with frequency, these frequencies of best coupling tend to be very low, in the range of airborne sound signals produced by elephants (Günther et al. 2004), but below those produced by small animals (Bennet-Clark 1998). In plants, where phase velocities increase with the square root of frequency (Cremer et al. 2005), the airborne sounds of many animals should be better coupled to the substrate. In some cases, however, modeling the ground surface as a homogenous solid may not be accurate. Ground surface layers are often made somewhat porous through the actions of weathering and biotic sources. The result is lowered wave conduction velocity in the uppermost ground layers and improved coupling between the air and substrate across sound frequencies (Arnott and Sabatier 1990; Bass et al. 1980; Sabatier et al. 1986). In nature, coupling between substrate and air can be much better than that predicted by current models (Fig. 6.3). Air-to-substrate coupling in plants is so good, in fact, that substrate vibration signals can propagate non-contiguous plant structures by radiating airborne sound waves, which in turn induce vibrations in nearby plants. Eriksson et al. (2011) demonstrated that the substrate vibration signals of leafhoppers

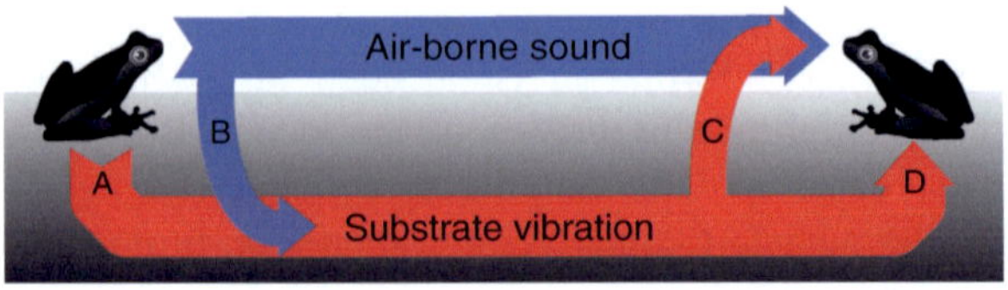


**Fig. 6.2** Paths of acoustic signal propagation. Vibrations generated by animal acoustic calls propagate to receivers through both substrate-borne- and airborne-communication channels. Substrate vibrations can be excited either by direct mechanical coupling between the signaler and the substrate (**a**), or they can be induced in the substrate through the action of airborne sound waves (**b**). Likewise, receivers may detect substrate vibrations by direct coupling with the substrate (**d**), or substrate vibrations may radiate into the air and be detected as airborne sound (**c**)

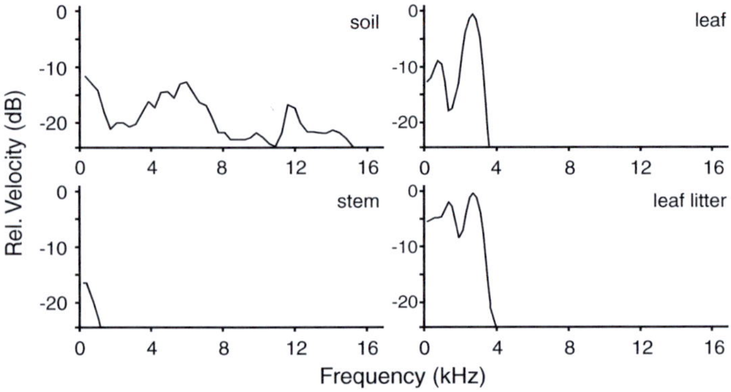


**Fig. 6.3** Substrate vibrations induced by airborne sound. Shown are frequency responses, as measured with a laser vibrometer, for four substrate types exposed to playback of airborne sound at 100 cm (equal amplitude tones 100 Hz–16 kHz): (1) damp soil, (2) the leaf of a small plant (*Philodendron sp.*), (3) a stem from the same plant, and (4) damp leaf litter over soil. All three plant structures strongly attenuate high frequencies. The soil has a broader frequency response, but for frequencies under 4 kHz was not as well coupled to the air as the leaf structures

(*Scaphoideus titanus*) elicit a behavioral response from conspecifics on non-contiguous plant cuttings separated by a gap of up to 6 cm.

Only a handful of studies have looked at the relative contributions of direct coupling and induction by airborne sound to the substrate components of bimodal acoustic signals. Wolf spiders (*S. ocreata*) respond defensively to airborne sound playback of predatory bird calls when on filter paper, but not on granite substrate, which is poorly coupled to air (Lohrey et al. 2009). These defensive responses ultimately lead to lower mating success when bird calls are present (Gordon and Uetz 2012). O'Connell-Rodwell et al. (2000) show a propagation velocity for the seismic component of Asian elephant rumbles that is considerably slower ($\sim$250 ms$^{-1}$) than the speed of sound in air ($\sim$340 ms$^{-1}$). They conclude that this seismic component is not induced in the substrate through the action of airborne sound and therefore must be the result of direct coupling via the animal's body. In a separate study, however, the seismic component of African Elephant (*L. africana*) rumbles was found to travel at considerably higher velocities ($\sim$410 $ms^{-1}$) and to interact with the airborne component (Günther et al. 2004). The nature of airborne-sound-induced vibrations in the substrate was also examined for the stridulatory songs of katydids (*T. cantans*). Airborne song components are limited to frequencies above 6 kHz, but plant-borne substrate vibrations excited by the same signal are broadband, with additional energy between 30 Hz and 5 kHz. To determine the relative contributions of direct coupling and airborne sound induction to the observed plant-borne vibrations, Keuper and Kühne (1983) compared substrate vibrations excited by a stridulating male perched on a stick to those excited by pre-recorded sound played through a nearby speaker. They found that low frequencies were excited by direct coupling, while higher frequencies, more characteristic of the airborne component, were excited through induction. A similar study investigating the role of acoustic induction in the generation of plant-borne vibrations by cicada (*Okanagana rimosa*) songs found that direct coupling to the plant was not necessary to excite strong substrate vibrations across the frequency range of the song, but that high-frequency components of a substrate signal were eliminated if the tymbals (airborne sound radiators) were destroyed (Stölting et al. 2002).

Just as pressure waves in fluid media can impart mechanical energy to solid substrates, biological signals traveling in the substrate can radiate as pressure waves in air and water (by acoustic reciprocity)(Fig. 6.2). Near- and far-field airborne sound radiating from vibrating substrates could be a useful communication channel for some species. Small animals unable to efficiently couple low-frequency body vibrations directly to the air could, for example, vibrate the substrate, relying on substrate structures with greater surface areas, such as nearby leaves, to radiate their signals into the air. Likewise, insects and arachnids possess a number of peripheral receptors that are sensitive to near-field particle movement in the air and may be particularly likely to utilize airborne vibrations induced by substrate movements (Barth 1998; Drosopoulos and Claridge 2005; Hergenröder and Barth 1983). Parasitoid wasps (*Sympiesis sericeicornis*), for example, may use near-field sound radiating off leaves to detect their leafminer hosts (Casas et al. 1998).

### 6.3.3 The Reception of Airborne Sound and Substrate Vibrations

Sensitivity to both airborne and substrate-borne vibration is a prerequisite for the evolution of bimodal sound/substrate vibration signals. While this condition is not universally met, and the exact proportion of animal species sensitive to both modalities is not known, the number of potential receivers for bimodal sound/ substrate vibration signals is enormous. Species from nearly every taxon examined have been found to be sensitive to substrate vibrations (Hill 2008), and a great number of aquatic and terrestrial animals can also detect pressure waves traveling through water or air (Bradbury and Vehrencamp 2011; Drosopoulos and Claridge 2005). Importantly, there is extensive overlap between the species sensitive to each of these modalities and therefore considerable opportunity for bimodal signals to evolve. Cocroft and Rodriguez (2005) estimate that just over 7 % of acoustically communicating insect species use exclusively airborne or waterborne sound signals. By contrast, more than 20 % of insect species employ a combination of airborne and substrate-borne signals. Given that bimodal acoustic signals have not been a focus of communication research and many bimodal signals may currently be assumed to function in only a single modality, this estimate could be low.

In some taxa, input from both sensory modalities is transduced by the same peripheral sensors or processed in the same areas of the central nervous system (see Sect. 6.3.3.2 below). We might expect bimodal sound/substrate vibration signals to be particularly common in these animals. Cross-modal sensitivity of the receiver's neurology will, in the very least, affect the perception of acoustic signals.

#### 6.3.3.1 Cross-Modal Sensitivity of Peripheral Sensors

Sensory receptors in many animals show cross-modal sensitivity to airborne and substrate-borne vibrations. This is not surprising given the fundamental similarities of acoustic waves in fluid and solid media and that in many cases sensors that transduce energy in one acoustic modality have evolved from predecessors tuned to the other modality. Arthropods employ a vast diversity of mechanoreceptors, with most species able to transduce acoustic waves using a variety of sensors (Drosopoulos and Claridge 2005; Hill 2008). The sensitivities of many of these receptors have not been fully mapped out, but it is clear that bimodal responsiveness to air and substrate vibration is a common occurrence throughout the taxon. One notable example is the bimodal response of insect subgenual organs. While the subgenual organ is often considered a specialized receptor for the transduction of substrate vibration, it has also been shown to respond to stimulation by far-field airborne sound in katydids and cockroaches (Drosopoulos and Claridge 2005; Shaw 1994). This pattern of cross-modal sensitivity may be, in part, due to the history of mechanoreceptor evolution. Across insects, it appears that organs sensitive to airborne sound have evolved from substrate vibration sensors (Shaw 1994). Indeed, sensitivity to substrate vibrations is phylogenetically more widespread and appears to predate the evolution of airborne sound sensitivity in several insect lineages (Cocroft 2005).

Insects are not the only group of arthropods known to possess bimodal mechanoreceptors. The three major classes of mechanoreceptors found in crustaceans capable of transducing acoustic waves (chordotonal organs, sensilla, and statocysts) are all sensitive both to waterborne or airborne vibration and to substrate vibration (Taylor and Patek 2010).

Whether bimodal mechanoreceptors are common in arachnids is unknown. Extensive research has documented the diversity and function of mechanoreceptors in arachnids, but this work has been mainly focused on the remarkable sensitivity of spiders and scorpions to substrate vibrations, and to a lesser extent near-field sound (Barth 1982; Brownell and Farley 1979). Spiders, at least, do respond behaviorally to far-field airborne sound (Barth 1982), but it is unclear whether the particular slit sensillae that transduce this energy are also sensitive to substrate vibrations.

The acoustic receptors of vertebrates also show cross-modal sensitivity. In contrast to the diversity of bimodal receptors found in arthropods, however, similar mechanisms involving bone conduction to the ear are responsible for all known cases of bimodal air and substrate vibration receptor sensitivity in vertebrates. In teleost fishes, the inner ear otoliths respond to head movements excited by acoustic wave particle motion in the water. Swim bladder pulsations excited by waterborne pressure waves are also coupled to the head through the body tissues, and in some teleost groups (e.g., Ostariophysi), this coupling is improved by a boney linkage, the Weberian ossicles (Popper et al. 2003). Thus, the inner ear responds to both the particle movement and pressure components of waterborne waves.

In frogs, a specialized structure, the opercularis muscle, connects the scapula directly to the operculum bone, which rests on the oval window of the inner ear and is responsible for exquisite sensitivity to substrate vibrations. Vibrations are then transduced by the saccule and amphibian papilla, structures that also transduce airborne sound stimuli (Narins et al. 2006).

Although snakes can use cutaneous sensors to detect ground movement, substrate vibrations are also coupled through the skull to the middle ear via bone conduction (Young 2010). Interestingly, because snakes lack the tympanic ears found in most terrestrial vertebrates, it appears that vibrations of the skull, induced by airborne sound, may be responsible for sensitivity to both acoustic modalities (Christensen et al. 2012).

It is not known whether birds can detect substrate vibrations through bone conduction to the ear or whether substrate vibration sensitivity in this group is limited to cutaneous mechanoreceptors (Dorward and McIntyre 1971). Some bird species, however, are very sensitive to extremely low-frequency airborne sound (Kreithen and Quine 1979; Warchol and Dallos 1989), increasing the likelihood that any substrate vibrations, which are typically of low frequency, that reach the ear will be transduced by the auditory receptors.

The mechanisms of substrate vibration reception in fossorial mammals are not totally resolved, but most species appear to receive this energy through bone conduction to the middle ear. The ear also shows sensitivity to airborne sound, although direct coupling of the ear to airborne vibrations is reduced in comparison to that found in other mammals (Mason and Narins 2001).

Finally, the inner ear of African elephants appears to be adapted to detect substrate vibrations through bone conduction and should be sensitive to both the airborne and substrate-borne components of elephant "rumble" vocalizations (Reuter et al. 1998).

#### 6.3.3.2 Bimodal Integration in Higher Neural Structures

Even in species with separate specialized acoustic receptors that respond to excitation in a single modality, airborne and substrate-borne vibration may be processed together at higher neural levels. In katydids, which possess specialized tympanal structures for the reception of airborne sound, acoustic neurons ascending to the head ganglia are without exception bimodal, responding to both airborne and substrate-borne vibrations (Rossler et al. 2006). This pattern of extensive integration of substrate vibration and auditory input extends to locusts, where a network of interneurons mediates excitatory or inhibitory interactions between neurons ascending from modality-specific sensors (Bickmeyer et al. 1992; Drosopoulos and Claridge 2005). Furthermore, cross-modal integration is not a phenomenon limited to arthropods. In snakes, cutaneous somatosensory receptors and auditory hair cells both project to the auditory midbrain (Hartline 1971; Young 2003). Likewise, in humans and macaques, sensory input from airborne sounds detected at the ears and substrate vibrations transduced via the somatosensory system converge in the auditory cortex, and similar cross-modal integration would be expected in other primates (Foxe et al. 2002; Schroeder et al. 2001).

The perceptual consequences of cross-modal sensitivity at the receptor level and multimodal integration of sensory input at higher levels are not well understood. In theory, because the transmission properties of each communication channel differ, airborne and substrate-borne signal components are likely to carry non-redundant sets of information about the signaler or signaling environment. Some of this information will not be available to receivers who lack the ability to assess signal components in each modality separately. By contrast, neural integration of sensory input from air and substrate vibration sources allows for a greater range of potentially informative comparisons of stimuli arriving through each communication channel (Hebets and Papaj 2005; Partan and Marler 2005). Further research is needed to elucidate the ways in which receiver neurology affects the structure and function of bimodal acoustic signals.

## 6.4 Implications for the Function and Evolution of Signaling Systems

With only a handful of examples of species known to communicate with bimodal acoustic signals, it is impossible to make broad conclusions about how these signals function and evolve in nature. We can, however, assemble a collection of intriguing and untested hypotheses that may serve as fertile ground for future biovibrations research.

### 6.4.1 The Evolution of Bimodal Acoustic Signals

Airborne and substrate-borne vibrations may be particularly likely to evolve jointly as components of multimodal signals. With the precondition that the signaler is in contact with some substrate, only these two modalities are unavoidably excited together. Evolution of an airborne acoustic signal insures that substrate vibrations will co-occur, setting the stage for this "unintended" component to be co-opted as part of a novel bimodal acoustic signal. Although perhaps less common (see Sect. 6.3.1.2), coincidental airborne sound produced by a signal that has evolved to function through substrate vibration can, likewise, be audibly intense and thereby subject to selection imposed by receivers. Furthermore, the abundance of peripheral receptors sensitive to both modalities and cross-modal processing at higher neural levels increases the likelihood that receivers attending to a signal that initially evolved through one acoustic modality can also detect components propagating though the other.

It is valuable to identify and understand bimodal acoustic signals, not just because unexplored signal components may mediate important aspects of behavioral interactions, but because selection acting on each component can affect the evolution of the entire signaling system, including components in the other modality, through functional trade-offs and pleiotropic interactions (Cooper and Goller 2004). When, for example, white-lipped frogs thump the ground with their vocal sacs while calling, it not only produces a vibrational signal in the substrate, but also introduces amplitude modulations into the airborne component of the call (Lewis et al. 2001).

### 6.4.2 Integration of Information Across Sensory Modalities

In some cases, both components of bimodal acoustic signals may be informative, with components assessed either as separate information sources or assessed relative to one another (Hebets and Papaj 2005; Partan and Marler 2005; Rowe 1999). Even when the airborne and substrate-borne components of a signal are generated through the same physiological mechanism, differences between the components in coupling to the signaling media and in propagation through the environment can imbue them with informative variation.

Receivers may exploit this variation in the localization and ranging of signal sources. The use of multi-sensor, or multi-sample, comparisons for the localization and ranging of visual, acoustic, or chemical signal sources is found in most animal taxa (Bradbury and Vehrencamp 2011; Popper and Fay 2005). Sand Scorpions (*Paruroctonus mesaensis*), for example, compare arrival times across an array of vibration sensors in their legs to localize moving prey at distances of up to 20 cm (Brownell and van Hemmen 2001). The types of substrate waves used for communication generally propagate at lower velocities than airborne sound (Virant-Doberlet et al. 2006). When the substrate component is assessed alone, this leads to greater arrival time and phase differences between spatially separated sensors than are seen with airborne sound and should facilitate localization. In addition, arrival time differences between airborne and substrate-borne components could facilitate ranging, without the constraint on sensors needing to be spread out spatially. Scorpions and spiders may use a similar unimodal strategy, comparing the arrival times of substrate-borne compressional and Rayleigh waves to determine distance to moving prey (Aicher and Tautz 1990; Brownell and Farley 1979). In theory, frequency dispersion or frequency-dependent attenuation of substrate vibrations could also be used for ranging (Elias et al. 2006; Michelsen et al. 1982), but there is currently no convincing evidence that receivers use frequency dispersion information in this way (Popper et al. 2003; Virant-Doberlet et al. 2006).

Because airborne and substrate-borne signal components can have different active spaces, the simple presence or absence of each modality may also communicate some information about a receiver's distance to a signal source. African elephants, for example, respond defensively to seismic presentation of alarm calls, but this response is reduced compared to that seen following playback of the airborne component (O'Connell-Rodwell et al. 2006). If the substrate vibrations generated by alarm calls travel further in the elephant's environment than their airborne counterparts (but see Günther et al. 2004), it is possible that receivers interpret the presence of the seismic vibrations in the absence of the airborne component as an indicator that the alarm call was issued from a distant signaler and thus corresponds to a non-imminent threat (O'Connell-Rodwell et al. 2006). This method of ranging may be particularly useful for territorial plant-dwelling animals who defend an area of contiguous substrate, where the active space of substrate vibration signals is well defined (Caldwell et al. 2010).

Differences in the propagation properties of airborne and substrate-borne vibration components could also aid receivers in the localization or assessment of signalers at different spatial scales (Hebets and Papaj 2005). Katydids (*Tettigonia cantans*) employ bimodal acoustic signals in this way. The airborne component of the male stridulatory signal attracts females from afar, and the plant-borne vibratory component aids in close range localization (Latimer and Schatral 1983). Similarly, some courting male wolf spiders in the genera *Lycosa* and *Schizocosa* couple papal stridulations to the leaf litter, producing substrate vibrations as well as airborne sound that radiates from the leaves. Rovner (1975) suggest that the airborne component of this bimodal strategy may function to communicate with distant females while the seismic vibrations become the dominant communication channel once a female is standing on contiguous substrate.

It is often suggested that substrate vibrations may be a relatively private information channel (Bell 1980; Henry 1994; Markl 1983), with a more limited active space and fewer potential eavesdroppers than visual or sound signals. With the diversity of animals sensitive to substrate vibrations, it is still very much an open question how commonly the concept of a private communication channel applies (Casas et al. 1998; Cocroft and Rodriguez 2005; Virant-Doberlet et al. 2011), but it is certainly the case that airborne and substrate-borne signal components have different, likely overlapping, sets of potential receivers. Moreover, in some cases, the two components may have adapted for communication with separate groups of receivers. This arrangement may be particularly common in lekking species, where signalers benefit from a signal that is effective in communicating both with potential mates at a distance and with nearby competitors. White-lipped frogs and prairie mole crickets appear to use bimodal acoustic signals in this way (Hill and Shadley 2001; Lewis et al. 2001).

### *6.4.3 Backup Signals*

Because both air- and substrate-borne vibrations are often generated by the same physiological mechanism, and both propagate as mechanical waves, they may be more likely to be redundant in information content than components in any other two modalities. Redundant signal components can, nevertheless, serve important communication functions by acting as backup signals and thereby improving transmission efficacy in noisy or variable signaling environments (Hebets and Papaj 2005; Partan and Marler 2005; Rowe 1999). White-lipped frogs use bimodal acoustic signals in this way. Males coordinate the timing of their sexual advertisement calls using the airborne component of calls from neighboring frogs. When this component is masked with noise, however, males instead adjust call timing based on the seismic component of neighbors' calls (Lewis et al. 2001).

Because solid substrates are variable and mechanically complex compared to fluid media (Cremer et al. 2005; Markl 1983; Michelsen et al. 1982), we might expect the unreliability of substrate components to drive the evolution of bimodally redundant acoustic calls. In some cases, however, constraints on sound propagation in air may also drive the evolution of redundant signals. Dense forest, for example, rapidly degrades airborne sound signals, and forest animals such as Asian elephants may benefit from informationally redundant seismic call components in areas where airborne signals are strongly attenuated by vegetation (O'Connell-Rodwell 2007).

Bimodal redundancy is not a foolproof strategy for mitigating the effects of environmental noise. By the same mechanisms that make simultaneous generation of airborne and substrate-borne signal components so likely, environmental noise can also be cross-modal. The receptivity of female wolf spiders (*S. ocreata*) to male substrate vibration courtship signals, for instance, is reduced in the presence of airborne synthetic noise (Gordon and Uetz 2012). As these spiders are not sensitive to airborne sound, it is vibrations induced in the substrate by the noise that are responsible for the reduced effectiveness of male signals. It is not clear whether airborne sounds mask the substrate signals of these spiders in nature. Playback of recorded cicada choruses did not have the same effect. The phenomenon of cross-modal masking might be of use to the agricultural industry, as vibrations induced in a plant substrate by airborne sound can mask the transmission of substrate-borne signals used by crop pests (Saxena and Kumar 1980).

### 6.4.4 Sensory Drive

The summed input of two redundant signal components can better stimulate a receiver's sensory system (Hebets and Papaj 2005; Partan and Marler 2005; Rowe 1999), and this may have significant implications for the function and evolution of bimodal acoustic signals. Whether call intensity encodes information about signaler quality or is a cue to signaler proximity, receivers show a general preference for higher amplitude acoustic signals (Castellano et al. 2000; Latimer and Sippel 1987; Ritschard et al. 2010). Signalers benefit by maximally exciting the sensory systems of receivers, and depending on signaling context, the most effective means of doing this may involve a multimodal strategy, simultaneously stimulating both sound and substrate vibration sensors (Rowe 1999).

The cross-modal sensitivity of mechanoreceptors in several taxa leaves open the door for the intriguing possibility that some bimodal acoustic signals evolve through sensory exploitation. Physiological or environmental constraints on the amplitude of unimodal airborne sound or substrate vibration signals can enforce signal honesty (Bennet-Clark 1998, Fitch and Hauser 2003). However, novel adaptations that better couple a signaler's call to a second transmission medium, either the surrounding fluid medium or the solid substrate, could in essence hijack preexisting receiver strategies that assess signaler quality or location using unimodal amplitude cues. This idea is as yet untested and would likely require a phylogenic analysis of acoustic call trait evolution to resolve.

### 6.4.5 Context-Dependent Shifts in Bimodal Signal Structure

In some behavioral contexts, it may be dangerous or ineffective to produce intense signals in a particular sensory modality (Elias and Mason 2010; Zuk and Kolluru 1998). Signalers would, therefore, benefit from the ability to facultatively alter the relative amplitude or other characteristics of bimodal acoustic signal components. Even when both airborne and substrate-borne vibrations are generated by the same mechanism, the characteristics of each component could be independently modified by adjusting the mechanical response of a signaler's resonant structures or by altering the coupling of each component to the environment.

Unintended receivers, such as predators, parasitoids, and conspecific rivals, can exert strong selection on acoustic signals (Bernal et al. 2006; Zuk and Kolluru 1998) and may shape the evolution of facultative shifts in the structure of bimodal acoustic signals. Cicadas (*Okanagana rimosa*), for instance, undergo periods of "silent singing" where they flex their tymbal organ without producing the high-amplitude airborne component of their advertisement call. This tactic may favor the substrate call component, reducing exposure to parasitoid flies (Stölting et al. 2002). Over an evolutionary timescale, a shift from airborne calls to substrate-borne signals does appear to be an adaptive strategy for avoiding airborne-sound-oriented predators.

Katydids from areas with high densities of predatory bats produce airborne calls with lower duty cycles and rely more heavily on substrate-borne tremulatory signals (Belwood and Morris 1987; Morris et al. 1994). Many animals exhibit lower amplitude airborne calls when they are in close proximity to intended receivers (Caldwell et al. 2010; Dabelsteen et al. 1998). Whether signalers are relying more heavily on substrate-borne signal components in these contexts is not known.

Another potential benefit of facultative changes in a signaler's relative investment into either airborne or substrate-borne signal components comes from the fact that the quality of signaling substrate can be highly variable depending on behavioral context. At times, receivers may not be in contact with contiguous substrate, or the substrate may transmit vibrational signals poorly (Elias and Mason 2010). Signalers would benefit from the ability to assess the current efficacy of communication in each channel and respond by investing more heavily in effective signal components. Similar strategies have been described in animals using other types of multimodal signals. Wolf spiders (*S. ocreata*), for example, use more visual signals on substrates where vibration transmission is poor (Gordon and Uetz 2011). In a related species (*S. rovneri*), courting males use behavioral feedback from responding females to determine whether vibrational signals are effectively propagating through the substrate and adjust relative signaling investment across modalities accordingly (Sullivan-Beckers and Hebets 2011).

Receivers too may benefit from the ability to behaviorally alter the coupling of sensory receptors to each acoustic communication channel. In species with peripheral mechanoreceptors that are sensitive to both airborne and substrate-borne vibrations, this would allow receivers to independently assess information from each sensory modality. The spider *Araneus sericatus* can do this by simply altering the positions of its legs, thereby damping the mechanical response of its exoskeleton (Finck 1981). Similarly, elephants appear to shift their weight forward when attending to seismic signals, thereby improving coupling between the "acoustic fat" of their feet and the substrate. Elephants also possess a sphincter-like muscle that can constrict around the ear canal, attenuating the ear's response to airborne sound (O'Connell-Rodwell and Wood 2010).

### *6.4.6 Costs of Multimodal Signaling*

While the use of bimodal airborne sound and substrate vibration signals may confer several adaptive advantages to signalers and receivers, there are some potential costs as well. First, as mentioned above, bimodal signals have the potential to reach a wider diversity of receivers and thus to expose signalers to increased rates of parasitism and predation. Bimodal acoustic signaling can also be energetically inefficient (Cremer et al. 2005; Partan and Marler 2005). A signal optimally tuned to couple well to either a solid or fluid medium, and propagate through it with minimal loss, will not function as efficiently in the other type of medium. Likewise, any signal tuned to propagate well in both modalities will not be optimally efficient in either. Another potential cost stemming from the inherently bimodal nature of acoustic signals is that cross-modal interactions can degrade signal quality. Because of the differing wave propagation characteristics of fluid and solid transmission media, interference between the two communication channels during signal propagation or reception may degrade frequency, temporal, and amplitude properties of the signal. A consequence of this is that even if an acoustic modality is not used by receivers as an information source, incidental vibrations in this modality excited by a calling animal may still affect the evolution of signal structure and signaling behavior by degrading signals in the other modality.

## 6.5 Implications for the Study of Acoustic Communication

There is a single methodological issue that rises above all others in the experimental study of bimodal airborne and substrate-borne signals. To document their existence and to determine how their components interact during communication requires the researcher to effectively isolate the two communication channels. The very property that makes these two modalities so likely to interact during communication, that they are mechanically intertwined and unavoidably excite one another, makes them difficult to properly control in an experimental setting. Effective methods for experimentally teasing apart the two acoustic modalities include altering a signaler's ability to produce signal components in one modality (Stölting et al. 2002), interfering with the reception of stimuli in one modality through manipulation of the receiver's sensory organs or by masking signal components with noise (Hergenröder and Barth 1983; Lewis et al. 2001), and carefully controlled recording and playback (O'Connell-Rodwell et al. 2007; Rado et al. 1998). It is worth noting, however, that in some cases, controlling each modality during experimentation is not necessary or behaviorally relevant, as receivers may never encounter isolated airborne or substrate-borne signal components in nature.

### *6.5.1 Recording*

Unintended cross-modal contamination of acoustic recordings is a common concern. This can be a significant issue when recording substrate vibrations in the presence of intense airborne sound using a laser Doppler vibrometer, which measures relative velocity between the sensor head and the substrate of interest (Arnott and Sabatier 1990, Caldwell pers. obs.). The thin housing surrounding the sensor head and the large surface area of these sensors (especially portable units) can cause them to vibrate in an airborne sound field. Care should be taken to insure that the sensor head is sufficiently distant from any intense airborne sound sources, and control measurements recorded with the laser focused on a non-vibrating target may be useful to assess sound-induced vibration of the vibrometer itself. The surface of the sensor head can also be covered with acoustic foam, but this will interfere with heat conduction away from the instrument.

Cross-modal contamination is less of an issue with accelerometers and geophones, which tend to be built of dense materials, have relatively small surface areas, and should not, therefore, be as strongly coupled to airborne sound (Fahy and Gardonio 2007). Furthermore, these two types of sensors are often buried in the substrate, precluding the direct action of airborne sound waves on their surfaces. Unnecessary exposure of instrument wiring to excitation by airborne sound should be avoided, however. Cables can also be taped down to damp vibrations propagating along their lengths.

### *6.5.2 Playback*

Unintended cross-modal excitations of vibrations can also complicate playback of acoustic signals. Mechanical shakers follow the same basic design as loudspeakers, but lack the speaker cone. Even without the impedance matching characteristics of a cone, shakers can produce sufficient airborne sound that they limit the range of useful substrate vibration presentation amplitudes (O'Connell-Rodwell et al. 2006) and the construction of a small sound attenuating enclosure surrounding the shaker may be necessary. Playback of airborne signals can pose similar problems. When a speaker and a test subject are resting on a common substrate, speaker vibrations can propagate through this substrate to the animal (Caldwell pers. obs.). Suspending playback speakers with vibrational damping material can reduce the amount of unintended energy transferred directly between the speaker and substrate, but cannot eliminate substrate vibrations induced by the action of airborne sound. Although less widely applicable, in some cases closed-field playback directly to an animal's peripheral sensors (Pinder and Palmer 1983) is an alternative to free-field playback and could reduce cross-modal noise. Regardless of the playback setup utilized, when feasible, it can also be helpful to determine response thresholds for animals exposed to signal components in each modality, so it is clear that unintended energy excited in the other modality during playback is not responsible for the response of test subjects (Lewis et al. 2001). Furthermore, collecting re-recordings of both airborne and substrate-borne vibrations excited by playback in both modalities is an important precaution, especially where airborne sound is likely to couple well with the substrate of interest, as it can during playback to plant-dwelling animals. If the facilities are available, re-recordings can be combined with active noise cancelation techniques to reduce unintended cross-modal noise (Ho and Narins 2006). For a more in depth discussion of vibration playback methods, see Cocroft et al. (Chap. 13, this volume).

## 6.6 Some Final Thoughts

This chapter has likely generated far more questions in the minds of readers than it has answered. That disparity mirrors the state of biovibrations as a field of scientific exploration. Several decades of research have now firmly established the importance of substrate-borne vibrations as an information source and a mode of communication nearly ubiquitous across animal taxa (Hill 2008). Now, we have the opportunity start to synthesize what we have learned, to identify patterns in the use of substrate-borne vibrations, and to begin to understand how they fit in among the full spectrum of other sensory modalities as parts of complete communication systems. Investigating the interactions between airborne sound and substrate-borne vibrations is a promising avenue for future study. It is not clear how common or important these cross-modal interactions are to communicating animals, but it is clear that these questions need to be answered.

**Acknowledgments** I would like to thank all those who have generously contributed to this chapter through discussions and logistical support. I am grateful to Rex Cocroft for his encouragement and for helping to shape the manuscript with his advice and comments. I would like to thank Peggy Hill for additional comments. I would also like to thank Karen Warkentin, John Christy, Rachel Page, and Alex Trillo for helping to develop many of the ideas presented here. Additional gratitude goes to Karen Warkentin and Mark Bee for the generous use of their equipment, Reinhard Lakes-Harlan for the use of his cicada recordings, and Tim Polnaszek for help in obtaining jay call recordings.

## References

Aicher B, Tautz J (1990) Vibrational communication in the fiddler crab, Uca pugilator. I. Signal transmission through the substratum. J Comp Physiol A 166:345–353

Arnott WP, Sabatier JM (1990) Laser-doppler vibrometer measurements of acoustic to seismic coupling. Appl Acoust 30:279–291

Barth FG (1982) Spiders and vibratory signals: sensory reception and behavioral significance. In: Witt PN, Rovner JS (eds) Spider communication mechanisms and ecological significance. Princeton University Press, Princeton, pp 66–122

Barth FG (1998) The vibrational sense of spiders. In: Hoy R, Popper AN, Fay R (eds) Comparative hearing: insects. Springer, New York, pp 228–278

Bass HE, Bolen LN, Cress D, Lundien J, Flohr M (1980) Coupling of airborne sound into the earth: frequency dependence. J Acoust Soc Am 67:1502–1506

Bell PD (1980) Transmission of vibrations along plant stems: implications for insect communication. J N Y Entomol Soc 88:210–216

Belwood JJ, Morris GK (1987) Bat predation and its influence on calling behavior in neotropical katydids. Science 238:64–67

Bennet-Clark HC (1987) The tuned singing burrow of mole crickets. J Exp Biol 128:383–409

Bennet-Clark HC (1998) Size and scale effects as constraints in insect sound communication. Philos T Roy Soc B 353:407–419

Bernal XE, Rand AS, Ryan MJ (2006) Acoustic preferences and localization performance of blood-sucking flies (*Corethrella Coquillett*) to tungara frog calls. Behav Ecol 17:709–715

Bickmeyer U, Kalmring K, Halex H, Mücke A (1992) The bimodal auditory–vibratory system of the thoracic ventral nerve cord in *Locusta migratoria* (Acrididae, Locustinae, Oedipodini). J Exp Zool 264:381–394
Bradbury JW, Vehrencamp SL (2011) Principles of animal communication. Sinauer Associates Inc, Sunderland
Brownell P, Farley RD (1979) Orientation to vibrations in sand by the nocturnal scorpion *Paruroctonus mesaensis*: mechanism of target localization. J Comp Physiol A 131:31–38
Brownell PH (1977) Compressional and surface waves in sand used by desert scorpions to locate prey. Science 197:479–482
Brownell PH, van Hemmen JL (2001) Vibration sensitivity and a computational theory for prey-localizing behavior in sand scorpions. Am Zool 41:1229–1240
Caldwell MS, Johnston GR, McDaniel JG, Warkentin KM (2010) Vibrational signaling in the agonistic interactions of red-eyed treefrogs. Curr Biol 20:1012–1017
Casas J, Magal C, Sueur J (2007) Dispersive and non-dispersive waves through plants: implications for arthropod vibratory communication. P Roy Soc Lond B Bio 274:1087–1092
Casas J, Bacher S, Tautz J, Meyhofer R, Pierre D (1998) Leaf vibrations and air movements in a leafminer-parasitoid system. Biol Control 11:147–153
Castellano S, Rosso A, Laoretti F, Doglio S, Giacoma C (2000) Call intensity and female preferences in the European green toad. Ethology 106:1129–1141
Christensen CB, Christensen-Dalsgaard J, Brandt C, Madsen PT (2012) Hearing with an atympanic ear: good vibration and poor sound-pressure detection in the royal python, *Python regius*. J Exp Biol 215:331–342
Cocroft R (2005) Vibrational communication facilitates cooperative foraging in a phloem-feeding insect. P Roy Soc Lond B Bio 272:1023–1029
Cocroft RB, Rodriguez RL (2005) The behavioral ecology of insect vibrational communication. Bioscience 55:323–334
Cokl A, Zorovic M, Millar JG (2007) Vibrational communication along plants by the stink bugs *Nezara viridula* and *Murgantia histrionica*. Behav Process 75:40–54
Collin SP, Marshall NJ (2003) Sensory processing in aquatic environments. Springer, New York
Cooper BG, Goller F (2004) Multimodal signals: enhancement and constraint of song motor patterns by visual display. Science 303:544–546
Cremer L, Heckl M, Petersson BAT (2005) Structure-borne sound: structural vibrations and sound radiation at audio frequencies. Springer, New York
Dabelsteen T, McGregor PK, Lampe H, Langmore N, Holland J (1998) Quiet song in song birds: an overlooked phenomenon. Bioacoustics 9:89–105
Dorward PK, McIntyre AK (1971) Responses of vibration-sensitive receptors in the interosseous region of the duck's hind limb. J Physiol 219:77–87
Drosopoulos S, Claridge MF (eds) (2005) Insect sounds and communication: physiology, behaviour, ecology, and evolution. CRC Press, Boca Raton
Elias DO, Mason AC (2010) Signaling in variable environments: substrate-borne signaling mechanisms and communication behavior in spiders. In: O'Connell-Rodwell CE (ed) The use of vibrations in communication: properties, mechanisms and function across taxa. Transworld, Kerala, pp 25–46
Elias DO, Lee N, Hebets EA, Mason AC (2006) Seismic signal production in a wolf spider: parallel versus serial multi-component signals. J Exp Biol 209:1074–1084
Eriksson A, Anfora G, Lucchi A, Virant-Doberlet M, Mazzoni V (2011) Inter-plant vibrational communication in a leafhopper insect. PLoS ONE 6:e19692
Fahy F, Gardonio P (2007) Sound and structural vibration: radiation, transmission and response. Academic Press, Oxford
Finck A (1981) The lyriform organ of the orb-weaving spider *Araneous sericatus*: vibration sensitivity is altered by bending the leg. J Acoust Soc Am 70:231–233
Fitch WT, Hauser MD (2003) Unpacking "honesty": vertebrate vocal production and the evolution of acoustic signals. In: Simmons AM, Fay RR, Popper AN (eds) Acoustic communication. Springer, New York, pp 65–137

Forrest TG (1994) From sender to receiver: propagation and environmental effects on acoustic signals. Amer Zool 34:644–654
Foxe JJ, Wylie GR, Martinez A, Schroeder CE, Javitt DC, Guilfoyle D, Ritter W, Murray MM (2002) Auditory-somatosensory multisensory processing in auditory association cortex: an fMRI study. J Neurophysiol 88:540–543
Gordon SD, Uetz GW (2011) Multimodal communication of wolf spiders on different substrates: evidence for behavioural plasticity. Anim Behav 81:367–375
Gordon SD, Uetz GW (2012) Environmental interference: impact of acoustic noise on seismic communication and mating success. Behav Ecol 23:707–714
Günther RH, O'Connell-Rodwell CE, Klemperer SL (2004) Seismic waves from elephant vocalizations: A possible communication mode? Geophys Res Lett 31
Hartline PH (1971) Physiological basis for detection of sound and vibration in snakes. J Exp Biol 54:349–371
Hebets EA, Papaj DR (2005) Complex signal function: developing a framework of testable hypotheses. Behav Ecol Sociobiol 57:197–214
Henry CS (1994) Singing and cryptic speciation insects. Trends Ecol Evol 9:388–392
Hergenröder R, Barth FG (1983) The release of attack and escape behavior by vibratory stimuli in a wandering spider (*Cupiennius salei* Keys). J Comp Physiol A 152:347–359
Hill PSM (2008) Vibrational communication in animals. Harvard University Press, Cambridge
Hill PSM, Shadley JR (2001) Talking back: sending soil vibration signals to lekking prairie mole cricket males. Am Zool 41:1200–1214
Hill PSM, Wells H, Shadley JR (2006) Singing from a constructed burrow: why vary the shape of the burrow mouth? J Orthopt Res 15:23–29
Ho C, Narins P (2006) Directionality of the pressure-difference receiver ears in the northern leopard frog, *Rana pipiens pipiens*. J Comp Physiol A 192:417–429
Holldobler B (1999) Multimodal signals in ant communication. J Comp Physiol A 184:129–141
Jardetzky WS, Press F (1952) Rayleigh-wave coupling to atmospheric compression waves. Bull Seismol Soc Am 42:135–144
Keuper A, Kuhne R (1983) The acoustic behaviour of the bushcricket *Tettigonia cantans* II. Transmission of airborne-sound and vibration signals in the biotope. Behav Process 8:125–145
Kreithen ML, Quine DB (1979) Infrasound detection by the homing pigeon: a behavioral audiogram. J Comp Physiol A 129:1–4
Latimer W, Schatral A (1983) The acoustic behaviour of the bushcricket *Tettigonia cantans* I. Behavioural responses to sound and vibration. Behav Process 8:113–124
Latimer W, Sippel M (1987) Acoustic cues for female choice and male competition in *Tettigonia cantans*. Anim Behav 35:887–900
Lewis ER, Narins PM (1985) Do frogs communicate with seismic signals. Science 227:187–189
Lewis ER, Narins PM, Cortopassi KA, Yamada WM, Poinar EH, Moore SW, Yu XL (2001) Do male white-lipped frogs use seismic signals for intraspecific communication? Am Zool 41:1185–1199
Lohrey AK, Clark DL, Gordon SD, Uetz GW (2009) Antipredator responses of wolf spiders (Araneae: Lycosidae) to sensory cues representing an avian predator. Anim Behav 77:813–821
Magal C, Scholler M, Tautz J, Casas J (2000) The role of leaf structure in vibration propagation. J Acoust Soc Am 108:2412–2418
Markl H (1983) Vibrational communication. In: Huber F, Markl H (eds) Neuroethology and behavioral physiology. Springer, Berlin, pp 332–353
Mason MJ, Narins PM (2001) Seismic signal use by fossorial mammals. Amer Zool 41:1171–1184
McComb K, Reby D, Baker L, Moss C, Sayialel S (2003) Long-distance communication of acoustic cues to social identity in African elephants. Anim Behav 65:317–329
Michelsen A, Fink F, Gogala M, Traue D (1982) Plants as transmission channels for insect vibrational songs. Behav Ecol Sociobiol 11:269–281

Morris GK (1980) Calling display and mating behaviour of *Copiphora rhinoceros* Pictet (Orthoptera: Tettigoniidae). Anim Behav 28:42–51
Morris GK, Mason AC, Wall P, Belwood JJ (1994) High ultrasonic and tremulation signals in neotropical katydids (Orthoptera: Tettigoniidae). J Zool 233:129–163
Narins PM, Reichman OJ, Jarvis JUM, Lewis ER (1992) Seismic signal transmission between burrows of the cape mole-rat, *Georychus capensis*. J Comp Physiol A 170:13–21
Narins PM, Feng AS, Fay RR, Popper AN (eds) (2006) Hearing and sound communication in amphibians. Springer, New York
Narins PM, Grabul DS, Soma KK, Gaucher P, Hödl W (2005) Cross-modal integration in a dart-poison frog. P Natl Acad Sci USA 102:2425–2429
O'Connell-Rodwell C, Wood J, Rodwell T, Puria S, Partan S, Keefe R, Shriver D, Arnason B, Hart L (2006) Wild elephant (*Loxodonta africana*) breeding herds respond to artificially transmitted seismic stimuli. Behav Ecol Sociobiol 59:842–850
O'Connell-Rodwell CE (2007) Keeping an "ear" to the ground: seismic communication in elephants. Physiology 22:287–294
O'Connell-Rodwell CE, Arnason BT, Hart LA (2000) Seismic properties of Asian elephant (*Elephas maximus*) vocalizations and locomotion. J Acoust Soc Am 108:3066–3072
O'Connell-Rodwell CE, Wood JD, Kinzley C, Rodwell TC, Poole JH, Puria S (2007) Wild African elephants (*Loxodonta africana*) discriminate between familiar and unfamiliar conspecific seismic alarm calls. J Acoust Soc Am 122:823–830
O'Connell-Rodwell CE, Wood JD (2010) Vibration generation, propagation and detection in elephants. In: O'Connell-Rodwell CE (ed) The use of vibrations in communication: properties, mechanisms and function across taxa. Transworld, Kerala, p 183
Partan SR, Marler P (2005) Issues in the classification of multimodal communication signals. Am Nat 166:231–245
Pinder AC, Palmer AR (1983) Mechanical properties of the frog ear: vibration measurements under free- and closed-field acoustic conditions. P Roy Soc Lond B Bio 219:371–396
Polajnar J, Svenšek D, Čokl A (2012) Resonance in herbaceous plant stems as a factor in vibrational communication of pentatomid bugs (Heteroptera: Pentatomidae). J Roy Soc Interface 9:1898–1907
Popper AN, Fay RR (eds) (2005) Sound source localization. Springer, New York
Popper AN, Fay R, Platt C, Sand O (2003) Sound detection mechanisms and capabilities of teleost fishes. In: Collin SP, Marshall J (eds) Sensory processing in aquatic environments. Springer, New York, pp 3–38
Press F, Ewing M (1951) Theory of air-coupled flexural waves. J Appl Phys 22:892–899
Rado R, Terkel J, Wollberg Z (1998) Seismic communication signals in the blind mole-rat (Spalax ehrenbergi): electrophysiological and behavioral evidence for their processing by the auditory system. J Comp Physiol A 183:503–511
Randall JA (2010) Drummers and stompers: vibrational communication in mammals. In: O'Connell-Rodwell CE (ed) The use of vibrations in communication: properties, mechanisms and function across Taxa. Transworld, Kerala, pp 99–120
Reuter T, Nummela S, Hemila S (1998) Elephant hearing. J Acoust Soc Am 104:1122–1123
Ritschard M, Riebel K, Brumm H (2010) Female zebra finches prefer high-amplitude song. Anim Behav 79:877–883
Rossler W, Jatho M, Kalmring K (2006) The auditory-vibratory sensory system in bushcrickets. In: Drosopoulos S, Claridge MF (eds) Insect sounds and communication: physiology, behaviour, ecology, and evolution. CRC Press, Boca Raton, pp 35–69
Rovner JS (1975) Sound production by nearctic wolf spiders: a substratum-coupled stridulatory mechanism. Science 190:1309–1310
Rowe C (1999) Receiver psychology and the evolution of multicomponent signals. Anim Behav 58:921–931
Sabatier JM, Bass HE, Bolen LN, Attenborough K (1986) Acoustically induced seismic waves. J Acoust Soc Am 80:646–649

Saxena KN, Kumar H (1980) Interruption of acoustic communication and mating in a leafhopper and a planthopper by aerial sound vibrations picked up by plants. Experientia 36:933–936

Schroeder CE, Lindsley RW, Specht C, Marcovici A, Smiley JF, Javitt DC (2001) Somatosensory input to auditory association cortex in the macaque monkey. J Neurophysiol 85:1322–1327

Shaw S (1994) Detection of airborne sound by a cockroach 'vibration detector': a possible missing link in insect auditory evolution. J Exp Biol 193:13–47

Stiedl O, Kalmring K (1989) The importance of song and vibratory signals in the behaviour of the bushcricket *Ephippiger ephippiger* Fiebig (Orthoptera, Tettigoniidae): taxis by females. Oecologia 80:142–144

Stölting H, Moore TE, Lakes-Harlan R (2002) Substrate vibrations during acoustic signalling in the cicada *Okanagana rimosa*. J Insect Sci 2:1–7

Sullivan-Beckers L, Hebets EA (2011) Modality-specific experience with female feedback increases the efficacy of courtship signalling in male wolf spiders. Anim Behav 82:1051–1057

Taylor JRA, Patek SN (2010) Crustacean seismic communication: Heard but not present? In: O'Connell-Rodwell CE (ed) The use of vibrations in communication: properties, mechanisms, and function across taxa. Transworld, Kerala, pp 9–24

Uetz GW, Roberts JA (2002) Multisensory cues and multimodal communication in spiders: Insights from video/audio playback studies. Brain Behav Evol 59:222–230

Uetz GW, Roberts JA, Taylor PW (2009) Multimodal communication and mate choice in wolf spiders: female response to multimodal versus unimodal signals. Anim Behav 78:299–305

Virant-Doberlet M, Cokl A (2004) Vibrational communication in insects. Neotrop Entomol 33:121–134

Virant-Doberlet M, Cokl A, Zorovic M (2006) Use of substrate vibrations for orientation. In: Drosopoulos S, Claridge MF (eds) Insect sounds and communication: physiology, behaviour, ecology, and evolution. CRC Press, Boca Raton, pp 81–97

Virant-Doberlet M, King RA, Polajnar J, Symondson WOC (2011) Molecular diagnostics reveal spiders that exploit prey vibrational signals used in sexual communication. Mol Ecol 20:2204–2216

Walker TJ, Figg DE (1990) Song and acoustic burrow of the prairie mole cricket, *Gryllotalpa major* (Orthoptera: Gryllidae). J Kans Entomol Soc 63:237–242

Warchol ME, Dallos P (1989) Neural response to very low-frequency sound in the avian cochlear nucleus. J Comp Physiol A 166:83–95

Whang A, Janssen J (1994) Sound production through the substrate during reproduction in the mottled sculpin, Cottus bairdi (Cottidae). Environ Biol Fishes 40:141–148

Young BA (2003) Snake bioacoustics: toward a richer understanding of the behavioral ecology of snakes. Q Rev Biol 78:303–325

Young BA (2010) Vibration detection in snakes. In: O'Connell-Rodwell CE (ed) The use of vibrations in communication: properties, mechanisms and function across taxa. Transworld, Kerala, pp 85–98

Zuk M, Kolluru GR (1998) Exploitation of sexual signals by predators and parasitoids. Q Rev Biol 73:415–438

# Chapter 7
# Vibrational Communication Networks: Eavesdropping and Biotic Noise

**Meta Virant-Doberlet, Valerio Mazzoni, Maarten de Groot, Jernej Polajnar, Andrea Lucchi, William O. C. Symondson and Andrej Čokl**

**Abstract** In nature, communication predominantly occurs in a group of several conspecific and/or heterospecific individuals within signaling and receiving range of each other, i.e., in a network environment. Vibrational communication in the context of sexual behavior has been, in the past, usually considered as a private communication channel, free of potential competitors and eavesdropping predators or parasitoids and consequently only rarely studied outside an emitter–receiver

M. Virant-Doberlet (✉) · M. de Groot · J. Polajnar · A. Čokl
Department of Entomology, National Institute of Biology, Večna pot 111, 1000 Ljubljana, Slovenia
e-mail: meta.virant@nib.si

M. de Groot
e-mail: maarten.degroot@gozdis.si

J. Polajnar
e-mail: jernej.polajnar@nib.si

A. Čokl
e-mail: andrej.cokl@nib.si

V. Mazzoni
Department of Sustainable Ecosystems and Bioresources, Fondazione Mach, Via E. Mach 1, 38010 San Michele all'Adige, Trentino, Italy
e-mail: valerio.mazzoni@fmach.it

*Present Address:*
M. de Groot
Slovenian Forestry Institute, Večna pot 2, 1000 Ljubljana, Slovenia

A. Lucchi
Department of Agriculture, Food and Environment, University of Pisa, Via del Borghetto 80, 56124 Pisa, Italy
e-mail: andrea.lucchi@unipi.it

W. O. C. Symondson
Cardiff School of Biosciences, Cardiff University, Sir Martin Evans Building Park Place, Cardiff CF10 3AX, UK
e-mail: symondson@cardiff.ac.uk

R. B. Cocroft et al. (eds.), *Studying Vibrational Communication*, Animal Signals and Communication 3, DOI: 10.1007/978-3-662-43607-3_7,

dyad. We provide an overview of work related to vibrational communication in the presence of (a) environmental (abiotic) noise, (b) other conspecific and/or heterospecific signalers (biotic noise), (c) rivals and (d) exploiters (predators and parasitoids). The evidence gathered in the last few years shows that arthropods relying on substrate-borne vibrations communicate within a rich and complex vibrational world and reveals diverse interactions and mechanisms. Considering vibrational communication from a network perspective may allow us in the future to identify sources of selection pressures that cannot be recognized in a communication dyad.

## 7.1 Introduction

Substrate-borne sound is mostly hidden from our weak human vibrational senses, and we can hear vibrational signals only with the aid of specialized equipment. The fact that in the field we cannot hear the vibrational signals of the species around us leads to the traditional perception that vibrational communication must be extremely rare. Moreover, since the techniques used to detect these signals often give the impression that we are spying on a private communication, another prevailing perception is that vibrational communication is limited exclusively to emitter–receiver dyads. Furthermore, since to us these signals appear extremely quiet and private, another general misconception is that they must be effective only as a short-range communication channel.

In the last decade, however, emerging evidence has fundamentally changed our perception and understanding of this communication modality. We now know that vibrational signaling is the most common and taxonomically widespread form of sound communication, probably used by more than 230,000 arthropod species (Barth 1998; Virant-Doberlet and Čokl 2004; Cocroft and Rodríguez 2005, Hill 2008). We realized that, even for smaller arthropods, the communication range of vibrational signals can extend up to several meters (Stewart and Zeigler 1984; McVean and Field 1996) and that it is not limited to a continuous substrate (Eriksson et al. 2011) (see also Chap. 8, this volume). Moreover, as predicted by Cocroft and Rodríguez (2005), in the last few years, an increasing number of studies revealed that animals relying on vibrational signaling communicate in a complex biotic landscape that contains other conspecific and heterospecific signalers (biotic noise), as well as rivals and exploiters (Fig. 7.1). Furthermore, environmental (abiotic) noise, which was previously regarded in laboratory studies only as an inconvenience, is now recognized as an important source of selection on vibrational communication.

Networks, where each signal can be received by several receivers and each receiver can receive signals from several emitters, can be considered as the most widespread context in which communication occurs (McGregor and Peake 2000; McGregor 2005; Peake 2005). While vibrational signals emitted by insects living in groups have always been related to a network context (reviewed in Kirchner 1997; Hill 2008; Cocroft and Hamel 2010), vibrational communication in sexual behavior

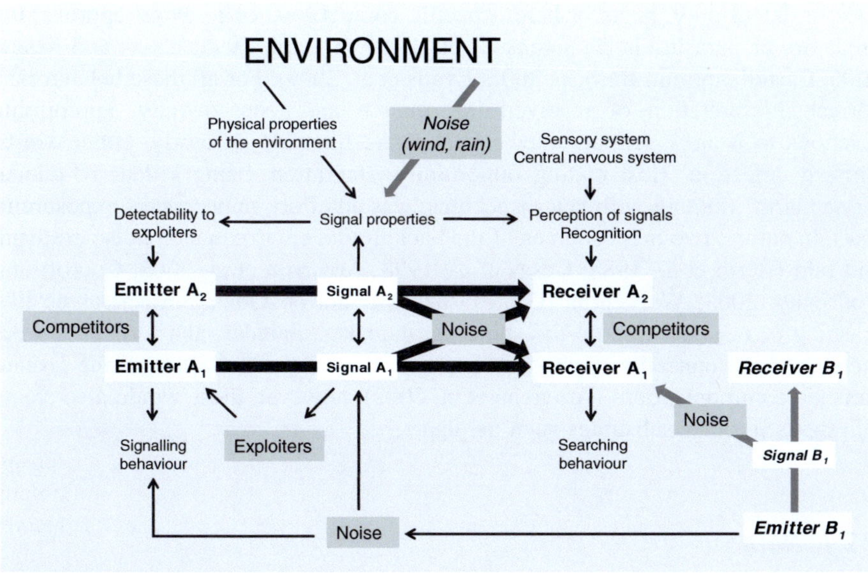


**Fig. 7.1** Basic components of the design of a sexual communication system within a network environment. Selection favors signals that experience less attenuation and degradation during the transmission through the environment. Signal properties and signaling behavior should evolve to increase the probability of correct responses from the intended receiver and minimize detection by the unintended receivers (*exploiters* predators and parasitoids) as well as to minimize the effects of biotic noise that arises from the signals of other individuals, conspecific (*A1*, *A2*) and heterospecific (*B1*), in the same habitat. Receiving mechanisms and the receiver's behavior should evolve to distinguish relevant signals from environmental noise (*wind, rain*) and to respond appropriately. The receiver's response strategy may depend on the presence of intrasexual competitors (*rivals*). In interactive communication, the same individual is the emitter and the receiver

has only rarely been considered other than in the context of emitter–receiver dyads. In this chapter, we wish to bridge this gap by focusing our review mainly on sexual communication in the presence of noise, as well as competitors, predators, and parasitoids. Considering vibrational communication from a network perspective may allow us to identify sources of selection pressures that cannot be recognized in a communication dyad. We start by reviewing the work on effects of abiotic noise on emission and detection of vibrational signals. Since there are very few studies dealing with the influence of environmental noise on vibrational communication, we also included other arthropod behaviors guided by vibrational signals.

## 7.2 Vibrational Channel and Abiotic Environmental Noise

In natural environments, background noise created by abiotic sources can affect any behavior that relies on information provided by substrate vibrations. Animals use vibrational signals not only in sexual communication, but also as incidental cues to detect prey, hosts or heterospecific competitors, or to avoid approaching predators or parasitoids (Pfannenstiel et al. 1995; Barth 1998; Casas and Magal 2006; Castellanos and Barbosa 2006; Evans et al. 2009). For all these behaviors, a correct interpretation of received information and, consequently, appropriate reactions to it are essential, since mistakes are likely to be costly, either due to 'missed detection' (lost mating opportunity, starvation, being killed) or due to 'false alarm' (mating with heterospecifics, wasted effort, unnecessary exposure to risk). In nature, two main sources of the background environmental noise are wind and rain (Barth et al. 1988; Casas et al. 1998; Arnarson et al. 2002; Cocroft and Rodríguez 2005; Warkentin 2005; Tishechkin 2007). Other noise sources like crashing of ocean waves on the shore, earthquakes, thunder, and tornadoes were mentioned in connection with vibrational signals transmitted through the ground (seismic communication) (Arnarson et al. 2002); however, these would also create vibrations in other substrates such as plants.

### *7.2.1 Rain*

Vibrations generated in plants by rain drops are composed of a short, high-amplitude, high-frequency irregular phase followed by a longer regular phase composed of diminishing low-frequency oscillations (up to 11 Hz), and the basic frequency of the regular phase is higher when a drop of water strikes the base of a leaf (Barth et al. 1988; Casas et al. 1998) (Fig. 7.2). The effect of rain on arthropod behavior, guided by vibrational signals or cues, has been directly studied only in connection with evaluation of predation risk by larvae of the moth *Semiothisa aemulataria* (Castellanos and Barbosa 2006). Caterpillars could distinguish between vibrational stimuli created by water drops and those induced by predators and did not respond to rain with defensive behavior. Observations indicate that parasitoid wasps exploiting incidental vibrations produced by their leafminer hosts do not forage in rain (Casas et al. 1998; Casas and Magal 2006); however, there is no evidence of a causal relationship between noise created by rain and reduced foraging. On the other hand, characteristic escape movements have been sometimes observed in their leafminer hosts when plants were watered (Casas and Magal 2006). It has been suggested that in temperate climate rain may not represent a source of selection on vibrational communication (Tishechkin 2007); however, in rainforests, where rain is frequent, vibrational signals used in intraspecific communication may evolve to stand out from background noise created by dripping foliage after rain (Cocroft and Rodríguez 2005).

The effect of rain on behavior that relies on information provided by vibrations has been studied in more detail in vertebrates. Embryos of the red-eyed tree frog *Agalychnis callidryas* use vibrational cues in snake attack to hatch prematurely and escape. However, vibrations caused by rain may represent 'false alarms' that

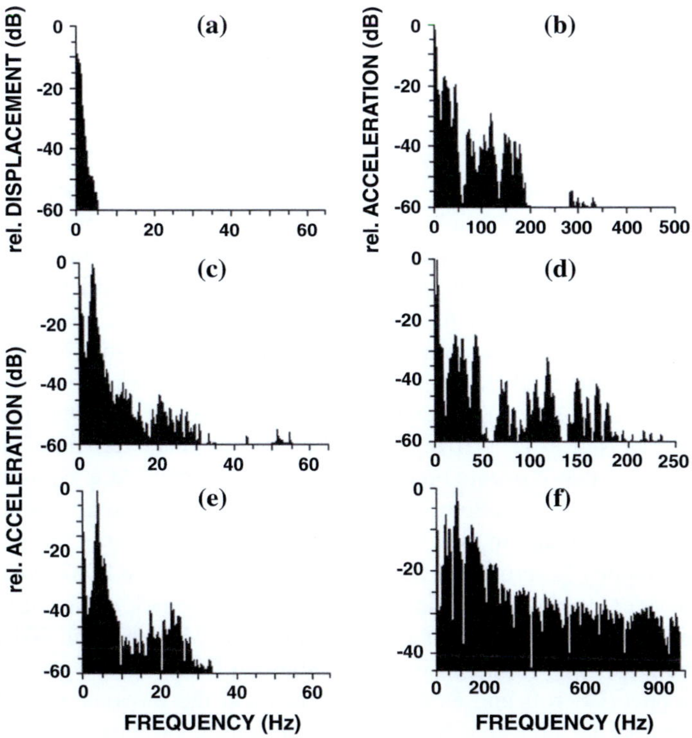


**Fig. 7.2** Frequency spectra of abiotic noise created in plants by wind and rain. Vibrations generated by wind recorded on a banana plant *Musa sapientum* (**a**, **b**) and bromeliad *Aechmea bractea* (**c**–**e**). Noise shown in (**b**) and (**d**) was recorded when leaves were colliding. Vibrations created in a bromeliad by falling water drops (**f**). Note differences in scales (Modified from Barth et al. 1988, Fig. 2, p. 197, with permission from Springer)

are potentially costly leading to unnecessary exposure to other predators (Warkentin 2005; Caldwell et al. 2010). These frogs lay egg clutches on vegetation hanging over ponds, and upon hatching, tadpoles fall into the water. In their environment in lowland wet forest of Central America, rainstorms are common and the temporal, amplitude, and frequency characteristics of vibrations produced by snake attack and by rain are partly overlapping (Warkentin 2005; Caldwell et al. 2010). To avoid premature hatching, embryos assess vibrational cues for up to several minutes (Warkentin et al. 2007) and have to combine information from multiple cues (duration, interval, frequency, and intensity pattern of perceived vibrations) to distinguish between lethal and benign sources (Warkentin et al. 2006; Caldwell et al. 2009; Caldwell et al. 2010).

### 7.2.2 Wind

For plant-dwelling arthropods, behaviorally relevant abiotic noise arises predominantly from wind (Barth et al. 1988; McVean and Field 1996; Casas et al. 1998; Cocroft and Rodríguez 2005; Tishechkin 2007; McNett et al. 2010). Vibrational noise generated by wind has primarily low dominant frequency (between 5–30 Hz); although when wind is strong enough that leaves are colliding, frequency bands extend up to 200 Hz (Barth et al. 1988) (Fig. 7.2). Frequencies up to 25 kHz were recorded in wind-induced noise; however, intensities decrease with increasing frequency (Casas et al. 1998; Cocroft and Rodríguez 2005). The amplitude and noise spectra can vary somewhat between structurally different plant species (Barth et al. 1988; Cocroft and Rodríguez 2005; Tishechkin 2007) and probably also between structurally and mechanically different parts of the same plant (McNett et al. 2010). Another important characteristic of noise generated by wind is that it occurs on two different time scales. At a given location, it may show predictable variation over the course of the day (usually wind speed is lower in the morning and in the evening) and unpredictable variation on a scale of seconds or minutes (Cocroft and Rodríguez 2005; Tishechkin 2007; McNett et al. 2010). However, at any single location, local wind velocity may differ greatly between open and sheltered areas (Tishechkin 2007; McNett et al. 2010).

Wind noise is likely to be an important source of selection in grasslands or in forest canopies (Cocroft and Rodríguez 2005), and the obvious adaptation is to restrict signaling to wind-free periods (Cocroft and Rodríguez 2005; Tishechkin 2007; McNett et al. 2010). In the treehopper *Enchenopa binotata* 'Ptelea,' most vibrational signaling in the field occurred during the morning and evening when wind speed was lowest (McNett et al. 2010) (Fig. 7.3b). Furthermore, in the same species, laboratory experiments revealed that, on a short time scale, males emitted significantly fewer calls during wind bursts than during wind-free gaps (Fig. 7.3c), while in the presence of wind females failed to respond to calls simulating a more distant male (McNett et al. 2010). Similarly, in the field, the planthopper *Criomorphus albomarginatus* emitted vibrational signals only in 30 s or longer lulls between wind-induced vibrations (Tishechkin 2007). It has been hypothesized that leaves, which are easily moved even in light winds, may have evolved to generate background noise that deters herbivorous insects that communicate using vibrational signals (Yamazaki 2011).

Field observations showed that females of parasitoid wasps do not fly when wind speed is higher than 0.6 m/s (Casas 1989); however, laboratory measurements of leaf vibrations indicate that under such conditions environmental noise may not interfere with vibrations created by a leafminer host (Casas et al. 1998; Casas and Magal 2006). The caterpillars of *S. aemulataria* did not respond to vibrational noise created by wind (Castellanos and Barbosa 2006). The web-building spider *Araneus diadematus* approached entangled prey more slowly when wind speed increased; however, it seems that wind affected the spider's locomotion directly and not its ability to detect and locate the prey (Turner et al. 2011).

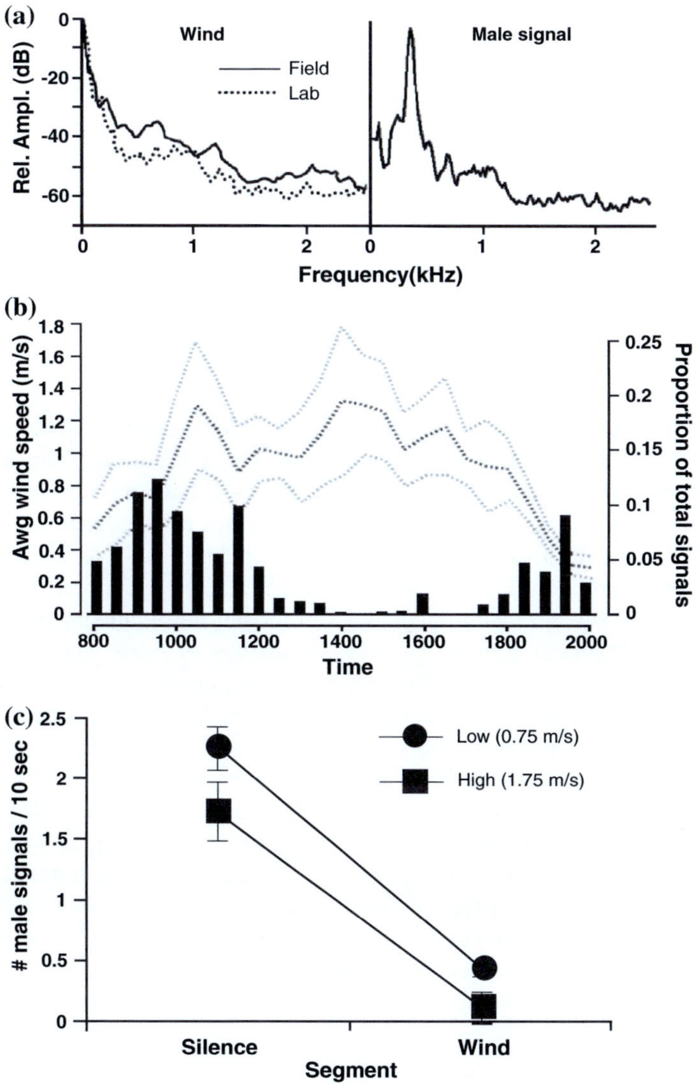


**Fig. 7.3** Effects of wind on vibrational signaling in *Enchenopa binotata* 'Ptelea.' **a** Amplitude spectra of noise induced in the plant *Ptelea trifoliata* in the field or generated in the laboratory (*left*) compared with male advertisement signal (*right*). **b** Diel variation in wind velocity (*dotted line*) and male signaling (*bars*) in the field. Average wind velocity together with 95 % confidence interval is shown. **c** Inhibition of male signaling by wind generated in the laboratory where two wind speeds were used (Modified from McNett et al. 2010, Figs. 1b, 4, 5, with permission from Springer)

Background noise generated by wind can, on the other hand, impair the ability of prey to detect vibrational cues emanating from an approaching predator. It has been shown that salticid spiders, preying on other spiders by invading their webs, exploit background noise created by wind as part of their predatory tactics (Wilcox et al. 1996; Cerveira et al. 2003). A similar hunting strategy was found in the araneophagic assassin bug *Stenolemus bituberus*, which was more likely to catch a web-building spider in the presence of wind (Wignall et al. 2011).

### 7.2.3 Anthropogenic Noise

An increasingly important problem, affecting animal communication in all environments, is low-frequency human-generated (anthropogenic) noise (Barber et al. 2010; Slabbekoorn et al. 2010). The main source of anthropogenic noise is human activity related to transportation (roads, railways, and air traffic). Most of the energy in anthropogenic noise is concentrated in the frequency spectrum below 2 kHz, and the affected area can extend more than 1 km away from the source (Slabbekoorn and Peet 2003; Halfwerk et al. 2011). It has been demonstrated that masking imposed by anthropogenic noise has a negative impact on vertebrate communication and chronic exposure to noise can ultimately restructure animal communities (Barber et al. 2010; Slabbekoorn et al. 2010). Currently, there is no information on whether anthropogenic vibrational noise affects emission and reception of vibrational signals; however, frequency of most vibrational signals used in communication is within the frequency range of anthropogenic noise, and interference is therefore likely. Indirect information, however, has been provided by studies using synthesized noise, which showed that in leafhoppers such vibrational noise reduced the level of male calling and interrupted established male–female duets (Hunt and Morton 2001; Mazzoni et al. 2009b). Moreover, the same effect has been achieved by broadcasting synthesized airborne sounds (Saxena and Kumar 1980). It has also been shown that in the terrestrial hermit crab *Coenobita clypeatus* anthropogenic acoustic noise can affect reaction to a visual threat, presumably due to distraction and attention limits (Chan et al. 2010a, b). However, the disruptive effect of noise can be exploited in pest management, as recently shown in the field (Eriksson et al. 2012).

## 7.3 Vibrational Communication Networks

In the last two decades, there has been a major paradigm shift in the understanding of animal communication by researchers working on airborne sound communication, namely that in nature, all communication occurs in a network environment, i.e., in a group of several animals within signaling and receiving range of each other (McGregor 1993, 2005; McGregor and Peake 2000) (Fig. 7.1). Several basic network structures have been recognized (Burt and Vehrencamp 2005; Otter and Ratcliffe 2005; Peake 2005): broadcast networks (one emitter sends information to several receivers); receiving network (one receiver receives information from several emitters); interactive networks (several emitters signaling interactively to one another); eavesdropping networks (to obtain information the receiver eavesdrops on the interaction in which he/she is not directly involved). One important element of communication networks in nature is also incidental receivers, i.e., heterospecific receivers present in the same environment that usually do not obtain any information from the emitter and to which signals represent only biotic noise. However, for incidental receivers, heterospecific signals may also be a source of 'false alarms' when they incorrectly interpret them as relevant information emitted by conspecifics. Any real-world communication network is likely to consist of any combination of the above-mentioned basic structures, adding more complexity (both additive and interactive) to the overall network structure (Burt and Vehrencamp 2005).

From potential interactions within the network, it emerges that signalers and receivers take on a range of costs and benefits that may have fundamental implications for communication strategies (McGregor and Peake 2000; Dabelsteen 2005; Peake 2005). In a network, conspecific signalers and target receivers may benefit from increased opportunity for mate choice (Otter and Ratcliffe 2005), while an eavesdropper benefits from low-cost gathering of information (Dabelsteen 2005). However, important effects of communication networks on conspecifics are also competition between the emitters for the receiver's attention and the receiver's problems of extracting relevant information emitted simultaneously from several sources and discriminating between them ('cocktail party problem'). Eavesdropping on conspecific interactions may limit attention and thus may increase predation risk, because vigilance is reduced (Peake 2005; Dabelsteen 2005). Another emerging property of the network environment is that signals can also be intercepted and exploited by rivals and predators or parasitoids. As a result, emitters suffer direct costs due to eavesdropping exploiters, while several conspecific receivers compete for resources represented by the emitter (rivalry). For the receivers, the presence of heterospecific signalers compounds the 'cocktail party problem' (Bee and Micheyl 2008) and increases the probability of false alarms, which are costly due to wasted effort and unnecessary exposure to risk and may ultimately result in mating with heterospecifics. When communication networks involve mating traits (mating signals and associated preferences for those signals, as well as mate-searching behavior), adaptation of these traits to the biotic landscape may ultimately lead to speciation (Rundle and Nosil 2005; Hoskin and Higgie 2010).

Here below, we provide an overview on what is known about vibrational communication in situations that relate to the network context. We start by reviewing the work on effects of biotic noise on emission and detection of vibrational signals, followed by studies on rivalry interactions, and conclude with research on eavesdropping by parasitoids and predators.

## 7.3.1 Biotic Noise

In intraspecific communication, noise is any signal that interferes with detection of conspecific signals and transmission of relevant information (Forrest 1994). In nature, interference often occurs as an incidental by-product from overlapping or alternating conspecific and/or heterospecific signals (Römer 1993; Brumm and Slabbekoorn 2005). Reproductive interference, due to signal jamming by heterospecific signals, is a common phenomenon in species communicating with airborne sounds (Gröning and Hochkirch 2008). In general, the most severe masking and interference effects result when noise and communication signals overlap in time and frequency domain, and, consequently, signals emitted by several conspecifics may present a number of challenges for the receiver.

Plants are probably the most common substrate for arthropods relying on vibrational communication. While problems of simultaneously detecting several conspecific and/or heterospecific signals, and discriminating between them, are present also in other signaling modalities (Gerhardt and Huber 2002; Greenfield 2002), potential impacts are particularly severe for species communicating via vibrational signals transmitted through plants. Although arthropods have a spatial array of vibration receptors located in all legs positioned in a circular arrangement on the substrate, which enables accurate localization of the source in two dimensions (Virant-Doberlet et al. 2006), plant-dwelling arthropods often perceive vibrational signals via one-dimensional plant stems and leaf stalks. In such situations, when all signals are reaching a receiver only from one side, signals from multiple sources may combine into a compound signal and the receiver may perceive them as emanating from a single source (Fig. 7.5a).

Although it has been shown in the field (Cocroft 2003; Cocroft and Rodríguez 2005) and in some laboratory studies (Ichikawa 1982; Hunt and Morton 2001; Virant-Doberlet and Žežlina 2007) that several conspecifics, as well as individuals of different insect species, can emit advertisement vibrational calls on the same plant, there are only a few studies that provide some insight into the role of biotic noise in vibrational communication. Indirect information about incidental interference from heterospecifics, in a vibrational communication system where partners form a precisely coordinated duet, has been provided by a study investigating the role of vibrational signals in reproductive isolation. In *Ribautodelphax* planthoppers, the results from a playback study in which conspecific and heterospecific female calls were presented simultaneously did not reveal any effect of biotic noise on the ability of males to recognize and locate conspecific females (de Winter and Rollenhagen 1990). By contrast, incidental interference between signalers in the southern green stink bug *Nezara viridula* showed a negative effect of biotic noise on mating behavior. Although in *N. viridula* both partners emit vibrational signals, termed songs (Čokl et al. 2000), a stereotyped duet is not formed during the initial advertising stage (de Groot et al. 2011b; Žunič et al. 2011) and a male searches for a continuously calling female. Vibrational contact between partners is initiated by the female calling song, which enables a male to identify the conspecific mate by species-specific temporal song pattern and to locate her on the plant (Čokl et al. 1999; Miklas et al. 2003; Hrabar et al. 2004; Žunič et al. 2011).

In *N. viridula,* the ability of males to recognize the song of a conspecific female, in the presence of incidental interference from other potential signalers, was tested in playback experiments. Male responsiveness was reduced when the conspecific female calling song was presented together with the female song of the closely related stink bug *Acrosternum hilare*, or with background noise from conspecific signals that obscured the species-specific temporal pattern of the *N. viridula* female song (Miklas et al. 2003; Polajnar and Čokl 2008; de Groot et al. 2010, 2011a) (Fig. 7.4). While the presence of two synchronized sources of *N. viridula* female calling song had a positive effect on male responsiveness (de Groot et al. 2010) (Fig. 7.4b), males perceived conspecific female songs emitted in alternation, from two sources, as a compound song with a signal repetition time outside the species-specific value (Fig. 7.5) and fewer males started searching (de Groot et al. 2011a). By contrast, searching activity was not significantly affected when the conspecific female song was presented together with the female signals of *A. hilare* that differ in frequency. However, when conspecific and heterospecific signals were overlapping, *N. viridula* males made orientation errors and the majority located the heterospecific source of lower frequency, which may be easier to locate (de Groot et al. 2011a).

Given the importance of vibrational signals for recognition and localization of a conspecific female, one would expect that the *N. viridula* communication strategy should evolve to resolve ambiguous situations when signals are emitted from several sources. The above-mentioned studies, however, suggest that males do not move to a receiving position that would enable spatial release from masking that would permit separation of two conspecific sources (Bee and Micheyl 2008), and therefore, the emitters (females) should take on the responsibility to assure a reliable transfer of information. To increase signal-to-noise ratio, females should either synchronize the emission, increase signal amplitude, shift to different frequencies within the species-specific range, or increase serial redundancy by emitting more signals or signing for a longer time (Brumm and Slabbekoorn 2005; Aubin and Jouventin 2002). In playback experiments, male responsiveness was not reduced when background biotic noise had lower amplitude (de Groot et al. 2010) (Fig. 7.4c, d). Although the treehopper *E. binotata* 'Ptelea' can change the amplitude of their advertisement signals (Sullivan-Beckers and Cocroft 2010), it is not clear whether *N. viridula* can also control the amplitude of emitted signals. Furthermore, on plants, the amplitude of vibrational signals oscillates during the transmission (Michelsen et al. 1982; Polajnar et al. 2012) and on a bean stem, differences between regularly repeated minima and maxima of the *N. viridula* female calling song range between 5 dB below and 10 dB above the value at the source (Čokl et al. 2007). While higher signal amplitude at the source would certainly increase the broadcast distance, it would not simultaneously assure higher signal-to-noise ratio at the position of the receiver. Rather than adjusting the amplitude, in the presence of a continuous background noise simulating

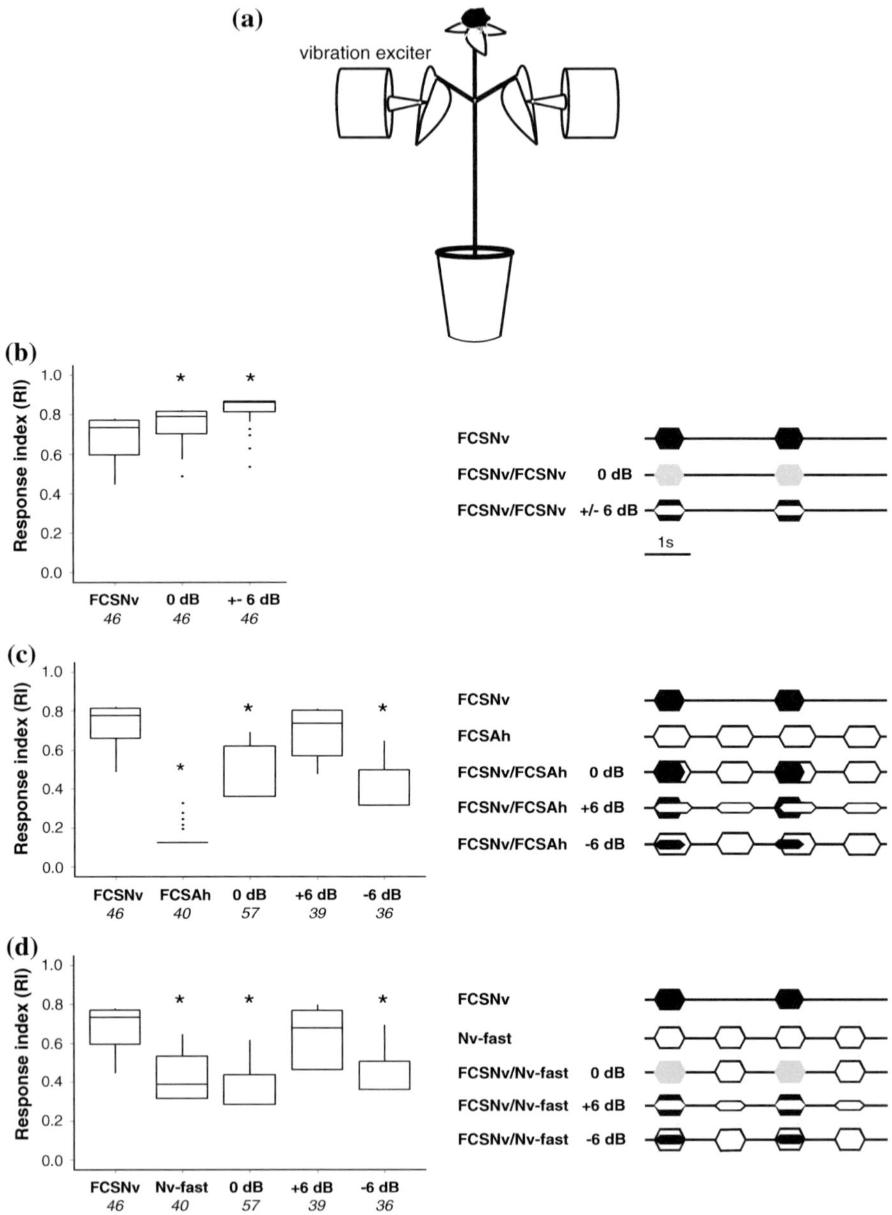
(a)
vibration exciter
(b)
1.0
0.8
0.6
0.4
0.2
0.0
Response index (RI)
FCSNv
46
0 dB
46
+- 6 dB
46
FCSNv
FCSNv/FCSNv 0 dB
FCSNv/FCSNv +/- 6 dB
1s
(c)
FCSNv
46
FCSAh
40
0 dB
57
+6 dB
39
-6 dB
36
FCSNv
FCSAh
FCSNv/FCSAh 0 dB
FCSNv/FCSAh +6 dB
FCSNv/FCSAh -6 dB
(d)
FCSNv
46
Nv-fast
40
0 dB
57
+6 dB
39
-6 dB
36
FCSNv
Nv-fast
FCSNv/Nv-fast 0 dB
FCSNv/Nv-fast +6 dB
FCSNv/Nv-fast -6 dB

◀ **Fig. 7.4** Effects of biotic noise and different signal-to-noise ratios on responsiveness of male *Nezara viridula*. **a** Experimental setup. *Nezara viridula* female calling song (FCSNv) was applied via vibration exciter to one side, while the other leaf was simultaneously vibrated with signals representing incidental interference from other conspecific signalers (**b**, **c**) or from closely related stink bug species *Acrosternum hilare* (**d**). Male responsiveness is represented by response index (*RI*), which describes the relative strength of male response in each treatment and incorporates the proportion of signaling males, signaling rate, and the proportion of searching males. Box plots show the median, lower and upper quartiles, 95 % confidence intervals, and outliers. Schematic presentations of the stimuli are given on the right. Conspecific female calling song applied to one side is shown in black, signals representing biotic noise in white, and two overlapping identical signals of equal amplitude are shown in gray. 0 dB: bilateral stimulation with signals of equal amplitude; +6 dB: bilateral stimulation in which FCSNv had 6 dB higher amplitude than signals representing biotic noise; −6 dB: bilateral stimulation in which FCSNv had 6 dB lower amplitude. *Asterisk* (*) Treatments in which the RI was significantly different ($P < 0.05$). All values were compared with unilateral stimulation with FCSNv, and differences in RI were tested with Kruskal–Wallis test following a multiple comparison test. Number of replicates are shown below each treatment (Modified from de Groot et al. 2010, Fig. 3, p. 3218, The Company of Biologists)

overlapping conspecific signals, *N. viridula* females shifted the dominant frequency of their calling song away from the background frequency (Polajnar and Čokl 2008). It has been suggested that this may be the main strategy for reducing incidental interference between emitters and presumably enables the receiver to perceive the presence of two sources due to frequency differences between two consecutive signals.

The obvious adaptation to avoid incidental interference between heterospecifics is time sharing of acoustic environment described in airborne sound communication of cicadas (Gogala and Riede 1995; Riede and Kroker 1995) and bush crickets (Römer et al. 1989). In *N. viridula*, males showed lower signaling activity in response to a conspecific female in the presence of heterospecific signals (Miklas et al. 2003; de Groot et al. 2010); however, the effect of heterospecific signals on emission of female calling songs has not been studied. Diel variation in vibrational signaling has been observed in the field (Cocroft 2003; McNett et al. 2010) as well as in laboratory studies in the absence of abiotic and biotic noise (Hunt and Nault 1991; Virant-Doberlet and Žežlina 2007; Mazzoni et al. 2009a); however, evidence of time sharing of the vibrational environment is lacking. It should be emphasized that the source of incidental interference with the vibrational communication channel could also be airborne sounds in the environment. *N. viridula* males responded to a human voice imitating the female calling song (Čokl et al. 1978), and the ambush bug *Phymata crassipes* emitted vibrational signals in alternation with human whistles (Gogala et al. 1984), while vibrations induced in a plant by musical sounds broadcast in the vicinity interfered with mating of the leafhopper *Amrasca devastans* and the planthopper *Nilaparvata lugens* (Saxena and Kumar 1980). The wolf spider *Schizocosa ocreata* perceived bird song as substrate vibrations (Lohrey et al. 2009), and it is likely that sources of airborne sounds, such as amphibian and insect choruses, contribute to background biotic noise.

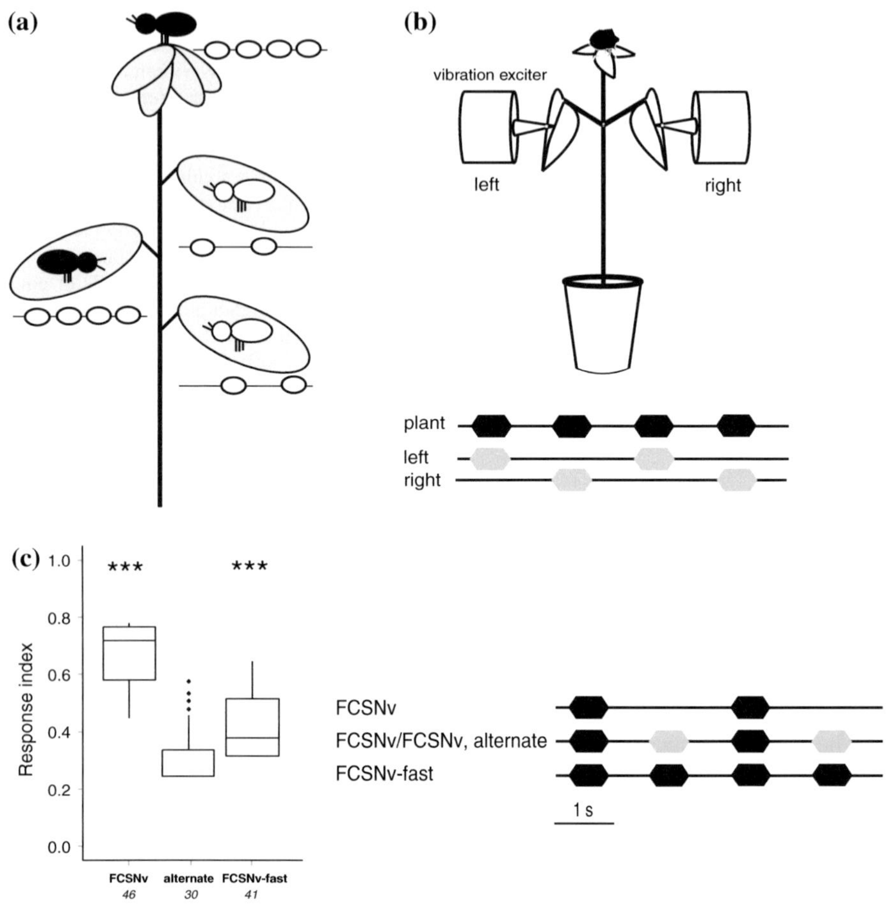


**Fig. 7.5** The effect of one-dimensional environment encountered on plants on reliability of perceived information and on male responsiveness in *Nezara viridula*. **a** A receiver positioned on a stem would perceive signals emitted from distant sources positioned on the same side of the receiver as a compound signal emanating from a single source. The same situation is encountered also by a receiver standing on a leaf, since all signals reach the receiver from one side, due to transmission via one-dimensional stem and petiole. Receivers are shown in *black*, emitters in *white*. **b** Experimental setup. *Nezara viridula* female calling song (FCSNv) was applied via vibration exciters to both sides in alternation, and the compound song had faster signal repetition that was outside species-specific value. **c** Male responsiveness is shown as response index (see Fig. 7.4). Box plots show the median, lower and upper quartiles, 95 % confidence intervals, and outliers. Schematic presentations of the stimuli are given on the right. In black and gray are shown FCSNv applied to different leaves. *Triple asterisk* (***) indicate significant difference ($P < 0.001$). The RI value for the bilateral treatment was compared with corresponding unilateral treatments (FCSNv and FCSNv-fast), and differences in RI were tested with Kruskal–Wallis test following a multiple comparison test. Number of replicates are shown below each treatment (Modified from de Groot et al. 2011a, Fig. 4, p. 1214, with permission from Oxford University Press)

### *7.3.2 Rivalry and Territoriality*

Many rivalry interactions are possible only in a network environment where advertising signals, aimed at the opposite sex of the same species, are received also by at least one another conspecific individual of either sex. Male–male interactions based on vibrational signals were found in many arthropods, and even within a species, a male can adopt several different rivalry tactics (Table 7.1). In addition, potential female rivalry has been described in a leafhopper (Percy and Day 2005). However, in studies of vibrational communication, rivalry itself has received little attention, and even when rivalry signals are mentioned, the exact behavioral context is often not clear. In particular, it is usually not clear whether the onset of rivalry behavior involves eavesdropping on intraspecific communication or whether only the presence or signaling of a potential rival is sufficient to trigger agonistic interactions. The simplest communication network consists of three individuals, and therefore, strictly speaking, exchange of rivalry signals between two males should be considered as an interactive signaler–receiver dyad. However, in many species, vibrational signals used in the later stages of pair formation, after partners have already established the contact, differ from calling signals used in the initial advertising stage, and in such cases, just the perception of a signal, typically emitted only in the presence of a receptive female, may be sufficient to induce male–male competition. Observed interactions relevant to a network context can be classified as (a) alternation of male calls, (b) exchange of rivalry vibrational signals that differ from signals used in interactions with females, (c) disruption by emission of jamming signals, and (d) silent approach to a duetting female (satellite behavior) (Table 7.1).

In the leafhopper *Graminella nigrifrons,* males form alternating choruses (Hunt and Morton 2001). Males coordinate emission of calling signals and avoid signal overlap, and chorus size may influence the call duration and calling rate of an individual male. Alternation of male calls has been observed also in other Auchenorrhyncha (Ossiannilsson 1949; Ichikawa 1982; Legendre et al. 2012); however, exchange of calling signals between more than two males has rarely been described (Virant-Doberlet and Žežlina 2007). In *G. nigrifrons*, alternating chorusing has been described as a competitive strategy in courtship disruption' although males alternate their calls also in the absence of a female (Hunt and Morton 2001). Very little is known about the effects of interference, from overlapping or alternating conspecific signals from chorusing males, on females. In *G. nigrifrons*, the female response does not decline significantly when male calls are overlapped (Hunt and Morton 2001). In the treehopper *Tylopelta gibbera*, two males alternate their signals and the female will duet with both of them (Legendre et al. 2012). In the wolf spider *Hygrolycosa rubofasciata*, males form synchronized choruses and females prefer closely synchronized males over loosely synchronized drumming clusters (Kotiaho et al. 2004). It has been suggested that in this species synchronous signaling may have evolved through female preference for the last signal.

**Table 7.1** Male–male interactions based on vibrational signals

| Arthropod taxa | Observed male–male interaction | References |
|---|---|---|
| **Crustacea** | | |
| Ocypodidae | Synchronized signaling | Popper et al. (2001), Clayton (2005) |
| **Araneae** | | |
| Ctenidae | Emission of rivalry signals | Schmitt et al. (1992) |
| | Vibrocryptic approach to a rival | Schmitt et al. (1992) |
| Lycosidae | Emission of rivalry signals | Fernandez-Montraveta and Schmitt (1994) |
| | Synchronized signaling | Kotiaho et al. (2004) |
| Sparassidae | Emission of rivalry signals | Henschel (2002) |
| **Insecta** | | |
| Tettigoniidae | Exchange of tremulatory signals (contest) | De Souza et al. (2011) |
| Cercopidae | Alternation of calls | Ossiannilsson (1949) |
| Cicadellidae | Alternation of calls | Ossiannilsson (1949), Tishechkin (2000), Hunt and Morton (2001) |
| | Emission of rivalry signals | Booij (1982), Heady et al. (1986), Nuhardiyati and Bailey (2005b) |
| | Disruption of male-female duet | Mazzoni et al. (2009a, b) |
| | Satellite behavior | Mazzoni et al. (2009a, b) |
| Membracidae | Emission of rivalry signals | Cocroft and McNett (2006) |
| | Alternation of calls | Cocroft and Rodríguez (2005), Legendre et al. (2012) |
| | Exchange of calls (contest) | Sullivan-Beckers and Cocroft (2010) |
| | Disruption of male-female duet | Miranda (2006), Legendre et al. (2012) |
| Delphacidae | Alternation of calls | Ossiannilsson (1949), Ichikawa (1982) |
| | Emission of rivalry signals | Ossiannilsson (1949) |
| | Emission of aggressive signals | Ichikawa (1982) |
| | Disruption of male-female duet | Claridge and Morgan (1993) |
| Flatidae | Alternation of calls | Virant-Doberlet and Žežlina (2007) |
| Pentatomidae | Emission of rivalry signals | Čokl et al. (2000), Blassioli Moraes et al. (2005), Bagwell et al. (2008), Čokl et al. (2011) |
| Chloropidae | Disruption of male calling | Kanmiya (2006) |
| Cypselosomatidae | Disruption of male calling | Kanmiya (2006) |

In the spider *Cupiennius getazi*, male vibrational courtship displays triggered male–male competition also in the absence of a female; however, her presence increased the likelihood of overt fighting (Schmitt et al. 1992). In the presence of a female, rival males emitted courtship and rivalry signals; however, they did not approach the responding female, but each other, and often approached the rival to attack without producing incidental vibrations during moving (conditional vibro-cryptic tactic).

In many species, a vibrational male–female duet is essential for recognition and successful localization of a female; however, duets appear vulnerable to disruption by eavesdropping competitors. In the leafhopper *Scaphoideus titanus,* rival males clearly eavesdrop on male–female interactions. Observations suggest that, in the absence of a receptive female, males do not interfere with each other's calling (Mazzoni et al., personal observation). However, rivals use alternative tactics, like emission of disruptive signals, to interrupt the ongoing male–female duet and/or satellite behavior (Mazzoni et al. 2009a, b) (Fig. 7.6). In this species, it seems that alternative tactics provide several benefits. Disruptive signals mask female replies and interrupt a duet and consequently prevent the courting male, which moves around only during a short period at the end of the duetting sequence, from searching. By contrast, the intruding male can move around all the time while the courting male is maintaining a duet. Furthermore, a silent satellite male orients specifically toward a responding female. Usually, multiple role reversals between courting and intruding male were observed, although sometimes males kept their respective roles. The cost of continuing duet disruption, however, was that 40 % of females stopped responding and moved away, and therefore, neither male gained from competition. Nevertheless, although in comparison with pairs, in trio situation (two males and a female) the number of copulations was significantly reduced, more copulations were achieved by intruding males. The only defensive tactic displayed by the courting male, after he became aware of a rival, was to simplify and shorten his courtship phrase, presumably to reduce time needed for localization of the female. The adaptive significance of different rivalry and defensive tactics is not entirely clear, but it is likely that they result from an intricate interplay between sexual selection and ecological selection arising from competitors.

The benefits of duet disruption are less clear in the treehopper *T. gibbera.* In trio situations, both males duet with the female and both produced masking signals mainly during the rival's advertisement signal, although emission during female response has also been observed (Legendre et al. 2012). Females responded less frequently to overlapped male signals, and, consequently, in trios, males needed more time to locate the female. Since the female duets with both males and mates with the first one that locates her, increasing the time a rival needs for localization is beneficial. However, since masking signals were produced during up to 82 % of advertisement signals, reducing the number of female responses, which are for both males essential for successful localization, both males may face costs associated with a longer search. Emission of jamming signals emitted during the courtship vibrational signal of another male has been proposed also in the

(a)

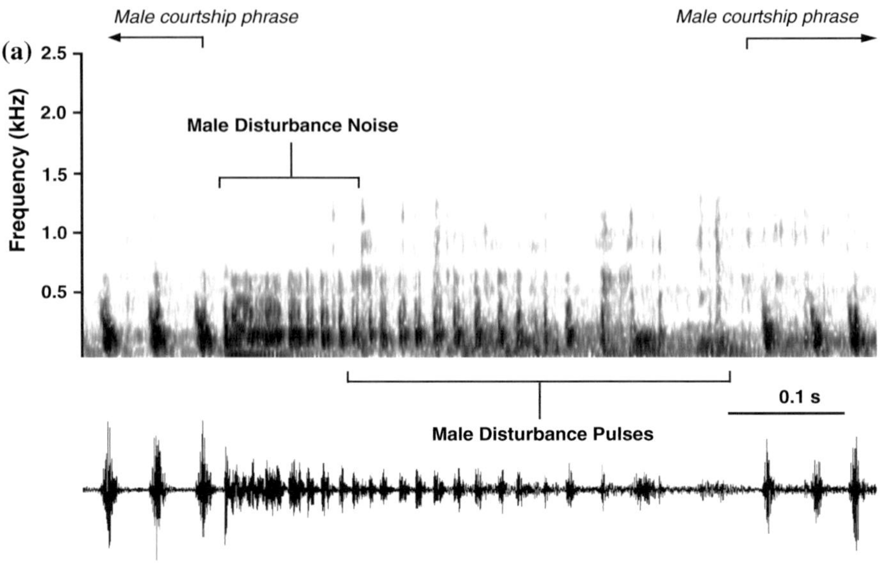


(b)

| | n | Type of behaviour | | |
|---|---|---|---|---|
| | | Females responding | Location | Copulation attempt |
| Pair | 24 | 100 | 100 | 100 |
| Trio | 15 | 60 | 73 | 20 |
| P | | 0.002 | 0.029 | <0.001 |

**Fig. 7.6** Male rivalry in *Scaphoideus titanus*. **a** Sonogram (*above*) and oscillogram (*below*) of parts of a male courtship phrase emitted by a courting male and male disturbance noise and disturbance pulses emitted by a rival are shown. **b** Table showing the effect of male rivalry on mating behavior. Percentage of females duetting at the end of the 20-min trial and percentages of males localizing the female and attempting a copulation are shown for experiments in which pairs and trios (two males and a female) were tested. n = number of pairs and trios tested. Chi-square test with Yates' correction was used to compare the results between pairs and trios (Modified from Mazzoni et al. 2009a, Fig. 4, Table 5, p. 409, with permission from Cambridge University Press)

treehopper *Ennya chrysura*. However, in this species, it was hypothesized that the function is to reduce the likelihood that the female would accept the courting male, because she becomes aware of the presence of another potentially preferred male (Miranda 2006).

In the planthopper *N. lugens,* male density-dependent agonistic interactions are triggered in the absence of females (Ichikawa 1982). When pairs of males were placed on a plant, males alternated their calls or emitted aggressive signals that could last for several minutes, which suppressed signaling in other males. In pairs consisting of males reared at higher male densities, rivals approached each other, emitted aggressive signals, and also even attacked each other while males that were reared singly avoided each other, never interrupted each other's calling, and often left the plant. Results of this study also suggest a role of vibrational signals in male hierarchy. In most cases, only one male in a pair emitted aggressive signals and this male never left the plant. Density-related changes in male signaling behavior may also be present in the treehopper *Vanduzea arquata* (Cocroft 2003).

In the vicinity of a female, male rivals of the treehopper *E. binotata* 'Ptelea' increase the amplitude of advertisement signals (Sullivan-Beckers and Cocroft 2010). Often, one male withdrew from such contests and the remaining one copulated with the female. The outcome of the contest was correlated with differences in male vibrational signals, although only the influence of whine length was significant.

In the bush cricket *Gnathoclita sodalis,* males exchange vibrational signals produced by tremulation to establish dominance (De Souza et al. 2011). The subordinate (usually smaller) male withdrew from the encounter, and the larger one kept the signaling perch. Both size and number of tremulations predicted the outcome of the contest. It has been shown in the bush cricket *Conocephalus nigripleurum* that body size is inversely related to the interval between tremulation pulses and that females discriminate between vibrational signals and orient toward those indicating a larger male (De Luca and Morris 1998). These studies suggest that the information about the size and dominance status of a male should be available also to the eavesdropping rivals that could thus obtain relative information about other males using less energy and with lower risk.

Incidental vibrational components of airborne advertisement calls are sometimes used by males to space themselves at appropriate distances. In the tettigoniid *Kawanaphila nartee* (Simmons and Bailey 1993) and in the prairie mole cricket *Gryllotalpa major* (Hill and Shadley 1997, 2001; Hill 1999) males respond to substrate-borne signals with behavior characteristic of male–male competition. Interestingly, males of the white-lipped frog *Leptodactylus albilabris* can also use the vibrational component of airborne calls emitted by neighboring males to adjust their own calling parameters (Lewis and Narins 1985; Lewis et al. 2001).

### *7.3.3 Eavesdropping by Parasitoids and Predators*

Parasitoid–host and predator–prey interactions are among the primary mechanisms underlying natural selection shaping animal communication systems (Zuk and Kolluru 1998; Haynes and Yeargan 1999). Arthropod predators and parasitoids are known to exploit incidental vibrational signals induced by moving or feeding prey (Pfannenstiel et al. 1995; Barth 1998; Meyhöfer and Casas 1999; Brownell and van Hemmen 2001). Somewhat surprisingly, more conspicuous and more informative vibrational signals used in pair formation have been, by contrast, traditionally regarded as a private channel not readily exploited by predators and parasitoids (Belwood and Morris 1987; Henry 1994; Zuk and Kolluru 1998; Lang et al. 2005; Römer et al. 2010). The evidence gathered in the last few years, however, clearly

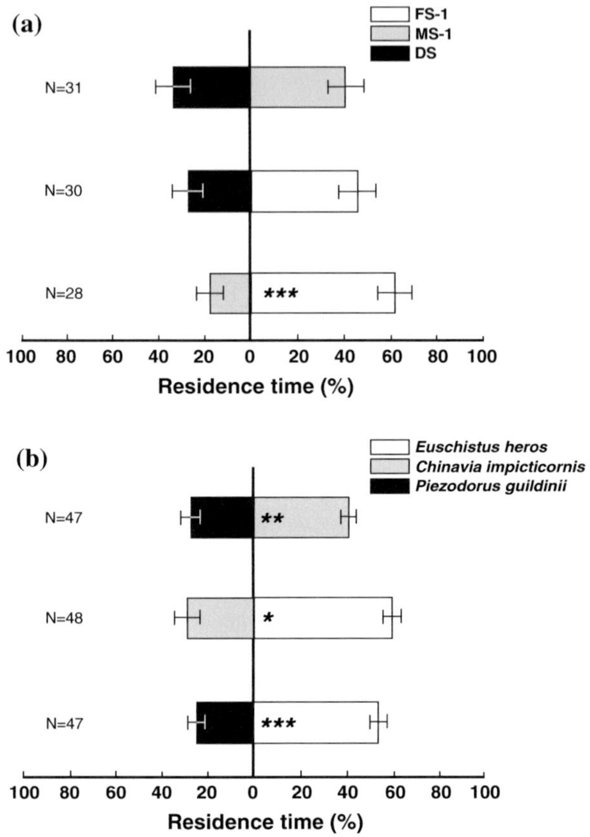


**Fig. 7.7** Eavesdropping by the parasitoid *Telenomus podisi* on vibrational signals of stink bugs. **a** Residence time (proportion of time spent in each vibrated area) (mean ± SE) of *T. podisi* females on bean leaves when stimulated with different combinations of the stink bug *Eschistus heros* vibrational songs. FS-1, female calling song; MS-1, male calling song; DS, duet song. **b** Residence time (mean ± SE) of *T. podisi* females on bean leaves when stimulated with different combinations of female calling songs of different stink bug species. * $P < 0.05$; ** $P < 0.01$; *** $P < 0.001$. Residence time was analyzed by generalized linear models with binomial distribution of errors and logit link function, using maximum likelihood estimation. N, number of replicates (Modified from Laumann et al. 2011, Figs. 4 and 5, p. 1180, with permission from Elsevier)

shows that exploiters eavesdrop on vibrational signals of their host or prey and rely on them when making foraging decisions.

The egg parasitoid *Telenomus podisi* (Hymenoptera: Scelionidae) is selectively attracted to species- and sex-specific female vibrational songs of the stink bug *Euschistus heros*, which is its preferred host (Laumann et al. 2007, 2011). *Telenomus podisi* females were attracted to this song also in the presence of female calling songs of other stink bug species (Fig. 7.7). The observed response appears to be innate, since the preference was observed in females that had no previous contact with stink bug vibrational signals. It was hypothesized that egg parasitoid searching behavior, based on female vibrational signals, may be related to higher probabilities of finding egg masses, provided that bugs mate on the same plant as they lay their eggs. The female song provides a good localization cue, since it is emitted continuously and with a regular repetition time by a stationary female (Blassioli Moraes et al. 2005).

Spiders are the most numerous groups of generalist predators in many terrestrial habitats (Wise 1993) and are major predators of insects like leafhoppers and planthoppers that communicate with vibrational signals (Döbel and Denno 1994; Fournier et al. 2008; Sanders et al. 2008). Spiders have a variety of specialized and highly sensitive receptors to detect substrate vibrations, and sensory-guided behaviors, such as prey detection, are commonly mediated by vibrational signals (Barth 1998; Uetz and Roberts 2002). There is anecdotal evidence that jumping spiders approach duetting leafhoppers and in playback experiments orient toward leafhopper vibrational signals (Narhardiyati and Bailey 2005a). Furthermore, it has been shown that vibrational signals incorporated within multimodal courtship display increase detectability to predators (Roberts et al. 2007). The jumping spider *Phidippus clarus* responded with orientation behavior faster to playback of courtship elements of the wolf spider prey species *Schizocosa ocreata* when vibrational signals were added to visual components.

To establish that exploitation by eavesdropping predators has influenced the evolution of vibrational signals and signaling behavior, however, it is not enough to observe that predation occurs, but also that predation during the mating period occurs frequently enough to be a significant source of mortality and that there is a positive relationship between vibrational signaling and the risk of predation. The tangle-web spider *Enoplognatha ovata* exploits vibrational communication of the leafhopper *Aphrodes makarovi* to obtain information about prey availability and uses this information to locate and capture prey (Virant-Doberlet et al. 2011). These spiders were the dominant predator taxon at the chosen field site where *A. makarovi* was present, and using molecular diagnostics, it was established that 25 % of *E. ovata* spiders were feeding on *A. makarovi* when signaling adults were present, while only 7 % were eating non-signaling nymphs (Fig. 7.8a). In microcosm experiments, one *E. ovata* killed on average four adult leafhoppers per week. Furthermore, although in microcosms the presence of *E. ovata* significantly reduced the numbers of both male and female *A. makarovi*, the spiders caught significantly more males than females (Fig. 7.8b). Male *A. makarovi* advertise themselves over longer periods, with calling signals that are structurally more complex than female replies and also have higher amplitude (de Groot et al. 2012). Males, therefore, face higher predation risk either due to greater detectability or possibly also due to higher mobility when searching for a replying female. Playback experiments, however, indicated that male vibrational signals play an important role in site selection, since *E. ovata* spent significantly more time on the plant in the presence of the male calling signal, but not when stimulated with female calls or incidental vibrational signals induced by walking leafhoppers (Virant-Doberlet et al. 2011) (Fig. 7.8c). However, there was a great variability in

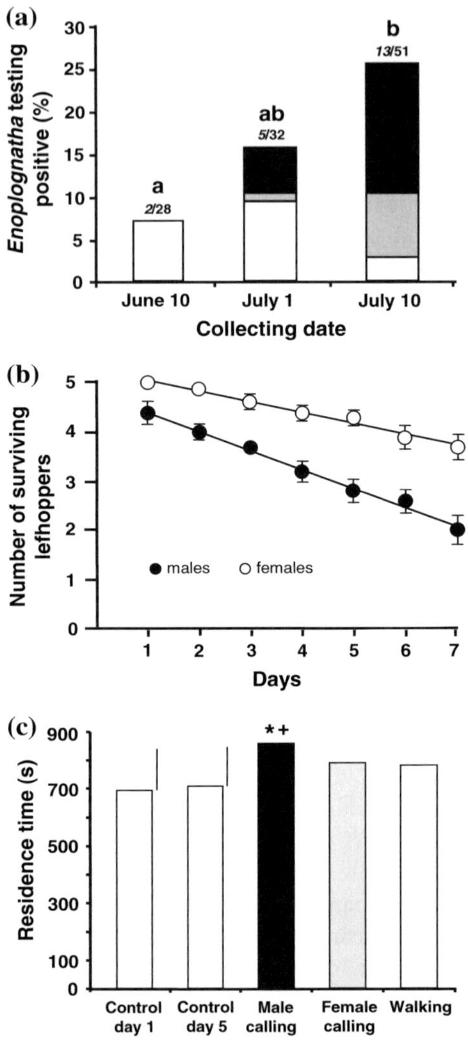
(a)
Enoplognatha testing positive (%)
30
25
20
15
10
5
0
a
2/28
ab
5/32
b
13/51
June 10
July 1
July 10
Collecting date
(b)
Number of surviving lefhoppers
5
4
3
2
1
0
males
females
1
2
3
4
5
6
7
Days
(c)
Residence time (s)
900
700
500
300
100
0
*+
Control day 1
Control day 5
Male calling
Female calling
Walking

◀ **Fig. 7.8** Exploitation of vibrational signals used in sexual communication of the leafhopper *Aphrodes makarovi* by *Enoplognatha* spiders. **a** Percentage of *Enoplognatha* spiders collected in field that tested positive for *Aphrodes* DNA. The numbers above bars are numbers of spiders testing positive (*in italics*) and numbers of spiders collected on each sampling date. Dates with different letters are significantly different for numbers testing positive (Fisher's exact test, $P < 0.05$). Shading in each bar represents the structure of *Aphrodes* population at the sampling site on collection date. Proportions of nymphs (*white*), adult females (*gray*), and adult males (*black*) are shown. **b** The effect of *E. ovata* spiders on the number of surviving male and female *Aphrodes* leafhoppers in microcosms. Mean values ± SE per microcosm (N = 10) are shown. In controls (not shown), the mean number of leafhoppers that died over 1 week was 0.7. The spiders significantly reduced leafhopper numbers. The presence of a spider significantly reduced the numbers of both males and female; however, the decline in numbers was significantly more rapid in males (general liner model analyses of covariance on raw data, $P < 0.001$). **c** The mean time that *E. ovata* spent on a plant in the presence and absence of vibrational stimuli. Eighteen adult female spiders were tested, each with three different types of leafhopper vibrational signals (male calling signal, female calling signal, and incidental signals induced by walking leafhoppers). Comparing playback treatments with controls (no stimulus conditions) showed that male calling signal significantly increased residence time (*asterisk*, *plus symbol* indicate significant differences with controls on days 1 and 5, respectively, $P < 005$). A priori application of the least significant difference (*LSD*), following two-way ANOVA, was used. Vertical bars represent LSDs (Modified from Virant-Doberlet et al. 2011, Figs. 3, 4, 5a, with permission)

spiders' response and while some spiders showed no response at all, others oriented and located the source of male vibrational signals. It has been suggested that field-collected spiders used in the playback experiments may have learned to associate *A. makarovi* vibrational signals with food availability and that the strength of this learnt association was related to experience with this particular prey.

The molecular analysis of gut contents of a range of spider species inhabiting the same nettle patch as *A. makarovi* showed very clearly that only a few spider species showed consistently high predation rates on this species (Virant-Doberlet et al. 2011). Wolf spiders of the genus *Pardosa* are major predators of *Prokelisia* planthoppers, and one spider can consume 70 planthoppers per day (Denno et al. 2002); however, wolf spiders from this genus were not feeding on *A. makarovi* in the field. Moreover, they did not kill them in microcosms, nor did they show any response to playback of leafhopper vibrational signals (Virant-Doberlet et al. 2011). For generalist predators, an essential element needed for optimal foraging is the ability to learn signals associated with the presence of a prey (Steidle and van Loon 2003). Cognitive abilities of spiders are well established (Wilcox and Jackson 1998; VanderSal and Hebets 2007), and learning in insects is also well documented (Dukas 2008). It has been shown that in jumping spiders a single encounter is sufficient for forming selective attention to specific features of a prey (Jackson and Li 2004). Moreover, antlions (Neuroptera: Myrmeleontidae) can associate a behaviorally neutral vibrational cue with the arrival of the prey and they learned to associate the cue with food after only two training sessions (Guillete et al. 2009; Hollis et al. 2011). Furthermore, it has been shown recently that learning plays an important role in parasitoid host-finding behavior (Hoedjes et al. 2011; Paur and

Gray 2011). The fact that experience and associative learning play an important role in exploiters' foraging decisions, and that foraging behavior can show geographic variation due to different selection pressures (Jackson and Carter 2001; Gray et al. 2007), emphasizes the difficulties encountered when studying exploitation of species- and sex-specific signals.

## 7.4 Concluding Remarks

Given the low number of studies and species in which sexual communication based on vibrational signals has been investigated, in more complex settings that more realistically represent the situation in nature, the diversity of interactions and mechanisms that have been revealed to date is surprising. Possibly, we should not be surprised, since again and again we find that there is a rich and complex vibrational world around us. In the last decade, our perception of substrate-borne sound communication has changed significantly, although it is probably still the least understood channel of communication. From the work reviewed in this chapter, it is evident that we have reached the stage where we can start asking important evolutionary questions. Sullivan-Beckers and Cocroft (2010) asked how selection arising from the transmission properties of the substrate, female choice, and male–male competition contribute to the evolution of male advertisement signals in the treehopper *E. binotata* 'Ptelea.' Although female preference was the most important cause of signal evolution, their results indicate that male–male competition may also be involved. Biotic landscape is likely to be a major driver of evolution in all its inhabitants, and by applying a network perspective, future studies should unravel many important interactions. It should be emphasized, however, that, in the field, the biotic landscape and communication networks within it change throughout the animal's lifetime as well as over the course of one season (Cocroft 2003). Mate-finding communication strategies may be shaped by population densities and operational sex ratio (Kvarnemo and Ahnesjo 1996; Heller et al. 1997; Kokko and Rankin 2006). Furthermore, not only does the set of species that spatially and temporally overlap change during the season, but due to their respective life cycles, the numbers of juveniles and adults of each species will also differ. The biotic landscape of species communicating with vibrational signals is an undiscovered country, and we have no doubt that by extending studies to a greater number of species living in different ecological conditions, future work will provide information not only on processes that affect evolution of vibrational communication, but are also central to our understanding of communication in general.

# References

Arnarson BT, Hart LA, O'Connel-Rodwell CE (2002) The properties of geophysical fields and their effect on elephants and other animals. J Comp Psychol 116:123–132

Aubin T, Jouventin P (2002) How to vocally identify kin in a crowd: the penguin model. Adv Stud Behav 31:243–277

Bagwell GJ, Čokl A, Millar JG (2008) Characterization and comparison of substrate-borne vibrational signals of *Chlorochroa uhleri*, *Clorochroa ligata* and *Chlorochroa sayi* (Heteroptera: pentatomidae). Ann Entomol Soc Am 101:235–246

Barber JR, Crooks KR, Fristrup KM (2010) The costs of chronic noise exposure for terrestrial organisms. TREE 25:180–189

Barth FG (1998) The vibrational sense in spiders. In: Hoy RR, Popper AN, Pay R (eds) Comparative hearing: insects. Springer, New York, pp 228–278

Barth FG, Bleckmann H, Bohnenberger J, Seyfarth E-A (1988) Spiders of genus *Cupiennius* Simon 1891 (Araneae, Ctenidae). II. On the vibratory environment of a wandering spider. Oecologia 77:194–201

Bee MA, Micheyl C (2008) Cocktail party problem: what is it? How can it be solved? And why should animal behaviorists study it? J Comp Psychol 122:235–251

Belwood JJ, Morris GK (1987) Bat predations and its influence on calling behaviour in Neotropical katydids. Science 238:64–67

Blassioli Moraes MC, Laumann RA, Čokl A, Borges M (2005) Vibratory signals of four Neotropical stink bug species. Physiol Entomol 30:175–188

Booij CJH (1982) Biosystematics of the *Muellerianella* complex (Homoptera, Delphacidae), interspecific and geographic variation in acoustic behaviour. Z Tierpsychol 58:31–52

Brownell PH, van Hemmen JL (2001) Vibration sensitivity and a computational theory for prey-localizing behavior in sand scorpions. Amer Zool 41:1229–1240

Brumm H, Slabbekoorn H (2005) Acoustic communication in noise. Adv Stud Behav 55:151–209

Burt JM, Vehrencamp SL (2005) Dawn chorus as an interactive communication network. In: McGregor PK (ed) Animal communication networks. Cambridge University Press, Cambridge, pp 320–343

Caldwell MS, McDaniel JG, Warkentin KM (2009) Frequency information in the vibration-cued escape hatching of red-eyed treefrogs. J Exp Biol 212:566–575

Caldwell MS, McDaniel JG, Warkentin KM (2010) Is it safe? Red-eyed treefrog embryos assessing predation risk use two features of rain vibrations to avoid false alarms. Anim Behav 79:255–260

Casas J (1989) Foraging behaviour of a leafminer parasitoid in the field. Ecol Entomol 14:257–265

Casas J, Magal C (2006) Mutual eavesdropping through vibrations in a host-parasitoid interaction: from plant biomechanics to behavioural ecology. In: Claridge MF, Drosopoulos S (eds) Insects sounds and communication: physiology, behaviour, ecology and evolution. Taylor & Francis, Boca Raton, pp 263–271

Casas J, Bacher S, Tautz J, Meyhofer R, Pierre D (1998) Leaf vibrations and air movements in a leafminer-parasitoid system. Biol Control 11:147–153

Castellanos I, Barbosa P (2006) Evaluation of predation risk by a caterpillar using substrate-borne vibrations. Anim Behav 72:461–469

Cerveira AM, Jackson RR, Guseinov EF (2003) Stalking decisions of web-invading araneophagic jumping spiders from Australia, Azerbaijan, Israel, Kenya, Portugal and Sri lanka: the opportunistic smokescreen tactics of *Brettus*, *Cocalus*, *Cyrba* and *Portia*. New Zeal J Zool 30:21–30

Chan AAY-H, Giraldo-Perez P, Smith S, Blumstein DT (2010a) Anthropogenic noise affects risk assessment and attention: the distracted prey hypothesis. Biol Lett 6:456–461

Chan AAY-H, Stahlman WD, Garlick D, Fast CD, Blumstein DT, Blaisdell AP (2010b) Increased amplitude and duration of acoustic stimuli enhance distraction. Anim Behav 80:1075–1079

Claridge MF, Morgan JC (1993) Geographical variation in acoustic signals of the planthopper *Nilaparvata bakeri* (Muir) in Asia: species recognition and sexual selection. Biol J Linn Soc 48:267–281

Clayton D (2005) Substrate (acoustic/vibrational) communication and ecology of the ghost crab *Ocypode jousseaumei* (Brachyura: Ocypodidae). Mar Freshwater Behav Physiol 38:53–70

Cocroft RB (2003) The social environment of an aggregating, ant-attended treehopper (Hemiptera: Membracidae: *Vanduzea arquata*). J Insect Behav 16:79–95

Cocroft RB, Hamel JA (2010) Vibrational communication in the "other insect societies": a diversity of ecology signals and signal functions. In: O'Connell-Rodwell CE (ed) The use of vibrations in communication: properties, mechanisms and function across taxa. Research Outpost, Kerala, pp 47–68

Cocroft RB, McNett GD (2006) Vibratory communication in treehoppers (Hemiptera: Membracidae). In: Claridge MF, Drosopoulos S (eds) Insects sounds and communication: physiology, behaviour, ecology and evolution. Taylor & Francis, Boca Raton, pp 305–317

Cocroft RB, Rodríguez RL (2005) The behavioural ecology of insect vibrational communication. Bioscience 55:323–334

Čokl A, Gogala M, Blaževič A (1978) Principles of sound recognition in three pentatomidae bug species (Heteroptera). Biol Vestn (Ljubljana) 2:81–94

Čokl A, Virant-Doberlet M, McDowell A (1999) Vibrational directionality in the southern green stink bug *Nezara viridula* (L.) is mediated by female song. Anim Behav 58:1277–1283

Čokl A, Virant-Doberlet M, Stritih N (2000) The structure and function of songs emitted by southern green stink bugs from Brazil, Florida, Italy and Slovenia. Physiol Entomol 25:196–205

Čokl A, Zorović M, Millar JG (2007) Vibrational communication along plants by the stink bugs *Nezara viridula* and *Murgantia histrionica*. Behav Process 75:40–45

Čokl A, Žunič A, Virant-Doberlet M (2011) Predatory bug *Picromerus bidens* communicates at different frequency levels. Cent Eur J Biol 6:431–439

Dabelsteen T (2005) Public, private or anonymous? Facilitating and countering eavesdropping. In: McGregor PK (ed) Animal communication networks. Cambridge University Press, Cambridge, pp 38–62

de Groot M, Čokl A, Virant-Doberlet M (2010) Effects of heterospecific and conspecific vibrational signal overlap and signal-to-noise ratio on male responsiveness in *Nezara viridula* (L.). J Exp Biol 213:3213–3222

de Groot M, Čokl A, Virant-Doberlet M (2011a) Species identity cues: possibilities for errors during vibrational communication on plant stems. Behav Ecol 22:1209–1217

de Groot M, Čokl A, Virant-Doberlet M (2011b) Search behaviour of two hemipteran species using vibrational communication. Cent Eur J Biol 8:756–769

de Groot M, Derlink M, Pavlovčič P, Prešern J, Čokl A, Virant-Doberlet M (2012) Duetting behaviour in the leafhopper *Aphrodes makarovi*. J Insect Behav 25:419–440

De Luca PA, Morris GK (1998) Courtship communication in meadow katydids: female preference for large male vibrations. Behaviour 135:777–793

De Souza LR, Kasumovic MK, Judge KA, Morris GK (2011) Communicating male size by tremulatory vibration in a Columbian rainforest katydid *Gnathoclita sodalis* (Orthoptera, Tettigoniidae). Behaviour 148:341–357

De Winter AJ, Rollenhagen T (1990) The importance of male and female acoustic behaviour for reproductive isolation in *Ribautodelphax* planthoppers (Homoptera: Delphacidae). Biol J Linn Soc 40:191–206

Denno RF, Gratton C, Peterson MA, Langellotto GA, Finke DL, Huberty AF (2002) Bottom-up forces mediate natural-enemy impact in a phytophagous insect community. Ecology 83:1443–1458

Döbel HG, Denno RF (1994) Predator-planthopper interactions. In: Denno RF, Prefect TJ (eds) Planthoppers and their management. Chapman & Hall, New York, pp 325–399

Dukas R (2008) Evolutionary biology of insect learning. Annu Rev Entomol 53:145–160

Eriksson A, Anfora G, Lucchi A, Virant-Doberlet M, Mazzoni V (2011) Inter-plant vibrational communication in a leafhopper insect. PLoS ONE 6(5):e19692. doi:10.1371/journal.pone.0019692
Eriksson A, Anfora G, Lucchi A, Lanzo F, Virant-Doberlet M, Mazzoni V (2012) Exploitation of insect vibrational signals reveals a new method of pest management. PLoS ONE 7(3):e32954. doi:10.1371/journal.pone.0032954
Evans TA, Inta R, Lai JCS, Prueger S, Wei Foo N, Wei'n FuE, Lenz M (2009) Termites eavesdrop to avoid competitors. Proc Roy Soc B 286:4035–4041
Fernandez-Montraveta C, Schmitt A (1994) Substrate-borne vibrations produced by male *Lycosa tarentula fasciiventris* (Araneae, Lycosidae) during courtship and agonistic interactions. Ethology 97:81–93
Forrest TG (1994) From sender to receiver: propagation and environmental effects on acoustic signals. Am Zool 34:644–654
Fournier V, Hagler J, Dane K, de León J, Groves J (2008) Identifying the predator complexes of *Homalodisca vitripennis* (Hemiptera: Cicadellidae): a comparative study of the efficacy of an ELISA and PCR gut content assay. Oecologia 157:629–640
Gerhardt HC, Huber F (2002) Acoustic communication in insects and anurans. University of Chicago Press, Chicago
Gogala M, Riede K (1995) Time sharing of song activity by cicadas in Temengor forest reserve, Hulu Perak and in Sabah, Malaysia. Malayan Nat J 48:297–305
Gogala M, Virant M, Blejec A (1984) Mocking bug *Phymata crassipes* (Heteroptera). Acoust Lett 8:44–51
Gray DA, Banuelos C, Walker SE, Cade WH, Zuk M (2007) Behavioural specialization among populations of the acoustically orienting parasitoid fly *Ormia ochracea* utilizing different cricket species as hosts. Anim Behav 73:99–104
Greenfield MD (2002) Signallers and receivers: mechanisms and evolution of arthropod communication. Oxford University Press, New York
Gröning J, Hochkirch A (2008) Reproductive interference between animal species. Q Rev Biol 83:257–282
Guillete LM, Hollis KL, Markatian A (2009) Learning in a sedentary insect predator: antlions (Neuroptera: Myrmeleodontidae) anticipate long wait. Behav Process 80:224–232
Halfwerk W, Hollerman LJM, Lessells CM, Slabbekoorn H (2011) Negative impact of traffic noise on avian reproductive success. J Appl Ecol 48:210–219
Haynes KF, Yeargan KV (1999) Exploitation of intraspecific communication systems: illicit signallers and receivers. Ann Entomol Soc Am 92:960–970
Heady SE, Nault LR, Shambaugh GF, Fairchild L (1986) Acoustic and mating behavior of *Dalbulus* leafhoppers (Homoptera: Cicadellidae). Ann Entomol Soc Am 79:727–736
Heller KG, von Helversen O, Sergejeva M (1997) Indiscriminate response behaviour in a female bushcricket: sex role reversal in selectivity of acoustic mate recognition. Naturwissenschaften 84:252–255
Henry CS (1994) Singing and cryptic speciation in insects. TREE 9:388–392
Henschel JR (2002) Long-distance wandering and mating by the dancing white lady spider (*Leucorchestris arenicola*) (Araneae, Sparassidae) across Namib dunes. J Arachnol 30:321–330
Hill PSM (1999) Lekking in *Gryllotalpa major*, the prairie mole cricket (Insecta: Gryllotalpidae). Ethology 105:531–545
Hill PSM (2008) Vibrational communication in animals. Harvard University Press, Cambridge
Hill PSM, Shadley JR (1997) Substrate vibration as a component of a calling song. Naturwissenschaften 84:460–463
Hill PSM, Shadley JR (2001) Talking back: sending soil vibration signals to lekking prairie mole cricket males. Am Zool 41:1200–1214
Hoedjes KM, Kruidhod HM, Higens ME, Dicke M, Vet LEM, Smid HM (2011) Natural variation in learning rate and memory dynamics in parasitoid wasps: opportunities for converging ecology and neuroscience. Proc R Soc B 278:889–897

Hollis KL, Cogswell H, Snyder K, Guillette LM, Nowbahari E (2011) Specialized learning in antlions (Neuroptera: Myrmeleontidae), pit-digging predators, shortens vulnerable larval stage. PLoS ONE 6(3):e17958. doi:10.1371/journal.pone.0017958

Hoskin CJ, Higgie M (2010) Speciation via species interactions: the divergence of mating traits within species. Ecol Lett 13:409–420

Hrabar N, Virant-Doberlet M, Čokl A (2004) Species specificity of male southern green stink bug *Nezara viridula* (L.) reactions to the female calling song. Acta Zool Sin 50:566–575

Hunt RE, Morton TL (2001) Regulation of chorusing in the vibrational communication system of the leafhopper *Graminella nigrifrons*. Am Zool 41:1222–1228

Hunt RE, Nault LR (1991) Roles of interplant movement, acoustic communication and phototaxis in mate-location behavior of the leafhopper *Graminella nigrifrons*. Behav Ecol Sociobiol 28:315–320

Ichikawa T (1982) Density-related changes in male-male competitive behavior in the rice brown planthopper *Nilaparvata lugens* (Stål) (Homoptera: Delphacidae). Appl Ent Zool 17:439–452

Jackson RR, Carter CM (2001) Geographic variation in reliance on trial-and-error signal derivation by *Portia labiata*, an araneophagic jumping spider from Philippines. J Insect Behav 14:799–827

Jackson RR, Li D (2004) One-encounter, search-image formation by araneophagic spiders. Anim Cogn 7:247–254

Kanmiya K (2006) Communication by vibratory signals in Diptera. In: Claridge MF, Drosopoulos S (eds) Insects sounds and communication: physiology, behaviour, ecology and evolution. Taylor & Francis, Boca Raton, pp 381–396

Kirchner WH (1997) Acoustical communication in social insects. In: Leherer M (ed) Orientation and communication in arthropods. Birkenhäuser Verlag, Basel, pp 273–300

Kokko H, Rankin DJ (2006) Lonely hearts or sex in the city? Density-dependent effects on mating systems. Phil Trans R Soc B 361:319–334

Kotiaho JS, Alatalo RV, Mappes J, Parri S (2004) Adaptive significance of synchronous chorusing in an acoustically signalling wolf spider. Proc R Soc Lond B 271:1847–1850

Kvarnemo C, Ahnesjo I (1996) The dynamics of operational sex ratios and competition for mates. TREE 11:404–408

Lang AB, Teppner I, Hartbauer M, Römer H (2005) Predation and noise in communication networks of neotropical katydids. In: McGregor PK (ed) Animal communication networks. Cambridge University Press, Cambridge, pp 152–169

Laumann RA, Blassioli Moraes MC, Čokl A, Borges M (2007) Eavesdropping on sexual vibratory signals of stink bugs (Hemiptera: Pentatomidae) by egg parasitoid *Telenomus podisi*. Anim Behav 73:637–649

Laumann RA, Čokl A, Lopes APS, Fereira JBC, Moraes MCB, Borges M (2011) Silent singers are not safe: selective response of a parasitoid to substrate-borne vibratory signals of stink bugs. Anim Behav 82:1175–1183

Legendre F, Marting PR, Cocroft RB (2012) Competitive masking of vibrational signals during mate searching in a treehopper. Anim Behav 83:361–368

Lewis ER, Narins PM (1985) Do frogs communicate with seismic signals? Science 227:187–189

Lewis ER, Narins PM, Cortopasi KA, Yamada WM, Poinar EH, Moore SW, Yu X-L (2001) Do male white lipped frogs use seismic signals for intraspecific communication. Am Zool 41:1185–1199

Lohrey AK, Clark DL, Gordon SD, Uetz GW (2009) Antipredator responses of wolf spiders (Araneae: Lycosidae) to sensory cues representing an avian predator. Anim Behav 77:813–821

Mazzoni V, Prešern J, Lucchi A, Virant-Doberlet M (2009a) Reproductive strategy of the Nearctic leafhopper *Scaphoideus titanus* Ball (Hemiptera: Cicadellidae). Bull Entomol Res 99:401–413

Mazzoni V, Lucchi A, Čokl A, Prešern J, Virant-Doberlet M (2009b) Disruption of the reproductive behaviour of *Scaphoideus titanus* by playback of vibrational signals. Entomol Exp Appl 133:174–185

McGregor PK (1993) Signalling in territorial systems: a context for individual identification, ranging and eavesdropping. Phil Trans R Soc Lond B 340:237–244
McGregor PK (2005) Introduction. In: McGregor PK (ed) Animal communication networks. Cambridge University Press, Cambridge, pp 1–6
McGregor PK, Peake TM (2000) Communication networks: social environments for receiving and signalling behaviour. Acta Ethol 2:71–81
McNett GD, Luan LH, Cocroft RB (2010) Wind-induced noise alters signaller and receiver behaviour in vibrational communication. Behav Ecol Sociobiol 64:2043–2051
McVean A, Field LH (1996) Communication by substratum vibration in the New Zealand tree weta *Hemiadenia femorata* (Stenopelmatidae: Orthoptera). J Zool (Lond) 239:101–122
Meyhöfer R, Casas J (1999) Vibrational stimuli in host location by parasitic wasps. J Insect Physiol 45:967–971
Michelsen A, Flemming F, Gogala M, Traue D (1982) Plants as transmission channels for insect vibrational songs. Behav Ecol Sociobiol 11:269–281
Miklas N, Čokl A, Renou M, Virant-Doberlet M (2003) Variability of vibratory signals and mate choice selectivity in the southern green stink bug. Behav Process 61:131–142
Miranda X (2006) Substrate-borne signal repertoire and courtship jamming by adults of *Ennya chrysura* (Hemiptera: Membracidae). Ann Entomol Soc Am 99:374–386
Narhardiyati M, Bailey WJ (2005a) Biology and natural enemies of the leafhopper *Balclutha incisa* (Matsumura) (Hemiptera: Cicadellidae). Austr J Entomol 44:104–109
Nuhardiyati M, Bailey W (2005b) Calling and duetting behavior in the leafhopper *Balclutha incisa* (Hemiptera: Cicadellidae: Deltocephalinae): opportunity for female choice? J Insect Behav 18:259–280
Ossiannilsson F (1949) Insect drummers. A study on the morphology and function of the sound-producing organ of Swedish Homoptera Auchenorrhyncha with notes of their sound production. Opusc Entomol Suppl 10:1–145
Otter KA, Ratcliffe L (2005) Enlightened decision: female assessment and communication networks. In: McGregor PK (ed) Animal communication networks. Cambridge University Press, Cambridge, pp 133–151
Paur J, Gray DA (2011) Individual consistency, learning and memory in a parasitoid fly *Ormia ochracea*. Anim Behav 82:825–830
Peake TM (2005) Eavesdropping in communication networks. In: McGregor PK (ed) Animal communication networks. Cambridge University Press, Cambridge, pp 13–37
Percy DM, Day MF (2005) Observations of unusual acoustic behaviour in two Australian leafhoppers (Hemiptera, Cicadellidae). J Nat Hist 39:3407–3417
Pfannenstiel RS, Hunt RE, Yeargan KV (1995) Orientation of a hemipteran to vibrations produced by feeding caterpillars. J Insect Behav 8:1–9
Polajnar J, Čokl A (2008) The effect of vibratory disturbance on sexual behaviour of the southern green stink bug *Nezara viridula* (Heteroptera, Pentatomidae). Cent Eur J Biol 3:189–197
Polajnar J, Svenšek D, Čokl A (2012) Resonance in herbaceous plant stems as a factor in vibrational communication of pentatomid bugs (Heteroptera: Pentatomidae). J R Soc Interface 9:1898–1907
Popper AN, Salmon M, Horch KW (2001) Acoustic detection and communication by decapod crustaceans. J Comp Psychol A 187:83–89
Riede K, Kroker A (1995) Bioacoustics and niche differentiation in two cicada species from Bornean lowland forest. Zool Anz 234:43–51
Roberts JA, Taylor PW, Uetz GW (2007) Consequences of complex signalling: predator detection of multimodal cues. Behav Ecol 18:236–240
Römer H (1993) Environmental and biological constraints for the evolution of long-range signalling and hearing in acoustic insects. Phil Trans R Soc Lond B 340:179–185
Römer H, Bailey W, Dadour I (1989) Insect hearing in the field. III. Masking by noise. J Comp Physiol A 164:609–620
Römer H, Lang A, Hartbauer M (2010) The signallers dilemma: a cost-benefit analysis of public and private communication. PLoS ONE 5(10):e13325. doi:10.1371/journal.pone.0013325

Rundle HD, Nosil P (2005) Ecological speciation. Ecol Lett 8:336–352
Sanders D, Nickel H, Grützner T, Platner C (2008) Habitat structure mediates top-down effects of spiders and ants on herbivores. Basic Appl Ecol 9:152–160
Saxena KN, Kumar H (1980) Interruption of acoustic communication and mating in a leafhopper and a planthopper by aerial sound vibrations picked up by plants. Experientia 36:933–936
Schmitt A, Schuster M, Barth FG (1992) Male competition in a wandering spider (*Cupiennius getazi*, Ctenidae). Ethology 90:293–306
Simmons LW, Bailey WJ (1993) Agonistic communication between males of a zaprochilinae katydid (Orthoptera: Tettigonnidae). Behav Ecol 4:364–368
Slabbekoorn H, Peet M (2003) Birds sing at a higher pitch in urban noise. Nature 424:267
Slabbekoorn H, Boutin N, van Opzeelan I, Coers A, ten Cate C, Popper AN (2010) A noisy spring: the impact of globally rising underwater sound level on fish. TREE 25:419–427
Steidle JLM, van Loon JJA (2003) Dietary specialization and infochemical use in carnivorous arthropods: testing a concept. Entomol Exp Appl 108:133–148
Stewart KW, Zeigler DD (1984) The use of larval morphology and drumming in Plecoptera systematics and further studies of drumming behavior. Ann Limnol 20:105–114
Sullivan-Beckers L, Cocroft RB (2010) The importance of female choice, male-male competition and signal transmission as causes of selection on male mating signals. Evolution 64:3158–3171
Tishechkin DYu (2000) Vibrational communication in Aphrodinae leafhoppers (Deltocephalinae auct., Homoptera: Cicadellidae) and related groups with notes on classification of higher taxa. Russ Entomol J 9:1–66
Tishechkin DYu (2007) Background noises in vibratory communication channels of Homoptera (Cicadinea and Psyllinea). Russ Entomol J 16:39–46
Turner J, Vollrath F, Hesselberg T (2011) Wind speed affects prey-catching behaviour in an orb web spider. Naturwissenschaften 98:1063–1067
Uetz GW, Roberts JA (2002) Multisensory cues and multimodal communication in spiders: insights from video/audio playback studies. Brain Behav Evol 59:222–230
VanderSal ND, Hebets EA (2007) Cross-modal effects on learning: a seismic stimulus improves color discrimination learning in a jumping spider. J Exp Biol 210:3689–3695
Virant-Doberlet M, Čokl A (2004) Vibrational communication in insects. Neotrop Entomol 33:121–134
Virant-Doberlet M, Žežlina I (2007) Vibrational communication of *Metcalfa pruinosa* (Say) (Hemiptera: Fulgoroidea: Flatidae). Ann Entomol Soc Am 100:73–82
Virant-Doberlet M, Čokl A, Zorović M (2006) Use of substrate vibrations for orientation: from behaviour to physiology. In: Claridge MF, Drosopoulos S (eds) Insects sounds and communication: physiology, behaviour, ecology and evolution. Taylor & Francis, Boca Raton, pp 81–97
Virant-Doberlet M, King RA, Polajnar J, Symondson WOC (2011) Molecular diagnostics reveal spiders that exploit prey vibrational signals used in sexual communication. Mol Ecol 20:2204–2216
Warkentin KM (2005) How embryos assess risk? Vibrational cues in predator-induced hatching of red-eyed treefrogs. Anim Behav 70:59–71
Warkentin KM, Caldwell MS, McDaniel JG (2006) Temporal pattern cues in vibrational risk assessment by embryos of the red-eyed treefrog *Agalychnis callidryas*. J Exp Biol 209:1376–1384
Warkentin KM, Caldwell MS, Siok TD, D'Amato AT, McDaniel JG (2007) Flexible sampling in vibrational assessment of predation risk by red-eyed treefrog embryos. J Exp Biol 210:614–619
Wignall AE, Jackson RR, Wilcox RS, Taylor PW (2011) Exploitation of environmental noise by an araneophagic assassin bug. Anim Behav 82:1037–1042
Wilcox RS, Jackson RR (1998) Cognitive abilities of araneophagic jumping spiders. In: Balda RP, Pepperberg IM, Kamil IC (eds) Animal cognition in nature. Academic Press, San Diego, pp 411–434

Wilcox RS, Jackson RR, Gentile K (1996) Spiderweb smokescreens: spider trickster uses background noise to mask stalking movements. Anim Behav 51:313–326

Wise DH (1993) Spiders in ecological webs. Cambridge University Press, New York

Yamazaki K (2011) Gone with the wind: trembling leaves may deter herbivory. Biol J Linn Soc 104:738–747

Zuk M, Kolluru GR (1998) Exploitation of sexual signals by predators and parasitoids. Q Rev Biol 73:415–438

Žunič A, Virant-Doberlet M, Čokl A (2011) Species recognition during substrate-borne communication in *Nezara viridula* (L.) (Pentatomidae: Heteroptera). J Insect Behav 24:468–487

# Chapter 8
# Active Space and the Role of Amplitude in Plant-Borne Vibrational Communication

**Valerio Mazzoni, Anna Eriksson, Gianfranco Anfora, Andrea Lucchi and Meta Virant-Doberlet**

**Abstract** Unlike airborne signals, substrate-borne vibrational signals are confined within the size and shape of their medium of communication, which in the case of small arthropods often coincides with the host plant. By following the substrate continuity, a vibrational signal creates a more or less complex active space network that enables communication between individuals. Due to the heterogeneity of plants, physical properties of the substrate can vary in the efficiency of signal transmission and in the diffusion of signals along the tissues. Under such circumstances, the identification and location of a potential partner may be a difficult task. Amplitude cues can be of great importance in orientation to the source of a vibrational signal by providing information about both direction and distance. As examples, we present two case studies on mating behavior of a leafhopper and a planthopper.

## 8.1 Introduction

The accurate detection and estimation of relevant ecological parameters allow animals to optimally exploit information collected from the environment. Signal emission and reception are constant actions used for survival and reproduction; in this regard, only the correct interpretation of sensory cues arising from relevant

V. Mazzoni (✉) · A. Eriksson · G. Anfora
Department of Sustainable Agro-Ecosystems and Bioresources, Fondazione Edmund Mach, via Mach 1, 38010 San Michele all'Adige, Trento, Italy
e-mail: valerio.mazzoni@iasma.it

A. Eriksson · A. Lucchi
Department C.D.S.L., Section of Agricultural Entomology, University of Pisa, via S. Michele degli Scalzi 2, 56124 Pisa, Italy

M. Virant-Doberlet
Department of Entomology, National Institute of Biology, Večna pot 111, 1000 Ljubljana, Slovenia

R. B. Cocroft et al. (eds.), *Studying Vibrational Communication*, Animal Signals and Communication 3, DOI: 10.1007/978-3-662-43607-3_8,

sources allows animals to fulfill their fundamental needs (Dall and Johnstone 2002; Schmidt et al. 2010).

In arthropods that produce and rely on plant-borne vibrational signals, the transmission medium (i.e., substratum) often coincides with the food source. The behavior of such species is thus strongly influenced by the physical and physiological characteristics of the host plant. In particular, signaling and signals are shaped according to substrate structure in order to best accomplish their goals (Čokl and Virant-Doberlet 2003; Cocroft and Rodriguez 2005; Hill 2008).

As in other communication systems, individuals have to extract fundamental information that can be summed up in three main questions: (1) who? (2) what? and (3) where? (Pollack 2000). In both intra- and interspecific interactions, vibrational signals have to intrinsically contain those features that allow the receiver to correctly identify the sender (who); otherwise, interference by either environmental noise or nontarget species is likely to occur. In some cases, as in mating communication, signal characteristics may transmit fitness cues, such as age, physical condition, strength, and size of the sender (what). This is particularly true for courtship signals, which indeed have evolved not necessarily to provide females with correct information about the sender, but rather to promote mating (Dawkins and Krebs 1978). In addition, in a world where many fundamental activities rely on the sense of hearing, it is important to use vibrations also to orient (where). What direction? What is the distance to the singing animal? Is the source on the same plant/leaf?

In this chapter, we provide an overview of the concept of active space of vibrational signaling and of the role of signal amplitude as an ecological parameter. The latter will be discussed both as a directional cue and as a behavioral trigger of plant-borne vibrational communication.

## 8.2 Signal Networks

### *8.2.1 Vibrational Signals*

Physiology, sexual selection, and selective pressure due to environmental factors, both biotic and abiotic, contribute to shaping the signaling characteristics of a species (Rendall et al. 2009). In plant-borne communication, the efficacy of transmission is strongly dependent on the combined characteristics of substrate and signal. The structure of vibrational signals is characterized by three principal parameters: frequency, amplitude, and temporal pattern (Čokl and Virant-Doberlet 2003; Cocroft and Rodriguez 2005). Even without taking into account the inter-individual variability of the source species, which can be significant (Cocroft et al. 2006), the situation is far from easy. The same signal performs rather differently, depending on the substrate through which it travels (e.g., Joyce et al. 2008). Thus, relevant variations can be found between plants or even different parts of the same plant. It is likely that signal specialization and complexity are correlated with parameters such as plant age or other host plant characteristics, including development or disappearance of organs and tissues and changes in turgor pressure. (Bell 1980). Indeed, the plant system is constantly changing and insect signaling should fit the physiology of feeding sites. Vibrational signals are therefore a compromise of different variables, which result in the species-specific signal repertoire (Cokl and Virant-Doberlet 2003; Cocroft et al. 2006; Virant-Doberlet et al. 2006).

From these premises, further hypotheses can be formulated, as for instance a different degree of signal specialization between feeding generalist and specialist species, or speciation elicited by the prolonged separation of different populations of the same species on different hosts (Sattman and Cocroft 2003; Cocroft et al. 2008, 2010; McNett and Cocroft 2008).

### *8.2.2 Signal Active Space*

A signal should accomplish the main aim for which it is emitted, and it consequently undergoes continuing selection. Mating signals are a typical example of this: in the first instance, they must be attractive enough to seduce a potential Partner; therefore, energetic efforts must be spent on them (Ryan 1988; Sullivan-Beckers and Cocroft 2009). Furthermore, the wider the signal's spatial range, the more effective it is in reaching potential listeners. Thus, the maximization of the signal active space can represent another crucial task to increase reproductive success. On the other hand, an important counter-indication is that such signals can be intercepted by both rival males and predators or parasitoids that exploit vibrational signals (Zuk and Kolluru 1998; Roberts et al. 2007; Virant-Doberlet et al. 2011).

In acoustics, the signal active space is defined as the three-dimensional spatial area in which the signal amplitude is sufficiently above the detection threshold of potential receivers to elicit a behavioral response. The factors that describe the active space are these: (1) amplitude of the signal at the source; (2) attenuation rate during transmission through the medium; (3) amplitude of background noise; and (4) threshold of receiver sensitivity (Brenowitz 1982). The fifth factor worth considering is the period of the day during which signaling occurs. This is important when signaling is restricted to a temporal window within the day and under certain weather conditions. In this sense, the factor "time" comes to constitute a fourth dimension of the signal active space.

The active space concept is often associated with another communication modality, olfaction, as a crucial factor in the use of pheromones as pest mating disruption tools (e.g., Grieshop et al. 2010; Ioriatti et al. 2011). Actually, parallels between airborne (including volatile pheromones and sound) and substrate-borne communication are numerous. However, there are some important differences. First of all, substrate-borne signals in plants travel in one (stems and rods)- or two (leaves)-dimensional space, whereas airborne signals spread from the source in three dimensions. Transmitted as bending waves, vibrational signals have a frequency pattern that rapidly changes according to the substrate structure (Michelsen et al. 1982). Therefore, the active space formed by vibrational signals can be characterized by an irregular pattern in which amplitude does not decrease monotonically (Čokl 1988; Čokl et al. 2007). It is likely that there are no two points of a plant with the identical "blend" of frequency/amplitude (Michelsen et al. 1982; Čokl 1988; Baurecht and Barth 1992). Despite these remarkable changes in structure, vibrational signals can propagate for several meters following the substrate continuity and receivers should be able to detect, identify, and trace them back to the source (Čokl and Virant-Doberlet 2003).

### 8.2.3 Active Space Networks

Numerous differences between airborne and substrate-borne signals lead to differences in the size and shape of the respective active space. A released odor or sound theoretically travels in a 3D space, progressively decreasing in intensity. Instead, vibrational signals that follow a continuous substrate are consequently limited to it. If we consider a simple stem, devoid of branches and leaves, the active space is reduced to two segments, with signals departing in opposite directions from the source. In small plants, like herbs, vibrations may reflect both at the root and at the top of the plant. Because of small internal damping, the reflected waves may travel up and down the stem several times. Such reflections change the pattern of the input signal in a way that the intensity of the vibratory signals does not decrease monotonically with distance (Čokl 1988; Miklas et al. 2001).

More branched plants, like most shrubs and trees, present a more complex system in which each bifurcation opens a new path for the signal and determines an active space network (ASN). The ASN can be defined as the cladogram-like substrate network covered by an active signal that spreads from a vibrational source.

In defining the ASN, one should consider the heterogeneity of the involved plant parts, such as trunks, stems, petioles and leaves, and how they contribute to attenuation of the signal. Even the same leaf can be characterized by relevant differences in signal amplitude and frequency, especially between lamina sectors separated by veins that can act as low-pass filters (Casas et al. 1998; Magal et al. 2000). Consequently, the ASN shape and effective range can be extremely variable and its limits difficult to predict. Furthermore, because of signal amplitude oscillations, active signal traits can alternate with others below the threshold of receiver sensitivity, thus creating a discontinuous framework in the ASN.

The ASN may also include the underground parts (Čokl and Virant-Doberlet 2003) and other plants (or materials) in direct contact with leaves or stems (Ichikawa and Ishii 1974) (Fig. 8.1). This means that in dense meadows, as well as in entangled vegetation of hedges, the ASN might be much larger than expected, by potentially including more than one focal plant.

**Fig. 8.1** Hypothetical active space network (*ASN*) created by vibrational signals emitted by an individual positioned on a leaf of a plant with complex geometry. The ASN can be rather heterogeneous: Areas at different signal intensity (*red*, *orange*, and *brown*) may alternate one other or with tissues where the signal is below the sensitivity threshold of the species (inactive signal, *white*). The ASN can also include the hypogeal parts (*A*), other plants in direct contact (*B*), and other surfaces not in contact but sufficiently close to the vibrating tissues to acquire active signal (*C*)

Last, but not least, plant tissues vibrated by the signal component transmitted through the air can belong to an ASN. In fact, for short ranges, e.g., 6 cm for the grapevine leafhopper *Scaphoideus titanus* (size 0.5 cm), the active signal can effectively pass from one leaf to another even without substrate continuity (Eriksson et al. 2011). This is a likely scenario when individuals are signaling on leaves from adjacent plants. Indeed, plants with overlapping leaves that are distanced by a few cm might be part of the same ASN even when not being connected by a common substrate.

### 8.2.4 Performances and Limits

Immediately after the emission from the source (i.e., the insect), the drop of signal intensity along the substrate (i.e., plant) can be relevant. For example, in a signal of the pentatomid bug *Nezara viridula*, a loss of approximately 10 db was measured on the surface of a *Hedera helix* plant, immediately below the legs of the singing animals (Amon and Čokl 1990). Damping of vibrations by the plant stem further reduces signal amplitude, especially when the stem is soft and flexible

(Michelsen et al. 1982). In *Cyperus* plants vibrated with *N. viridula* signals, a difference up to 20 dB was found between nodes and internodes (Cokl 1988). In a stinging nettle, a drop in intensity from the leaf vibrated by a female calling song of the planthopper *Hyalesthes obsoletus* to the main stem was 15 dB on average (Mazzoni unpublished data).

Indeed, studies investigating in detail the ASN of any species are lacking, and usually, only the maximal distance of active signal from the source is reported. The largest effective range has been observed in relatively large arthropods (size >1 cm) like spiders and large insects, whose signals may travel up to several meters (McVean and Field 1996; Barth 2002; Römer et al. 2010). Male and female stone flies were even able to communicate at 8 m distances along a wooden rod (Stewart and Zeigler 1984). In contrast, in small arthropods (<1 cm), the documented communication range is from a few centimeters up to 2 m (Keuper and Kühne 1983; Cocroft 1996; Čokl and Virant-Doberlet 2003).

The effective scale of most vibrational interactions often fits well with the specific substrate context (i.e., host plant size and habitus). For example, Römer et al. (2010) measured the active range of the katydid *Docidocercus gigliotosi* tremulatory signals, which could be perceived at distances up to 4 m; in this study, the species active signal range was detected throughout all plant tissues so that the ASN coincided with the whole host plant, the Bromeliaceae *Aechmea magdalenae*. In the planthopper *Nilaparvata lugens*, males were observed to respond to female signals up to 70 cm on rice plants (Ichikawa and Ishii 1974); however, again this is more or less the whole plant. In some *Eupteryx* leafhopper species, courtship songs were detectable throughout the nettle plant with no apparent decrease in amplitude (Stiling 1980).

The ASN does not necessarily correspond to the whole host plant, in that at some parts of the plant, the signal could be non-active. Male calling songs of the leafhopper *S. titanus*, which were adjusted to the same intensity level as emitted by live males and played back to potted grapevine plants, were detectable throughout the plant by laser vibrometer (Eriksson et al. 2012). However, not all measured points (some of the more distant leaves and the main stem) had amplitude values above the sensitivity threshold of the species.

Remarkably, the ASN can be increased by the multitude of singing individuals: In some treehoppers, the aggregated nymphs produced simultaneous calls in response to predatory risk (Cocroft 1996, 2001). This signaling spreads through the group in a chain reaction, generating a group signal that is longer and higher in amplitude than an individual signal.

On the contrary, background vibrations due to any environmental factor may reduce the ASN range in terms of loss of signal activity. This means either that the original signal is no longer detectable by the receiver when being masked by interference noise or that the presence of alien elements may prevent reliable recognition of the source and, consequently, the expression of the behavioral response. In addition, the possible reduction of the emitter's motivation may shorten the duration of emissions or result in a shift of the dominant frequency by the calling insect (Polajnar and Čokl 2008). The use of specific signals to disrupt or jam the courtship of another male has been described in the treehopper *Tylopelta gibbera* (Legendre et al. 2012), as well as in the leafhopper *S. titanus* (Mazzoni et al. 2009b): In *S. titanus*, two distinct "disturbing" elements were described, called "disturbance noise" and "disturbance pulses," respectively. The first was emitted by rival males to cover the female reply signal during a courtship duet, causing the immediate interruption of the duet, while the second was used as an element of interference: the rival emitted his own pulses in alternation with the courting male. In laboratory conditions, when placing two males together with one receptive female, this strategy resulted in 87 % definite interruption of courtship duets.

Little is known about interference of heterospecific vibrational signals emitted by other insects signaling on the same plant or of background abiotic noise caused, for example, by wind (Cocroft and Rodrıguez 2005). However, it has been demonstrated that the use of playback of male disturbance noise or white noise with a flat spectrum over a 0–8000 Hz frequency band was effective in interrupting the communication and consequently the mating process, between male and female *S. titanus* (Mazzoni et al. 2009a). Recently, it has been demonstrated that the use of male disturbance noise, played back into grapevine plants (also in the open field) through vineyard-supporting wires, was effective in dramatically reducing the mating rate of the species (Eriksson et al. 2012).

## 8.3 What Is the Role of Signal Intensity?

Signals belonging to different sensory modalities can be received and interpreted interactively in order to grasp most of the available information. Obviously, according to the context, one of the senses can play the main role. Similarly, although we can assume that frequency, amplitude, and temporal patterns of the vibrational signal are all important for providing fundamental information, it is still conceivable that each of them might be devoted to perform primarily a specific task. Frequency would therefore be useful mainly for best tuning with the transmission medium in order to maximize the active space by minimizing attenuation effects (Čokl and Virant Doberlet 2003). The temporal pattern would mainly be an identification tool, but also a provider of fitness information in terms of rhythm and repetition time (De Luca and Morris 1998; Pollack 2000). For instance, by stimulating *N. viridula* males either with heterospecific female signal or with various conspecific signals with altered temporal parameters, their responsiveness was significantly reduced (De Groot et al. 2010).

In contrast, the role of amplitude in vibrational communication is still much debated. Taken alone, deprived of specific frequency and temporal features, amplitude does not contain essential information for recognition (Baurecht and Barth 1992). Sometimes it reflects body size (Otten et al. 2001; Ulyshen et al. 2011), but unless two individuals are very close to each other, amplitude cannot represent a crucial cue of fitness, since relevant, and often unpredictable, intensity variation may occur, while the signal is traveling (McVean and Field 1996).

**Fig. 8.2** Without the ability to orient, a male of *Scaphoideus titanus* would face a difficult challenge in reaching the female on a branch with $n$ leaves and $n - 1$ bifurcations. Since the female may dwell on any leaf, the probability of her occurrence on each of them is 1/n. According to the binomial distribution, the probability that the male finds the female on its first attempt is also 1/n. On the other hand, the probability of meeting on a certain leaf depends on the structure of the branch and becomes lower with the distance. In this example, with 10 bifurcations, a male would need, respectively, 8 and 17 attempts to have 50 % and 80 % chance to find the female on any leaf

Michelsen et al. (1982) first suggested that insects communicating with vibrational signals might be able to judge the distance and direction to the singing animal because of amplitude cues. When localizing a source, it is important to be as fast as possible for two main reasons: (1) reducing energetic costs and (2) avoidance of eavesdroppers, either inter- or intraspecifics, and also considering that in some species the female will mate with the first male to arrive. Proper orientation is important since the lack of any directional cue may entail a huge loss of time by letting, for example, a searcher go randomly up and down the ASN paths. If no directional information is available at each bifurcation point, the chance of correct choice would be reduced to 50 %, and the probability of accidentally choosing the right path in a branch with $n$ bifurcations would be 1 out of $2^n$, which would make the search an extremely difficult task (Fig. 8.2). Furthermore, the awareness of the effective or plausible distance from the source would allow the receiver to restrict the area of search to a certain part of the substrate and also to adjust the mating behavior according to the perceived distance from the emitter.

### *8.3.1 Directionality*

Directional searching driven by substrate-borne vibrations, also termed vibrotaxis (Meyhöfer and Casas 1999), has been reported for different arthropod taxa (reviewed in Virant-Doberlet et al. 2006). Nevertheless, what would allow the directionality is still an object of discussion. Theoretically, directionality can be achieved either by evaluating arrival time and amplitude differences of signals sampled at spatially separated receptors (Michelsen 1998) or by using amplitude gradients along the substrate by means of successive comparisons (Čokl 1985) (the latter could be defined as vibroklinotaxis). There is an additional hypothesis, proposed by Cocroft et al. (2000), which suggests that special mechanical responses of the body to vibrations are used as directional cues. Individuals would receive directional information because the same insect body would vibrate differently according to the signal direction. However, this has been suggested only for *Umbonia crassicornis*, a treehopper species, and more investigations are required to consider its applicability to other taxa.

The use of time delay cues between the arrivals of signals at spatially separated receptors would make the searching process a straightforward task; however, there is a neurobiological barrier. For example, *N. viridula* has a maximum leg span of 1 cm, but propagation velocities of their signals on a bean plant are between 40 m/s and 80 m/s, thus creating a time-of-arrival difference between 0.125 and 0.250 ms, which is on the borderline of the lowest behaviorally determined threshold found in scorpions (0.2 ms; Brownell and Farley 1979). Indeed, *N. viridula* was observed to stretch the legs at the branching point over a 2 cm distance, which would increase the time delay to 0.25–0.50 ms (Čokl and Virant-Doberlet 2003; Virant-Doberlet et al. 2006). On the other hand, for smaller insects such as Auchenorrhyncha species (many of them under 0.5 cm) and most parasitoid wasps, there would not be enough distance between receptor organs to enable the nervous system to process the directional cue.

Saxena and Kumar (1984) proposed phonoklinotaxis (=vibroklinotaxis) as a possible explanation for female location by males in a leafhopper species, *Amrasca devastans*. They observed that males used to make “irregular running in swift jerky steps in different directions for varying distances, often revising their course, punctuated by intermittent brief halts” (1984, p. 81). Vibroklinotaxis has been mainly neglected because of the irregular weakening of the signal with distance travelled (Hunt and Nault 1991; McVean and Field 1996; Virant-Doberlet et al. 2006). Tishechkin (2007) observed that psyllid males, instead of choosing the shortest way to a female, often crawl on the plant in various directions before coming into close contact with the sexual partner. As a result, he assumed that psyllids cannot directly locate the source of vibrational signals due to the lack of any proper gradient of amplitude from the source. Nevertheless, this is the same explanation previously given by Saxena and Kumar (1984) to support exactly the opposite statement. By definition, vibroklinotaxis should be quite an inefficient strategy, characterized by “trial and error” with many wrongly chosen routes caused by the weak relationship between target angle and turn angle (Oldfield 1980). Such a “random” searching behavior has been observed in several other planthoppers and leafhoppers (e.g., Hunt and Nault 1991; Gillham 1992; Mazzoni et al. 2010); however, this does not imply the real use of amplitude gradients. Nevertheless, vibroklinotaxis should enable a better than purely random chance to localize the source, even if still lower than a straight directional search. A confirmation of this comes from a recent paper (Legendre et al. 2012) in which the distance walked by treehopper males in search of receptive females was measured. On average, the path walked was 4–5 times longer than the minimum required to reach the female. Still, this length was significantly shorter than for a pure search by chance. In another study (Mazzoni et al. unpublished data), males of *H. obsoletus* stimulated with playback of female calling songs in a two choice test showed a marked preference for a specific signal type containing higher-frequency components. When simultaneously stimulated with a more and a less attractive female signal, males always oriented toward the attractive signal, indicating a non-random search, driven only by vibrational cues.

In parasitoid wasps, vibrotaxis has been suggested as the mode of searching behavior, especially for those species that use their ovipositor to search for hosts buried within plants (Meyhöfer and Casas 1999; Fellowes et al. 2005). Studies on leaf miners proved that low-energy signals from larval movements do not propagate much further than the mined area delimited by veins and that a searching wasp consequently may detect the exact location of the prey, primarily by signal intensity cues. In this case, it is interesting to note that the wasp may be "disrupted" by the occurrence in the same leaf of several mines whose larvae alternately emit vibrations altogether (Casas et al. 1998; Kocarek 2009).

Finally, in another study, female signals played back into plant tissues allowed males of the treehopper *U. crassicornis* to locate the source when the amplitude gradients were increased with proximity. On the contrary, they failed when the pattern was reversed (Swatek et al. 2011).

### 8.3.2 Distance

The perception of an active signal triggers a behavioral response that is not necessarily directional. If the response is proportional to the intensity of the stimulus, it belongs to the "kinetic" category, and if the stimulus is vibrational, it is vibrokinesis (Meyhöfer and Casas 1999). In sensu lato, very different types of behavioral response could be included in this definition. For example, the spider *Cupiennius salei* can distinguish either opportunity or danger from amplitude cues of certain frequency by increasing the reaction speed in approaching or withdrawing from a vibrational source (Hergenröder and Barth 1983). Larvae of the moth *Semiothisa aemulataria* respond to differences in the amplitude of the high-frequency components of the substrate-borne vibrations of predatory stinkbugs and wasps by producing silk threads of different lengths (Castellanos and Barbosa 2006). In the Colorado potato beetle, *Leptinotarsa decemlineata*, the length of recovery from induced quiescence depends mostly upon substrate displacement caused by the stimulus of danger (Acheampong and Mitchell 1997). The parasitoid *Sympiesis sericeicornis* responds to vibrational cues emitted by the leaf miner *Phyllonorycter malella* while feeding by increasing rates of turning in its vicinity (Meyhöfer et al. 1994, 1997).

There is another type of behavioral response where the perceived signal intensity does not influence the behavioral response in terms of quantity but rather of quality. In this case, the response will be "chosen" among a number of distinct and discrete types. Many interesting examples of it can be found in the mating behavior of some Hemiptera. In that case, copula is often preceded by two phases, both mediated by vibrational signals: partner identification and search. Once identification has been achieved, one of the partners, usually the male, starts searching. When localization has been accomplished, the basic calling signal is often replaced by more specific courtship or precopula signals, functional in the final step of the mating approach. Such signals are usually calling signals, furnished with additional sound elements and/or produced at faster rates. Their emission occurs when the male arrives within a certain range from the female (Čokl and Virant-Doberlet 2003; Tishechkin 2006). In most of the pentatomid species investigated, the male courtship song pulse train originates from single male calling song pulses, whose repetition rate increases, leading finally to their fusion in a pulse train of species-specific temporal characteristics (Čokl and Virant-Doberlet 2003). In *Acrosternum hilare*, the male produces two different songs, each associated with a different phase of mating behavior (Čokl et al. 2001). Distinct courtship songs were also described from treehoppers (Hunt 1994) and leafhoppers of the subfamilies Agalliinae (Shaw 1976), Typhlocybinae (Shaw et al. 1974; Shaw and Carlson 1979; Saxena and Kumar 1984; Gillham 1992), Macropsinae (Claridge and Nixon 1986) and Deltocephalinae of the tribes Fieberiellini and Platymetopinii (Tishechkin 2006). The courtship signal is often also characterized by distinct harmonic elements in addition to typical broadband pulses. This was described from land bug species (Gogala and Razpotnik 1974), but also from several Deltocephalinae leafhoppers (Michelsen et al. 1982; Mazzoni et al. 2009b).

Why is this strategy used? A prompt recognition between the sexes of a species at an early stage in mating behavior is important for many reasons (Bell 1990); however, spending a lot of energy signaling when a partner is still far away is also costly. In fact, the emission of long and elaborate signals from the beginning would make the actual search much slower, and consequently, energetic costs and eavesdropping risks would also increase. The emission of courtship songs, which express the highest level of species specificity, will determine the final outcome in terms of mating choice; for this reason, it is absolutely important to concentrate the maximum effort only after the partner has been reliably located. Furthermore, additional components typical of courtship songs, especially harmonic sounds at frequencies above 500 Hz (i.e., in *S. titanus* courtship signal, Mazzoni et al. 2009b), would be really useful only if emitted in the range of a few cm, because frequencies outside the 50–500 Hz range are destined to attenuate rather rapidly (Čokl et al. 2003).

Taking into account everything mentioned above, can we state that qualitative and quantitative responses to a signal according to its amplitude imply awareness in the receiver of the source distance? The reported mating strategies would be possible only if males were to some degree aware of their distance from a female;

consequently, they can adjust their behavior and use their most seductive and energetic songs only when they are sufficiently confident to be in the vicinity of the female. Obviously, to talk of awareness in the case of insects could seem excessive; the question should more reasonably be reduced to selective pressures that eventually defined this multi-step behavior in a way to maximize the chance of reproductive success.

What would allow the distance evaluation? A candidate cue is the time interval between arrivals of different frequencies due to the dispersive propagation of bending waves; this implies a progressive evolution of spectrum with distance, especially as concerns the dominant frequency (Michelsen et al. 1982; McVean and Field 1996; Cocroft et al. 2006). The distance estimation could possibly also be done by comparing amplitudes of the signal in the two-dimensional array around the stem (Virant-Doberlet et al. 2006). At any rate, what is clear is that there is a rather strong correlation between amplitude signal and type of behavioral response as reported in the case studies described below.

## 8.4 Two Case Studies

### *8.4.1 The Leafhopper Scaphoideus titanus*

*Scaphoideus titanus* is a small insect (4–5 mm as adult), sap feeder, grapevine specialist, originated from North America, but since the 1960s is also present in Europe. The species is known as a vector for a dangerous phytoplasmas disease, flavescence dorée. This disease can cause mortality of infected grapevine plants after only a few years, and consequently, the EU authorities are demanding chemical treatments for its control. Because of its economic relevance, scientists have been studying the biology of *S. titanus*; in this context, intraspecific communication and mating behavior been considered relevant topics in order to establish sustainable pest management.

In a first study (Mazzoni et al. 2009b), the mating strategy at a single leaf range was described. Receptive females responded to the male calling song, the latter made of regularly repeated pulses, called male pulses of type 1 (MP1). Immediately after the first female reply, the male calling song turned into a courtship song, including additional vibrational elements such as a second type of pulse (male pulse of type 2 (MP2)) and a harmonic sound (called "buzz") with dominant frequency of 560 Hz and a fundamental of 280 Hz.

But what happens when males and females are placed on two different leaves? In another experiment (Polajnar et al. submitted), pairs were placed on leaves of the same grapevine shoot (base of petioles 10 cm distant) (Fig. 8.3a). Again, males started with the calling song, but after the female had replied, the mating process proceeded differently. Primarily, the male calling song did not turn into a courtship song after the vibrational contact with the female was established. Following the

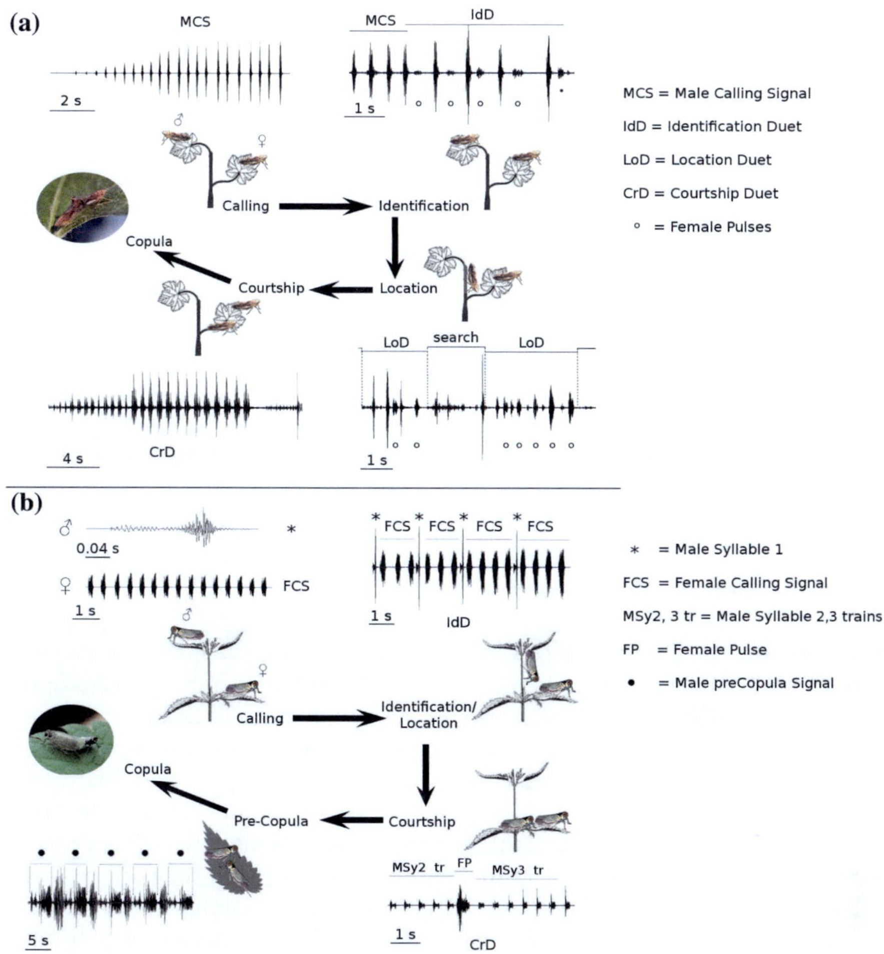


**Fig. 8.3** Mating behavior of *Scaphoideus titanus* (**a**) and *Hyalesthes obsoletus* (**b**). Different behavioral phases are characterized by a particular between-partners distance and specifically related signals

first perception of female signals, the male brusquely but shortly interrupted his call, by adjusting the pulse repetition time to a value that was approximately twice the normal duration. This initial duet, characterized by a very irregular rhythm of male emission, was termed the identification duet. It lasted for a few minutes, during which a male made random short walks without leaving the leaf. Then, rather abruptly, males shifted toward another behavior that was either (1) call-fly or (2) a proper directional search (Fig. 8.4). The call-fly is characterized by alternation of male calling songs (normally one or two in rapid succession) and short flights to other plant parts (or also on the cage walls of the experimental arena). It can go on for several minutes until males either again perceive a female

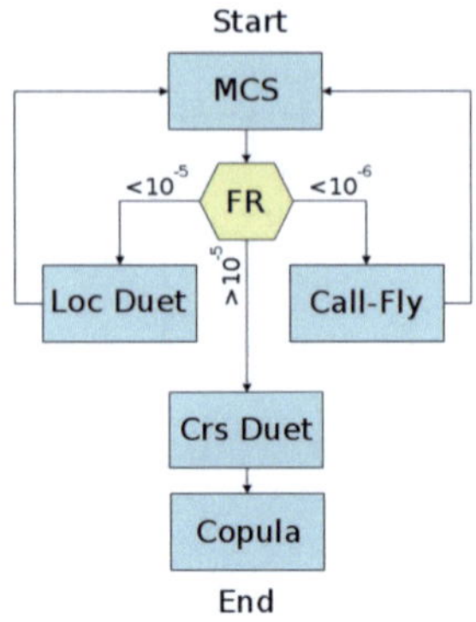


**Fig. 8.4** In *Scaphoideus titanus*, the mating behavior is tightly associated with the intensity of the partner's signal. When the perceived signal intensity was below $1 \times 10^{-6}$ mm/s, males usually adopted the call-fly strategy, whereas above this value, they went into location duet with the female. The proper courtship duet was observed only when the perceived signal intensity was above $1 \times 10^{-5}$ mm/s, which was reached when pairs were on the same leaf, even if not in visual contact. For this reason, in first experiments (Mazzoni et al. 2009b) where pairs were placed on the same leaf, only the Crs Duet was observed. MCS male calling signal; FR female's reply; Loc Duet location duet; Crs Duet courtship duet

pulse or give up. The directional search is the way males approach a receptive female by alternating short calls with rapid walks. While searching, males emit a short series of MP1 at rather regular pulse repetition rates. In fact, pulse interval after the female reply was significantly shorter than in the identification duet and slightly longer than in calling/courtship songs. This male–female interaction, called the location duet, was maintained by the male from the petiole of the starting leaf to the petiole of the leaf with the female. Eventually, and once on the female leaf, males immediately started emitting the courtship song.

In a third experiment (Eriksson et al. 2011), a male and a female were placed on two different shoots, without physical contact and an increasing gap between parallel leaves, starting from 0.5 cm. As well as in the experiments described above, directional search and call-fly were observed, but with a clear prevalence of the first type of behavior up to a leaf–leaf gap of 1 cm. After identification, males established a location duet with the female while walking toward the leaf margin. In this way, they passed directly onto the female leaf simply by walking, and once there, they immediately started the courtship duet. On the contrary, only call-fly was observed when the gap measured 2–6 cm; then, starting from 7 cm, no male–female communication was ever recorded.

By analyzing the values of signal intensity (as velocity of substrate vibration at the dominant frequency), it was found that within a 1-cm gap, the intensity of female signals was above $1 \times 10^{-6}$ m/s; the same value was measured from the distant leaf in the previous experiment. Indeed, in both cases, pairs established location duets. On the other hand, from 2 to 6 cm of gap, the intensity dropped to levels between $5 \times 10^{-7}$ and $1 \times 10^{-6}$ mm/s. Finally, when signals were measured from the same leaf as the source, the intensity values were constantly above $1 \times 10^{-5}$ mm/s.

These experiments suggest that the mating and, in particular, the singing behavior of *S. titanus* are in good part regulated by the perceived intensity of the partner's signal. Here would reside the difference between performing call-fly, directional search, or courtship. The question is whether this choice depends upon certain thresholds of signal intensity that, when reached, let a male extract sufficient directional cues or even be confident enough to stay in the vicinity of the female. It follows that certain behavioral responses are tightly associated with specific ranges of intensity.

### 8.4.2 The Planthopper *Hyalesthes obsoletus*

*Hyalesthes obsoletus* belongs to the Cixiidae family and is characterized by a nymphal age spent on roots of dicotyledon herbs like nettle and bindweed. Adults (body length 4–5 mm) feed on aboveground parts of the same plants, but occasionally also dwell on woody plants, such as grapevine to which they can transmit a dangerous phytoplasma disease known as "stolbur." The mating behavior of *H. obsoletus* shows some similarities with *S. titanus* but also several important differences. Experiments (Mazzoni et al. 2010) revealed that both genders can start communication with a calling signal that is followed by a "recognition (=identification) duet." The latter is performed as long as individuals are on different parts of the plant (for initial position in the experiment, they were placed on different leaves of stinging nettle plants 20 cm tall) (Fig. 8.3b). In this early stage of mating behavior, males stimulated the emission of female pulse trains by emitting a specific signal (called male syllable 1 (MSy1)). Later, when males were in the proximity of the female's leaf, females stopped emitting their calls, while males emitted courtship signals. Courtship pulse trains were made of two other types of syllables (MSy2 and MSy3) to which females responded with single pulses (and not trains as before). Finally, when pairs were in close vicinity, within a range of 2–3 cm on the same leaf, males emitted a precopula signal (made of MSy1 combined with a further syllable, MSy4), while females completely stopped signaling until the copulation attempt.

As well as in *S. titanus*, a well-defined behavioral sequence, in this case composed of recognition—location—courtship—precopula—copula, is a species characteristic. Again, the question is whether vibrational perception, and in particular the intensity parameter, may play a major role in this planthopper.

During the tests, when males came to the branching point between the stem and two-side leaves (stinging nettle has opposite leaves), one of these hosting the female, the chance to walk either to the right or to the wrong petiole was more or less the same (a similar observation was also made by Hunt and Nault (1991) for the leafhopper *Graminella nigrifrons*). But more importantly, when a male walked along the wrong petiole, he always turned back to the stem and did not enter the leaf lamina. In contrast, individuals that chose the right petiole continued to the leaf and finally located the female. This experiment was carried out both with living females and with playbacks of female calling song (Mazzoni and Virant-Doberlet unpublished data), and in both cases, it gave identical results. It follows that orientation is based on vibrational signals and not on other possible cues (olfaction and vision). It is vibrotaxis, but which one? Is it depending upon amplitude gradients? Males walking along the wrong petiole turned back before entering the wrong leaf; however, measured substrate velocity was higher in this petiole than in the main stem from which the male came. This would rather suggest that the change in walking direction was based on the perception of some form of directional cue. More experiments are needed to resolve this question.

Finally, by measuring the signal intensity over the whole stinging nettle plant, a 15-dB loss in the intensity of the female signal was found immediately outside her leaf. This difference was probably large and abrupt enough to trigger a switch of behavior by the approaching males.

## 8.5 Conclusions

Mating success is the final aim for all sexually reproducing species. However, before copulation can occur, it is crucial to find a partner. Consequently, efficient orientation and correct interpretation of all cues that can enable partner location are the key to mating success. Orientation includes not only the concept of directionality, but also the concept of "awareness" of the distance from the signaling source. On this topic, there are still many open questions that need to be answered. In particular, for small animals such as arthropods with body size under 1 cm, there is still open discussion about which cues are principally involved. For example, is the amplitude the crucial parameter that allows directionality? Can small species perceive and interpret differences in amplitude? Is vibroklinotaxis the way to achieve directionality?

A discussion could be based on the following four statements:

1. Many insects vibrate their host plants with amplitude 10–30 times above the threshold of receptor sensitivity (Michelsen et al. 1982).
2. These insects are also able to perceive a wide range of intensities, from the threshold of sensitivity to the highest intensity in close proximity of the source (Michelsen et al. 1982; Eriksson et al. 2011).
3. The higher the amplitude of the perceived signal, the higher the directional accuracy (Swatek et al. 2011; Legendre et al. 2012).
4. A constant monotone gradient is not likely to be found in plant tissues (Michelsen et al. 1982).

Taken together, these four points suggest that the vibrational environment can be highly variable and that animals able to perceive and interpret substrate-borne vibrations can theoretically exploit a large range of signal amplitudes. The potential information provided by a signal is often in direct proportion to the amplitude perceived by a receiver. Parameters that define a signal and make it sufficiently specific for source identification may not necessarily be sufficient to allow its location. However, the highest signal amplitudes are supposed to be found in proximity to the source; in fact, in the ASN of complex structures like those of shrubs or trees, or even of physically touching neighboring plants, there are relevant drops in signal intensity that correspond to crossing of the signal from one part to another. It follows that a certain polarization of signal intensity in the plant structures closer to the source should characterize any ASN. This was confirmed by different studies: the signal of the green stinkbug *N. viridula* transmitted through two adjacent *Cyperus* plants, connected by roots, was attenuated at about 0.1 dB/cm between apical points on neighboring stems, but only 0.04 dB/cm on the same stem (for a total distance of 80 cm) (Čokl and Virant-Doberlet 2003). Even more interestingly, simply by passing from a petiole to the main stem, signal intensity can abruptly drop more than 10 dB (Stritih et al. 2000). This demonstrates that relevant differences in amplitude can occur over short distances. Then, it is conceivable that abrupt and marked differences, rather than small progressive decreases, in signal intensity might be the cues that indicate distance to the searcher approaching or distancing himself from the source. In this way, a substantial reduction in intensity would elicit a backward turn on the stem, and a sufficiently strong amplitude could trigger a certain behavioral adaptation. As observed both in *S. titanus* and in *H. obsoletus*, males went on walking up and down the main stem, from the base to the top, while emitting signals to induce female calling. However, once males entered the leaf with the female, where male–female reciprocal perception of signals was at least 10–15 dB higher than previously, courtship maneuvers were immediately activated.

Actually, what exactly triggers a transition from one behavior to another, from simple to elaborate signals, and if all these are elicited exclusively by the perceived intensity of the signal emitted by the duetting partner, is still to be definitely demonstrated. Indeed, a complementary role of signal frequency and temporal pattern is conceivable. At any rate, it is evident that the mechanisms of vibrational communication are very different from olfaction. The latter works in terms of regular intensity attenuation and linear dose-dependent response, whereas in substrate-borne vibrations (at least of small species), it would rather be a sort of "ladder scheme," made of different steps or ranges of signal intensity that can be associated with corresponding behaviors.

Directionality represents an even more complicated question. A completely random search would be an inefficient strategy. If the ASN is complex and includes branching points, the location of a duetting partner becomes problematic. However, although many authors stated that "their insects" were randomly searching, indeed, in almost all the cases, a proper statistical analysis of the accuracy of directional decisions was not done. On the contrary, when performed (Legendre et al. 2012), analysis demonstrated that location was significantly more accurate than predicted by chance, and in addition, that likelihood to make the right decision was lower with increasing distance from the source. It is interesting to note that in the same study, the ability of the treehopper to make a correct directional decision was significantly lower at distances larger than 30 cm. Although not reported in the paper, it is possible that over that distance, signal amplitude fell below a certain threshold of insect sensitivity, where the directional information was no longer, or perhaps only to a minor degree, available.

In conclusion, although there are several possible explanations both for directionality and, more generally, for orientation, many questions still remain open.

## References

Acheampong S, Mitchell BK (1997) Quiescence in the Colorado potato beetle, *Leptinotarsa decemlineata*. Entomol Exp Appl 82:83–89

Amon T, Čokl A (1990) Transmission of the vibratory song of the bug Nezara viridula (Pentatomidae, Heteroptera) on the Hedera helix plant. Scopolia (Suppl. 1):133–141

Barth FG (2002) A spider's world. Senses and behaviour. Springer, Berlin

Baurecht D, Barth FG (1992) Vibratory communication in spiders. J Comp Physiol A 171:231–243

Bell PD (1980) Transmission of vibrations along plant stems: implications for insect communication. J New York Entomol S 88:210–216

Bell WJ (1990) Searching behaviour patterns in insects. Annu Rev Entomol 35:417–467

Brenowitz EA (1982) The active space of red-winged blackbird song. J Comp Physiol A 147:511–522

Brownell P, Farley RD (1979) Orientation to vibrations in sand by the nocturnal scorpion *Paruroctonus mesaensis*: mechanism of target localization. J Comp Physiol A 131:31–38

Claridge MF, Nixon GA (1986) *Oncopsis flavicollis* (L.) associated with tree birches (Betula): a complex of biological species or a host plant utilization polymorphism? Biol J Linn Soc 27:381–397

Casas J, Bacher S, Tautz J, Meyhöfer R, Pierre D (1998) Leaf vibrations and air movements in a leafminer-parasitoid system. Biol Control 11:147–153

Castellanos I, Barbosa P (2006) Evaluation of predation risk by a caterpillar using substrate-borne vibrations. Anim Behav 72:461–469

Cocroft RB (1996) Insect defense vibrational signals. Nature 382:679–680

Cocroft RB (2001) Vibrational communication and the ecology of group-living, herbivorous insects. Amer Zool 41:1215–1221

Cocroft RB, Rodriguez RL (2005) The behavioral ecology of insect vibrational communication. Bioscience 55:323–334

Cocroft RB, Tieu TD, Hoy RR, Miles RN (2000) Directionality in the mechanical response to substrate vibration in a treehopper (Hemiptera: Membracidae: *Umbonia crassicornis*). J Comp Physiol A 186:695–705

Cocroft RB, Shugart HJ, Konrad KT, Tibbs K (2006) Variation in plant substrates and its consequences for insect vibrational communication. Ethology 112:779–789

Cocroft RB, Rodriguez RL, Hunt RE (2008) Host shifts, the evolution of communication, and speciation in the *Enchenopa binotata* species complex of treehoppers. In: Tilmon KJ (ed) Specialization, speciation and radiation: the evolutionary biology of herbivorous insects. University of California Press, Berkeley, pp 88–100

Cocroft RB, Rodriguez RL, Hunt RE (2010) Host shifts and signal divergence: mating signals covary with host use in a complex of specialized plant-feeding insects. Biol J Linn Soc 99:60–72

Čokl A (1988) Vibratory signal transmission in plants as measured by laser vibrometry. Period Biol 90:193–196
Čokl A, Otto C, Kalmring K (1985) The processing of directional vibratory signals in the ventral nerve cord of *Locusta migratoria*. J Comp Physiol A 156:45–52
Čokl A, McBrien HL, Millar JG (2001) Comparison of substrate-borne vibrational signals of two stink bug species, *Acrosternum hilare* and *Nezara viridula* (Heteroptera: Pentatomidae). Ann Entomol Soc Am 94:471–479
Čokl A, Virant-Doberlet M (2003) Communication with substrate-borne signals in small plant-dwelling insects. Annu Rev Entomol 48:29–50
Čokl A, Zorovic M, Millar JG (2007) Vibrational communication along plants by the stink bugs *Nezara viridula* and *Murgantia histrionica*. Behav Process 75:40–54
Dall SRX, Johnstone RA (2002) Managing uncertainty: information and insurance under the risk of starvation. Phil Trans R Soc Lond B 357:1519–1526
Dawkins R, Krebs JR (1978) Animal signals: information or manipulation. In: Krebs JR, Davies NB (eds) Behavioural ecology: an evolutionary approach. Blackwell Scientific, Oxford, pp 282–309
De Groot M, Čokl A, Virant-Doberlet M (2010) Effects of heterospecific and conspecific vibrational signal overlap and signal-to-noise ratio on male responsiveness in *Nezara viridula* (L.). J Exp Biol 213:3213–3222
De Luca PA, Morris GK (1998) Courtship communication in meadow Katydids: female preference for large male vibrations. Behaviour 135:777–793
Eriksson A, Anfora G, Lucchi A, Virant-Doberlet M, Mazzoni V (2011) Inter-plant vibrational communication in a leafhopper insect. PLoS ONE 6(5):e19692
Eriksson A, Anfora G, Lucchi A, Lanzo F, Virant-Doberlet M, Mazzoni V (2012) Exploitation of insect vibrational signals reveals a new method of pest management. PLoS ONE 7(3):e32954
Fellowes MDE, Van Alphen M, Jervis MA (2005) Foraging behaviour. In: Jervis MA, Kidd NAC (eds) Insects as natural enemies: a practical perspective, 2nd edn. Springer, Dordrecht, pp 1–71
Gillham MC (1992) Variation in acoustic signals within and among leafhoppers species of the genus *Alebra* (Homoptera, Cicadellidae). Biol J Linn Soc 45:1–15
Gogala M, Razpotnik R (1974) An oscillographic-sonagraphic method in bioacoustical research. Biol Vestn 22:209–216
Grieshop MJ, Brunner JF, Jones VP, Bello NM (2010) Recapture of codling moth (Lepidoptera: Tortricidae) males: influence of lure type and pheromone background. J Econ Entomol 103:1242–1249
Hergenröder R, Barth FG (1983) The release of attack and escape behavior by vibratory stimuli in a wandering spider (*Cupiennius salei* Keys.) J Comp Physiol 152:347-358
Hill PSM (2008) Vibrational communication in animals. Harvard University Press, Cambridge
Hunt RE (1994) Vibrational signals associated with mating behavior in the treehopper *Enchenopa binotata* Say (Hemiptera: Membracidae). J New York Entomol Soc 102:266–270
Hunt RE, Nault LR (1991) Roles of interplant movement, acoustic communication, and phototaxis in mate-location behavior of the leafhopper *Graminella nigrifrons*. Behav Ecol Sociobiol 28:315–320
Ichikawa T, Ishii S (1974) Mating signal of the brown planthopper, *Nilaparvata lugens* Stål (Homoptera: Delphacidae): vibration of the substrate. Appl Ent Zool 9:196–198
Ioriatti C, Anfora G, Tasin M, De Cristofaro A, Witzgall P, Lucchi A (2011) Chemical ecology and management of *Lobesia botrana* (Lepidoptera: Tortricidae). J Econ Entomol 104:1125–1137
Joyce AL, Hunt RE, Bernal JS, Vinson SB (2008) Substrate influences mating success and transmission of courtship vibrations for the parasitoid *Cotesia marginiventris*. Entomol Exp Appl 127:39–47
Keuper A, Kühne R (1983) The acoustic behaviour of the bushcricket *Tettigonia cantans* II. Transmission of airborne-sound and vibration signals in the biotope. Behav Process 8:125–145

Kocarek P (2009) Sound production and chorusing behaviour in larvae of *Icosium tomentosum*. Cent Eur J Biol 4:422–426

Legendre F, Marting PR, Cocroft R (2012) Competitive masking of vibrational signals during mate searching in a treehopper. Anim Behav 83:361–368

Magal C, Schöller M, Tautz J, Casas J (2000) The role of leaf structure in vibration propagation. J Acoust Soc Am 108:2412–2418

Mazzoni V, Lucchi A, Čokl A, Prešern J, Virant-Doberlet M (2009a) Disruption of the reproductive behaviour of *Scaphoideus titanus* by playback of vibrational signals. Entomol Exp Appl 133:174–185

Mazzoni V, Prešern J, Lucchi A, Virant-Doberlet M (2009b) Reproductive strategy of the Nearctic leafhopper *Scaphoideus titanus* Ball (Hemiptera: Cicadellidae). Bull Entomol Res 99:401–413

Mazzoni V, Lucchi A, Ioriatti C, Doberlet-Virant M, Anfora G (2010) Mating behavior of *Hyalesthes obsoletus*. Ann Entomol Soc Am 103:813–822

McNett GD, Cocroft RB (2008) Host shifts favor vibrational signal divergence in *Enchenopa binotata* treehoppers. Behav Ecol 19:650–656

McVean A, Field LH (1996) Communication by substrate vibration in the New Zealand tree weta, *Hemideina femorata* (Stenopelmatidae: Orthoptera). J Zool Lond 239:101–122

Meyhöfer R, Casas J (1999) Vibratory stimuli in host location by parasitic wasps. J Insect Physiol 45:967–971

Meyhöfer R, Casas J, Dorn S (1994) Host localization by a parasitoid using leafminer vibrations: characterizing the vibrational signals produced by the leafmining host. Physiol Entomol 19:349–359

Meyhöfer R, Casas J, Dorn S (1997) Vibration-mediated interactions in a host-parasitoid system. Proc R Soc Lond B 264:261–266

Michelsen A (1998) Biophysics of sound localization in insects. In: Hoy RR, Popper AN, Fay RR (eds) Comparative hearing: insects. Springer, New York, pp 18–62

Michelsen A, Fink F, Gogala M, Traue D (1982) Plants as transmission channels for insect vibrational songs. Behav Ecol Sociobiol 11:269–281

Miklas N, Stritih N, Čokl A, Virant-Doberlet M, Renou M (2001) The influence of substrate on male responsiveness to the female calling song in *Nezara viridula*. J Insect Behav 14:313–332

Oldfield BP (1980) Accuracy of orientation in female crickets, *Teleogryllus oceanicus* (Gryllidae): dependence on song spectrum. J Comp Physiol 141:93–99

Otten H, Wäckers F, Battini M, Dorn S (2001) Efficiency of vibrational sounding in the parasitoid *Pimpla turionellae* is affected by female size. Anim Behav 61:671–677

Polajnar J, Čokl A (2008) The effect of vibratory disturbance on sexual behaviour of the southern green stink bug *Nezara viridula* (Heteroptera, Pentatomidae). Cent Eur J Biol 3:189–197

Polajnar J, Eriksson A, Rossi Stacconi MV, Lucchi A, Anfora G, Virant-Doberlet M, Mazzoni V (submitted) The process of pair formation mediated by substrate-borne vibrations in a small insect

Pollack G (2000) Who, what, where? Recognition and localization of acoustic signals by insects. Curr Opin Neurobiol 10:763–767

Rendall D, Owren MJ, Ryan MJ (2009) What do animal signals mean? Anim Behav 78:233–240

Roberts JA, Taylor PW, Uetz GW (2007) Consequences of complex signaling: predator detection of multimodal cues. Behav Ecol 18:236–240

Ryan MJ (1988) Energy, calling and selection. Amer Zoologist 28:885–898

Römer H, Lang A, Hartbauer M (2010) The signaller's dilemma: a cost–benefit analysis of public and private communication. PLoS ONE 5(10):e13325

Sattman DA, Cocroft RB (2003) Phenotypic plasticity and repeatability in the mating signals of *Enchenopa* treehoppers, with implications for reduced gene flow among host-shifted populations. Ethology 109:981–994

Saxena KN, Kumar H (1984) Acoustic communication in the sexual behaviour of the leafhopper, *Amrasca devastans*. Physiol Entomol 9:77–86

Schmidt KA, Dall SRX, Van Gils JA (2010) The ecology of information: an overview on the ecological significance of making informed decisions. Oikos 119:304–316

Shaw KC (1976) Sounds and associated behavior of *Agallia constricta* and *Agalliopsis novella* (Homoptera: Auchenorrhyncha: Cicadellidae). J Kans Entomol Soc 49:1–17

Shaw KC, Carlson OV (1979) Morphology of the tymbal organ of the potato leafhopper *Empoasca fabae* Harris (Homoptera: Cicadellidae). J Kans Entomol Soc 52:701–711

Shaw KC, Vargo A, Carlson OV (1974) Sounds and associated behavior of some species of *Empoasca*. J Kans Entomol Soc 47:284–307

Stewart KW, Zeigler DD (1984) The use of larval morphology and drumming in *Plecoptera* systematics, and further studies of drumming behaviour. Ann Limnol 20:105–114

Stiling PD (1980) Competition and coexistence among *Eupteryx* leafhoppers (Hemiptera: Cicadellidae) occurring on stinging nettles (*Urtica dioica*). J Anim Ecol 49:793–805

Stritih N, Virant Doberlet M, Čokl A (2000) Green stink bug Nezara viridula detects differences in amplitude between courtship song vibrations at stem and petiolus. Pflug Arch Eur J Phy 439 (Suppl.):R190–R192

Sullivan-Beckers L, Cocroft RB (2009) The importance of female choice, male–male competition, and signal transmission as causes of selection on male mating signals. Evolution 11:3158–3171

Swatek CA, Gibson JS, Cocroft RB (2011) Use of an amplitude gradient during vibration localization by a small plant-dwelling insect. Ecological Society of America Annual Meeting (abstract). Available online from: http://eco.confex.com/eco/2011/webprogram/Paper30908.html. Cited 8 August 2013

Tishechkin DY (2006) Acoustic characters in the classification of higher taxa of Auchenorrhyncha (Hemiptera). In: Drosopoulos S, Claridge MF (eds) Insect sound and communication. Physiology, behaviour, ecology and evolution. Taylor and Francis, Boca Raton, pp 319–329

Tishechkin DY (2007) New data on vibratory communication in jumping plant lice of the families Aphalaridae and Triozidae (Homoptera, Psyllinea). Entomol Rev 87:394–400

Ulyshen MD, Mankin RW, Chen Y, Duan JJ, Poland TM, Bauer LS (2011) Role of emerald ash borer (Coleoptera: Buprestidae) larval vibrations in host-quality assessment by *Tetrastichus planipennisi* (Hymenoptera: Eulophidae). J Econ Entomol 104:81–86

Virant-Doberlet M, King RA, Polajnar J, Symondson WOC (2011) Molecular diagnostics reveal spiders that exploit prey vibrational signals used in sexual communication. Mol Ecol 20:2204–2216

Virant-Doberlet M, Čokl A, Zorović M (2006) Substrate vibrations for orientation: from behaviour to physiology. In: Drosopoulos S, Claridge MF (eds) Insect sound and communication. Physiology, behaviour, ecology and evolution. Taylor and Francis, Boca Raton, pp 81-97

Zuk M, Kolluru GR (1998) Exploitation of sexual signals by predators and parasitoids. Q Rev Biol 73:415–438

# Chapter 9
# Mutual Behavioral Adjustment in Vibrational Duetting

**Rafael L. Rodríguez and Flavia Barbosa**

**Abstract** Animal communication often involves a back-and-forth of influence between the sexes. Not only do males produce signals to court females–females often reply back, as is the case in many plant-feeding insects. Here, we explore the behavioral complexity that arises from these interactions. We examine the potential for substrate-borne vibrational duetting insects to serve as case studies of the evolution and evolutionary consequences of mutual influence between the sexes, including mutual mate choice. Female mate choice on the basis of male signals has been documented in several species of insects that communicate via substrate-borne vibration, but it is less clear how often males modify their behavior (up to and including male mate choice) on the basis of variation in female vibrational signals. We assessed the potential for the signals of one sex to influence the behavior of the other sex with a literature review in which we compared the signals used by males and females in vibrational duetting. We found that female signals were at least as long and variable as male signals, although male signals often had more components than female signals. Thus, it seems likely that female vibrational duetting behavior is involved in proximate and evolutionary dynamics involving mutual influence and stimulation between the sexes.

## 9.1 Introduction

Animal communication often involves reciprocal interactions between participants. This means that the signaler–receiver framework that is familiar to behavioral ecologists may often, in fact, involve a two-way avenue of influence.

R. L. Rodríguez (✉) · F. Barbosa
Behavioral and Molecular Ecology Group, Department of Biological Sciences, University of Wisconsin–Milwaukee, Lapham Hall, 3209 N. Maryland Ave, Milwaukee, WI 53201, USA
e-mail: rafa@uwm.edu

R. B. Cocroft et al. (eds.), *Studying Vibrational Communication*, Animal Signals and Communication 3, DOI: 10.1007/978-3-662-43607-3_9, 

Consider a female engaged in mate choice among courting males: Her behavior and body posture may offer cues about her inclinations and potential decisions, and males may use such cues to adjust their own efforts. Thus, in courtship interactions, males seek to influence female mating decisions, and females may in turn influence (and be selected to influence) male behavior in ways that range from subtle effort adjustments to outright mutual mate choice. A growing literature indicates that such dynamics may be widespread (e.g., Patricelli et al. 2002; Peretti et al. 2006; Rodríguez et al. 2012)—and, furthermore, that they are not limited to instances of sex role reversal (Gwynne 1991; Andersson 1994). Instead, back-and-forth exchanges of stimulation and influence occur even in 'traditional' mating systems characterized by strong female mate choice (West-Eberhard 1983; Amundsen 2000; Bonduriansky 2001; Sæther et al. 2001; Clutton-Brock et al. 2006; Rillich et al. 2009; Rubenstein and Lovette 2009; Elias et al. 2010).

Back-and-forth interactions between the sexes can have strong consequences for the action of sexual and social selection, and for how these evolutionary forces contribute to divergence and speciation (Clutton-Brock 2007, 2009). For example, males and females may attend to different features in each other's signals, so that each sex's behavior may either counter or reinforce the patterns of assortative mating that would arise from the behavior of the other sex (Henry and Wells 2006; Kozak et al. 2009; Noh and Henry 2010; Rodríguez et al. 2012). Understanding the nature and evolutionary consequences of male–female reciprocal influence during communication is therefore of high interest. It is also challenging, because it requires asking whether and how the behavior of one sex modifies the behavior of the other. Here, we showcase communication with substrate-borne vibrational signals as ideal for this task. In many species that use substrate-borne vibrational communication, male–female interactions involve signal exchanges that can be easily observed and experimented upon.

Many animals use substrate-borne vibrational communication in diverse aspects of their lives, such as foraging, defense against predators, pair formation, and courtship during mating (Cocroft and Rodríguez 2005; Hill 2008). Pair formation with substrate-borne signals often takes place by means of male–female signal exchanges, or duets (Henry 1994; Cokl and Virant-Doberlet 2003; Virant-Doberlet and Cokl 2004; Cocroft and Rodríguez 2005; Uhl and Elias 2011). As an example of a mating system with vibrational duetting, consider the process of pair formation in *Enchenopa* treehoppers.

### *9.1.1 Vibrational Duetting in the* **Enchenopa Binotata** *Species Complex (Hemiptera: Membracidae)*

*Enchenopa* treehoppers are specialized phloem-feeders that spend their entire life cycle on one or a few individuals of their host plant species (Wood 1993; Cocroft et al. 2008). When males reach sexual maturity, they search for mates by flying

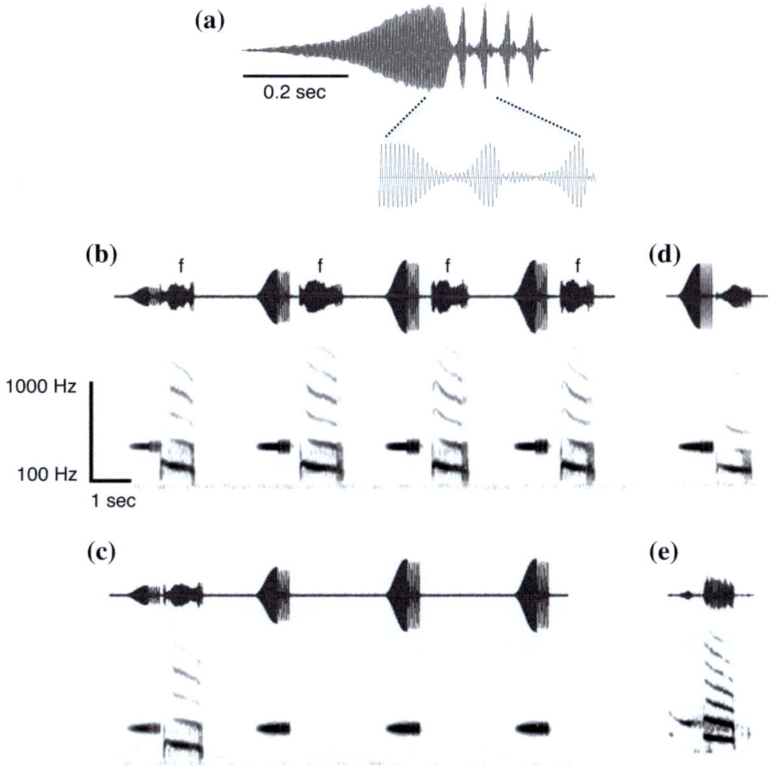


**Fig. 9.1** Example of "typical' vibrational duetting, as used in pair formation by *Enchenopa* treehoppers (Hemiptera: Membracidae). We emphasize the variability of female response signals and the experimental tractability of vibrational duetting. **a** Male advertisement signals have two components: a 'whine' consisting of a pure tone that slightly sweeps down in frequency, which is followed by a second component consisting of a series of 'pulses'. **b–e** Panels show the waveform (energy over time) and sonogram (frequency over time) of an interaction between a live female and either a live male or a synthetic vibrational playback stimulus. **b** A female duet with a synthetic playback consisting of a bout of four stimuli. This playback was attractive to the female, and she produced response signals (labeled "f") after each of the playback stimuli. **c** The same female, now duetting with a less attractive stimulus, produced only one response signal (to the first of the four stimuli in the playback). **d** Excerpt of another interaction of the female with a less attractive playback, to which she responds with a shorter signal. **e** Excerpt from a prolonged duetting interaction between a live male–female pair; after a few minutes, the male had decreased his signal amplitude and the female had increased hers

from plant to plant and producing advertisement signals (Fig. 9.1a). If a receptive female is attracted by the male, she produces her own response signals, thereby establishing a duet that continues until mating begins (Rodríguez and Cocroft 2006; Fig. 9.1b–e). Thus, by selectively duetting with some males and not others, females can influence the likelihood of being located by those males that they find attractive. Female mate preferences are an important source of selection on male signals in *Enchenopa* treehoppers (Rodríguez et al. 2004, 2006; Cocroft et al. 2010; Sullivan-Beckers and Cocroft 2010). For example, changes in mate preferences among species are associated with changes in signal features, and the level of signal-preference correspondence increases with the strength of the mate preference (Rodríguez et al. 2006). Female duetting behavior, besides expressing mate preferences and influencing male searching behavior, also influences male signaling behavior. Like male signals, female duetting signals vary in frequency and length among species (Rodríguez and Cocroft 2006). Moreover, females are not only more likely to duet in response to attractive male signals, but also produce more and longer response signals when duetting with males producing attractive stimuli (Rodríguez et al. 2004, 2012). This variation in female duetting signals influences the likelihood that males will themselves signal when presented with playbacks of female duetting responses: Males are more likely to signal when they perceive the duetting signals produced by females that were responding to stimuli that they find attractive (Fig. 9.2) and are relatively little influenced by species identity cues in female signals (Rodríguez et al. 2012).

### 9.1.2 Vibrational Duetting and Two-Way Avenues of Influence Between the Sexes

We suggest that vibrational duetting offers excellent opportunities to study the influence of two-way male–female interactions on sexual selection. Insect vibrational duetting often entails female signals that are comparable in length to the males' (see below). This offers a striking contrast with most insect air-borne acoustic duetting, which is characterized by extremely short female response signals (Bailey 2003). Very short female responses may be the product of selection stemming from the high risk of predation that comes with air-borne acoustic communication (Belwood and Morris 1987; Conner and Corcoran 2012; Bailey 2003). There are, of course, exceptions, such as prolonged in-flight duets in mosquitos (Cator et al. 2009). Nevertheless, the general contrast in female signal length between vibrational and acoustic insect duets suggests that predation risk may be lower for females in species using vibrational signals. This may in turn allow female duetting behavior to be shaped by additional sources of selection, such as expressing the female's own mate preferences, and influencing male behavior.

The above is not to say that there is no risk in vibrational signaling. Eavesdropping on vibrational signals is very likely. A great diversity of vibrationally sensitive predators and parasitoids occurs on the substrates used by vibrational animals (Cocroft and Rodríguez 2005; Hill 2008; Cocroft 2011). And although research on this topic is only beginning, there already are demonstrations of vibrational eavesdropping by predators (Laumann et al. 2007; Cocroft 2011; Virant-Doberlet et al. 2011). There are also interesting examples of vibrational

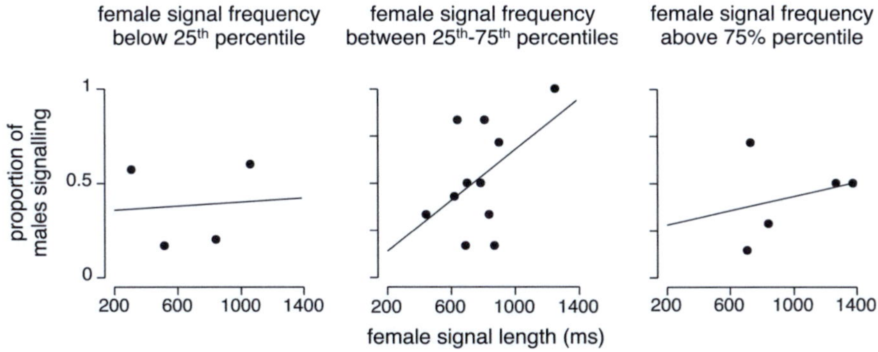


**Fig. 9.2** Example of how the expression of female mate preferences by means of duetting may influence male behavior. *Enchenopa* males are more likely to signal if they are presented with the duetting signals of females that were responding to attractive stimuli—i.e., if they are presented with longer duetting signals—an effect that was especially pronounced for signals of intermediate frequencies (*middle panel*) (Modified with permission from Rodríguez et al. 2012)

eavesdropping to avoid natural enemies or competitors (Djemai et al. 2004; Catania 2008; Evans et al. 2009; Mitra et al. 2009).

Even if we contemplate risks to vibrational signaling, predation risk due to vibrational eavesdropping may usually be lower for females, for various reasons. For example, female signaling efforts may generally be lower, as females often respond only to a fraction of the signals produced by males (Fig. 9.1; Rodríguez et al. 2004, 2006, 2012). Further, the costs of mate searching may be borne mostly by the male, who both signals and moves about; such conspicuous combinations of movement and signaling will likely provide more cues for many predators than the female's signaling. Also, male signals are often higher in amplitude than female signals and thus more detectable by natural enemies.

An interesting possibility is that the dynamism of duetting interactions may help males and females modulate the level of risk that they incur. For example, in *Enchenopa* treehoppers, we have observed that in prolonged duetting interactions, the male may diminish the amplitude of his signals, once the duet is established and he searches for the female, while the female may increase hers (e.g., Fig. 9.1e). Another example involves an observation of a prolonged duet in *Vanduzea arquata* treehoppers (RB Cocroft, pers. comm.; see also Cocroft 2003). In this species, the male taps his head on the substrate at the beginning of the signal, which consists of a downsweep harmonic series. The female signal is relatively simple, resembling that of *Enchenopa* females (see Fig. 9.1), but in the instance of this prolonged duet, the male was having difficulty locating the female, and eventually, she began tapping *her* head as well. These anecdotes suggest that females may have the ability to (and may be selected to) strike a balance between providing cues of their location to males and avoiding informing natural enemies of their location.

We take, then, the length of female response signals in vibrational duetting as an indication of the likelihood that they have been selected to influence male behavior. The potential for such influence arises from two lines of reasoning: (1) Greater length facilitates the assessment of signal features (cf. Bailey and Hammond 2003). When female signals are very short clicks, as is often the case in airborne acoustic communication, the only opportunity for variation in female behavior to influence male behavior lies in the timing of female signals relative to male signals (Bailey 2003). By contrast, as signal length increases, additional features may come into play, while timing can continue to be important. For example, in some species with long female signals, length, frequency, and relative timing are species specific (Rodríguez and Cocroft 2006). Additionally, as mean signal length increases, variation in signal length may become important (e.g., Rodríguez et al. 2012). Also, greater signal length facilitates assessing frequency content (Fletcher 1992). Such consequences of signal length on the assessment of signal features might then allow those features to influence the behavior of receivers. Further, longer signals may contain more components (see definition below) and thus increase the potential for signal features to influence the behavior of the receiver, a hypothesis that we test in this chapter. (2) In air-borne acoustic communication, male signals show a stronger signature of selection (reduced variability) with increasing length (Reinhold 2011). This pattern may be confounded by a positive relationship between signal length and variability and perceptual acuity (Reinhold 2011), but is also consistent with the rationale that longer signals may be more likely to be selected on the basis of their influence on receivers.

In this chapter, we use a literature review to ask the following questions: (1) Are female vibrational duetting signals at least as long and variable as male signals? (2) Do longer signals contain more distinct components, and does this vary between the sexes? (3) What is the relationship between signal length and other signal traits and trait variabilities, and does this vary between the sexes? *Ceteris paribus*, female signals should be more variable than male signals, for two reasons: First, female signals express mate preferences and this requires within-individual variation (e.g., *Enchenopa* females vary in the likelihood of response and in the number and length of signals according to male attractiveness; Rodríguez et al. 2004, 2012). Second, females typically have lower variance in reproductive success than males (Shuster and Wade 2003; for *Enchenopa,* see Sullivan-Beckers and Cocroft 2010), so that the signature of selection on female signals should be weaker, and the signals should be more variable. Note that we interpret trait variability in two different ways: Greater variability suggests a greater potential for the presence of features that may influence receiver behavior; lower variability is consistent with stronger past selection (cf. Rodríguez et al. 2006; Reinhold 2011).

## 9.2 Methods

### 9.2.1 Methods: Review of Studies on Vibrational Duetting

We conducted a literature search through the SciVerse Scopus database and our own reference collection. We focused on signal length and its variability, and on the dominant frequency of signals and its variability—the only other trait that was widely reported. We measured trait variability with the between-individual coefficient of variation (CV = 100 × standard deviation/mean). We included studies that addressed vibrational duetting and that provided at least enough information to allow comparing at least one of the above traits between the sexes. We found 15 studies on 32 species in 19 genera and 11 insect families that met these criteria (Appendix). When a species had more than one signal type, we excluded only those signal types stated by the authors not to involve male–female duetting; this yielded 1–4 signal types per species. Whenever we were able to determine from the waveforms shown in each paper, we counted the number of signal components, defining a distinct component whenever we observed a change in the amplitude modulation of the signal (e.g., in Fig. 9.1a, we would identify two components for the male signals and one component for the female signals). This criterion is based on the rationale that a researcher would require different traits to describe each component—and that receivers would likely perceive each component differently. For example, the 'whine' in Fig. 9.1a is best described in terms of its length, amplitude envelope, and frequency, whereas the 'pulses' component is best described in terms of the number, rate, and duration of the pulses (Cocroft et al. 2008, 2010). For this analysis, we included all signal types. Note that with our operational definition of 'component,' we are avoiding the related issue of complexity, because defining and quantifying complexity is very difficult, if not impossible (Eberhard 2007; Healy and Rowe 2007). Consider two signal types, one with a single component (say, a 'whine'); the other with two components (say, a 'whine' and 'pulses'). Can the latter be said to be twice as complex as the former? A counter is easy to articulate: e.g., if the pulses represent a higher rate of amplitude modulation (as in Fig. 9.1a), the signal with two components may not be twice as complex as the signal with one component. However, it can be objectively stated that there are two components, each potentially offering different features for receivers.

### 9.2.2 Statistical Analysis

We asked about sex differences in signal length, dominant frequency, and their CVs. We used a model that included species and signal type (nested within species) as random effects, and sex as a fixed effect. Including species and signal type in the model accounts for lack of independence of data points within species. Since our purpose here was descriptive, we did not attempt to account for potential phylogenetic bias in the data.

To ask about the relationship between signal length and the number of signal components, we used a model similar to the above, except that we excluded signal type. Here, phylogenetic bias may confound the results, since the relationships are represented by variation among species (see below). However, this analysis suggested that potential relationships with signal length are an artifact of two outlier and closely related species (see below), and we did not attempt further corrections for hierarchical species relatedness.

We also asked about sex differences in the relationship between signal length and dominant frequency and the CVs for length and frequency. Here, the potential for phylogenetic bias is relevant, since the relationships are represented by variation among species (Freckleton et al. 2002). Addressing this problem without phylogenies is challenging. Our approach was to run the analysis with the full data set, and after averaging across progressively higher taxonomic levels (see below). The basic model had the following explanatory variables: signal length ($\log_{10}$ transformed), sex, and the sex $\times$ signal length interaction to test for sex differences in the relationships. We conducted one analysis with all the data points; i.e., one data point for each male and female trait for all signal types and species. We then repeated the analysis averaging across signal types within species (which yielded one data point per sex per species); within genera (one data point per sex per genus); and within families (one data point per sex per family). Significance tests with family means as data points have very low statistical power with $n = 11$ families, but we report them for completeness. This is a blunt corrective, but patterns remaining throughout the progression of runs are robust to non-independence of data points due to hierarchical species relatedness (e.g., Eberhard 1994). We performed all statistical analyses in JMP (v. 7.0.1; SAS Institute, Cary, NC, USA).

## 9.3 Results

### *9.3.1 Comparisons Between the Signals of Males and Females*

Female signals used in vibrational duetting were at least as long as male signals, if not longer; and they were also at least as variable, if not more. Signal length varied by an order of magnitude across species and showed a skewed distribution, with most species having short signals (Fig. 9.3a, b). Distributions were similar for males and females, with a small and marginally significant difference: Female signals were longer by an average (obtained from the least square means of the statistical model) of 0.9 s, likely due to the presence of more extreme values in females (Table 9.1; Fig. 9.3a, b). The CV in signal length also tended to be higher

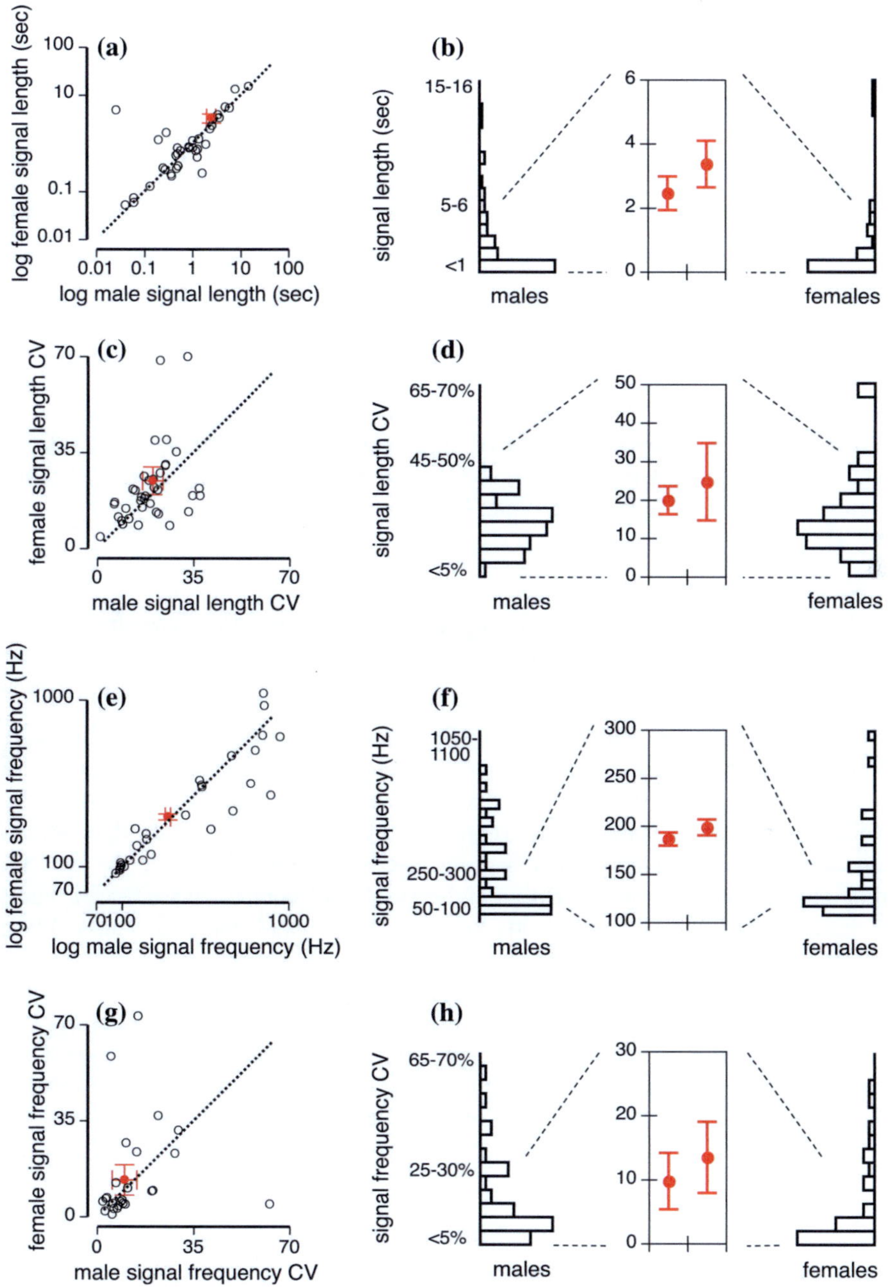


**Fig. 9.3** Comparison of the features of male and female vibrational duetting signals in 26 species of vibrational insects (see Appendix). We compare signal length, signal dominant frequency, and their variability (CV). *Black symbols*: raw data, with multiple signal types for some species. *Red symbols*: least square means (±95 % confidence interval) for each trait and sex, obtained from a statistical model that included species and signal type as random terms (see Sect. 9.3.2). The panels on the left column (**a**, **c**, **e**, **g**) show scatter plots of female traits against male traits. Dotted lines indicate a 1:1 relationship. Points above the 1:1 line indicate higher values for females than for males. The panels on the right column (**b**, **d**, **f**, **g**) show histograms for the same data, with insets detailing the position of the least square means

**Table 9.1** Signal length and its variability for male and female vibrational duetting signals in 26 species of vibrational insects (see Appendix)

| Term | Signal length | | | Signal length CV | | |
|---|---|---|---|---|---|---|
| | *F* | df | *P* | *F* | df | *P* |
| Species | 4.01 | 14, 26.583 | **0.001** | 1.46 | 14, 24.915 | 0.20 |
| Signal type | 4.15 | 28, 20 | **0.0008** | 1.66 | 28, 18 | 0.13 |
| Sex | 3.60 | 1, 20 | **0.072** | 2.10 | 1, 18 | 0.16 |

Species and signal type (nested within species) were random effects. *Bold*: significant or marginally significant terms

for females, but not significantly so (Table 9.1; Fig. 9.3c, d). We also found a small but significant sex difference in the dominant frequency of signals, with females having slightly higher-pitched signals (by 12 Hz; Table 9.2; Fig. 9.3e, f). The CV in dominant frequency tended to be higher in female signals, but not significantly so (Table 9.2; Fig. 9.3g, h).

### 9.3.2 Relationship Between Signal Length and the Number of Signal Components

Our analysis revealed two patterns. First, sex had a significant effect, with males having on average more signal components than females (least square mean ± SE for males = 1.4 ± 0.08; for females = 1.2 ± 0.08; Table 9.3). Although this difference is small, male signals more often had two components than female signals, and only male signals had three components (Fig. 9.4). Second, the relationship between signal length and the number of components was marginally significant and differed significantly between the sexes (Table 9.3). This is likely an artifact of the two species that had three signal components for male signals (Fig. 9.4), which were both in the same family (Cicadellidae; see Appendix). Similarly, the two species for which female signals had two components (Fig. 9.4) are also both in the same family (Pentatomidae; see Appendix). Nevertheless, there does appear to be a minimum signal length below which it may not be feasible to have more than a single component (ca. 0.15 ms for males).

### 9.3.3 Relationship Between Signal Length and Signal Frequency and Trait Variability

Patterns related to signal length were robust to analysis at different taxonomic levels. Longer signals tended to be lower in pitch (Fig. 9.5)—a pattern that has been noted previously in male signals in vibrational insects (among various insect

**Table 9.2** Signal dominant frequency and its variability for male and female vibrational duetting signals in 26 species of vibrational insects (see Appendix)

| Term | Signal frequency | | | Signal frequency CV | | |
|---|---|---|---|---|---|---|
| | *F* | df | *P* | *F* | df | *P* |
| Species | 21.93 | 12, 20.476 | **<0.0001** | 71.30 | 11, 1.2568 | **0.056** |
| Signal type | 8.95 | 21, 17 | **<0.0001** | 0.20 | 19, 14 | 0.99 |
| Sex | 4.82 | 1, 17 | **0.042** | 0.99 | 1, 14 | 0.34 |

Species and signal type (nested within species) were random effects. *Bold*: significant or marginally significant terms

**Table 9.3** Signal length and its relationship with the number of signal components across 22 species of vibrational insects (see Appendix)

| Term | F | df | P |
|---|---|---|---|
| Species | 2.21 | 19, 36 | **0.02** |
| Signal length | 3.79 | 1, 36 | **0.06** |
| Sex | 7.80 | 1, 36 | **0.008** |
| Signal length × sex | 5.71 | 1, 36 | **0.022** |

Species was a random effect. *Bold*: significant or marginally significant terms

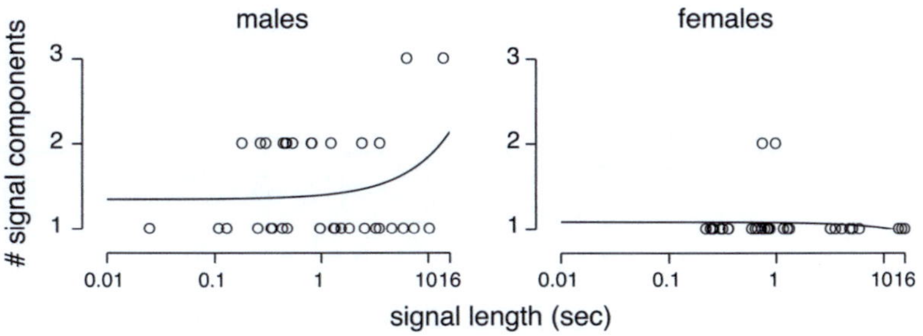


**Fig. 9.4** Relationship between signal length and the number of signal components in 22 species of vibrational insects (see Appendix). For males (*left*) but not for females (*right*), the number of signal components increased with signal length (fitted lines; Table 9.3). However, the relationship for males is likely an artifact of the two species with three components, which were in the same family (Cicadellidae; see Appendix). Note that no signals under ca. 0.15 ms had more than a single component. Fitted lines appear curved because of the log transformation for the *x*-axis

orders and within the family Membracidae: Cocroft and De Luca 2006; within the *E. binotata* complex: Cocroft et al. 2010). This relationship did not differ between the sexes, and it was significant or marginally significant in all analyses except for family means (Table 9.4).

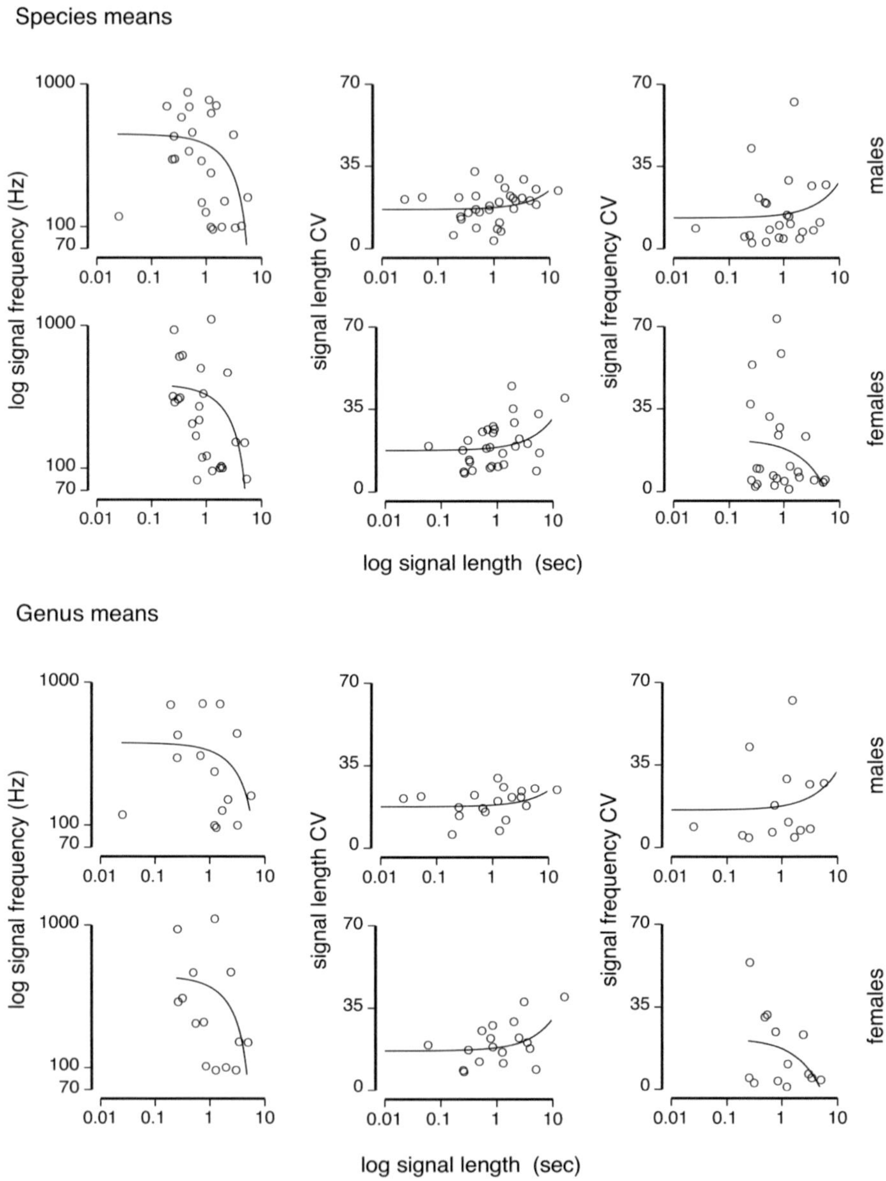


**Fig. 9.5** Relationship between signal length and dominant signal frequency and their variability (CV) across 26 species of vibrational insect (see Appendix). We show two of the four analyses in Table 9.4. The *top panel* of six figures shows the data used in the analysis with species means. The *bottom panel* of six figures shows the data used in the analysis with genus means. At both levels of analysis, signal frequency decreased with signal length for males and females, and the CV in signal length increased with signal length for males and females. By contrast, the CV in signal frequency tended to increase with signal length for males and to decrease for females, non-significantly in both cases

Longer signals also tended to have higher CVs for signal length (Fig. 9.5). This relationship was significant or marginally significant at all levels of analysis except for family means (Table 9.4). There was weak evidence that this relationship differed between the sexes: It was apparently steeper for females (Fig. 9.5), and the sex × signal length interaction had an $F$-ratio = 2.04 in the species means test (Table 9.4). We interpret this as indicative of a small difference in the steepness of the relationship.

The relationship between signal length and the CV for signal frequency was less clear cut. For males, longer signals tended to have higher CVs for frequency; for females, longer signals tended to have lower CVs for frequency (Fig. 9.5). These relationships were not significant, but the slope was always negative (Table 9.4), suggesting that the model weighed the pattern for females more heavily, perhaps because it was steeper than for males (Fig. 9.5). The sex × signal length interaction was not significant, but its $F$-ratio was 2.01 and 2.57 at the species means and genus means levels (Table 9.4). We interpret this as weak evidence that the CV of signal frequency varied with signal length and that the pattern differed between the sexes.

## 9.4 Discussion

In insect substrate-borne vibrational duetting, female signals were longer and slightly higher in pitch than male signals, and as variable if not slightly more in both length and frequency. Under the expectation that length and variability increase the likelihood that signal features will influence receiver behavior and perhaps mate choice, our findings suggest a high likelihood that female signals will influence male behavior in duetting interactions. That is to say, similar length and variability in male and female duetting signals suggest that both sides of the duetting interaction offer similar amounts (if not more for females) of features that can influence the behavior of the other sex. Further, the duetting signals of both sexes offer similar amounts of variation (if not more in females) for selection to act upon. Another finding, however, suggests that male duetting signals may offer more features: Although most male and female signals had a single component, multicomponent signals were more common in males. This preliminary comparison suggests the question of why male duetting signals would more often evolve multiple components than female duetting signals. It will be interesting to pursue this question in future research. We note, however, that more components do not necessarily mean stronger influence on the opposite sex, or stronger sexual selection. For example, of two case studies of substrate-borne vibrational duetting involving strong sexual selection, one features sexually dimorphic signals with two-component male signals and single-component female signals (*Enchenopa*

**Table 9.4** Relationship between signal length and dominant signal frequency and variability (CV) in signal length and frequency across 26 species of vibrational insects (see Appendix)

| | Signal frequency | | | | Signal length CV | | | | Signal frequency CV | | | |
|---|---|---|---|---|---|---|---|---|---|---|---|---|
| Term | *F* | df | *P* | *b* | *F* | df | *P* | *b* | *F* | df | *P* | *b* |
| *Analysis with all data points* | | | | | | | | | | | | |
| Signal length | 4.49 | 1, 70 | **0.038** | −0.16 | 2.76 | 1, 83 | **0.10** | 3.30 | 0.84 | 1, 64 | 0.36 | −3.98 |
| Sex | 0.56 | 1, 70 | 0.46 | | 0.88 | 1, 83 | 0.35 | | 0.10 | 1, 64 | 0.75 | |
| Interaction | 0.60 | 1, 70 | 0.44 | | 0.37 | 1, 83 | 0.55 | | 1.07 | 1, 64 | 0.30 | |
| *Analysis with species means* | | | | | | | | | | | | |
| Signal length | 9.06 | 1, 46 | **0.0042** | -0.30 | 8.09 | 1, 57 | **0.0062** | 5.41 | 0.38 | 1, 44 | 0.54 | -3.49 |
| Sex | 0.46 | 1, 46 | 0.50 | | 0.71 | 1, 57 | 0.40 | | 0.16 | 1, 44 | 0.69 | |
| Interaction | 1.47 | 1, 46 | 0.23 | | 2.04 | 1, 57 | 0.16 | | 2.01 | 1, 44 | 0.16 | |
| *Analysis with genus means* | | | | | | | | | | | | |
| Signal length | 3.22 | 1, 25 | **0.085** | -0.23 | 6.18 | 1, 33 | **0.018** | 4.95 | 0.56 | 1, 23 | 0.46 | −4.87 |
| Sex | 0.02 | 1, 25 | 0.90 | | 0.07 | 1, 33 | 0.80 | | 0.18 | 1, 23 | 0.68 | |
| Interaction | 1.23 | 1, 25 | 0.28 | | 1.50 | 1, 33 | 0.23 | | 2.57 | 1, 23 | 0.12 | |
| *Analysis with family means* | | | | | | | | | | | | |
| Signal length | 2.12 | 1, 10 | 0.18 | -0.22 | 1.84 | 1, 16 | 0.19 | 2.36 | 0.13 | 1, 10 | 0.73 | −3.45 |
| Sex | 0.04 | 1, 10 | 0.85 | | 0.10 | 1, 16 | 0.76 | | 0.03 | 1, 10 | 0.87 | |
| Interaction | 0.02 | 1, 10 | 0.88 | | 1.30 | 1, 16 | 0.27 | | 0.13 | 1, 10 | 0.76 | |

Length and frequency were $\log_{10}$ transformed for the analysis. The sex × signal length interaction term tests for sex differences in these relationships. To address the problems of non-independence of data within species and of possible phylogenetic bias, we repeated the analysis four times: all data points, averaging across signal types within species, averaging across species within genus, and averaging across genus within families. Bold: significant or marginally significant terms treehoppers; see Sect. 9.2.1), while the other involves sexually monomorphic, largely single-component signals (*Chrysoperla* green lacewings; Wells and Henry 1998; Henry et al. 2013). For future research, whether the duetting signals of one sex or the other are more influenced by the action of sexual selection may best be answered by comparing the relative rates of divergence in male versus female signals. Such comparisons would ideally be informed by data on the relative rates of divergence in the mate preferences of each sex (when present) as well as by the relative strength of those preferences (cf. Rodríguez et al. 2006, 2013).

Another goal of this chapter was to assess the potential for a signature of past selection to be manifest in relation to signal length in insect vibrational duetting. The rationale here is that, because longer signals have a higher likelihood of presenting traits and variation for selection to act upon, they should exhibit lowered trait variability. We found a robust pattern for variability in signal length to increase with signal length. The pattern for variability in signal frequency was weaker and more complex: It tended to increase with signal length for males and to decrease (as predicted) for females. Thus, overall, there was no evidence that longer duetting signals are more strongly selected.

The most robust pattern that we found in relation to the length of duetting signals was a negative association with signal frequency: In both sexes, longer signals tended to have lower pitch. This result suggests that a third variable may be involved: If body size is associated with signal length, it might explain the correlation with signal frequency, because larger insects may produce lower-pitched signals. The relationship between size and frequency in the substrate-borne vibrational channel is less pronounced than in the air-borne acoustic channel, but it is present (Greenfield 2002; Cocroft and Rodríguez 2005; Cocroft and De Luca 2006; Cocroft et al. 2010). However, we are not aware of evidence indicating that larger vibrational insects produce longer signals and thus leave this possibility open for further research.

There are two potential sources of bias in our literature review that should be taken into account in drawing conclusions from the above patterns. It seems likely that the literature underreports variability in female signals: Female duetting behavior varies in part as a function of the male signals with which the females are interacting (e.g., Rodríguez et al. 2004, 2012; Henry and Wells 2006). Thus, researchers interested in describing the features of female signals may often select their sample from those signals given in response to average or 'standard' males—at least one of us has done so (Rodríguez and Cocroft 2006). Indeed, this is recommended for the purpose of providing trait means and assessing species differences. Nevertheless, to the extent that this excludes female signals given in response to unattractive or extremely attractive males, it will undersample variability in female duetting behavior. In short, our review may underestimate variability in female signals. We thus consider that our tentative conclusion that there is strong potential for female duetting to influence the back-and-forth dynamics of sexual selection on both sexes is likely robust and worthy of further attention and research.

As a final point of discussion, we emphasize that we are likely to underestimate the diversity of ways in which a back-and-forth duet between the sexes may influence the dynamics and outcomes of sexual selection. For example, our focus on vibrational duets excluded cases in which the exchange of signals is multimodal; e.g., when one sex initiates the interaction with a chemical signal that then establishes a vibrational duet (Virant-Doberlet and Cokl 2004); or when the female response is both visual and vibrational (Sullivan-Beckers and Hebets 2011). Further diversity may be found among birds, where air-borne acoustic duetting is common. Here, the dynamics are different, since bird duets often involve the synchronization of breeding or coordinated territory defense against intruders, but there is the potential for mate choice on the basis of how well signaling is coordinated (Slater and Mann 2004; Mann et al. 2009). Finally, we have concentrated on interactions involved in mate location and pair formation, but there is evidence that male–female mutual influence may often continue beyond pair formation (Eberhard 1994; Rodríguez 1998; Patricelli et al. 2002, Peretti et al. 2006).

In conclusion, the study of mutual mate choice/influence is likely to reshape how behavioral ecologists view the process and outcome of sexual selection. Vibrational duetting insects offer extremely tractable case studies to elucidate these patterns and processes.

**Acknowledgments** We thank Rex Cocroft and Gerlinde Höbel for discussion and constructive comments to the manuscript. Funding was provided in part by NSF grant IOS–1120790 to RLR and KD Fowler-Finn, and by University of Wisconsin–Milwaukee Research Grant Initiative grant 101x197 to Gerlinde Höbel.

## A.1 Appendix

Mean and coefficient of variation for signal length and dominant frequency for males and females of the vibrational duetting species used in the analysis. A dash indicates that data were not available from the references given

| Taxon | Sex | Call type | Call length | | Dominant frequency | | Number of components | References |
|---|---|---|---|---|---|---|---|---|
| | | | Mean (s) | CV | Mean (Hz) | CV | | |
| **Coleoptera** | | | | | | | | |
| **Scarabaeidae** | | | | | | | | |
| *Aphodius ater* | F | – | 0.06 | 19.2 | – | – | – | Hirschberger and Rohrseitz (1995) |
| | M | – | 0.05 | 21.7 | – | – | – | Hirschberger and Rohrseitz (1995) |
| **Hemiptera** | | | | | | | | |
| **Aleyrodidae** | | | | | | | | |
| *Aleurothrixus floccosus* | F | Iriomote Island | 0.28 | 21.57 | 300.6 | 1.99 | 1 | Kanmiya and Sonobe (2002) |
| | M | Iriomote Island | 0.25 | 12.42 | 294.6 | 2.27 | 1 | 2002 |
| | F | Ishigaki Island | 0.31 | 12.53 | 308.1 | 2.92 | – | Kanmiya and Sonobe (2002) |
| | M | Ishigaki Island | 0.22 | 21.58 | 292.6 | 5.53 | – | Kanmiya and Sonobe (2002) |
| **Cicadellidae** | | | | | | | | |
| *Aphrodes makarovi* | F | – | 15.42 | 39.56 | – | – | 1 | de Groot et al. (2011) |
| | M | – | 13.43 | 24.49 | – | – | 3 | de Groot et al. (2011) |
| *Graphocephala atropunctata* | F | – | 0.51 | 25.24 | 203 | 31.53 | – | Percy et al. (2008) |
| | M | – | 1.17 | 19.66 | 235 | 28.93 | – | Percy et al. (2008) |
| *Homalodisca liturata* | F | – | 1.23 | 11.38 | 95 | 10.52 | – | Percy et al. (2008) |
| | M | – | 1.26 | 7.143 | 95 | 10.52 | – | Percy et al. (2008) |
| *Scaphoideus titanus* | F | Call 1 | – | – | 150.5 | 32.29 | 1 | Mazzoni et al. (2009) |
| | F | Call 2 | – | – | 137.7 | 40.74 | – | Mazzoni et al. (2009) |
| | M | Call 1 | 6.2 | 30.64 | 158.8 | 27.01 | 3 | Mazzoni et al. (2009) |
| | M | Call 2 | 4.6 | 19.56 | – | – | – | Mazzoni et al. (2009) |
| **Cixiidae** | | | | | | | | |
| *Hyalesthes obsoletus* | F | – | 0.25 | 7.69 | 287 | 53.65 | 1 | Mazzoni et al. (2010) |
| | M | Syllable 1 | 0.17 | 11.36 | 576 | 41.14 | 2 | Mazzoni et al. (2010) |
| | M | Syllable 2 | 0.26 | 14.44 | 292 | 53.42 | 2 | Mazzoni et al. (2010) |
| | M | Syllable 3 | 0.29 | 14.96 | 447 | 34.67 | 2 | Mazzoni et al. (2010) |
| | M | Syllable 4 | 0.23 | 13.04 | 379 | 40.63 | – | Mazzoni et al. (2010) |

(continued)

(continued)

| Taxon | Sex | Call type | Call length | | Dominant frequency | | Number of components | References |
|---|---|---|---|---|---|---|---|---|
| | | | Mean (s) | CV | Mean (Hz) | CV | | |
| **Flatidae** | | | | | | | | |
| *Metcalfa pruinosa* | F | Call | 2.34 | 22.22 | 463 | 23.11 | – | Virant-Doberlet and Žežlina (2007) |
| | M | Call | 2.35 | 20 | 442 | 27.61 | – | Virant-Doberlet and Žežlina (2007) |
| | M | Phrase pulse train 1 | 4.31 | 31.55 | 414 | 27.05 | – | Virant-Doberlet and Žežlina (2007) |
| | M | Phrase pulse train 2 | 2.39 | 12.55 | 446 | 25.11 | – | Virant-Doberlet and Žežlina (2007) |
| **Membracidae** | | | | | | | | |
| *Enchenopa binotata* 'Celastrus' | F | – | 0.7 | 18.7 | 215 | 5.4 | 1 | Rodriguez and Cocroft 2006 |
| | M | – | 0.52 | 15.43 | 451 | 7.97 | 2 | Rodriguez and Cocroft (2006) |
| *Enchenopa binotata* 'Cercis' | F | – | 0.79 | 24.8 | 118 | 26.9 | 1 | Rodriguez and Cocroft (2006) |
| | M | – | 0.79 | 18.05 | 146 | 9.9 | 2 | Rodriguez and Cocroft (2006) |
| *Enchenopa binotata* 'Ptelea' | F | – | 0.61 | 18.2 | 167 | 6.6 | 1 | Rodriguez and Cocroft (2006) |
| | M | – | 0.45 | 16.54 | 333 | 2.64 | 2 | Rodriguez and Cocroft (2006) |
| *Enchenopa binotata* 'Viburnum rufidulum' | F | – | 0.84 | 26.3 | 329 | 58.4 | 1 | Rodriguez and Cocroft (2006) |
| | M | – | 0.78 | 16.41 | 285 | 4.48 | 2 | Rodriguez and Cocroft (2006) |
| **Pentatomidae** | | | | | | | | |
| *Acrosternum hilare* | F | Song | 0.64 | 26.17 | 82 | 2.44 | 1 | Cokl et al. (2001) |
| | M | Call 1 | 0.94 | 0.53 | 120 | 6.66 | 2 | Moraes et al. (2005) |
| | M | Song 1 | 2.31 | 20.09 | – | – | 1 | Moraes et al. (2005) |
| *Acrosternum impicticorne* | F | Call 1 | 0.81 | 4.19 | 133 | 3.01 | 1 | Moraes et al. (2005) |
| | F | Call 2 | 1.13 | 16.82 | 109 | 5.51 | 1 | Moraes et al. (2005) |
| | M | Call 2 | 0.94 | 5.75 | 130 | 1.54 | 1 | Moraes et al. (2005) |
| *Euschistus heros* | F | Call 1 | 0.95 | 19.30 | 145 | 4.82 | 2 | Moraes et al. (2005) |
| | F | Call 2 | 5.66 | 21.15 | 157 | 4.46 | 1 | Moraes et al. (2005) |
| | M | Call 1 | 1.75 | 34.05 | 137 | 8.76 | 1 | Moraes et al. (2005) |
| | M | Call 2 | 4.41 | 13.20 | 136 | 9.56 | 1 | Moraes et al. (2005) |
| | M | Call 3 | 0.05 | 16.66 | 175 | 2.86 | – | Moraes et al. (2005) |

(continued)

(continued)

| Taxon | Sex | Call type | Call length | | Dominant frequency | | Number of components | References |
|---|---|---|---|---|---|---|---|---|
| | | | Mean (s) | CV | Mean (Hz) | CV | | |
| *Nezara viridula* | F | Calling song broad band | 0.13 | 35.15 | 109 | – | – | Cokl et al. (2000) |
| | F | Calling song narrow band | 1.66 | 21.91 | 99 | – | – | Cokl et al. (2000) |
| | F | Courtship song | 3.91 | 30.19 | 91 | – | 1 | Cokl et al. (2000) |
| | M | Calling song broad band | 0.12 | 28.09 | 109 | – | – | Cokl et al. (2000) |
| | M | Calling song narrow band | 0.26 | 36.5 | 96 | – | – | Cokl et al. (2000) |
| | M | Courtship song | 3.11 | 24.08 | 90 | – | 1 | Cokl et al. (2000) |
| *Piezodorus guildinii* | F | Call 1 | 4.91 | 13.06 | 168 | 5.95 | 1 | Moraes et al. (2005) |
| | F | Call 2 | 4.63 | 4.32 | 131 | 1.52 | 1 | Moraes et al. (2005) |
| | M | Call 1 | 0.02 | 20.83 | 117 | 8.54 | 1 | Moraes et al. (2005) |
| *Thyanta custator* accerra | F | Song 1 | 0.34 | 19.28 | 102 | 3.92 | 1 | McBrien et al. (2002) |
| | F | Song 2 | 3.08 | 69.91 | 97 | 12.37 | 1 | McBrien et al. (2002) |
| | M | Song 1 | 0.46 | 36.77 | 97 | 7.21 | 1 | McBrien et al. (2002) |
| | M | Song 2 | 2.44 | 32.28 | 95 | 6.31 | 1 | McBrien et al. (2002) |
| | M | Song 3 | 9.89 | 30.55 | 98 | 8.16 | 1 | McBrien et al. (2002) |
| | M | Song 4 | 0.12 | 17.46 | 99 | 9.09 | 1 | McBrien et al. (2002) |
| *Thyanta pallidovirens* | F | Song 1 | 0.21 | 30.62 | 100 | 7 | 1 | McBrien et al. (2002) |
| | F | Song 2 | 3.41 | 39.37 | 105 | 4.76 | 1 | McBrien et al. (2002) |
| | M | Song 1 | 0.34 | 24.18 | 101 | 2.97 | 1 | McBrien et al. (2002) |
| | M | Song 2 | 3.36 | 20.27 | 96 | 5.21 | 1 | McBrien et al. (2002) |
| *Thyanta perditor* | F | Call 1 | 13.41 | 14.49 | | | 1 | Moraes et al. (2005) |
| | F | Call 2, long | 1.46 | 43.09 | 84 | 4.76 | 2 | Moraes et al. (2005) |
| | F | Call 2, short | 0.71 | 40.84 | 83 | 4.82 | – | Moraes et al. (2005) |
| | M | Call 1 | 7.1 | 9.80 | 122 | 12.29 | 1 | Moraes et al. (2005) |
| | M | Call 2 | 3.39 | 29.20 | 109 | 11.01 | 2 | Moraes et al. (2005) |
| | M | Call 3 | 2.11 | 21.80 | 70 | 10 | – | Moraes et al. (2005) |

(continued)

(continued)

| Taxon | Sex | Call type | Call length | | Dominant frequency | | Number of components | References |
|---|---|---|---|---|---|---|---|---|
| | | | Mean (s) | CV | Mean (Hz) | CV | | |
| **Triozidae** | | | | | | | | |
| *Aacanthocnema dobsoni* | F | – | 0.24 | 8.33 | 925 | 4.64 | 1 | Percy et al. (2006) |
| | M | – | 1.48 | 25.67 | 698 | 62.17 | 1 | Percy et al. (2006) |
| *Schedotrioza apicobystra* | F | – | 0.75 | 10.66 | 499 | 23.64 | 1 | Percy et al. (2006) |
| | M | – | 1.19 | 10.92 | 615 | 13.65 | 2 | Percy et al. (2006) |
| *Schedotrioza distorta* | F | – | 0.23 | 17.39 | 315 | 36.82 | 1 | Percy et al. (2006) |
| | M | – | 0.33 | 15.15 | 576 | 21.52 | 1 | Percy et al. (2006) |
| *Schedotrioza marginata* | F | – | 0.34 | 8.82 | 614 | 9.44 | 1 | Percy et al. (2006) |
| | M | – | 0.46 | 8.69 | 683 | 19.18 | 2 | Percy et al. (2006) |
| *Schedotrioza multitudinea* | F | – | 0.3 | 13.33 | 602 | 9.63 | 1 | Percy et al. (2006) |
| | M | – | 0.43 | 32.55 | 869 | 19.56 | 2 | Percy et al. (2006) |
| *Schedotrioza*, sp. nov. | F | – | 0.7 | 10 | 268 | 73.13 | – | Percy et al. (2006) |
| | M | – | 1.09 | 8.256 | 765 | 14.25 | – | Percy et al. (2006) |
| *Trioza*, sp. nov. | M | – | 0.18 | 5.55 | 690 | 4.92 | – | Percy et al. (2006) |
| *Trioza*, sp. nov. | F | – | 1.18 | 16.10 | 1096 | 0.82 | – | Percy et al. (2006) |
| Neuroptera | | | | | | | | |
| **Chrysopidae** | | | | | | | | |
| *Chrysoperla nipponensis* type B | M | Site1 | 2.14 | 16.82 | – | – | – | Henry et al. (2009) |
| *Chrysoperla nipponensis* type B | F | Site1 | 1.99 | 19.09 | – | – | – | Henry et al. (2009) |
| *Chrysoperla nipponensis* type A | F | Site2 | 5.4 | 16.29 | – | – | – | Henry et al. (2009) |
| *Chrysoperla nipponensis* type A | M | Site2 | 5.43 | 18.60 | – | – | – | Henry et al. (2009) |
| Orthoptera | | | | | | | | |
| **Gryllacrididae** | | | | | | | | |
| *Hadrogryllacris* sp, | F | – | 0.8 | 27.5 | – | – | 1 | Field and Bailey (1997) |
| undescribed species | M | – | 0.45 | 22.22 | – | – | 1 | Field and Bailey (1997) |

## References

Amundsen T (2000) Why are female birds ornamented? Trends Ecol Evol 15:149–155
Andersson M (1994) Sexual selection. Princeton University Press, Princeton
Bailey WJ (2003) Insect duets: underlying mechanisms and their evolution. Physiol Entomol 28:157–174
Bailey WJ, Hammond TJ (2003) Duetting in insects—does call length influence reply latency? J Zool 260:267–274
Belwood JJ, Morris GK (1987) Bat predation and its influence on calling behavior in neotropical katydids. Science 238:64–67
Bonduriansky R (2001) The evolution of male mate choice in insects: a synthesis of ideas and evidence. Biol Rev 76:305–339
Catania KC (2008) Worm grunting, fiddling, and charming—humans unknowingly mimic a predator to harvest bait. PLoS ONE 3:e3472
Cator LJ, Arthur BJ, Harrington LC, Hoy RR (2009) Harmonic convergence in the love songs of the dengue vector mosquito. Science 323:1077–1079
Clutton-Brock TH (2007) Sexual selection in males and females. Science 318:1882–1885
Clutton-Brock TH (2009) Sexual selection in females. Anim Behav 77:3–11
Clutton-Brock TH, Hodge SJ, Spong G, Russell AF, Jordan NR, Bennett NC, Sharpe LL, Manser MB (2006) Intrasexual competition and sexual selection in cooperative mammals. Nature 444:1065–1068
Cocroft RB (2003) The social environment of an aggregating, ant-tended treehopper (Hemiptera: Membracidae: *Vanduzea arquata*). J Insect Behav 16:79–95
Cocroft RB (2011) The public world of insect vibrational communication. Mol Ecol 20:2041–2043
Cocroft RB, De Luca P (2006) Size–frequency relationships in insect vibratory signals. In: Drosopoulos S, Claridge MF (eds) Insect sounds and communication: physiology, behaviour, ecology and evolution. Taylor & Francis Group, Boca Raton, FL, pp 99–110
Cocroft RB, Rodríguez RL (2005) The behavioral ecology of insect vibrational communication. Bioscience 55:323–334
Cocroft RB, Rodríguez RL, Hunt RE (2008) Host shifts, the evolution of communication and speciation in the *Enchenopa binotata* complex of treehoppers. In: Tilmon K (ed) Specialization, speciation, and radiation: the evolutionary biology of herbivorous insects. University of California Press, Berkeley, pp 88–100
Cocroft RB, Rodríguez RL, Hunt RE (2010) Host shifts and signal divergence: mating signals covary with host use in a complex of specialized plant-feeding insects. Biol J Linn Soc 99:60–72
Cokl A, Virant-Doberlet M, Stritih N (2000) Temporal and spectral properties of the songs of the southern green stink bug *Nezara viridula* (L.) from Slovenia. Eur J Physiol 439:R168–R170
Čokl A, McBrien HL, Millar JG (2001) Comparison of substrate-borne vibrational signals of two stink bug species, *Acrosternum hilare* and *Nezara viridula* (Heteroptera: Pentatomidae). Ann Entomol Soc Am 94:471–479
Cokl A, Virant-Doberlet M (2003) Communication with substrate-borne signals in small plant-dwelling insects. Annu Rev Entomol 48:29–50
Conner WE, Corcoran AJ (2012) Sound strategies: the 65-million-year-old battle between bats and insects. Annu Rev Entomol 57:21–39
de Groot M, Čokl A, Virant-Doberlet M (2011) Search behaviour of two hemipteran species using vibrational communication. Cent Eur J Biol 8:756–769
Djemai I, Casas J, Magal C (2004) Parasitoid foraging decisions mediated by artificial vibrations. Anim Behav 67:567–571
Eberhard WG (1994) Evidence for widespread courtship during copulation in 131 species of insects and spiders, and implications for cryptic female choice. Evolution 48:711–733

Eberhard WG (2007) Miniaturized orb–weaving spiders: behavioural precision is not limited by small size. Proc R Soc B 274:2203–2209

Elias DO, Botero CA, Andrade MCB, Mason A, Kasumovic MM (2010) High resource valuation fuels “desperado” fighting tactics in female jumping spiders. Behav Ecol 21:868–875

Evans TA, Inta R, Lai JCS, Prueger S, Foo NW, Fu EW, Lenz M (2009) Termites eavesdrop to avoid competitors. Proc R Soc B 276:4035–4041

Field LH, Bailey WJ (1997) Sound production in primitive Orthoptera from Western Australia: sounds used in defence and social communication in *Ametrus* sp. and *Hadrogryllacris* sp. (Gryllacrididae: Orthoptera). J Nat Hist 31:1127–1141

Fletcher NH (1992) Acoustic systems in biology. Oxford University Press, New York

Freckleton RP, Harvey PH, Pagel M (2002) Phylogenetic analysis and comparative data: a test and review of evidence. Am Nat 160:712–726

Greenfield MD (2002) Signalers and receivers. Oxford University Press, New York

Gwynne DT (1991) Sexual competition among females: what causes courtship–role reversal? Trends Ecol Evol 6:118–121

Healy SD, Rowe C (2007) A critique of comparative studies of brain size. Proc R Soc B 274:453–464

Henry CS (1994) Singing and cryptic speciation in insects. Trends Ecol Evol 9:388–392

Henry CS, Wells MLM (2006) Testing the ability of males and females to respond to altered songs in the dueting green lacewing, *Chrysoperla plorabunda* (Neuroptera: Chrysopidae). Behav Ecol Sociobiol 61:39–51

Henry CS, Mochizuki A, Nakahira K, Haruyama N, Nomura M (2009) Courtship songs of *Chrysoperla nipponensis* (Neuroptera: Chrysopidae) delineate two distinct biological species in eastern Asia. Ann Entomol Soc America 102:747–758

Henry CS, Brooks SJ, Duelli P, Johnson JB, Wells MM, Mochizuki A (2013) Obligatory duetting behaviour in the *Chrysoperla carnea*-group of cryptic species (Neuroptera: Chrysopidae): its role in shaping evolutionary history. Biol Rev 88:787–808

Hill PSM (2008) Vibrational communication in animals. Harvard University Press, Cambridge

Hirschberger P, Rohrseitz K (1995) Stridulation in the adult dung beetle *Aphodius ater* (Coleoptera: Aphodiidae). Zoology (Jena) 99:97–102

Kanmiya K, Sonobe R (2002) Records of two citrus pest whiteflies in Japan with special reference to their mating sounds (Homoptera: Aleyrodidae). Appl Entomol Zool 37:487–495

Kozak G, Reisland M, Boughman JW (2009) Sex differences in mate recognition and conspecific preference in species with mutual mate choice. Evolution 63:353–365

Laumann RA, Moraes MCB, Cokl A, Borges M (2007) Eavesdropping on sexual vibratory signals of stink bugs (Hemiptera: Pentatomidae) by the egg parasitoid *Telenomus podisi*. Anim Behav 73:637–649

Mann NI, Dingess KA, Barker FK, Graves JA, Slater PJB (2009) A comparative study of song form and duetting in neotropical *Thryothorus* wrens. Behaviour 146:1–43

Mazzoni V, Prešern J, Lucchi A, Virant-Doberlet M (2009) Reproductive strategy of the Nearctic leafhopper *Scaphoideus titanus* Ball (Hemiptera: Cicadellidae). Bull Entomol Res 99:401–413

Mazzoni V, Lucchi A, Ioriatti C, Doberlet-Virant M, Anfora G (2010) Mating behavior of *Hyalesthes obsoletus*. Ann Entomol Soc Am 103:813–822

McBrien HL, Çokl A, Millar JG (2002) Comparison of substrate-borne vibrational signals of two congeneric stink bug species, *Thyanta pallidovirens* and *T. custator* accerra (Heteroptera: Pentatomidae). J Insect Behav 15:715–738

Mitra O, Callaham MA Jr, Smith ML, Yack JE (2009) Grunting for worms: seismic vibrations cause *Diplocardia* earthworms to emerge from the soil. Biol Lett 5:16–19

Moraes MCB, Laumann RA, Cokl A, Borges M (2005) Vibratory signals of four Neotropical stink bug species. Physiol Entomol 30:175–188

Noh S, Henry CS (2010) Sexually monomorphic mating preferences contribute to premating isolation based on song in European green lacewings. Evolution 64:261–270

Patricelli GL, Uy JAC, Walsh G, Borgia G (2002) Male displays adjusted to female’s response. Nature 415:279–280

Percy DM, Taylor GS, Kennedy M (2006) Psyllid communication: acoustic diversity, mate recognition and phylogenetic signal. Invertebr Syst 20:431–445
Percy DM, Boyd EA, Hoodle MS (2008) Observations of acoustic signaling in three sharpshooters: *Homalodisca vitripennis*, *Homalodisca liturata*, and *Graphocephala atropunctata* (Hemiptera: Cicadellidae). Ann Entomol Soc Am 101:253–259
Peretti A, Eberhard WG, Briceño RD (2006) Copulatory dialogue: female spiders sing during copulation to influence male genitalic movements. Anim Behav 72:413–421
Reinhold K (2011) Variation in acoustic signalling traits exhibits footprints of sexual selection. Evolution 65:738–745
Rillich J, Buhl E, Schildberger K, Stevenson PA (2009) Female crickets are driven to fight by the male courting and calling songs. Anim Behav 77:737–742
Rodríguez RL (1998) Possible female choice during copulation in *Ozophora baranowskii* (Heteroptera: Lygaeidae). J Insect Behav 11:725–741
Rodríguez RL, Cocroft RB (2006) Divergence in female duetting signals in the *Enchenopa binotata* species complex of treehoppers (Hemiptera: Membracidae). Ethology 112:1231–1238
Rodríguez RL, Sullivan LE, Cocroft RB (2004) Vibrational communication and reproductive isolation in the *Enchenopa binotata* species complex of treehoppers (Hemiptera: Membracidae). Evolution 58:571–578
Rodríguez RL, Ramaswamy K, Cocroft RB (2006) Evidence that female preferences have shaped male signal evolution in a clade of specialized plant–feeding insects. Proc R Soc B 273:2585–2593
Rodríguez RL, Haen C, Cocroft RB, Fowler-Finn KD (2012) Males adjust signaling effort based on female mate–preference cues. Behav Ecol 23:1218–1225
Rodríguez RL, Boughman JW, Gray DA, Hebets EA, Höbel G, Symes LB (2013) Diversification under sexual selection: the relative roles of mate preference strength and the degree of divergence in mate preferences. Ecol Lett 16:964–974
Rubenstein DR, Lovette IJ (2009) Reproductive skew and selection on female ornamentation in social species. Nature 462:786–790
Sæther SA, Fiske P, Kålås JA (2001) Male mate choice, sexual conflict and strategic allocation of copulations in a lekking bird. Proc R Soc Lond B 268:2097–2102
Shuster SM, Wade MJ (2003) Mating systems and strategies. Princeton University Press, Princeton
Slater PJB, Mann NI (2004) Why do the females of many bird species sing in the tropics? J Avian Biol 35:289–294
Sullivan-Beckers L, Cocroft RB (2010) The importance of female choice, male–male competition, and signal transmission as causes of selection on male mating signals. Evolution 64:3158–3171
Sullivan-Beckers L, Hebets EA (2011) Modality-specific experience with female feedback increases the efficacy of courtship signalling in male wolf spiders. Anim Behav 82:1051–1057
Uhl G, Elias DO (2011) Communication. In: Herberstein ME (ed) Spider behaviour: flexibility and versatility. Cambridge University Press, Cambridge, pp 127–189
Virant-Doberlet M, Cokl A (2004) Vibrational communication in insects. Neotrop Entomol 33:121–134
Virant-Doberlet M, Žežlina I (2007) Vibrational communication of *Metcalfa pruinosa* (Say) (Hemiptera: Fulgoroidea: Flatidae). Ann Entomol Soc Am 100:73–82
Virant-Doberlet M, King RA, Polajnar J, Symondson WOC (2011) Molecular diagnostics reveal spiders that exploit prey vibrational signals used in sexual communication. Mol Ecol 20:2204–2216
Wells MLM, Henry CS (1998) Songs, reproductive isolation, and speciation in cryptic species of insects. In: Howard DJ, Berlocher SH (eds) Endless forms. Oxford University Press, New York, pp 217–233
West-Eberhard MJ (1983) Sexual selection, social competition, and speciation. Q Rev Biol 58:155–183
Wood TK (1993) Speciation of the *Enchenopa binotata* complex (Insecta: Homoptera: Membracidae). In: Lees DR, Edwards D (eds) Evolutionary patterns and processes. Academic Press, New York, pp 299–317

# Chapter 10
# Communication Through Plants in a Narrow Frequency Window

**Andrej Čokl, Maja Zorović, Alenka Žunič Kosi, Nataša Stritih and Meta Virant-Doberlet**

**Abstract** Different vibration-producing mechanisms determine frequency characteristics of insect communication signals transmitted through plants. Plant-dwelling stinkbugs (Heteroptera: Pentatomidae) communicate with the substrate-borne component of the vibratory emissions produced by tremulation of the whole body, tremulation of the abdomen or wings, and percussion. The main characteristics of signals produced by the mechanisms mentioned are their dominant component around 100 Hz, a different number of higher harmonics generally not exceeding 1,000 Hz and different degrees of frequency modulation. Higher-frequency components of the low-amplitude percussion signals are strongly attenuated during transmission through herbaceous plants. Stinkbug non-species-specific, low-frequency and narrow-band signal characteristics are tuned with plants as their main communication media. Herbaceous plants act as low-pass filters that optimally transmit signals of dominant frequency around 100 Hz and attenuate those above 600 Hz. Narrow-band low-frequency stinkbug vibratory signals are transmitted through stems with regularly repeated peaks of velocity minima and maxima caused by resonance. The stinkbug sensory system with underlying neuronal network effectively codes the inner frequency structure of signals produced by

A. Čokl (✉) · M. Zorović · A. Ž. Kosi · N. Stritih · M. Virant-Doberlet
Department of Entomology, National Institute of Biology,
Večna pot 111, 1000 Ljubljana, Slovenia
e-mail: andrej.cokl@nib.si

M. Zorović
e-mail: maja.zorovic@nib.si

A. Ž. Kosi
e-mail: alenka.zunic-kosi@nib.si

N. Stritih
e-mail: natasa.stritih@nib.si

M. Virant-Doberlet
e-mail: meta.virant-doberlet@nib.si

R. B. Cocroft et al. (eds.), *Studying Vibrational Communication*,
Animal Signals and Communication 3, DOI: 10.1007/978-3-662-43607-3_10,

different mechanisms. Communication through a narrow-band frequency range on the one hand efficiently increases signal-to-noise ratio but on the other hand does not allow evolution of signals with species-specific frequency character.

## 10.1 Introduction

Communication with signals transmitted as substrate vibrations is one of several channels through which insects exchange the information in the field. In most cases, signals of different modalities interact and contribute to complex multimodal communication systems. Substrate-borne sound communication has been confirmed in a variety of insect groups (Cocroft and Rodríguez 2005) that developed various strategies to optimize information exchange in different environment conditions (Drosopoulos and Claridge 2006). Looking for definitions and rules of any process or phenomena in biology is not easy and evolution hates it: for every tidy definition a biologist constructs, evolution will have provided some exception (Bradbury and Vehrencamp 1998). We started to think about the chapter with the idea of reviewing the role of frequency in insect vibrational communication; however, soon we realized that it would be an impossible task taking into account the lack of data for different insect groups, which prevents synthesis in a general and realistic frame. Finally, we decided "to stay at home" and write a chapter on vibratory communication of stinkbugs, emphasizing advantages and disadvantages of their choice to use a private and narrow frequency channel.

An important reason to choose Heteroptera, and specifically the Pentatomidae, is also their worldwide economically important pest status, which triggered numerous studies on species geographical distribution, general biology, ecology, feeding habits, seasonal variation and other topics (Panizzi et al. 2000). Such a wide interest made it easy for us to get colleagues and friends in different parts of the world and in very different surroundings to share curiosity-driven research on insect communication. We dedicate the chapter to all of them and especially to those who have shown us that vibrational communication, at least in stinkbugs, is crucially modified by species-specific chemical ecology. Interaction of these two communication channels with different signal modalities (Čokl and Millar 2009) represents a big challenge for future investigations in insect biology.

During long-term research, data on the biology of stinkbugs have been upgraded with studies on communication, including investigations of the anatomy and function of vibration-producing mechanisms, sensory organs and underlying neuronal networks, multimodal communication on plants and the role of plants in transmission, and diversity and information content of signals (for review see Čokl and Virant-Doberlet 2003; Virant-Doberlet and Čokl 2004; Virant-Doberlet et al. 2006; Čokl et al. 2006a; Čokl 2008). Most studies have been directly or indirectly connected with mating behaviour since effective exchange of information significantly contributes to mating success and species survival. The picture is far from complete, and the recent investigations of communication in predatory stinkbugs of the pentatomid subfamily Asopinae (Čokl et al. 2011; Laumann et al. 2013), together with the data on leafhoppers that exchange vibratory signals through the air (Eriksson et al. 2011), have opened new horizons in Pentatominae (Kavčič et al. 2013).

Although broad-band vibratory signalling was presumed advantageous for plant-dwelling insects (Michelsen et al. 1982), stinkbugs were found to communicate with narrow-band low-frequency signals using different mechanisms. In this chapter, we present frequency characteristics of stinkbug vibratory signals and their role during transmission through plants, the tuning of sensory organs and the underlying neuronal system with signal spectral properties, and finally the advantages and disadvantages of communication through a narrow band-pass frequency window in noisy surroundings.

## 10.2 Frequency of Signals Produced by Different Mechanisms

Insects produce vibratory communication signals by tremulation (vibration) of the whole body or of its parts, by stridulation or by percussion (Ewing 1989). All these mechanisms have been described in different groups of Heteroptera (Gogala 2006).

Tremulation of the abdomen is a widely used sound-producing mechanism of Heteroptera in families Pentatomidae, Acanthosomatidae, Plataspidae, Cydnidae, Scutelleridae and Alydidae (Gogala 2006). The morphological and functional basis of the abdomen tremulation mechanism was described in *Nezara viridula* (Maluf 1932; Kuštor 1989; Amon 1990). *N. viridula* signals are produced by dorso-ventral vibration of the abdomen without touching the substrate. Signal velocity ranged between 0.1 and 1 mm/s when measured on the thorax and varied between 6 dB above and 2 dB below these values when measured on the leaf immediately below the signaller (Čokl et al. 2007).

Communication signals produced by abdomen tremulation have been recorded and described in the calling, courtship and rivalry context of mating behaviour in more than 20 stinkbug species (Čokl 2008). Mates recognize these signals by their species-specific temporal characteristics (Žunič et al. 2011), which have been shown in *N. viridula* to vary extensively among geographically isolated populations (Čokl et al. 2000). The authors have shown that the songs of populations from Brazil, Slovenia, Florida and Italy have the same basic pattern but different pulse duration and repetition rates, which occur mainly in the male and female calling songs. The basic difference in the pattern of the first female song of the Australian *N. viridula* population (Ryan et al. 1996) needs reinvestigation.

Contrary to inter-individual and inter-population differences in temporal parameters, all the stinkbug signals produced by abdomen tremulation that have been analysed so far show similar frequency characteristics: a narrow dominant frequency peak below 200 Hz and a different number of higher harmonics

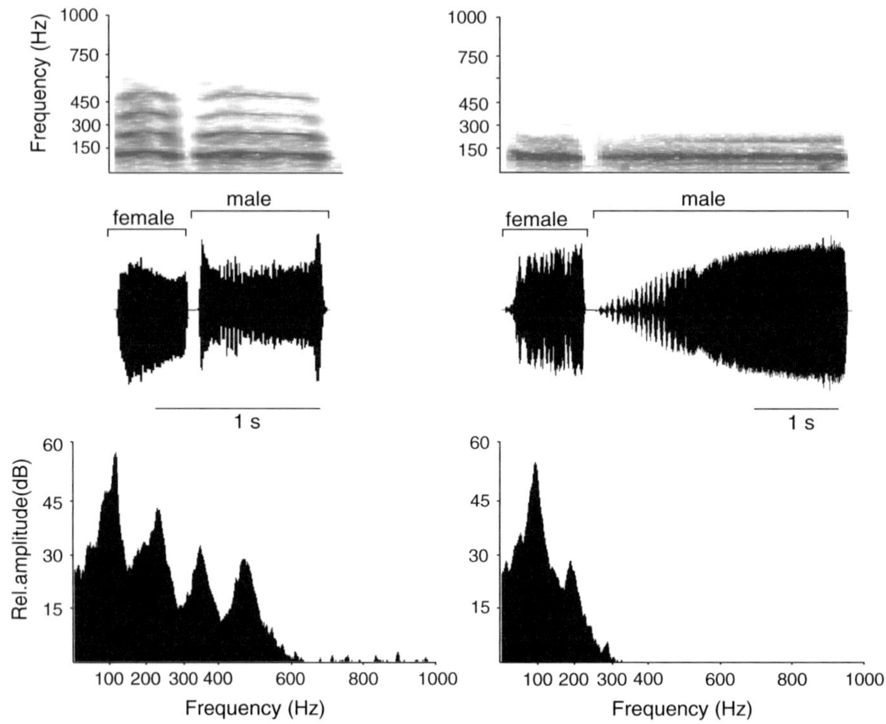


**Fig. 10.1** Female–male duet of *Holcostethus abbreviatus* (*left*) and *N. viridula* (*right*) with sonograms (*above*), oscillograms and frequency spectra (*below*)

generally not exceeding 600 Hz (Fig. 10.1) (Čokl 2008). The lowest mean dominant frequency for signals recorded on a non-resonant substrate was 62 Hz in the *Chlorochroa ligata* male courtship song (Bagwell et al. 2008) and the highest one was 175 Hz, determined in the *Eushistus heros* third male song (Blassioli-Moraes et al. 2005). Inter-individual and inter-population differences in the dominant frequency of the same song type have been documented, but they all take place in the frequency range of 20 Hz around the characteristic mean value. The mean dominant frequency of the species song of the stink bug, *N. viridula*, varied in geographically isolated populations from Brazil, Florida, Italy and Slovenia between 80 and 125 Hz (Čokl et al. 2000). Female *N. viridula*, for example, also actively change the dominant frequency within the same calling song sequence as a response to disturbance by producing a pure tone spectrally similar to that of its own signals in the species-characteristic range of variation between 70 and 130 Hz (Polajnar and Čokl 2008). The main difference in the frequency characteristics of pentatomine stink bug signals has been demonstrated in the number of higher harmonics and in the presence or absence of frequency modulation (FM). Narrower dominant frequency peaks are generally accompanied by a higher number of higher harmonic peaks, such as in the *Murgantia histrionica* first male song recorded on a non-resonant substrate. Here, seven spectral peaks in the range up to 1,000 Hz follow the narrow dominant frequency peak found around 110 Hz (Čokl et al. 2007).

FM is present in signals produced by abdomen tremulation in several pentatomine stink bug species such as *Piezodorus lituratus* (Gogala and Razpotnik 1974), *Chlorochroa uhleri*, *C. ligata* and *C. sayi* (Bagwell et al. 2008), *Eushistus conspersus* (McBrien and Millar 2003) and *Eushistus* (*Chinavia*) *heros* (Blassioli-Moraes et al. 2005). FM is expressed at the dominant and higher-frequency level and frequency sweeps may reach values up 70 Hz/s, as shown in the *E. heros* male rival song (Blassioli-Moraes et al. 2005). The extent of FM may differ also among the pulses of the same pulse train. The female *N. viridula* calling song pulse train is composed of the frequency-modulated pre-pulse followed by a long pulse of steady-frequency characteristics (Čokl et al. 2000). While the male responsiveness decreased significantly during stimulation with signals that were lacking the pre-pulse, variation of the pre-pulse velocity or frequency had no significant effect (Žunič et al. 2011). To our knowledge, the only study of the role of FM for song recognition in stink bugs has been conducted in *Piezodorus lituratus* (Gogala and Razpotnik 1974), for which the playback experiments demonstrated that signals of the same time characteristics that differed in the FM patterns triggered different male responses (Gogala 2006).

Significantly different frequency spectra, compared to signals produced by abdomen tremulation, are characteristic of vibratory signals produced by stridulation. This sound- and vibration-producing mechanism is widely present in several Heteroptera families such as Cydnidae, Thaumastellidae, Parastrahiidae, Tessaratomidae, Piesmatidae, Colobathristidae, Largidae, Coreidae, Scutelleridae, Alydidae and Aradidae (Gogala 2006) but has not been reported yet in Pentatomidae. Stridulatory signals are characterized by broad-band spectra with the dominant frequency above 0.5 kHz and distinct frequency peaks extending up to 10 kHz. Spectra of species- and sex-specific stridulatory signals of underground-living burrower bugs *Scaptocoris castanea* and *S. carvalhoi*, for example, have the dominant frequency peak between 0.5 and 1 kHz and distinct peaks up to 7 kHz; signal velocity ranges between 0.01 and 0.06 mm/s when recorded on the soil at less than 5 mm distance from the source (Čokl et al. 2006b). FM is present in stridulatory signals as well as in tremulation; Gogala (1984) reported in *Enoplops scapha* (Coreidae) a fluctuation of the dominant frequency between 1.5 and 11 kHz. Stridulatory signals are emitted as distress or disturbance calls or are combined in several Cydnidae species with signals produced in the behavioural contexts of calling and courtship by abdomen tremulation (Gogala 1984, 2006).

Less attention was paid among Heteroptera to species non-specific (or less species-specific) vibratory signals produced by tremulation of the whole body and by percussion, as described, for example, in meadow katydids (De Luca and Morris 1998) and Heelwalkers (Mantophasmatodea: Eberhard and Picker 2008), respectively. Signals produced by these mechanisms were described for the first time in the song repertoire of the predatory stinkbugs (Fig. 10.2) (Asopinae) (Žunič et al. 2008; Čokl et al. 2011; Laumann et al. 2013) and recently also

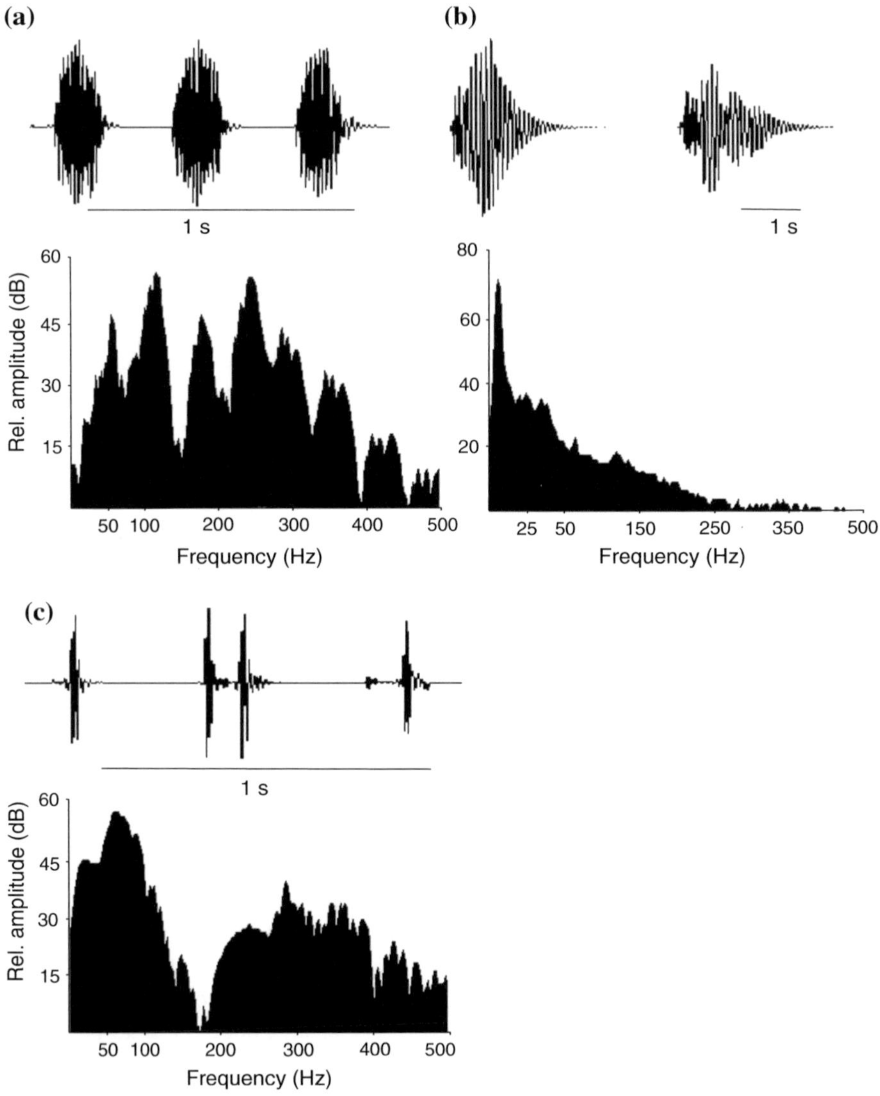


**Fig. 10.2** Oscillograms (*above*) and frequency spectrograms (*below*) of vibratory signals of *Podisus maculiventris* recorded on plumbago plant. **a** Male calling signals produced by abdomen vibration, **b** tremulatory signals and **c** percussion signals

confirmed in the subfamily Pentatominae (Kavčič et al. 2013). Spectra of the rhythmically repeated low-velocity (0.1–0.5 mm/s) percussion signals, recorded on a non-resonant substrate, are characterized by about 100-Hz dominant peak and a strong higher-frequency component between 1,500 and 3,000 Hz; the high-frequency component is reduced in signals recorded on the plant. In plants, tremulation of the whole body induces low-frequency signals with velocity ranging

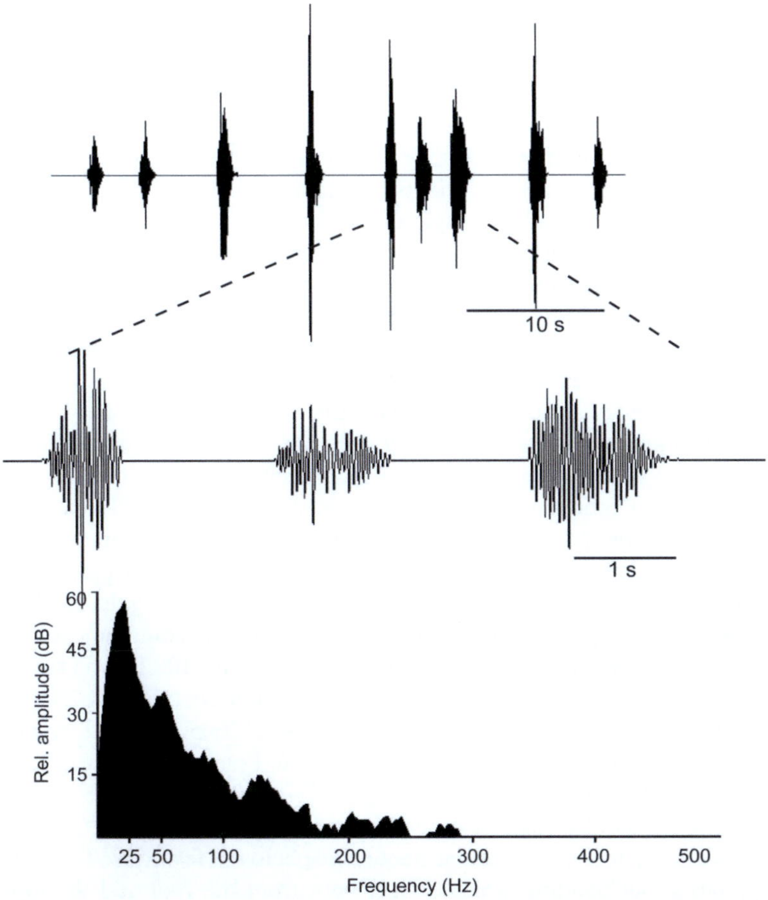


**Fig. 10.3** Oscillograms of tremulatory signals of *P. maculiventris* on two time scales with frequency spectra

between 1 and 6 mm/s and the dominant frequency below 200 Hz (Fig. 10.3). The pulse train of the male predatory stinkbug *Podisus maculiventris* is composed of a high-amplitude whole body tremulatory pulse and lower-amplitude species-specific pulses produced by abdomen tremulation (Žunič et al. 2008). Such a pulse train carries information on the presence of a potential mate at distances, where the velocity of the abdomen tremulation signals decreases below the receiver's sensory system threshold level.

We can conclude that despite different vibration-producing mechanisms and variations among populations and individuals, the plant-dwelling Heteroptera emit signals with the main energy produced in the narrow and low frequency range. Although the broad-band higher-frequency stridulatory signals represent an important part of the song repertoire in many Heteroptera, they are often combined with signals emitted in the frequency range below 200 Hz.

## 10.3 Plants and Insects

Plants represent the most natural surrounding for most insects, giving them shelter, food and substrate for reproduction (Schoonhoven et al. 1998). The southern green stink bug *N. viridula* and other highly polyphagous plant-dwelling Pentatomidae that feed on different mono- and dicots (Kiritani et al. 1965; Todd 1989; Panizzi 1997) serve as an instructive example of efficient insect–plant interactions during communication when mating.

Plants create a special communication environment for insects; dense vegetation prevents optical contact at larger distances and highly attenuates high-frequency signals. Furthermore, local air currents decrease the efficiency of chemical signals to guide searching mates to the signal source. Most insects are physically too small to emit low-frequency airborne sound efficiently (Markl 1983; Bennet-Clark 1998). For example, one-cm insects cannot emit airborne sound efficiently below 10 kHz and have to choose between flying out of their shelters into the free acoustic field where they attract predators, or to use the substrate (plant)-borne component of their emitted vibratory signals. Substrate-borne vibrational communication is indeed the prevalent form of signalling among different forms of mechanical communication in insects (Greenfield 2002; Cocroft and Rodríguez 2005).

Studies of insect–plant interactions during vibrational communication started 30 years ago with the pioneering work of Michelsen et al. (1982), who set the basis for following investigations of this phenomenon in insects (Cocroft 2000; Miklas et al. 2001; Miles et al. 2001; Čokl et al. 2004, 2005; McNett et al. 2006; Rodríguez et al. 2006; Casas et al. 2007; Čokl 2008; Polajnar et al. 2012) and other arthropods (Barth 2002). Michelsen et al. (1982) confirmed that insects communicate through plants by dispersive bending waves (Cremer et al. 1973) with little vibrational energy lost by friction at frequencies below a few kHz. They proposed the hypothesis that bending waves reflect both from the root and the top of the plant, creating standing wave conditions with the waves travelling up and down the plant's stem several times. They also demonstrated a non-monotonous decrease of the vibrational signal amplitude with the increasing distance from the source and about 20-dB higher attenuation of artificially induced vibrations at frequencies above 200 Hz, compared with those around 100 Hz when measured in a *Thesium bavarum* plant at the distance of 17 cm from the source. Finally, they postulated that it would not be a good strategy to use pure tone vibrational signals for communication through plants, particularly because it would be difficult to locate the emitter, and hypothesized that in the case of a broad-band spectrum, some portion of the frequency band will always get through to the listening insect.

Casas et al. (2007) quantified the wave propagation modes in rush stems (*Juncus effusus*) over the wide range of frequencies used by arthropods. The non-dimensionalized analysis has shown that mechanical signals propagate not only as dispersive but also as non-dispersive waves and that an arthropod can use non-dispersive waves either by producing high-frequency signals or by choosing large stems. The authors concluded that non-dispersive waves represent an advantage

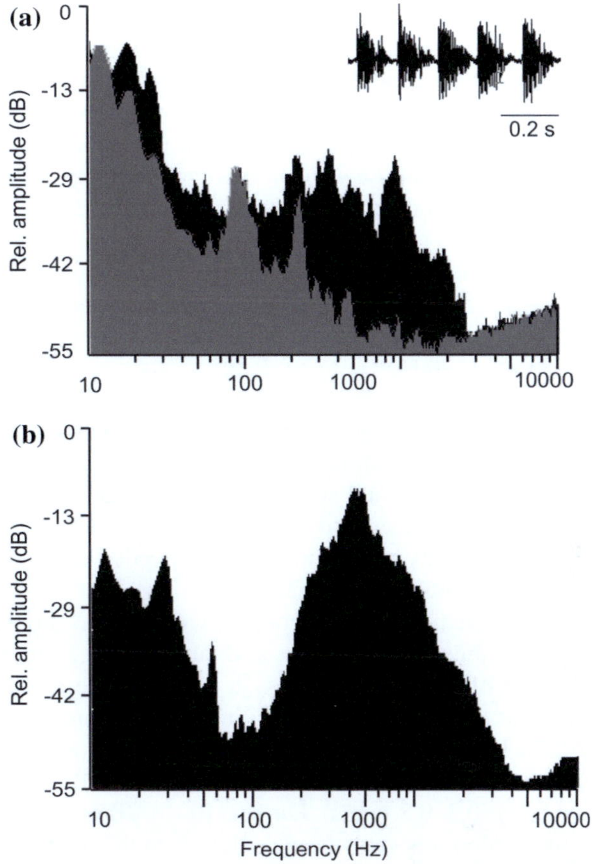


**Fig. 10.4** Spectral properties of vibratory signals emitted by burrower bugs on soybean roots and recorded from a plant: **a** 2 cm (*black*) and 28 cm (*grey*) above the soil, **b** spectrogram of male *Scaptocoris castanea* signals recorded from the soil 5 mm from the source (from Čokl et al. 2006b, with permission)

over dispersive ones in terms of signal integrity and give several examples of insects that emit high-frequency vibrational signals (Keuper and Kühne 1983; Gogala 1984; Devries 1991; White et al. 1993; Stölting et al. 2002; Elias et al. 2003; Evans et al. 2005).

Low-pass filter properties of plants were later confirmed in several insect–plant models, indicating that broad-band signals are not always convenient for longer-distance communication. The high-frequency components of the pedipalpal signals of the wandering spider *Cupiennius salei* (>1 kHz) are far more strongly attenuated by transmission through a plant than the frequency components below 200 Hz (over 30 dB difference) (Barth 1998). Burrower bug species *S. castanea* and *S. carvalhoi* (Heteroptera, Cydnidae) that live, feed and mate under soil on the soybean roots emit species- and sex-specific stridulatory signals that are transmitted from the roots onto the whole plant (Čokl et al. 2006b). Stridulatory signals recorded from the plant lose the higher-frequency component with increasing distance from the soil: the broad peak above 0.4 kHz disappears in recordings from the stem at 29 cm above the ground and only the low-frequency components remain in the stridulatory signal (Fig. 10.4). Similar low-pass filtering above 400 Hz has been demonstrated for percussion signals produced by predatory bug *P. maculiventris* on the plumbago plant (Žunič et al. 2008).

Low-pass filtering properties of plants cannot be generalized as a rule. McNett and Cocroft (2008) demonstrated that in *Cercis canadensis*, the host of one of the species inside the *Enchenopa binotata* treehopper species complex, the stems act as low-pass, and petiole as high-pass, filters. *E. bipunctata* Cercis were found to communicate primarily through the stems, and their signal frequency matched the frequency of least attenuation characteristic of the stem of the Cercis plant. Low-pass filtering has been shown for the stinkbug *M. histrionica* signals and its host plants when measured in the leaf but not when measured in the woody stem (Čokl et al. 2007).

Experiments in the laboratory with artificially induced pure tone vibrations (Čokl 1988) demonstrated predictable non-monotonous amplitude variation with distance when transmitted through the plant's stem. Such a relationship was not shown for broad-band signals of "small cicadas" species *Euscelis variegates* and *E. lineolatus* transmitted through bean (Michelsen et al. 1982). Regular and predictable velocity variation with distance during vibratory signal transmission has been also confirmed in natural conditions (Čokl et al. 2007). Female calling song of the southern green stink bug *N. viridula* is composed of sequences of regularly repeated pulse trains lasting several minutes, emitted spontaneously even in the absence of a male. Therefore, it was possible to measure velocity of signals at different distances from the calling animal along the plant (Fig. 10.5) (Čokl et al. 2007). The velocity of leaf recorded signals at a distance less than 5 mm from the source ranged between 6 dB above and 2 dB below the values recorded from the body (0.3–0.8 mm/s). Further, 10–14 dB of attenuation occurred on sedge with transmission of signals from the leaves to the stem. Attenuation of 0.06–0.1 dB/cm during transmission through the sedge stem is comparable to damping values measured for low-frequency vibrations transmitted through a banana plant (Barth 2002). Signal velocity did not decrease linearly but with regularly repeated velocity maxima and minima about every 20 cm along the 70 cm stem length. The amplitude of velocity maxima decreased with distance from 7.0 to 0.3 dB above the value at the reference (top of the stem). The maximum difference between neighbouring velocity maxima and minima peaks was 19.4 dB.

The distance between velocity peak maxima (and minima) decreases with increasing signal frequency. Consequently, the peak-to-peak distance of the dominant frequency is twice that of its first harmonic close to the source, and at a distance from the source, where the dominant frequency velocity peak is at its minimum, the first harmonic velocity peak may dominate in the spectrum, being at its maximum. Such a situation has been shown in the *M. histrionica* signal spectra with the narrow dominant frequency peak and several higher harmonic ones (Čokl et al. 2007). Changes of FM spectra during transmission need to be investigated on plants in natural conditions at different distances from the source. Because of decreasing distance between velocity peaks with increasing vibration frequency, and because of different propagation velocities of different frequencies, we can

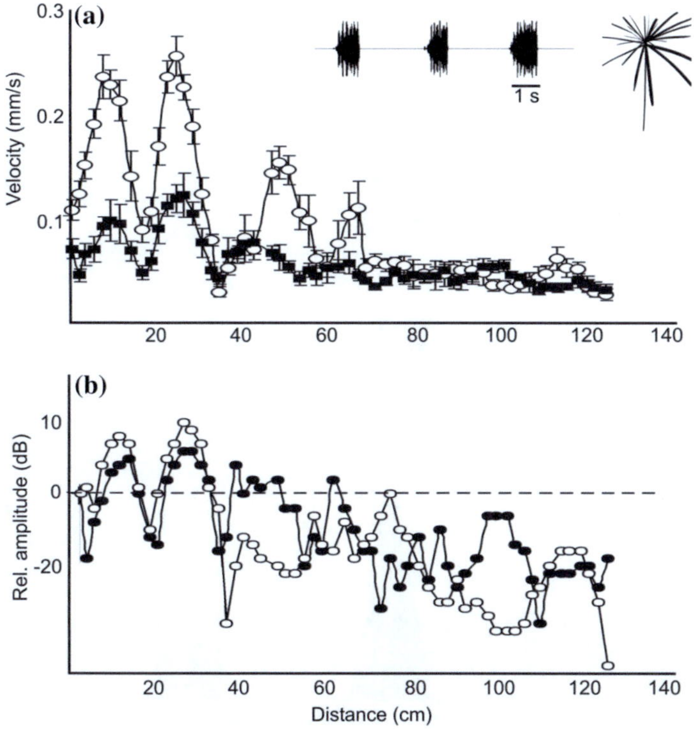


**Fig. 10.5** **a** Velocity variation with distance of *N. viridula* female calling songs transmitted through a sedge stem shown in absolute (mean ± SD) values and **b** dominant frequency spectral peak amplitude difference (re. value at the reference point) variation with distance. Different *symbols* mark signals emitted by different females (from Čokl et al. 2007, with permission)

expect different amplitude modulation patterns of FM signals. Such significant differences have been observed in 1-s-long FM vibratory signals (40–160 Hz) induced artificially in cyperus or bean and recorded at distances of 20–25 and 50 cm from the source (Fig. 10.6) (A. Čokl, unpublished data).

In a recent study, Polajnar et al. (2012) described the cyclic variation of signal amplitude with the distance as observed in natural conditions (Čokl et al. 2007). The wavelength of this variation along the model plant's stem proved to be frequency-dependent, which indicated the role of reflections from end points, already predicted by Michelsen et al. (1982). The authors (Polajnar et al. 2012) confirmed that resonance causes the observed variation by comparing measured values with the physical model of bending vibration in a fixed-free elastic rod with given material parameters. Eigenfrequencies of comparable systems, i.e. rod-like plant structures, are therefore predicted to influence the choice of signalling frequency on a particular plant, for example, least force is required to drive the system at a given velocity, and therefore least mechanical power, when the input matches resonance frequency. Frequency dependence of the amplitude variation is

**Fig. 10.6** Oscillograms and frequency spectra of the frequency-modulated (*FM*) signal (input) recorded on cyperus and bean plants at different distances from the stimulation point. **a** 20 cm on cyperus and 25 cm on bean, **b** 50 cm on cyperus and bean

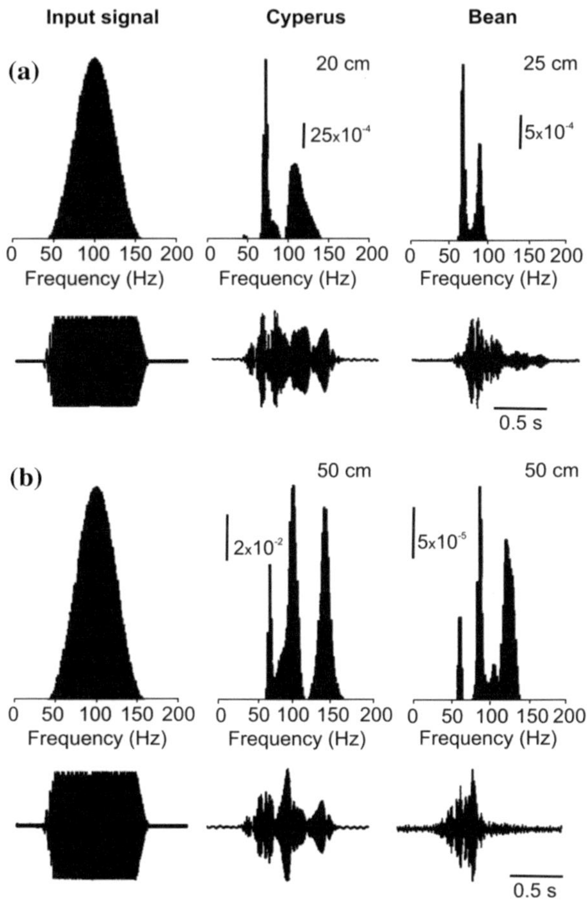


continuous, but vibrating with a frequency not corresponding to a resonance peak is by definition less efficient. However, little is known about the range of variation of resonance properties as experienced by the signallers, which depends on geometry and elasticity of individual stems. Systematic data need to be obtained before making conclusions about the actual importance of resonance in vibrational communication.

The hypothesis on standing wave conditions is based on reflected waves travelling up and down the plant several times. Such conditions cannot be expected in smaller insects, like treehoppers and their generally woody host plants, in which little if any reflection has been recorded (Cocroft and McNett 2006). We can conclude that communication signals produced by stinkbugs are well suited for transmission through their host plants that act as low-pass filters and provide predictable non-monotonous velocity variation with distance from the emitter.

## 10.4 Neural Coding of Frequency

The success of vibratory communication of small plant-dwelling insects through plants is on the one hand due to tuning of plant resonant peaks with spectral frequencies of songs and on the other hand due to tuning of these peaks to best sensitivity of individual receptor organs/cells (Čokl et al. 2005). The interneurons at the central processing level reflect the frequency tuning classes of receptor neurons. As in other areas of vibratory communication research, comparative investigations in Heteroptera are lacking; therefore, we will mainly summarize the results of the study on the model stinkbug species *N. viridula*.

### *10.4.1 Morphology and Function of Sensory Organs and Vibratory Receptor Neurons*

Debasieux (1938) described the leg vibratory sense organs in *Pyrrhocoris apterus* and Michel et al. (1983) those in the legs of the southern green stink bug *N. viridula*. Devetak et al. (1978) showed in burrower bugs (Cydnidae) that detection of higher-frequency signals takes place in the tibia and Čokl (1983) was the first to investigate and describe the functional properties of the leg vibratory organs at the single neuron level in *N. viridula*. Comparative morphological and physiological investigations in other groups of land bugs are needed.

In the last 5 years, a significant step forward has been achieved in investigations of the receptor and higher-order ventral cord neurons in *N. viridula*, which serves as the model for other stink bugs. The axons of the femoral chordotonal organ run together with those of the subgenual organ sensory cells in a sensory nerve and join with the main leg nerve in the posterior part of the femur close to the femur/trochanter joint. Data on the central projections and final arborizations of the receptor neurons sensitive to vibration of the legs (Fig. 10.7a) are summarized in Čokl et al. (2006a). Each receptor neuron terminates at the ipsilateral site of the appropriate part of the ventral cord [prothoracic ganglion and the meso- or metathoracic part of the central ganglion (CG)]. Receptor neurons most sensitive to frequencies below 100 Hz (see below) terminate with the main branch anteriorly, close to the ganglion midline, and have side branches diverging anteriorly from the main one immediately after entering the ganglion. Receptor neurons most sensitive to middle (about 200 Hz) or higher (500–1,000 Hz) frequencies (see below) enter the ganglion in the posterior third of the leg nerve and finally terminate in a dense arborization in the median half of the ipsilateral side of the ganglion.

Singlecell recordings from each leg nerve at its entrance to the ganglion revealed two groups of receptor neurons differing in their frequency/intensity sensitivity (Čokl 1983). The low-frequency neurons respond best between 40 and 70 Hz with peak velocity sensitivity between 0.03 and 0.06 mm/s (Fig. 10.7b).

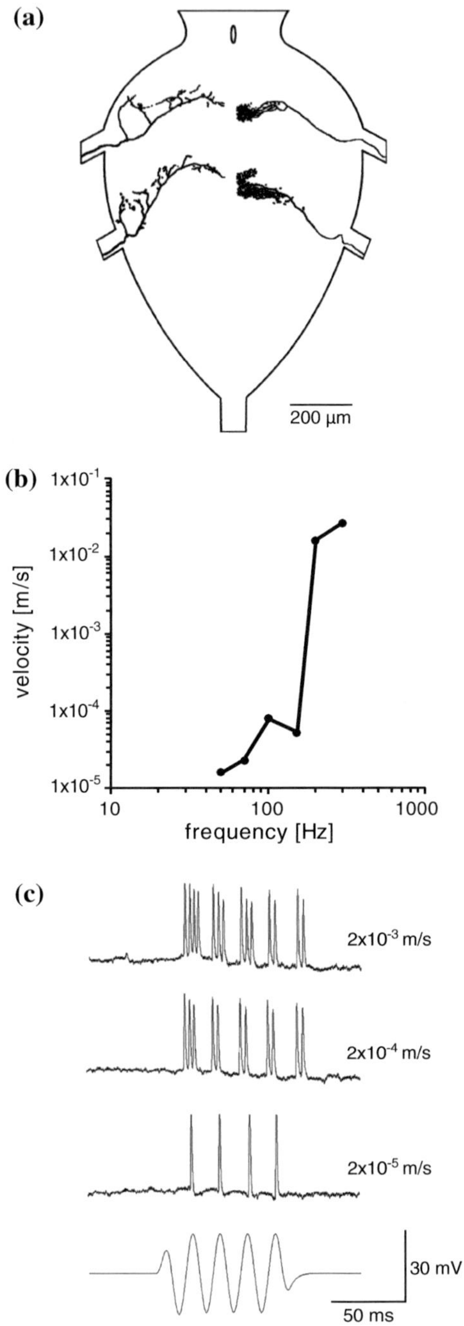


**Fig. 10.7** **a** Dorsal view of the central ganglion of *N. viridula* showing central projections of low (*left*)- and middle (*right*)-frequency receptor neurons. **b** Threshold curve of the low-frequency receptors and **c** their responses to 50 Hz stimulus of different intensities (partly redrawn from Čokl et al. 2006a, with permission)

The low-frequency neurons' threshold curves follow the line of equal displacement values (around $10^{-7}$ m) in the range above best frequency sensitivity. Due to technical limits of the experimental set-up, responses of neurons have not been tested with stimuli of frequencies below 30 Hz; however, strong responses at this frequency indicate that threshold may lie even lower. In the frequency range below 120 Hz, the low-frequency neurons respond in a phase-locked manner (Fig. 10.7c) with action potentials coupled to the upward leg movement (Type 1), downward leg movement (Type 2) or the peak leg displacement in both directions (Type 3). Responses of the low-frequency neurons originate in excitation of the joint chordotonal organs and campaniform sensilla and show many similarities with those described in locusts and bush crickets (Kühne 1982).

Although there are several types of low-frequency receptor neurons, only two neurons were identified in each leg nerve with best sensitivity above 100 Hz. The middle-frequency receptor neuron is tuned to frequencies around 200 Hz with threshold sensitivity of 0.01–0.02 mm/s velocity, and the higher-frequency receptor neuron to frequencies between 500 and 1,000 Hz with threshold velocity sensitivity of 0.002–0.003 mm/s (Čokl 1983). Threshold curves of the latter two receptor neurons follow the line of equal acceleration values (around $10^{-2}$ m/s$^2$) in the range below the best frequency and the line of equal displacement values ($2–6 \times 10^{-8}$ m for the middle-frequency receptor neuron and $5–7 \times 10^{-10}$ m for the higher-frequency receptor one) in the range above the best frequency. Neurons code stimulus duration in a tonic way, with responses of the middle-frequency receptor neuron outlasting the stimulus duration at 200 Hz. These two receptor neurons originate in the subgenual organ.

Regular tapping of the substrate by antennae, displayed during mating behaviour in *N. viridula* (Ota and Čokl 1991), indicate that they are not only the receptor site for chemical, touch and temperature stimuli, but can also act as an additional vibratory sensory organ. Jeram and Čokl (1996) and Jeram and Pabst (1996) described 12 campaniform sensilla on the antennae and within them the Johnston's organ with the centrally positioned chordotonal organ. Johnston's organ consists of 45 scolopidia distributed around the periphery of the distal part of the third antennal segment, while the central organ consists of seven mononematic scolopidia, which comprise one or two sensory cells. Axons of the 17 Johnston's organ sensory cells are joined in one of the antennal nerves with those of four scolopidia of the central chordotonal organ, while the other three axons of the central organ run together with 28 axons of the Johnston's organ in the other antennal nerve. Best sensitivity to vibrations of the proximal flagellar segment lies around 50 Hz at threshold velocity value of 2 mm/s (Jeram 1993, 1996; Jeram and Čokl 1996). Back-fill staining of the antennal nerve has shown that mechanosensory fibres terminate at the ipsilateral side of the subaesophageal and prothoracic ganglion, with some of them ending in the abdominal part of the CG.

Airborne sound of frequencies between 100 and 600 Hz and intensities between 50- and 60-dB SPL inhibits vibrational communication in *N. viridula* (Čokl 1984). Both subgenual receptor cells are most sensitive to sound around 250 Hz with threshold around 50-dB SPL. The probable site of excitation is cuticle that is induced to vibrate by airborne stimuli. Although the anatomy of Heteroptera leg inner structures, together with relevant mechanosensory organs, is significantly different from that of cockroaches, other inputs to the subgenual organ as described by Shaw (1994) cannot be excluded. The intensity of the airborne component of signals produced by abdomen vibration and recorded 1 cm above the calling bug was below the threshold of the subgenual organ receptor cells. Nevertheless, other mechanoreceptors present on and in different parts of the insect body may detect vibrations induced by airborne sound at the relevant distance between the sender and emitter. Hair sensilla are involved in detection of low-frequency body tremulations in *Cupiennius salei* Keys spiders (Barth 2002). The wing beats of a fly introduces a periodic pulsation and produces oscillations of the air at frequencies around 100 Hz. Air movements are also generated by insects crawling on a plant and such spatially confined signals are characterized by low frequencies (<20 Hz) and low particle velocities (<2 cm/s) (Gnatzy and Kämper 1990). A spider's trichobotrium response follows the time course of the air movements caused by a fly flying 5 cm from the spider. The response is proportional to the logarithm of the mean flow velocity and the degree of turbulence in a broad range from 1 to 1,000 mm/s at frequencies below 150 Hz. The greatest distance at which a flying fly elicited a response in a trichobotrium of a spider standing on a plate (leaf) was 55 cm. Because the airborne stimulus is directional and the higher-frequency components, typical of the prey signal, are eliminated by distance, the range of detecting airborne signals of a flying fly in natural conditions does not exceed 30 cm. Such detailed experiments have not been conducted in *N. viridula* or in any other stink bug species. However, the presence of both trichobotria at the abdominal edge and wing "buzzes"—recorded in many pentatomide species on plants—indicates that at relevant distances the airborne signals produced by approaching or "buzzing" mates (prey or predator) might also be registered by the very sensitive hair sensilla-like trichobotria. This would add substantial information from a vibrational component on the position of the emitter of communication signals. This is a promising direction for future research of multimodal communication in Heteroptera.

The maximum sensitivity of the leg vibro-receptor system of *N. viridula* in the frequency range under 70 Hz lies 35 dB below the level of stink bug vibratory emissions measured at the source on a typical substrate (0.1–1 mm/s), while the maximum sensitivity in the frequency range between 70 and 200 Hz lies about 40 dB below the same level. The former is attributed to joint chordotonal organs and campaniform sensilla and the latter to the very sensitive subgenual organ. Comparative studies of vibrational receptors in different groups of Heteroptera are needed, and investigations have to be extended to include lower frequencies and infrasound, together with studies of vibratory communication through the air, as shown recently by Eriksson et al. (2011).

### *10.4.2 Morphology and Function of Vibratory Interneurons*

Higher-order ventral cord neurons sensitive to vibration of the legs within Heteroptera have been studied only in *N. viridula*. First, four different types were identified electrophysiologically (Čokl and Amon 1980) and later in detailed studies Zorović et al. (2004, 2008) described the morphology and function of ten different types of vibration-sensitive interneurons in thoracic ganglia of the species. Based on their gross morphology, they were divided into four categories (Fig. 10.8). Five different types form the ascending L-shaped interneuron CG-AC group with the cell body located in the metathoracic neuromere of the CG and with a contralaterally ascending axon. Another group of ascending interneurons, CG-AB, described as dorsal unpaired median neurons (DUM), has the cell body in the mesothoracic neuromere of the CG and two bilaterally ascending axons projecting through the prothoracic ganglion. The authors also described three types of local interneurons (CG-L) of the central ganglion, with gross morphology similar to that of the cricket omega cells (Huber et al. 1989). Only one type of vibration-sensitive interneuron with a descending axon has been identified (PTG-DC). Its cell body lies in the prothoracic ganglion and the contralaterally descending neuron branches at the midline of the meso- and meta-thoracic neuromeres. Based on their frequency tuning, the neurons were divided into low- and middle-frequency units. The low-frequency neurons show best sensitivity around 50 Hz with acceleration threshold at $10^{-1}$ m/s$^2$ (0.1–1 mm/s velocity) and the middle-frequency neurons around 200 Hz with acceleration threshold at $5 \times 10^{-3}$ m/s$^2$ (about 0.005 mm/s velocity). Broad-band neurons and neurons tuned to higher frequencies have not been yet identified.

Recently, Zorović (2011) investigated the temporal selectivity for *N. viridula* vibratory signals in four types of ascending vibration-sensitive interneurons. The mean spike repetition rate of responses revealed preference for pulse durations below 600 ms and no selectivity for pulse interval duration. On the other hand, the maximum spike rate of the responses showed selectivity either for short pulse duration and long-interval duration or no selectivity at all. In all neurons tested, the author found weakest responses when long pulses were combined with short intervals. None of the response arrays showed a receiver preference for either constant period or duty cycle. The neurons' vibratory song temporal pattern selectivity matched vibratory signals of the male *N. viridula*. The mechanism thus enables temporal filtering for the conspecific vibratory signals at the early level of signal processing. Distinct and regular oscillations of the membrane potential following the responses to vibratory stimuli, found in one of the neurons, matched the temporal structure of the male calling song, indicating a potential resonance-based mechanism for signal recognition.

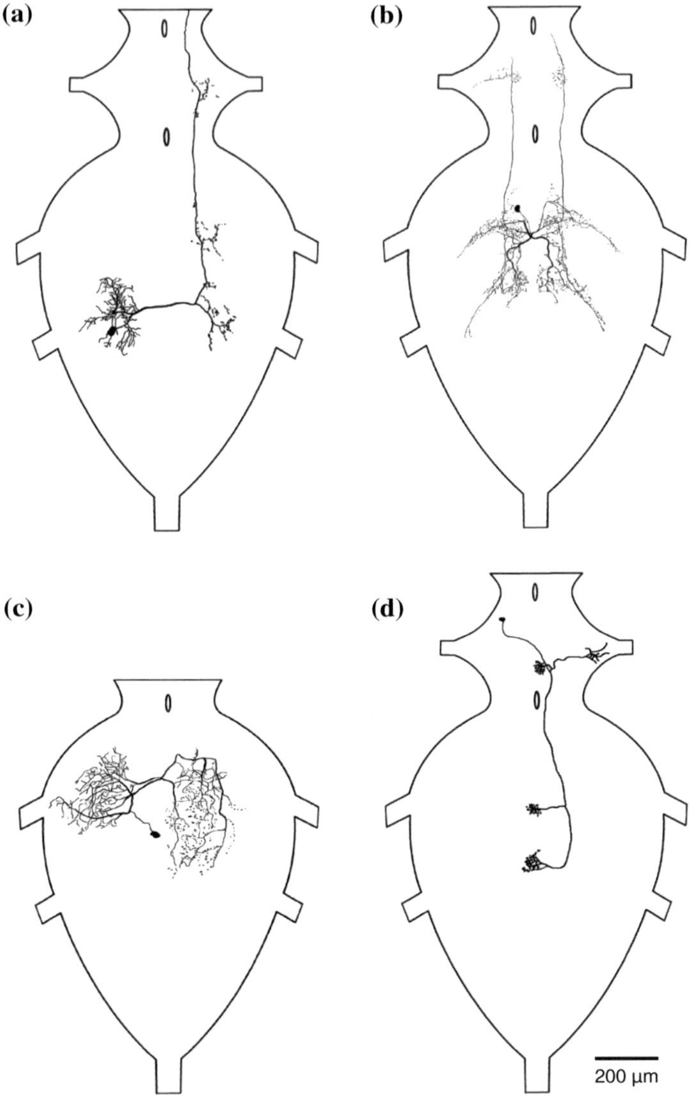


**Fig. 10.8** Whole-mount drawings of four morphologically distinct groups of vibratory interneurons in thoracic ganglia of *N. viridula*. **a** CG-AC, **b** CG-AB, **c** CG-L, **d** PTG-DC (partly redrawn from Zorović et al. 2008, with permission)

## 10.5 Signal Frequency and Environmental Noise

In the field, communication takes place in noisy surroundings, which decreases the signal-to-noise ratio for any modality involved. When decision demands complex processing of multimodal information, as demonstrated for example in stinkbug close range courtship (Borges et al. 1987), chemical, optical and mechanical noise decreases the efficiency of information exchange. On plants, substrate-borne signal communication prevails in the calling phase of stinkbug mating behaviour, and environmental noise significantly decreases effectiveness of vibrational directionality, preventing meeting of mates on a plant. The use of narrow-band signals tuned both to the mechanical properties of the medium and the properties of the receiver might be an advantage due to filtering of noise of frequencies below and above the frequency band used during communication.

The main abiotic noise is produced by wind and rain drops (Cocroft and Rodríguez 2005). Most energy of noise produced by wind is present at low frequency (below 30 Hz) (Barth et al. 1988; McVean and Field 1996; Barth 1997; Casas et al. 1998). In the higher frequency range abiotic noise is produced by falling water drops that induce vibrations of a leaf with a high-frequency irregular and low-frequency regular phase (Barth et al. 1988; Casas et al. 1998). Vibrations induced in a leaf by falling water drops are similar to signals produced by tremulation of the body (Žunič et al. 2008; Čokl et al. 2009, 2011); the initial high-amplitude irregular phase with frequencies extending up to several kHz is followed by a lower-amplitude regular phase of frequencies below 20 Hz (Fig. 10.3). Velocity of body tremulation signals measured on bean and plumbago close to the source ranged between 2 and 15 mm/s, and decreased by 1.5–0.3 dB/cm in bean and by 0.9–0.3 dB/cm in plumbago (Čokl et al. 2009).

Biotic noise and its effect on vibrational communication have been described in detail (see Chap. 7, this volume). Spectra of vibratory noise that is produced in a living environment range from a few to several thousand Hz. The advantage of band-pass communicating insects is the filtering of noise below and above the effective frequency range, and one of the main disadvantages is that there are not many possibilities for evolution of species specificity based on frequency parameters, as this relies principally on the signal temporal characteristics (Fig. 10.9) (Žunič et al. 2011).

During evolution, insects developed numerous strategies to increase signal-to-noise ratio in the field. Masking of the signals may be avoided by communication in the wind-free period of the day or in silent windows that appear unpredictably (Greenfield 1994; McNett et al. 2010). To our knowledge, there is no evidence that insects would increase the amplitude of vibratory signals above the noise level as shown for example in vertebrates (Brumm and Slabbekorn 2005). Nevertheless, the combination of body tremulation produced with high amplitude, and species-specific pulse produced by abdomen tremulation with low amplitude, in the male song pulse train of *P. maculiventris* (Žunič et al. 2008) might indicate that long distance signalling is combined with species specificity to optimize communication in different conditions.

Short-term frequency changes within the narrow range in the presence of noise have been demonstrated in frogs (Howard and Young 1998), birds (Manabe 1997; Slabbekoorn and Peet 2003) and recently also in the southern green stinkbug *N. viridula* (Polajnar and Čokl 2008). The main energy of Pentatominae species-specific communication signals produced by abdomen tremulation lies outside the

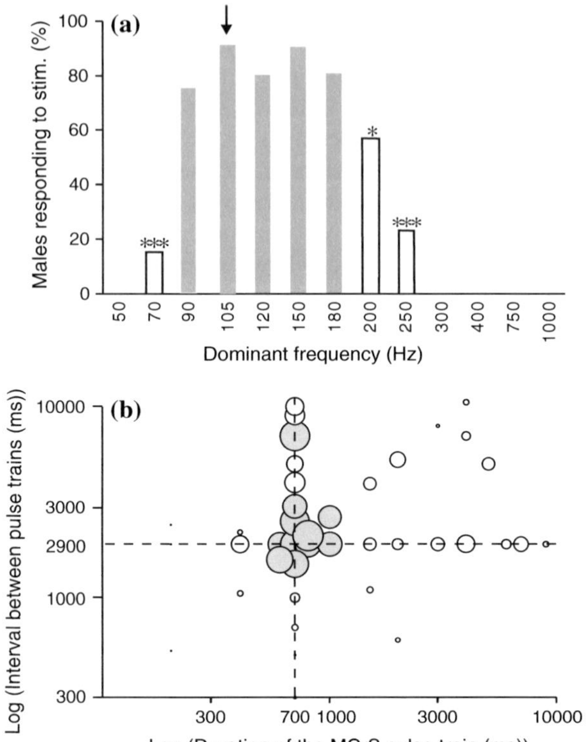


**Fig. 10.9** Preferences of male *N. viridula* for different frequency and time characteristics of synthesized conspecific female calling song reproduced on a non-resonant substrate. **a** Vibratory responses of males to stimuli of constant time characteristics and different dominant frequency. The magnitude of responses to stimuli of frequencies below 90 and above 180 Hz is significantly lower (*white bars*) than the one to the reference stimulus. **b** Male responsiveness to synthesized female calling song of constant frequency and different time characteristics. The *circles* show the percentage of males responding to specific pulse train duration and intervals combination. *Open circles* show significantly lower values compared to the reference. The reference stimulus is characterized by the frequency and time values of the natural female calling song (edited from Žunič et al. 2011, with permission)

frequency range of abiotic noise. In the presence of pure tone disturbance vibration around 100 Hz, the female *N. viridula* contrasts its calling song signals by shifting the dominant frequency away from the frequency of the disturbance. All the dominant frequency variations above or below the disturbance frequency occur within the species-characteristic range between 70 and 130 Hz. Žunič et al. (2011) demonstrated that male *N. viridula* respond to an artificial female calling song of constant natural temporal characteristics in the effective frequency range between 90 and 180 Hz with best responses at stimulus frequency of 105 Hz. Compared with the effective range of temporal parameters, the effective range of frequency was broader, and no correlation was observed between the emission of the male song and manipulation of the stimulus values.

We can conclude that communication through plants with narrow-band signals of main energy around 100 Hz decreases the effect of low- and high-frequency noise produced by abiotic and biotic factors. Similarities between temporal and frequency characteristics of signals produced by falling water drops and whole body tremulation appear to represent only a minor constraint in plant-borne communication, due to higher or lower repetition rate regularity of the latter. Because all known vibratory communication signals in Heteroptera run in the narrow frequency range, we may conclude that recognition of signals relies on different species-specific temporal parameters.

## References

Amon T (1990) Electrical brain stimulation elicits singing in the bug *Nezara viridula*. Naturwissenschaften 77:291–292

Bagwell GJ, Čokl A, Millar JG (2008) Characterization and comparison of the substrate-borne vibrational signals of *Chlorochroa uhleri*, *C. ligata* and *C. sayi*. Ann Entomol Soc Am 101:235–246

Barth FG (1997) Vibrational communication in spiders: adaptation and compromise at many levels. In: Lehrer M (ed) Orientation and communication in arthropods. Birkhauser Verlag, Basel, pp 247–272

Barth FG (1998) The vibrational sense in spiders. In: Popper RR, Fay AN (eds) Handbook of auditory research. Comparative hearing in insects. Springer, New York, pp 228–278

Barth FG (2002) A spider's world: senses and behavior. Springer, Berlin

Barth FG, Bleckmann H, Bohnenberger J, Seyfarth EA (1988) Spiders of the genus *Cupiennius* Simon 1891 (Aranea, Ctenidae), II: on the vibratory environment of a wandering spider. Oecologia 77:194–201

Bennet-Clark HC (1998) Size and scale effects as constraints in insect sound communication. Philos Trans R Soc B 353:407–419

Blassioli-Moraes MC, Laumann RA, Čokl A, Borges M (2005) Vibratory signals of four neotropical stink bug species. Physiol Entomol 30:175–188

Borges M, Jepson PC, Howse PE (1987) Long-range mate location and close-range courtship behaviour of the green stink bug, *Nezara viridula* and its mediation by sex pheromone. Entomol Exp Appl 44:205–212

Bradbury JW, Vehrencamp SL (1998) Principles of animal communication. Sinauer Associates Inc., Sunderland

Brumm H, Slabbekorn H (2005) Acoustic communication in noise. Adv Stud Behav 35:151–209

Casas J, Bacher S, Tautz J, Meyhöfer R, Pierre D (1998) Leaf vibrations and air movements in a leafminer–parasitoid system. Biol Control 11:147–153

Casas J, Magal C, Sueur J (2007) Dispersive and non-dispersive waves through plants: implications for arthropod vibratory communication. Proc R Soc B Biol 274:1087–1092

Cocroft RB (2000) Directionality in the mechanical response to substrate vibration in a treehopper (Hemiptera: Membracidae: *Umbonia crassicornis*). J Comp Physiol A 186:695–705

Cocroft RB, McNett GD (2006) Vibrational communication in treehoppers (Hemiptera: Membracidae). In: Drosopoulos S, Claridge MF (eds) Insect sounds and communication: physiology, behaviour, ecology and evolution. CRC Taylor & Francis, Boca Raton, pp 2305–2317

Cocroft RG, Rodríguez RL (2005) The behavioral ecology of insect vibrational communication. Bioscience 55(4):323–334

Čokl A (1983) Functional properties of vibroreceptors in the legs of *Nezara viridula* (L.) (Heteroptera, Pentatomidae). J Comp Physiol 150:261–269

Čokl A (1984) Problems in sound communication in land bug species *Nezara viridula* L. (Heteroptera, Pentatomidae). In: Kalmring K, Elsner N (eds) Acoustic and vibration communication in insects. Paul Parey, Hamburg, pp 163–168

Čokl A (1988) Vibratory signal transmission in plants as measured by laser vibrometry. Period Biol 90:193–196

Čokl A (2008) Stink bug interaction with host plants during communication. J Insect Physiol 54:1113–1124

Čokl A, Amon T (1980) Vibratory interneurons in the central nervous system of *Nezara viridula* L. (Pentatomidae, Heteroptera). J Comp Physiol 139:87–95

Čokl A, Millar JG (2009) Manipulation of insect signalling for monitoring and control of pest insects. In: Ishaaya I, Horowitz AR (eds) Biorational control of arthropod pests: application and resistance management. Springer, Dordrecht, pp 279–316

Čokl A, Virant-Doberlet M (2003) Communication with substrate-borne signals in small plant-dwelling insects. Annu Rev Entomol 48:29–50

Čokl A, Virant-Doberlet M, Stritih N (2000) The structure and function of songs emitted by the southern green stink bugs from Brazil, Florida, Italy and Slovenia. Physiol Entomol 25:196–205

Čokl A, Prešern J, Virant-Doberlet M, Bagwell GJ, Millar JG (2004) Vibratory signals of the harlequin bug and their transmission through plants. Physiol Entomol 29:372–380

Čokl A, Zorović M, Žunič A, Virant-Doberlet M (2005) Tuning of host plants with vibratory songs of *Nezara viridula* L. Heteroptera: Pentatomidae). J Exp Biol 208:1481–1488

Čokl A, Virant-Doberlet M, Zorović M (2006a) Sense organs Involved in the vibratory communication of bugs. In: Drosopoulos S, Claridge MF (eds) Insect sounds and communication: physiology, behaviour, ecology and evolution. CRC Taylor & Francis, Boca Raton, pp 71–80

Čokl A, Nardi C, Bento JMS, Hirose E, Panizzi AR (2006b) Transmission of stridulatory signals of the burrower bugs, *Scaptocoris castanea* and *Scaptocoris carvalhoi* (Heteroptera: Cydnidae) through the soil and soybean. Physiol Entomol 31:371–381

Čokl A, Zorović M, Millar JG (2007) Vibrational communication along plants by the stink bugs *Nezara viridula* and *Murgantia histrionica*. Behav Process 75:40–54

Čokl A, Žunič A, Millar JG (2009) Transmission of *Podisus maculiventris* tremulatory signals through plants. Centr Eur J Biol 4:585–594

Čokl A, Žunič A, Virant-Doberlet M (2011) Predatory bug *Picromerus bidens* communicates at different frequency levels. Centr Eur J Biol 6:431–439

Cremer I, Heckl M, Ungar EE (1973) Structure-borne sound, structural vibrations and sound radiation at audio frequencies. Springer, Berlin

De Luca PA, Morris GK (1998) Courtship communication in meadow katydids: female preference for large male vibrations. Behaviour 135:777–794

Debasieux P (1938) Organes scolopidiaux des pattes d'insectes. Cellule 47:77–202

Devetak D, Gogala M, Čokl A (1978) A contribution to the physiology of vibration receptors in the bugs of the family Cydnidae. Biol Vestnik 26:131–139

Devries PJ (1991) Call production by myrmecophilous riodinid and lycaenid butterfly caterpillars (Lepidoptera): morphological, acoustical, functional, and evolutionary patterns. Am Mus Novit 3025:1–23

Drosopoulos S, Claridge MF (2006) Insect sounds and communication: physiology, behaviour, ecology and evolution. CRC Taylor & Francis, Boca Raton, pp 1–532

Eberhard M, Picker MD (2008) Vibrational communication in two sympatric species of Mantophasmatodea (Heelwalkers). J Insect Behav 21:240–257

Elias DO, Mason AC, Maddison WP, Hoy RR (2003) Seismic signals in a courting male jumping spider (Araneae: Salticidae). J Exp Biol 206:4029–4039

Eriksson A, Anfora G, Lucchi A, Virant-Doberlet M, Mazzoni V (2011) Inter-plant vibrational communication in a leafhopper insect. PLoS ONE 6(5):e19692. doi:10.1371/journal.pone.0019692

Evans TA, Ali JCS, Toledano E, McDowall L, Rakotonarivo S, Lenz M (2005) Termites assess wood size by using vibration signals. Proc Natl Acad Sci USA 102:3732–3737

Ewing AW (1989) Mechanisms of sound production. In: Ewing AW (ed) Arthropod bioacoustics. Edinburgh University Press, Edinburgh, pp 17–57

Gnatzy W, Kämper G (1990) Digger wasp against crickets: II. An airborne signal produced by a running predator. J Comp Pysiol 167:551–556

Gogala M (1984) Vibration producing structures and songs of terrestrial Heteroptera as systematic character. Biol Vestnik 32:19–36

Gogala M (2006) Vibratory signals produced by Heteroptera–Pentatomorpha and Cimicomorpha. In: Drosopoulos S, Claridge MF (eds) Insect sounds and communication: physiology, behaviour, ecology and evolution. CRC Taylor & Francis, Boca Raton, pp 275–296

Gogala M, Razpotnik R (1974) A method of oscillographic sonagraphy for bio-acoustic research. Biol Vestnik 22:209–216

Greenfield MD (1994) Cooperation and conflict in the evolution of signal interactions. Annu Rev Ecol Syst 25:97–126

Greenfield MD (2002) Signalers and receivers: mechanisms and evolution of arthropod communication. Oxford University Press, Oxford

Howard RD, Young RJ (1998) Individual variation in male vocal trials and female mating preferences in *Bufo americanus*. Anim Behav 55:1165–1179

Huber F, Moore TE, Loher W (1989) Cricket behaviour and neurobiology. Cornell University Press, Ithaca

Jeram S (1993) Anatomical and physiological properties of antennal mechanoreceptors of the bug species *Nezara viridula* (L.) (Pentatomidae, Heteroptera). MSc thesis, University of Ljubljana, Ljubljana

Jeram S (1996) Structure and function of Johnston's organ in the bug species *Nezara viridula* (L.) (Pentatomidae, Heteroptera). PhD thesis, University of Ljubljana, Ljubljana

Jeram S, Čokl A (1996) Mechanoreceptors in insects: Johnston's organ in *Nezara viridula* (L.) (Pentatomidae, Heteroptera). Pflügers Archiv 431(Suppl.):R281

Jeram S, Pabst AM (1996) Johnston's organ and central organ in *Nezara viridula* (L.) (Heteroptera, Pentatomidae). Tissue Cell 28:227–235

Kavčič A, Čokl A, Laumann RA, Moraes MCB, Borges M (2013) Tremulatory and abdomen vibration signals enable communication through air in the stink bug *Eushistus heros*. PLoS ONE 8(2):1–10

Keuper A, Kühne R (1983) The acoustic behaviour of the bushcricket *Tettigonia cantans*. II. Transmission of airborne-sound and vibration signals in the biotope. Behav Process 8:125–145

Kiritani KN, Kimura K, Nakasuji F (1965) Imaginal dispersal of the southern green stink bug, *Nezara viridula* L., in relation to feeding and oviposition. Jpn J Appl Entomol Zool 9:291–297

Kühne R (1982) Neurophysiology of the vibration sense in locusts and bushcrickets: response characteristics of single receptor units. J Insect Physiol 28:155–163

Kuštor V (1989) Activity of muscles of the vibration producing organ of the bug *Nezara viridula*. MSc thesis, University of Ljubljana, Ljubljana

Laumann RA, Kavčič A, Moraes MCB, Borges M, Čokl A (2013) Reproductive behaviour and vibratory communication of the neotropical predatory stink bug *Podisus nigrispinus*. Physiol Entomol 38:71–80

Maluf NSR (1932) The skeletal motor mechanism of the thorax of the "stink bug" *Nezara viridula* L. Bull Soc R Ent Egypt 16:161–203

Manabe K (1997) Various modifications of vocalization by operant conditioning. Biomed Res 18:125–132

Markl H (1983) Vibrational communication. In: Huber F, Markl H (eds) Neuroethology and behavioural physiology. Roots and growing points. Springer, Berlin, pp 332–353

McBrien HL, Millar JG (2003) Substrate-borne vibrational signals of the consperse stink bug (Hemiptera: Pentatomidae). Can Entomol 135:555–567
McNett GD, Cocroft RB (2008) Host shifts favour vibrational signal divergence in *Enchenopa binotata* treehoppers. Behav Ecol 19:650–656
McNett GD, Miles RN, Homentcovschi F, Cocroft RB (2006) A method for two-dimensional characterization of animal vibrational signals transmitted along plant stems. J Comp Physiol A 192:1245–1251
McNett G, Luan LH, Cocroft RG (2010) Wind-induced noise alters signaller and receiver behavior in vibrational communication. Behav Ecol Sociobiol 64:2043–2051
McVean A, Field LH (1996) Communication by substratum vibration in the New Zealand tree weta, *Hemideina femorata* (Stenopelmatidae: Orthoptera). J Zool 239:101–122
Michel K, Amon T, Čokl A (1983) The morphology of the leg scolopidial organs in *Nezara viridula* (L.) (Heteroptera, Pentatomidae). Rev Can Biol Exp 42:130–150
Michelsen A, Fink F, Gogala M, Traue D (1982) Plants as transmission channels for insect vibrational songs. Behav Ecol Sociobiol 11:269–281
Miklas N, Stritih N, Čokl A, Virant-Doberlet M, Renou M (2001) The influence of substrate on male responsiveness to the female calling song in *Nezara viridula*. J Insect Behav 14:313–332
Miles RN, Cocroft RB, Gibbons C, Batt D (2001) A bending wave simulator for investigating directional vibration sensing in insects. J Acoust Soc Am 110:579–587
Ota D, Čokl A (1991) Mate location in the southern green stink bug *Nezara viridula* (Heteroptera: Pentatomidae) mediated through substrate-borne signals on ivy. J Insect Behav 4:441–447
Panizzi AR (1997) Wild hosts of pentatomids: ecological significance and role in their pest status on crops. Annu Rev Entomol 42:99–122
Panizzi AR, McPherson JE, James DG, Javahery M, McPherson RM (2000) Stink bugs (Pentatomidae). In: Schaefer CW, Panizzi AR (eds) Heteroptera of economic importance. CRC Press, Boca Raton, p 828
Polajnar J, Čokl A (2008) The effect of vibratory disturbance on sexual behaviour of the southern green stink bug *Nezara viridula* (Heteroptera, Pentatomidae). Centr Europ J Biol 3:189–197
Polajnar J, Svenšek D, Čokl A (2012) Resonance in herbaceous plant stems as a factor in vibrational communication of pentatomid bugs (Heteroptera: Pentatomidae). J R Soc Interf 9:1898–1907
Rodríguez RL, Ramaswamy K, Cocroft RB (2006) Evidence that female preferences have shaped male signal evolution in a clade of specialized plant-feeding insects. Proc R Soc B Biol 273:2585–2593
Ryan MA, Čokl A, Walter GH (1996) Differences in vibratory communication between a Slovenian and Australian population of *Nezara viridula* (L.) (Heteroptera: Pentatomidae). Behav Process 36:183–193
Schoonhoven LM, Jermy T, van Loon JJA (1998) Insect–plant biology: from physiology to evolution. Chapman & Hall, London, p 409
Shaw SR (1994) Detection of airborne sound by a cockroach vibration detector: a possible missing link in insect auditory evolution. J Exp Biol 193:13–47
Slabbekoorn H, Peet M (2003) Birds sing at a higher pitch in urban noise. Nature 424:267
Stölting H, Moore TE, Lakes-Harlan R (2002) Substrate vibrations during acoustic signalling in the cicada *Okanagana rimosa*. J Insect Sci 2:1–7
Todd JW (1989) Ecology and behaviour of *Nezara viridula*. Annu Rev Entomol 34:273–292
Virant-Doberlet M, Čokl A (2004) Vibrational communication in insects. Neotrop Entomol 33:121–134
Virant-Doberlet M, Čokl A, Zorović M (2006) Use of substrate vibrations for orientation: from behaviour to physiology. In: Drosopoulos S, Claridge MF (eds) Insect sounds and communication: physiology, behaviour, ecology and evolution. CRC Taylor & Francis, Boca Raton, pp 81–98
White PR, Birch MC, Church S, Jay S, Rowe E, Keenlyside JJ (1993) Intraspecific variability in the tapping behavior of the deathwatch beetle, *Xestobium rufovillosum* (Coleoptera, Anobiidae). J Insect Behav 6:549–562

Zorović M (2011) Temporal processing of vibratory communication signals at the level of ascending interneurons in *Nezara viridula* (Hemiptera: Pentatomidae). PLoS ONE 6:1–8. doi:10.1371/journal.pone.0026843

Zorović M, Čokl A, Virant-Doberlet M (2004) The interneural network underlying vibrational communication in the green stink bug *Nezara viridula* (L.). In: Processing of the female calling song with application for orientation and species recognition. The 7th congress of the international society for neuroethology, program and abstracts, University of Southern Denmark, Odense, p 201

Zorović M, Prešern J, Čokl A (2008) Morphology and physiology of vibratory interneurons in the thoracic ganglia of the southern green stinkbug *Nezara viridula* (L.). J Comp Neurol 508:365–381

Žunič A, Čokl A, Virant-Doberlet M, Millar JG (2008) Communication with signals produced by abdominal vibration, tremulation, and percussion in *Podisus maculiventris* (Heteroptera: Pentatomidae). Ann Entomol Soc Am 101:1169–1178

Žunič A, Virant-Doberlet M, Čokl A (2011) Species recognition during substrate-borne communication in *Nezara viridula* (L.) (Pentatomidae: Heteroptera). J Insect Behav 24:468–487

# Part III
# Practical Issues in Studying Vibrational Communication

# Chapter 11
# Physical Aspects of Vibrational Communication

**Axel Michelsen**

**Abstract** Thirty years ago, we found that insect vibrational songs may travel as bending waves through the stems of various plants. It was already known that other kinds of waves were involved when ants or scorpions detect vibrations through soil or sand, and we anticipated that several other kinds of waves would be involved in different substrates. This review summarizes the progress made since our study and points out some problems that need scrutiny: the energetic costs of communicating through different substrates, how vibrations propagate in plants and soils, discrimination between attenuations due to the substrate and those due to geometric spreading, and whether we can be sure that we record the kinds of waves sensed by the animals.

## 11.1 Introduction

About thirty years ago, we found that insect vibrational songs may travel as bending waves through the stems of various plants (Michelsen et al. 1982). A few years later, I cooperated with Martin Lindauer in the studies of vibrational signalling in honeybees, especially the tooting and quacking signals of young queens (Michelsen et al. 1986). I then left the field and turned my attention to other matters like sound emission, the mechanisms of directional hearing, and the dances of honeybees. This invited chapter thus gives me an opportunity to return to old ground and find out what has happened since my time.

Most investigators of vibrational communication are, of course, more interested in biological problems than in the physical aspects. The costs of various sorts of communication are important for animals, however, and a direct way to determine

A. Michelsen (✉)
Department of Biology, University of Southern Denmark, 5230 Odense, Denmark
e-mail: A.Michelsen@biology.sdu.dk

R. B. Cocroft et al. (eds.), *Studying Vibrational Communication*,
Animal Signals and Communication 3, DOI: 10.1007/978-3-662-43607-3_11,

**Table 11.1** Power outputs in some sound and vibrational calls

| Insect | Body length (mm) | Emission | Power output | Source of data |
|---|---|---|---|---|
| Buried leaf-cutting ant (*Atta*) | 9–12 | Omnidirectional vibration | 10–20 nW | Markl (1968) |
| Cydnic bug (*Seherus*) | 6–7 | Bidirectional vibration | 30 nW | Michelsen et al. (1982) |
| Grasshopper (*Omocestus*) | 14 | Directional sound | 52 nW | Michelsen and Elsner, (1999) |
| Grass cicada (*Tympanistalna*) | 15 | Omnidirectional sound | 280 nW | Michelsen and Fonseca (2000) |
| Cricket (*Gryllus campestris*) | 30 | Directional sound | 60 μW | Bennet-Clark (1970) |
| Bladder cicada (*Cystosoma*) | 50 | Omnidirectional sound | 350 μW | Mac Nally and Young (1981) |
| Mole cricket (*Gryllotalpa*) | 45 | Omnidirectional sound | 1.2 mW | Bennet-Clark (1970) |

the costs is to determine the kind of waves and to estimate the input impedance of the substrate for these waves (the *relative* costs of different kinds of communication can be determined with other methods like measuring the production of $CO_2$, see Römer et al. 2010). A big surprise in our 1982 study was how cheap vibrational signalling could be: one microgram of muscle may be sufficient for communicating through a plant stem. In the most efficient small producers of airborne sound, the power output is of similar magnitude (Table 11.1).

This is an example of a "counter-intuitive" finding. Our experience as humans is that it does not cost us much work to communicate by means of airborne sounds. We also expect to work harder, if we try to communicate with substrate-borne vibrations. Most insects are so small that they can only emit sounds efficiently at high frequencies, and their vibration receptors are generally more sensitive than ours.

## 11.2 Biologists and Physics

Some insight into physics is required when biologists are studying physical aspects of biology, but in many cases, a superficial understanding is sufficient. When we began our studies on vibrational signals in plants 30 years ago, I had much experience in acoustics, but no in depth of education in physics. I thought that I could start by setting up a small experiment in order to "learn by doing". So, I replaced a loudspeaker in my laboratory with a small vibration table attached to a plant stem, sent an electrical impulse to the vibration table, and measured the vibration in the plant with a laser vibrometer. Much to my surprise, I found that the vibration impulse in the plant seemed to arrive at some distance from the vibration table before it had started! This experience convinced me that it would be better to start by reading a book about vibrations, and I was then introduced to strange phenomena like phase and group velocities of bending waves.

In contrast to airborne sounds, where all components travel with the same velocity, in bending waves, the high frequencies travel faster than the low frequencies. One consequence is that the signals change shape and perhaps also duration when they travel. A frequency sweep that begins at a high frequency and ends at a low frequency will increase its duration when it travels (and vice versa). When working with airborne sounds, one often selects a characteristic part of the signal (e.g. a maximum of amplitude) and looks for this part in other recordings. This is not a wise strategy when working with bending waves. The strategies used by the animals for distinguishing signals are thus likely to be different from those used by animals communicating with sounds.

A major problem confronting biologists interested in biophysical problems is to understand how physicists work, especially when they make theoretical models and try to predict reality by means of calculations. Many models are based on a too small number of simple assumptions about the system under study. This was especially true earlier, when calculations had to be performed with pencil and paper. For example, the first models of soil assumed that the grains were of the same size, spherical, and made of the same material. One can learn much from such a simple model, but one cannot expect the results and predictions to be realistic for real soils, in which grains of all possible shapes and sizes allow far more contact between each grain and its neighbours.

Different mathematical models of the same phenomenon may rest on different assumptions. I had my first experience with this more than 40 years ago when I tried to understand the physics of the locust ear (Michelsen 1971). One of the many problems was how much of the sound received by the ear is reradiated because the excited, vibrating eardrum acts as a loudspeaker. By searching the literature, I found two mathematical solutions to this problem, but unfortunately, the results differed by a factor of five! My acoustical mentors said that both models appeared to be reasonable and based on realistic assumptions, so it was not possible to say which model was closest to reality. This may be an extreme example, but it convinced me that it always is a good idea to study the assumptions behind the mathematical expressions very carefully and not to trust theoretical results without (if possible) testing them in experiments. It is also a good idea, even for very experienced investigators, to use a professional physicist as mentor when entering new fields.

## 11.3 Plants as Transmission Channels for Vibration Signals

In 1974, two pioneers in this field, Ichikawa and Ishii, demonstrated that male planthoppers may respond to the abdominal vibrations of virgin females situated on the stem of the same rice plant. With very simple experiments, they also

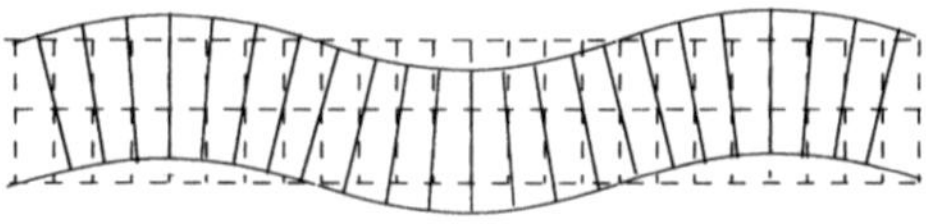


**Fig. 11.1** Kinematic form of a bending (flexural) wave, in which the particle displacements have components both normal and parallel to the direction of wave propagation (Reprinted from Fahy 2001 with permission from Elsevier)

showed that the vibration signals may spread to other plants, if leaves from the plants touch each other. At the same time, evidence was found for "substrate-borne sound communication" in cydnid bugs (Gogala et al. 1974).

In 1973, I obtained a grant to allow a Danish company to build a prototype of a transportable laser vibrometer (earlier versions were huge and stationary). After a few years, we had learned to use it (despite its very unstable character) for measurement of ear drum vibrations in various insects. Matija Gogala now suggested that we should use this technique for measuring vibrations in plants, which were too tiny to support the heavy accelerometers of that time.

Our set-up was quite elaborate with a function generator for generating the signals, a spectrum analyser, and a signal averager (with no less than 4 k of 20 bit computer memory, which was very expensive in 1980!). A special feature was the method for causing the plants to vibrate: we glued a small, 25 mg magnet to the plant, and caused it to vibrate with an electromagnet 1 cm away. In this manner, we hoped to reduce restricting the movements of the plant. (I am not aware whether this concern was justified. It would be interesting to compare this technique with a direct physical contact between plant and vibrator).

The main findings of our investigation were that the vibration waves travelling through plants with very different physical properties (ranging from soft bean plants to reeds and maples) were bending waves (Fig. 11.1). We suggested that a part of the wave energy might be transformed into a longitudinal wave at branching junctions of the plants, but this possibility does not seem to have been investigated. We also measured the mechanical properties (the moment of inertia, Young's modulus of elasticity, and the mass per unit length) and used the figures for calculating expected group velocities of bending waves, which appeared to be close to the measured values (Fig. 11.2). The measurements and calculations were straightforward, and the results could be used for calculation of the input impedance of the plants and thus of the muscular power necessary for producing the vibrations. We calculated the power used by *Sehirus* bugs that send vibrational songs through *Thesium* plants and found that the animal only needed to use one microgram of muscle. We anticipated that future investigators would make similar measurements and calculations for other animals and plants, but so far no one seems to have followed this line of research.

We found that the bending waves could spread to branches of the plant that originate at the root, and we observed strong reflections of the bending waves, both

| Plant | $I$ | $E$ | $m'$ | Calculated | | Measured | |
|---|---|---|---|---|---|---|---|
| | | | | 200 Hz | 2 kHz | 200 Hz | 2 kHz |
| *Vicia faba* | $2 \cdot 10^{-10}$ | $4 \cdot 10^{6}$ | $9 \cdot 10^{-3}$ | 39 | 122 | 36 | 120 |
| *Galeobdolon vulgare* | $3 \cdot 10^{-12}$ | $2 \cdot 10^{8}$ | $5 \cdot 10^{-3}$ | 42 | 132 | 45 | 143 |
| *Thesium bavarum* | $6 \cdot 10^{-13}$ | $10^{9}$ | $3 \cdot 10^{-3}$ | 47 | 150 | 40 | 162 |
| *Phragmites communis* | $4 \cdot 10^{-11}$ | $3 \cdot 10^{8}$ | $8 \cdot 10^{-3}$ | 78 | 246 | 75 | 220 |
| *Phalaris arundinacea* | $14 \cdot 10^{-13}$ | $2 \cdot 10^{9}$ | $2 \cdot 10^{-3}$ | 77 | 243 | 74 | – |
| *Acer pseudoplatanus* | $2 \cdot 10^{-10}$ | $6 \cdot 10^{9}$ | $5 \cdot 10^{-1}$ | 88 | 279 | 95 | – |

**Fig. 11.2** Mechanical properties and calculated and measured bending wave group velocities (in m/s) in six species of plants with very different mechanical properties. I: surface moment of inertia (in $m^4$); E: Young's modulus of elasticity (in $N/m^2$); m′: mass per unit length (in kg/m) (From Michelsen et al. 1982)

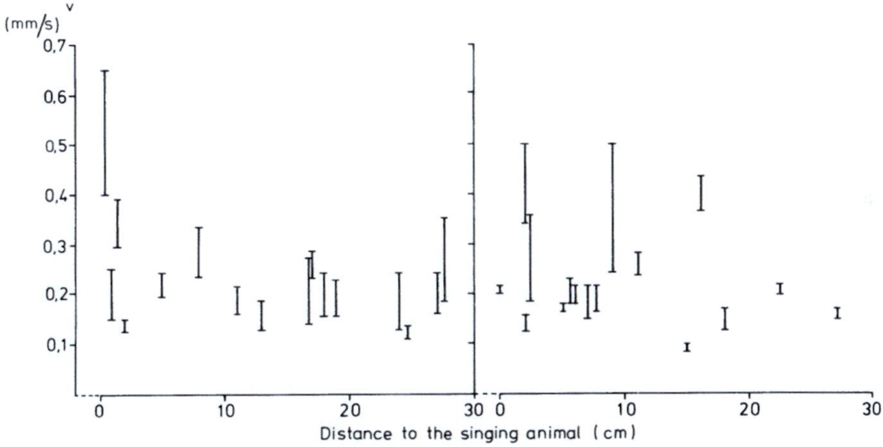


**Fig. 11.3** Maximum peak-to-peak vibration velocities of songs recorded on bean plants at various distances from singing males of the small cicadas *Euscelidius variegatus* (*left*) and *Euscelis lineolatus* (*right*) (From Michelsen et al. 1982)

at the soil and at the top of the plants. As a result, maxima and minima of vibration amplitude occurred along the stem, and the amplitude of the vibrations did not seem to be a reliable cue to the distance to the singing animal (Fig. 11.3).

As already mentioned, Ichikawa and Ishii (1974) observed that the vibrations may spread to other plants if leaves from the plants touch each other. In a recent paper by Eriksson et al. (2011), it is shown that 220–250 Hz signals can spread to other plants if large leaves are close neighbours. In their experiment, wine leaves with areas of about 60 $cm^2$ were mounted parallel to each other with about 30 $cm^2$ overlap and a variable distance of 0.5–11 cm. At all distances, vibrations generated by a singing male on one leaf caused vibrations in the other leaf. At distances of 0.5–1 and 11 cm, the vibrations of the second leaf were about 20 and 40 dB down, respectively. At distances up to 6 cm, a singing male on one leaf could release a response from a female on the other leaf. It remains to be seen whether this mechanism is sufficiently robust to play a role outdoors, where the songs may be masked by noise- and wind-induced vibrations.

A very likely explanation for a part of the variation in vibration amplitude shown in Fig. 11.3 was published by McNett et al. (2006). They had access to two laser vibrometers and used them for studying the two-dimensional properties of plant stem motion. The laser vibrometers were placed in an orthogonal position, which allowed the investigators to calculate the motion in space. Single points on the surface of the stem were found to move in an ellipse with the major axis of the ellipse corresponding to the maximum amplitude of vibration. Measurements with a single transducer will therefore often underestimate the amplitude of bending waves. Furthermore, the direction of the major axis may change when the signal travels. This may have contributed to the large variation of vibration amplitudes with distance in Fig. 11.3. The issue is complicated further by the fact that the insects detect the vibrations by means of six subgenual organs with different orientations and sensitivities (see Sect. 11.5 and Fig. 11.7). The paper has so far been cited in six publications, but the promising potential of the method has not yet been utilized.

In our experiments, we observed reflections of the bending waves both at the top of plants and at the ground. Depending on the amount of damping in the particular plants, the direct and (often several times) reflected waves could add to very long-lasting vibrations. This raised the question of how the animals determined the direction to the origin of the waves (that is, to the signalling animal). Elegant experiments performed with a bending wave simulator, which allows the investigator to map the passive body movements of an animal subjected to bending waves, showed that the body oscillations depend on the direction of the wave (Cocroft et al. 2000). In addition, discussions during the 2011 DGaaE meeting in Berlin showed that reflections from top or ground do not seem to be very common.

A recent publication is of particular interest for our understanding of the physics of vibrational communication through plants. Casas et al. (2007) demonstrated that plant stems that are mechanically excited within a broad frequency range may carry dispersive vibration waves at low frequencies and non-dispersive waves at higher frequencies (Fig. 11.4). The dispersive waves travel with a velocity that is proportional to the square root of frequency whereas the non-dispersive waves travel with a constant velocity. At higher frequencies, simple bending is no longer possible, because the wavelength approaches the diameter of the stem. The dispersive waves are now named corrected bending waves (see, e.g. Cremer et al. 2005). The physical mechanisms have also become more complicated. In bending waves, the movements are caused by an interaction between potential and kinetic energy, but in corrected bending waves, rotational energy and shear deformations also play a role. At still higher frequencies, the waves approach the behaviour of non-dispersive surface (Rayleigh) waves. We considered it unlikely that surface waves could occur in the stems of the plants investigated (Michelsen et al. 1982), but apparently they can occur above 5 kHz in stems of 3–4 mm diameter. Unfortunately, Casas et al. use the name “non-dispersive bending waves” for the surface waves, which is likely to cause much confusion (2007).

The experiments by Casas et al. (2007) left several questions unanswered. The vibrations were released by dropping a metal ball on the plant stem, but the weight

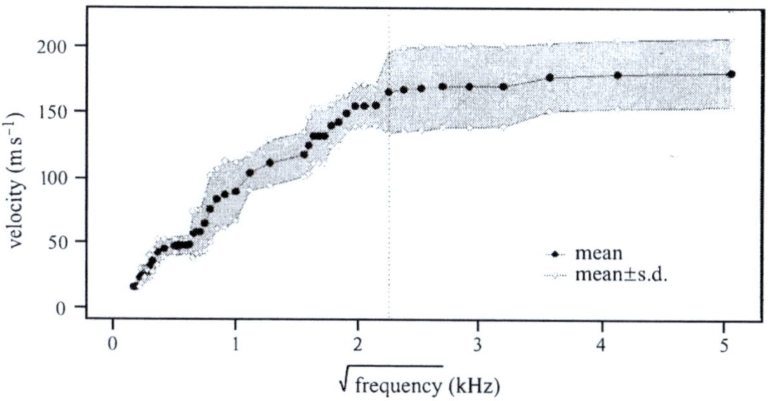


**Fig. 11.4** Both dispersive and non-dispersive waves are released when a falling metal ball hits a plant stem. (Redrawn from Casas et al. 2007 with permission from the Royal Society)

of the ball was not varied. It is therefore not known whether the vibrations were within the mechanically linear range of the stems and thus whether the findings were valid for smaller or larger releasing forces. The authors point out that the non-dispersive waves would be much better suited for communication, because the signals would be less distorted when they travel. The animals could obtain this advantage by singing on thicker stems, but further studies of the energetic aspects are needed in order to find out whether this strategy has any drawbacks. For example, what is the attenuation with distance, and how well are the signals transmitted to thinner branches?

The same technique had earlier been used for characterizing vibrations of leaves (Magal et al. 2000). Again, the investigators did not investigate the degree of linearity. In addition, the authors did not report the measured attenuations of the vibration signals after crossing over the leaf lamina and large or small veins as dB values, but as percentage energy losses. Energy losses of 40–80 % were thus reported for crossing the midvein at the apex and leaf base, respectively. These figures sound dramatic, but in a linear system, they would only correspond to 2.2 and 7 dB attenuations of vibration amplitude.

In these studies, a brief impulse of energy from a falling metal ball released the bending waves that were favoured by the mechanics of the stems and leaves. Several investigators of tapping or drumming signals have argued that such stimuli must cause compression waves. For example, Henry (2006) argued that drumming causes compression waves, whereas tremulation (shaking) produces bending waves. Some investigators have referred to the paper by Henry and supported the idea that different kinds of motion cause different kinds of waves, but changed the message (for example, Eberhard et al. 2010 claim that tapping the abdomen onto the plant substrate generates compression and transversal waves). These claims may sound logical, but in the world of physics, the claims should be based on the magnitudes of the input impedances of plants for the different kinds of waves.

## 11.4 Vibration Signals in Soil

More than 40 years ago, Hubert Markl investigated the transmission of vibration signals through soil (Markl 1968). This pioneering study addressed several important problems, including the physics of stridulation. Leaf-cutting ants stridulate when they are in trouble, and stridulatory vibrations from buried ants attract nest mates and cause them to start digging. Markl observed that the airborne stridulatory sounds and the vibrations in the soil were a maximum at 20–60 and 1–5 kHz, respectively, and that these spectra did not overlap. Markl interpreted the dominance of high frequencies in the airborne sound to be caused by an increase in the sound radiation with frequency. He further listed three mechanisms, which might attenuate the high frequencies when the vibrations were transmitted through the soil (friction, heat transfer, and multiple scatterings). The total effect of the three processes may be expressed as an attenuation coefficient (dB/cm).

Markl also proposed an explanation for the relationship between the physical processes during stridulation and the resulting vibration or sound, but unfortunately the equipment of that time did not allow experimental tests of his proposal. He further estimated the power necessary for producing the observed vibrations in the soil (see Table 11.1). Finally, he outlined the basic principles for the propagation of the vibration signals through the soil with different properties (see Sect. 11.4.1), but he was unable to observe the theoretically expected differences. The reason for this may have been his efforts to ensure a perfect contact between the soil and the very large and heavy accelerometers that were available in the 1960s. The impact of his papers would probably have been much larger, if they had been published in English. His measurements should be repeated with modern light-weight accelerometers and laser vibrometers.

In another classical study, Brownell and Farley (1979) showed that desert scorpions can detect the distance and direction to moving prey on or below the surface of sand. The movements gave rise to two kinds of waves: longitudinal compression waves with a propagation velocity of about 150 m/s and Rayleigh surface waves with a velocity of about 50 m/s. The vibrations are in the direction of wave propagation in the compression waves, but vertical and almost transverse to the direction of wave propagation in the surface waves. The compression waves affect tarsal hair receptors, whereas the surface waves affect basitarsal compound slit sensilla, which are located above the joint of the tarsus and the basitarsus of each leg. The vertical movements compress the slits and activate two sensory neurons behind each slit. Both types of receptors can respond to movements with amplitudes of a few Å.

It was suggested that the scorpions use the time difference between the responses of the two types of receptors for determining the distance to the prey. The responses allow the scorpion to detect and to determine the direction of insects moving more than 15 cm away, which is also the distance at which a buried sand swimming lizard can detect vibrations from its insect prey (Hetherington 1989). The directional localization of the prey by scorpions can be mimicked by a model, in which second-order neurons are excited by input from one tarsus and inhibited by input from a triad of legs opposite to it (Stürzl et al. 2000).

Before the discovery of the prey detection of scorpions, only one example of vibrational wave propagation over sandy free surfaces had been discovered, namely the "song of dunes": dunes of sorted and very dry sand may make loud booming sounds (about 100 Hz at 100 dB SPL) when they avalanche (Carus-Wilson 1891). Over the years, several explanations were proposed for this phenomenon before it finally was found to be caused by coherent elastic waves originating inside the avalanche and spreading to neighbouring areas, travelling a few centimetre below the free surface with a velocity of about 40 m/s (Andreotti 2004).

In 1990, Aicher and Tautz published a very thorough study of the signal transmission through wet sand, as a part of a study of vibrational communication in fiddler crabs, which live on beaches with wet sand. The males of these crabs drum the ground with a large claw (*chela*) and produce three kinds of waves: love waves, Rayleigh waves, and surface pressure waves with velocities of 50–60, 70–80, and 150–160 m/s, respectively. The signals were recorded with accelerometers and a laser vibrometer. This paper contains a detailed discussion of the different kinds of waves and the directions of vibration and propagation when the waves are travelling inside a three-dimensional medium and when the waves are at the boundary of two media. In the Materials and Methods section, the possible experimental errors connected with the methods and instruments are discussed so carefully that the text should be compulsory reading for newcomers to the field of vibrational communication. The vibrational directions of small volumes of sand were measured by fastening small (4 mm times 4 mm) pieces of very thin gold foil to the sand, thus improving the reflection of light from the laser vibrometer.

### *11.4.1 Theories and Data for Attenuation of Vibrational Signals in Soil*

The literature on wave propagation in soil is difficult to read for non-physicists (like me), partly because the behaviour of real grain particles differ much from the simple assumptions in the basic theories and also because there are often large differences in opinion between physicists who argue by means of mathematical expressions. For a fairly simple explanation of the development of the ideas from the pioneering work by Heinrich Hertz (1882) on contact between elastic solid particles of simple shape until the present time, see the anonymous article "Contact mechanics" in Wikipedia. The two papers by Bonneau et al. (2007, 2008) are reasonably simple introductions to the wave problems.

In Sect. 11.2, it was mentioned that the first theoretical models of soil heavily underestimated the amount of contact between the grains in soil. Another important factor is the content of air and water in the soil. In soils saturated with air, the grains constitute a stiff frame, and a slow compression wave travels mainly in the pores. In contrast, in soils saturated with water, a fast compression wave travels mainly in the solid frame (Albert 1993). These two extremes are relatively simple, but real soils may be somewhere in between.

The elasticity of the sand is due to the compression caused by gravity, which forces the grains towards their neighbours and causes slight elastic deformations of the grains, the so-called elastic skeleton, through which the sound may propagate. There has been much disagreement among physicists about the detailed mechanisms, partly because of simplifications introduced in various theoretical calculations. The stiffness of the elastic skeleton increases with depth, and this causes an increase in the velocity of sound with depth. Sound waves travelling parallel to the surface, therefore, tend to bend (refract) upwards (similar to sound waves travelling parallel to ground on sunny days when a higher temperature exists close to ground). Some physicists have claimed that the sound in the soil disappears completely when it reaches the surface (Jaeger and Nagel 1977). However, there is now evidence for an almost total reflection similar to the reflection of sound waves at the open end of an open organ pipe. The vibrational waves close to the surface of soil are thus guided by the free surface and by the increasingly rigid material at some distance below the surface (Andreotti 2004; Gusev et al. 2006). The latter reference also contains an explanation for the surprisingly low propagation speed.

Most experimental data of the attenuation of vibrational signals in soil are presented as dB rms values obtained at various distances to the source. Several factors may contribute to the decrease in signal amplitude with distance. Waves travelling in soil lose energy by the interaction between the wave and the soil, and their amplitudes also decrease because the waves are spreading to larger volumes of soil. The frictional losses are generally proportional to the distance travelled (dB/m), whereas the geometric losses depend on the geometry. When vibrations travel evenly in all directions in the soil, the attenuation due to geometric spreading is 6 dB/dd, where dd means distance doubled. This is similar to the simple spherical spreading loss of sound in homogeneous air. However, the layered nature of soils may restrict the spreading to a layer near the surface, and now the (cylindrical) spreading has a loss of only 3 dB/dd. Finally, vibrations travelling in a channel may have a geometric spreading loss close to zero.

The calculations of the values of dd (distance doubled) should start outside the near field (the space around the source that is affected within one oscillation period). The border between the near field and the far field depends on frequency, and changes in the vibration spectrum with distance to the source may thus complicate the calculations.

Obviously, most reports of measured attenuation with distance do not contain sufficient information to allow the readers to correct the published attenuations for the effects of geometric spreading in order to arrive at the attenuation with distance caused by the medium. Figure 11.5 shows attenuations of playback signals in leaf litter, clay and pine litter at distances up to 16 cm from a mini-shaker (Elias et al. 2010). The data leave no doubt that it is very sensible that wolf spiders prefer to generate their vibration signals on leaf litter and not on clay or pine litter.

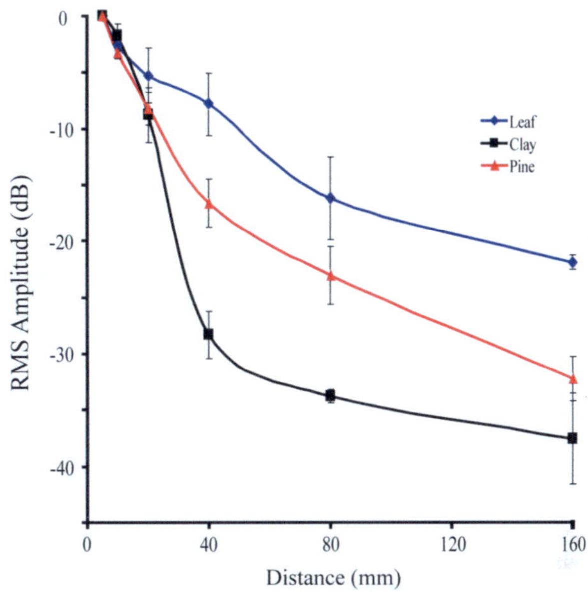


**Fig. 11.5** Attenuations of the vibratory signals of wolf spiders through three different substrates (From Elias et al. 2010 with permission from the editors of Zoology)

However, a close look at the data shows that most of the attenuation occurred over the first 2–4 cm. The attenuation from 4 to 16 cm was close to 6 dB/dd in leaf litter and pine litter, whereas it was 9 dB/dd in clay. The 6 dB/dd expected for spherical spreading leaves no room for any attenuation due to the medium in the case of leaf litter and pine litter. These attempts to understand the reported findings were also complicated by the fact that most of the reported dB values were averages from several measurements and that the scatter of the data was indicated with the same length of a vertical bar above and below the average. The averages were thus probably averages of dB values, and the true averages were not available.

A similar situation is found in the paper on transmission of stridulation signals in burrower bugs by Čokl et al. (2006), who reported that the signals were transmitted through the soil with high attenuation and that they only could be detected less than 6 cm from the source. Attenuations of 5.5 dB/0.5 cm and 15–19 dB/3.5 cm were observed and said to correspond to 11 and 4–5 dB/cm, respectively, without taking into account the geometric spreading losses.

It occurs to me that this area of research would benefit much, if someone would perform model studies that would allow future investigators to analyse their data in such a way that geometric spreading losses and attenuations caused by the substrates were reported separately. Perhaps, it may be possible to analyse the data with a factor analysis with the parameters dB/distance and dB/dd.

Before we leave the field of vibrations in soils, it should be mentioned that the output from accelerometers is not always related to vibrations (Fig. 11.6). Various environmental effects may affect the output, but the most important factor is the mounting of the accelerometer and its cable. Serridge and Licht (1987) provide a thorough introduction to the many problems.

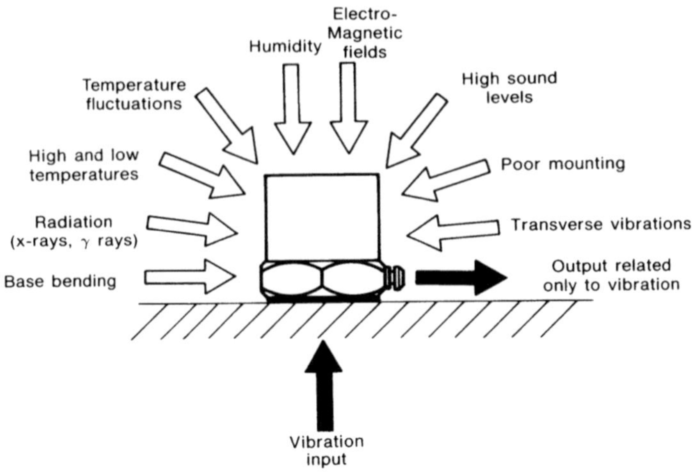


**Fig. 11.6** Many extraneous inputs can result in non-vibration related outputs from an accelerometer. Most of them can be eliminated or reduced by a good design of the accelerometer, but the mounting of the accelerometer and its cable is critical, also in well-designed accelerometers (Reprinted from Serridge and Licht 1987. Figure courtesy of Brüel and Kjaer sound and vibration measurement A/S)

## 11.5 Which Waves Do the Investigators and the Animals Detect?

In most studies of vibrational communication, the vibrations are measured with a laser vibrometer or an accelerometer, and the most sensitive direction of the measuring devices is perpendicular to the surface of the plant or the soil. These methods are sufficient for waves in which the major direction of vibration is perpendicular to the surface. In other kinds of waves, the major direction of vibration is in the plane of the surface, either in the direction of wave propagation or perpendicular to the direction of wave propagation.

For example, pure longitudinal waves with vibrations only in the direction of wave propagation only exist inside solids whose dimensions are several wavelengths in all directions. In rod-shaped objects, we find quasi-longitudinal waves (Fig. 11.7), in which a contraction of the cross section occurs in addition to the axial extension, and vice versa (see, e.g. Cremer et al. 2005). The exact magnitude of the contractions depends on several parameters, but as a rule of thumb, the ratio between the greatest lateral displacement (l) to the greatest longitudinal displacement (L) is close to the ratio between the diameter of the stem (d) and the wavelength of the longitudinal wave (λ).

For example, let us consider a rod of soft wood with a diameter of 5 mm, which carries a quasi-longitudinal wave with a velocity of 500 m/s. At 1,000 Hz, the longitudinal wavelength is 500 mm, which is 100 times larger than the diameter of the rod. The lateral displacement, which is the signal measured by our laser

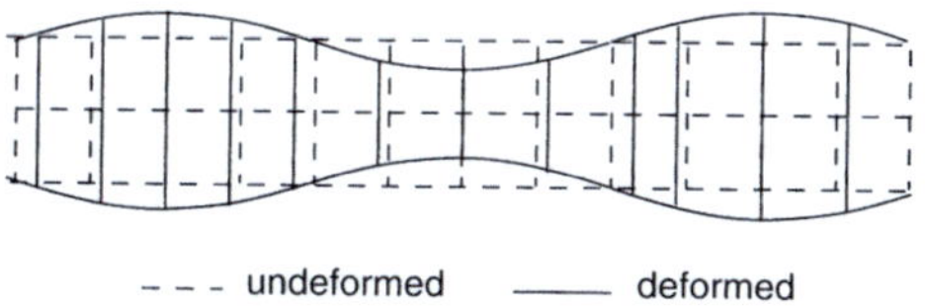


**Fig. 11.7** Textbook illustrations of quasi-longitudinal waves may create an unrealistic impression of the relationship between the lateral and the longitudinal deformations (Reprinted from Fahy 2001 with permission from Elsevier)

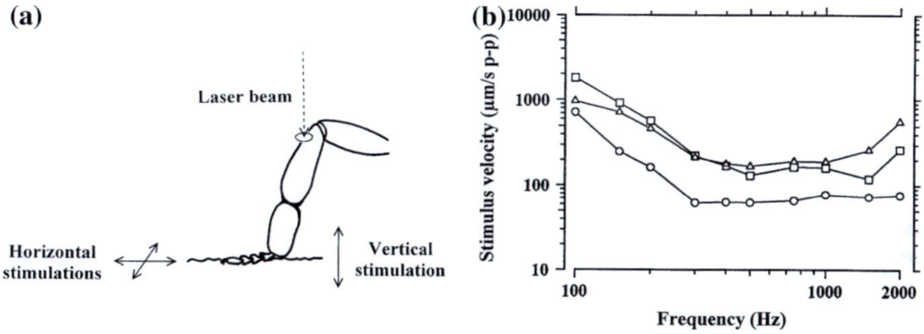


**Fig. 11.8** The sensitivity to substrate vibrations of the subgenual organ in the middle leg of a freely standing honeybee. **a** The experimental set-up. The vibrations were measured with laser vibrometry. **b** Electrophysiological thresholds of summated potentials from the nerve recorded in the femur. *Circles* vertical stimulus: *Triangles* horizontal stimulus perpendicular to the axial direction of the femur; *Squares* horizontal stimulus parallel to the femur. The two *upper curves* are only significantly different from each other at 1.5 kHz, but significantly different from the *lower curve* at all frequencies (from Rohrseitz and Kilpinen 1997. Reprinted with permission from Elsevier)

vibrometers or accelerometers, is therefore only about 1 % of the longitudinal displacement.

At the end of last century, some of my co-workers discovered that it is possible to measure the amplitudes and phase angles of the vibrations of subgenual organs in newly emerged honeybees, and they were thus able to analyse the biophysics of this seismic receptor (Storm and Kilpinen 1998). In addition, with electrophysiological methods, Rohrseitz and Kilpinen (1997) measured the thresholds for vibrations in two directions in the plane of the substrate and in the direction perpendicular to the plane of the substrate (Fig. 11.8). They found that the subgenual organ is about 10 dB more sensitive to vibrations perpendicular to the substrate than to vibrations in the plane of the substrate (Fig. 11.8).

If we assume that this finding can be applied to other insects, then the insects respond to vibrations much like three-directional accelerometers. Unfortunately, although such instruments are available, they are generally too heavy to be used on substrates like plant leaves or stems. If we now return to the situation where a stem carries a weak quasi-longitudinal wave, then the vibrations perpendicular to the surface (which are those recorded by a laser vibrometer or an unidirectional accelerometer) will be much smaller than the vibrations in the plane of the surface. In conclusion, an insect may respond to vibrations in the substrate, which activates its sense organs, but may be almost invisible for our most popular instruments.

How can one measure the vibrations in the plane of the substrate? My favourite method (used in measurements of the mechanics of the eardrums of small cicadas) is to use an atomic force cantilever as a "go-between". The cantilevers, which carry a sharpened tip at their ends, are 500 μm long, 20 μm broad, but only 2 μm thick. They are therefore extremely compliant to bending forces. One of the 500 × 20 μm surfaces is coated with metal and an excellent mirror for laser light. When such a cantilever is positioned almost perpendicular to the surface and attached at its tip, reflected laser light can reveal the vibrations in the plane of the surface.

**Acknowledgments** I am grateful to my colleague Ole Næsbye Larsen and to the editors of the book for comments on the manuscript.

## References

Aicher B, Tautz J (1990) Vibrational communication in the fiddler crab, *Uca pugilator*. I. Signal transmission through the substratum. J Comp Physiol A 166:345–353

Albert DG (1993) A comparison between wave propagation in water-saturated and air-saturated porous materials. J Appl Phys 73:28–36

Andreotti B (2004) The song of dunes as a wave-particle mode locking. Phys Rev Lett 93:238001–238004

Bennet-Clark HC (1970) The mechanism and efficiency of sound production in mole crickets. J Exp Biol 52:619–652

Bonneau L, Andreotti B, Clément E (2007) Surface elastic waves in granular media under gravity and their relation to booming avalanches. Phys Rev E 75:016602

Bonneau L, Andreotti B, Clément E (2008) Evidence for Rayleigh-Hertz surface waves and shear stiffness anomaly in granular media. Phys Rev Lett 101:118001

Brownell P, Farley RD (1979) Detection of vibrations in sand by tarsal sense organs of the nocturnal scorpion, *Paruroctonus mesaensis*. J Comp Physiol A 131:23–30

Carus-Wilson C (1891) The production of musical notes from non-musical sand. Nature 44:322

Casas J, Magal C, Sueur J (2007) Dispersive and non-dispersive waves through plants: implications for arthropod vibratory communication. Proc R Soc B 274:1087–1092

Cocroft RB, Tieu TD, Hoy RR, Miles RN (2000) Directionality in the mechanical response to substrate vibration in a treehopper (*Hemiptera: Membracidae: Umbonia crassicornis*). J Comp Physiol 186:695–705

Čokl A, Nardi C, Mauricio J, Bento S, Hirose E, Panizzi AR (2006) Transmission of stridulatory signals of the burrower bugs, *Scaptocoris* castanea and *Scaptocoris carvalhoi* (Heteroptera: Cydnidae) through the soil and soybean. Physiol Entomol 31:371–381

Cremer L, Heckl M, Petersson BAT (2005) Structure-borne sound: structural vibrations and sound radiation at audio frequencies. Springer, Berlin

Eberhard MJB, Lang D, Metcher B, Pass MD, Picker MD, Wolf H (2010) Structure and sensory physiology of the leg scolopidial organs in Mantophasmatodea and their role in vibrational communication. Arthropod Struct Dev 39:230–241

Elias DO, Mason AC, Hebets EA (2010) A signal-substrate match in the substrate-borne component of a multimodal courtship display. Curr Zool 56:370–378
Eriksson A, Anfora G, Lucchi A, Virant-Doberlet M, Mazzoni V (2011) Inter-plant vibrational communication in a leafhopper insect. PloS ONE 6(5):e19692. doi:10.1371/journal.pone.0019692
Fahy F (2001) Foundations of engineering acoustics. Academic Press, Amsterdam
Gogala M, Čokl A, Drašlar K, Blaževic A (1974) Substrate-borne sound communication in Cydnidae (Heteroptera). J Comp Physiol 94:25–31
Gusev VE, Aleshin V, Tournat V (2006) Acoustic waves in an elastic channel near the free surface of granular media. Phys Rev Lett 96:214301
Henry CH (2006) Acoustic communication in neuropterid insects. In: Drosopoulos S, Claridge MF (eds) Insect sounds and communication—physiology, behaviour, ecology and evolution. Taylor and Francis, London, pp 153–166
Hertz H (1882) Ueber die Berührung fester elastischer Körper. J Reine Angew Math 92:156–171
Hetherington TE (1989) Use of vibratory cues for detection of insect prey by the sandswimming lizard *Scincus scincus*. Anim Behav 37:290–297
Ichikawa T, Ishii S (1974) Mating signal of the brown planthopper, *Nilaparvata lugens* Stål (Homoptera: Delphacidae): vibration of the substrate. Appl Ent Zool 9:196–198
Jaeger HM, Nagel SR (1977) Dynamics of granular material. Amer Sci 85:540–545
Mac Nally R, Young D (1981) Song energetics of the bladder cicada, *Cystosoma saundersii*. J Exp Biol 90:185–196
Magal C, Schöller M, Tautz J, Casas J (2000) The role of leaf structure in vibration propagation. J Acoust Soc Am 108:2412–2418
Markl H (1968) Die Verständigung durch Stridulationssignale bei Blattschneiderameisen. ii. Erzeugung und Eigenschaften der Signale. Z Vergl Physiol 60:103–150
McNett GD, Miles RN, Homentcovschi D, Cocroft RB (2006) A method for two-dimensional characterizations of animal vibrational signals transmitted along plant stems. J Comp Physiol A 192:1245–1251
Michelsen A (1971) Physiology of the locust ear. Z Vergl Physiol 71:49–128
Michelsen A, Fink F, Gogala M, Traue D (1982) Plants as transmission channels for insect vibrational songs. Behav Ecol Sociobiol 11:269–281
Michelsen A, Kirchner WH, Andersen BB, Lindauer M (1986) The tooting and quacking vibration signals of honeybee queens: a quantitative analysis. J Comp Physiol A 158:605–611
Michelsen A, Elsner N (1999) Sound emission and the acoustic far field of a singing acridid grasshopper (*Omocestus viridulus*). J Exp Biol 202:1571–1577
Michelsen A, Fonseca P (2000) Spherical sound radiation patterns of singing grass cicadas, *Tympanastalna gastrica*. J Comp Physiol A 186:163–168
Rohrseitz K, Kilpinen O (1997) Vibration transmission characteristics of the legs of freely standing honeybees. Zoology 100:80–84
Römer H, Lang A, Hartbauer M (2010) The signaller's dilemma: a cost-benefit analysis of public and private communication. PLoS ONE 5:e13325
Serridge M, Licht TR (1987) Piezoelectric accelerometer and vibration preamplifier handbook. Brüel & Kjaer
Storm J, Kilpinen O (1998) Modelling the subgenual organ of the honeybee, *Apis mellifera*. Biol Cybern 78:175–182
Stürzl W, Kempter R, van Hemmen JL (2000) Theory of arachnid prey localization. Phys Rev Lett 84:5668–5671

# Chapter 12
# The Role of Wave and Substrate Heterogeneity in Vibratory Communication: Practical Issues in Studying the Effect of Vibratory Environments in Communication

**Damian O. Elias and Andrew C. Mason**

**Abstract** The substrate-borne sensory modality is inherently more complex than other modalities mainly due to the availability of many possible signaling channels, each with potentially distinct physical properties that may affect transmission between senders and receivers. In addition, in any given signaling channel, multiple wave types and patterns of propagation are possible leading to a combinatorial expansion of signal parameters that must be considered in analyses of vibratory sensory ecology. In the context of substrate-borne communication, animals have adapted to variation in signaling environments in a variety of ways including the evolution of distinct signaling strategies, multiple signal production mechanisms, and context-dependent behavior. This rich diversity is a subject of growing interest, but also presents major experimental challenges for scientists. In this review, we survey the literature of vibratory sensory ecology and discuss issues relating directly to the measurement of transmission characteristics in substrates and the use of artificial and natural substrates in behavioral experiments. We suggest that the vibratory sensory modality is an ideal study system for questions on sensory ecology and urge further research integrating mathematical models, carefully measured behavioral recordings, and comparative analyses.

D. O. Elias (✉)
Department of Environmental Science, Policy and Management,
University of California, Berkeley, Berkeley, CA 94720, USA
e-mail: doelias@berkeley.edu

A. C. Mason
Integrative Behaviour and Neuroscience Group, University of Toronto Scarborough,
Toronto, ON, M1C 1A4, Canada

R. B. Cocroft et al. (eds.), *Studying Vibrational Communication*,
Animal Signals and Communication 3, DOI: 10.1007/978-3-662-43607-3_12,

## 12.1 Introduction

Since Shannon (1949), information theory has influenced studies of communication. Shannon's formal framework defined the components of a communication system as a sender exchanging information (encoded in a signal) with a receiver via a channel through which the signal must propagate. A key feature of this framework is the recognition that signaling channels are inherently noisy (i.e., signals are modified in unpredictable ways by properties of the channel), and the information content of signals is therefore uncertain. Consequently, in studies of biological systems, a major focus has been sender–receiver coevolution (Alexander 1962; Butlin and Ritchie 1989; Boake 1991; Endler 1992; Endler and Basolo 1998; Wiley and Shaw 2010), which largely deals with common effects on senders and receivers during information transfer. Several later theories (e.g., sensory exploitation, sensory bias) also considered the influence of preexisting sensory mechanisms on the evolution of signals and communication (Ryan 1990; Proctor 1992; Arak and Enquist 1993; Ryan and Rand 1993; Boughman 2002; Arnqvist 2006) and consequences of divergent interests in senders and receivers (e.g., antagonistic coevolution, chase-away selection; Arak and Enquist 1993; Holland and Rice 1998; Rowe and Arnqvist 2002; Arnqvist and Rowe 2005; Rowe et al. 2005; Arnqvist 2006; Parker 2006; Peretti and Aisenberg 2011). A common theme among these diverse approaches to animal communication has been the implicit assumption that the main effect of the signaling channel is to corrupt signal content via the introduction of noise. It has long been recognized that natural environments impose substantial and at times severe constraints on the ability of animals to transmit, detect, and process information (Aylor 1972a, b; Morton 1975; Michelsen et al. 1982; Bleckmann and Barth 1984; Bleckmann and Rovner 1984; Fleishman 1988). An understanding of the physical environment and its effects on animal physiology, behavior, and evolution is thus essential to understanding fundamental biological processes.

Alternatively, some views of sensory processing (particularly in the context of navigation and orientation) have focused on the role of predictable structure in the sensory environment as a driver of economical design in sensory systems. Wehner (1987) introduced the idea of "matched filters" in which spatial patterning of sensory stimuli (e.g., polarized light) may be reflected in a complementary filter in the sensory system, so that navigation and orientation problems may be solved by simple alignment of the sensory system with a spatial pattern in the environment, rather than by a more complex neural computation. This view is also extended to the central nervous system in the concept of "sensory focusing" (Bernays and Wcislo 1994), which suggests that dedicated feature detectors can reduce costs of information processing at higher levels. More broadly, these issues have come together in the last two decades under the banner of sensory ecology—the study of how animals identify, process, and respond to information about their environment (Dusenbery 1992; Chittka and Thomson 2001; Dangles et al. 2009; Stevens 2012). Much of the thought in this emerging field was fueled by the need to understand how information is modified, filtered, and/or distorted by often heterogeneous and chaotic natural environments. Several recent reviews have sought to synthesize sensory ecology research on different sensory modalities (Bell and Cardé 1984; Endler 1992; Kroodsma and Miller 1996; Archer et al. 1999; Barth and Schmid 2001; Hill 2001; Barth 2002; Gerhardt and Huber 2002; Greenfield 2002; Cardé and Millar 2004; Dicke and Takken 2006; Hill 2008; Elias and Mason 2011) and to identify broad generalities between different modalities (Bernays and Wcislo 1994; Dukas 1998; Endler and Basolo 1998; Boughman 2002; Dangles et al. 2009; Stevens 2012).

In this review, we focus on one aspect of the sensory ecology of communication that is particularly relevant to substrate-borne vibrations. It is often the case that the signaling channel does not simply introduce random noise into signals, but rather imposes systematic biases on the range and quality of signals that can be transmitted, and this results in selection for signals that are the best match to the filtering properties of the channel. We argue that this is particularly important for vibrational signals propagating in solid substrates because of the inherent heterogeneity of substrate signaling environments and their effects on mating behavior, signal evolution, and species diversification. We argue that substrate-borne signaling environments are distinctly diverse, multi-dimensional, and heterogeneous relative to visual, airborne, chemical, and electrical signalling environments. While transmission in airborne, chemical, and visual environments can be complex and vary as a function of abiotic conditions such as temperature, humidity, time of day and weather, average channel characteristics in these scenarios are relatively stable (Endler 1993; Bradbury and Vehrencamp 1998). This is a consequence of the fact that signals in these modalities typically propagate in a single medium (air or water) and usually do not traverse boundaries between media with different physical properties. In substrate-borne signaling, however, typical natural environments have many surfaces available (e.g., plant, litter, soil, rocks) and any surface can be a potential signaling channel. Many of these potential signaling channels are contiguous, and signals routinely traverse boundaries between similar (i.e., between contiguous leaves on the ground) and dissimilar (i.e., between rock and leaf) channels. For example, in leafhoppers, vibratory signals can be transmitted between plants that are not physically connected (Eriksson et al. 2011), and in the burrower bug *Scaptocoris carvalhoi*, vibratory signals produced in the soil can be detected on the leaves and stems of close by nearby plants (Čokl et al. 2006). In this review, we will discuss research on the transmission of substrate-borne vibrations in natural environments, the effect of substrates on mating behavior, and strategies for effective signal transmission. We argue that adaptations to signaling environments ("sensory drive") routinely lead to specialization and consequently signaling and/or species divergence and that studying vibration sensory ecology will be particularly informative as to the role of the environment in driving the evolution of sensory systems, behavior, and speciation. Below we discuss three types of variation/heterogeneity that are key components in substrate-borne sensory ecology: (1) variation in the physical form of transmission, (2) heterogeneity within a single transmission path (intrinsic variation), and

(3) diversity of possible channels for information transmission. Later we discuss practical experimental methods for studying substrate-borne environments with particular emphasis on the study of vibratory communication.

## 12.2 Variation in the Form of Information Transmission (Wave Heterogeneity)

Substrate-borne information is transmitted as waves that can come in a variety of different forms that are defined by, among other things, differences in the direction of energy propagation, in speeds of propagation, their dispersive properties, and in their attenuation properties (Achenbach 1973; Cremer et al. 1973; Markl 1983; Casas et al. 2007; Hill 2008). Substrate-borne waves include pure longitudinal, quasi-longitudinal, transverse, torsional, and bending waves (Achenbach 1973; Cremer et al. 1973; Markl 1983; Hill 2008). Substrate-borne information can be transmitted with one or multiple wave types, with each wave type having potentially different physical properties (Morris 1980; Hill 2008). The major determinates of the type of waves that can be transmitted are (1) the geometry and material properties of the conducting substrate and (2) the physical coupling of the source to the conducting substrate (i.e., are they on the surface of the substrate or are they surrounded by the substrate) (Achenbach 1973; Michelsen et al. 1982; Markl 1983; Barth et al. 1988; Magal et al. 2000; Tautz et al. 2001; Casas et al. 2007; Hill 2008). In general, four broad categories of substrate-borne environments have been described in the literature: (1) thin long rod-like structures, i.e., structures with a small diameter relative to their length, with vibration propagation primarily in one dimension. This category is exemplified by the stems and branches of many plants. (2) Thin plate-like structures with propagation in two dimensions. This category is exemplified by some leaves and leaf litter. (3) Substrates in which vibration may propagate in three dimensions. This category is exemplified by the ground (e.g., soil, sand), rocks, thick tree trunks, etc. (4) Organism-manufactured substrates such as spider webs and bee combs, which may combine properties of the first two categories above.

In thin rod-like structures, such as plant stems, Michelsen et al. (1982) demonstrated that, of all the wave types possible (quasi-longitudinal, transverse, torsional, and bending or flexural waves), bending waves propagated with the least attenuation. In fact, some signals were observed to travel up and down the plant several times (Michelsen et al. 1982; Barth et al. 1988; Čokl et al. 2007). Under experimental conditions, it was common to observe standing waves in the plant, and this has been hypothesized to allow signal propagation over long distances (Field and Bailey 1997; Čokl and Virant-Doberlet 2003; Čokl et al. 2007; Čokl 2008) and to ameliorate the effects of disturbance noise (Polajnar and Čokl 2008). Standing waves in plants would theoretically distort any directional or amplitude information (Michelsen et al. 1982; Miklas et al. 2003; Čokl et al. 2007). Animals, however, could compensate by either producing broad band, harmonic, or frequency-modulated signals

(Michelsen et al. 1982; Virant-Doberlet and Čokl 2004; Čokl et al. 2005; Čokl 2008) or producing signals at amplitudes that are insufficient to create standing wave conditions (Cocroft et al. 2006). Reflections at the tips of plants could also distort the temporal pattern of signals as was demonstrated in green stinkbugs (Miklas et al. 2003). Although standing and reflected waves clearly occur in plants under some conditions, behavioral trials have demonstrated no impediment to localization or mating, even though the exact mechanism of how this occurs is unknown. It is possible, however, that some animals can adjust signal amplitudes to reduce the impact of reflected and standing waves (see plant-borne signaling chapters, this volume).

While bending waves propagate with minimal attenuation, their propagation is also dispersive (speed of propagation is frequency-dependent), which has several potential shortcomings for information transmission since signal structure changes as the wave propagates along a stem (Casas et al. 2007). Given that the structure of signals is important in species identification and mate choice of many animals that communicate on plants (Henry 1994; Henry et al. 1999b; Virant-Doberlet and Čokl 2004; Cocroft et al. 2006; McNett and Cocroft 2008; Cocroft et al. 2010), Casas et. al (2007) suggested that animals could use non-dispersive waves either by signaling at higher frequencies or signaling on larger-diameter plants. Many animals include either percussive elements (with high-frequency energy) and/or high-frequency stridulations in their substrate-borne signals (Keuper and Kuhne 1983; Gogala 1985; Field and Bailey 1997; Parri et al. 1997b; Rohrig et al. 1999; Elias et al. 2003; Evans et al. 2005; Elias et al. 2006b; Miranda 2006; Hebets et al. 2008; Zunic et al. 2008; Evans et al. 2009) suggesting that at least some species may use multiple wave types. Unfortunately, in order to adequately address these questions, multiple measurement points using two or more devices are needed, and this has been rarely done (Magal et al. 2000; Cocroft et al. 2006; McNett et al. 2006). Many more studies like this are needed to understand if/how animals use multiple wave types to guide behavior. Careful behavioral studies, aimed at measuring the limits of signal discrimination (relative to the properties of available wave types), in combination with mathematical models describing wave propagation are needed as a first step to identify possible systems in which animals selectively attend to different wave components.

To date, we are aware of no studies explicitly examining the types of waves transmitted through plate-like structures such as broad leaves or leaf litter, although these substrates are routinely used by animals that can detect substrate-borne energy (Brownell 1977; Uetz and Stratton 1982; Stratton and Uetz 1983; Scheffer et al. 1996; Parri et al. 1997a; Kotiaho et al. 1998; Rivero et al. 2000; Parri et al. 2002; Uetz and Roberts 2002; Elias et al. 2004, 2006a, b; Gibson and Uetz 2008; Hebets et al. 2008; Hoefler et al. 2009; Uetz et al. 2009; Sivalinghem et al. 2010; Elias et al. 2012). Theoretically, plate-like structures will be able to transmit the greatest diversity of wave types (pure longitudinal, quasi-longitudinal, transverse, torsional), but given that most animals are not entirely surrounded by the substrate (but see Evans et al. 2005; Čokl et al. 2006; Evans et al. 2007; Fertin and Casas 2007; Evans et al. 2009), pure longitudinal wave signals are probably not used. It is likely that most animals use surface or plate waves (Cremer et al. 1973). Surface (or Rayleigh) waves occur in relatively thick solid substrates of a depth of at least one wavelength (Cremer et al. 1973). Close to the surface, the majority of compression occurs in the transverse direction, and particle motion is intermediate between the longitudinal and transverse direction (Cremer et al. 1973; Markl 1983). If the solid is relatively thin (the material is not more than a few wavelengths thick), then waves can propagate on both sides of the plate-like structure. These plate waves are known as Lamb waves and are very similar to Rayleigh waves in that the majority of compression occurs in the transverse direction and particle motion is intermediate between the longitudinal and transverse direction (Cremer et al. 1973). Rayleigh and Lamb waves are sometimes used interchangeably.

Another common substrate category is substrates that are thick relative to the wavelengths of vibrations produced. This includes substrates used by many animals that communicate on the ground, whether it be soil, sand, etc. Note that this does not include animals that signal on leaf litter, as the waves in those habitats are probably transmitted via leaf fragments with a plate-like structure (see above). This also likely includes large rod-like structures such as tree trunks (i.e., in which the diameter is much greater than the relevant wavelengths). Some pioneering work in substrate-borne sensory ecology was conducted by Brownell and colleagues who demonstrated that scorpions use surface Rayleigh waves in sand for prey localization and mating behavior (Brownell 1977; Brownell and Farley 1979; Brownell and Van Hemmen 2000). Rayleigh waves propagate for long distances and are thought to be important to a variety of organisms, including vertebrates (Brownell 1977; Brownell and Farley 1979; Aicher et al. 1983; Aicher and Tautz 1990; Brownell and Van Hemmen 2000; Mason and Narins 2001; Narins 2001; Randall 2001; O'Connell-Rodwell 2007) such as elephants, which are thought to communicate via Rayleigh waves over distances of several kilometers (O'Connell-Rodwell et al. 2000, 2001; O'Connell-Rodwell 2007). Pure longitudinal waves also occur in these substrates but are probably only used by animals that are completely surrounded by the substrate (e.g., antlions—Devetak et al. 2007; Fertin and Casas 2007; Devetak, Chap. 16, this volume; termites, Rohrig et al. 1999; mole crickets, Hill and Shadley 1997). Transverse surface waves (Love waves), where particle motion occurs in the transverse direction, are also possible. Scorpions have been shown to be able to detect this energy (Brownell 1977). Love waves may be produced by scorpions that use percussive glancing tail blows (Brownell and Van Hemmen 2000) and by the percussive claw strikes of fiddler crabs (Aicher and Tautz 1990).

The final substrate type is manufactured substrates and includes such structures as combs and webs (Herberstein and Tso 2011). Webs are special cases and can be thought of as stretched elastic fibers where transverse, pure longitudinal, and torsional waves are all possible (Masters and Markl 1981; Frohlich and Buskirk 1982; Klarner and Barth 1982; Masters 1984; Masters et al. 1986; Barth 1998). Bending waves do not occur in webs because silk lacks bending stiffness (Barth 1998). In webs, longitudinal waves propagate with less attenuation than transverse waves, and behaviorally, spiders react to lower amplitudes of longitudinal waves than to transverse waves (Masters and Markl 1981; Klarner and Barth 1982; Masters 1984).

Most of our understanding of the form of energy propagation in substrate-borne environments comes from a small set of seminal papers (Brownell 1977; Masters and Markl 1981; Frohlich and Buskirk 1982; Michelsen et al. 1982; Masters 1984; Aicher and Tautz 1990; Cocroft et al. 2000; Magal et al. 2000; Casas et al. 2007). Many more studies are needed on this important topic. In addition, it is unclear if/how/and to what extent animals can use different wave types to assess their environment. Each wave type may have very different physical properties, and animals may use different waves to acquire different information. For example, the dispersive property of bending and Rayleigh waves may be used to estimate distance from the source (Brownell 1977; Brownell and Farley 1979; Frohlich and Buskirk 1982; Michelsen et al. 1982; Brownell and Van Hemmen 2000; Cocroft et al. 2000; Čokl et al. 2009), while non-dispersive waves can be used to assess temporal and spectral structure in mating signals (Casas et al. 2007). Many animals that communicate in substrate-borne environments use multiple signals produced using multiple mechanisms (Dierkes and Barth 1995; Elias et al. 2003, 2006b; Hebets et al. 2008; Zefa et al. 2008; Zunic et al. 2008), and it is possible that different signals may be most adequately detected and/or decoded using different wave types.

## 12.3 Properties of Different Substrates and Within-Substrate Variation

Substrates impose severe constraints on information transmission. Each substrate has its own attenuation and filtering properties, and these can constrain wave propagation and, in the case of courtship signals, substantially modify and distort the spectral and temporal structure of signals. For signals, natural selection is expected to drive the evolution of signaling behavior that maximizes effective transmission (Endler 1992, 1993; Endler and Basolo 1998; Boughman 2002). For animals that signal on one or few substrates throughout their lifetime, this process is expected to drive signal–substrate matching, that is, signal properties (temporal and/or spectral) that are matched to the transmission properties of their environment (Cocroft et al. 2008; Joyce et al. 2008; Cocroft et al. 2010; Cocroft 2011). For example, in parasitoid wasps that use vibrational sounding (echolocating in solid surfaces) to locate hosts, species that target hosts found in only one substrate, find them faster and more efficiently (Fischer et al. 2003). One of the best examples is communication in the *Enchenopa binotata* species complex (Cocroft 2001; Cocroft and Rodriguez 2005). The *E. binotata* species complex consists of 11 plant specialists found sympatrically throughout North America (Wood 1993). Each member is adapted to feed and live on their plant host (Wood and Keese 1990; Wood et al. 1990). *E. binotata* planthoppers use substrate-borne signals transmitted along the stems of plants to mediate mating (Hunt 1994; Cocroft and Rodriguez 2005). Male mating displays consist of pure-tone signals followed by a series of pulses (Cocroft et al. 2008). Females show strong preferences for the signal frequencies produced by conspecific males (Rodriguez and Cocroft 2006; Rodriguez et al. 2006). The frequency of the pure tone in two of the species for which signal transmission was studied, transmits best in their host plant, strongly suggesting that host plant shifts favor divergence in male courtship songs and that the host plant channel drives the evolution of particular signaling characteristics (McNett and Cocroft 2008; Rodriguez et al. 2008; Cocroft et al. 2010; Sullivan-Beckers and Cocroft 2010). Similar adaptations in female response calls of *E. binotata* to their host plant environment have also been demonstrated (Rodriguez and Cocroft 2006).

Another well-studied system is that of the green stinkbug, *Nezara viridula*. Both males and females of this species duet using tonal, low-frequency vibratory signals on plant stems (Čokl and Virant-Doberlet 2003). Duets, and in particular female mating calls, are thought to mediate localization (Ota and Čokl 1991; Čokl et al. 2000; Čokl and Virant-Doberlet 2003). Once males and females locate each other, males continue to produce a courtship song that is thought to mediate species identification and possibly mate choice (Čokl et al. 2000; Miklas et al. 2003). When experiments were conducted on artificial substrates (a non-resonant speaker membrane), males differentiated between different female calling songs, but this was not the case when choice was measured on their natural host plant substrates (Miklas et al. 2001). This was potentially caused by the distortion of the species-specific temporal pattern due to reflections at the tips of the plant, which were substantial (Michelsen et al. 1982; Miklas et al. 2001). When signals produced on a speaker membrane (animal-induced vibrations alone, unmodified by transmission through plant) were compared to signals on a natural host plant (animal vibrations plus any plant-induced modification), large differences were observed (Čokl et al. 2005). The signal's fundamental frequency was significantly altered when the signaling animal was coupled to a natural (plant) substrate (Čokl et al. 2005). The propagated signal on the plant more closely matched the tuning characteristics of the bug's vibration receptors leading Čokl et al. (2005) to suggest that *N. viridula* used the resonance characteristics of the plant to extend the range of their calls. In this scenario, the animal sender excites resonances in the plant channel using narrowband body vibrations tuned to the plant's mechanical properties, which can then propagate for long distances (over several meters) due to the resonance characteristics of the plant (Čokl et al. 2005). In a later study, Polajnar et al. (2012) demonstrated that resonance properties of plants were indeed responsible for amplitude variations of transmitted signals. This signaling strategy is very efficient given that *N. viridula* are found on a variety of host plants (Panizzi 1997) and the filtering properties would be expected to vary between different plant species.

For manufactured substrates, signal–substrate matching is expected to be prevalent, and animals may take advantage of particular features of these substrates. For example, honey bees use particular areas of their comb with small open cells to produce substrate-borne vibrations (Sandeman et al. 1996). Small open cells facilitate transmission of substrate-borne signals (Sandeman et al. 1996), and dances done in these cells recruit 3X more workers (Tautz 1996). Honey bees appear to take advantage of the particular architecture of combs where cells at specific distances induce abrupt phase reversals in vibration signals (Tautz et al. 2001). Bees tend to straddle combs in such areas, and it has been suggested that this results in an amplification of the substrate-borne vibrations produced by bees (Tautz et al. 2001). In another study in the wasp *Asteloeca ujhelyii*, it was shown that substrate-borne vibrations propagated with high efficiency in nests, and the nest appeared to amplify specific frequencies in animal signals (Nascimento et al. 2005).

In addition to the strong effects on signals that are imposed by specific substrates, heterogeneity within substrates will further diversify the range of conditions under which signals must propagate. For example, Magal et al. (2000) investigated the propagation of waves along leaves. Different parts of the leaf transmitted vibrations differently, with the leaf lamina transmitting vibrations better than the leaf veins (Magal et al. 2000). Major veins acted as low-pass filters, suggesting that animals on different sides of a vein may receive different vibratory information (Magal et al. 2000; Čokl 2008). In one study on harlequin bugs, which signal on cabbage, it was shown that attenuation through veins was low relative to other areas on the cabbage plant (Čokl et al. 2004). This suggests that selection for information transfer may not only drive signal–substrate matching but also behaviors and signaling attributes may be matched to particular regions within a substrate. This parallels some observations in acoustic signaling systems, such as the singing from specific perch heights in the crickets *Anurogryllus arboreus* (Paul and Walker 1979; Arak and Eiriksson 1992), the use of baffles in some tree crickets (Prozesky-Schulze et al. 1975; Forrest 1982), and singing from naturally resonant cavities in some ground crickets (Bennet-Clark 1987; Walker and Figg 1990; Daws et al. 1996) and frogs (Lardner and bin Lakim 2002; Penna 2004). In studies of *Enchenopa,* species were observed not only to specialize on host plants but also to specialize on specific regions on the plants (McNett and Cocroft 2008). *Enchenopa binotata* "Cercis" signal on stems while *E. binotata* "Ptelea" signals on both stems and leaf petioles (McNett and Cocroft 2008). Experiments examining signaling on different plant areas demonstrated that signals were matched not only to their particular substrate but also to the sections of plants on which the animals were typically found (McNett and Cocroft 2008).

While environmental effects on signal evolution are expected, this was not supported in studies of green lacewings in the genus *Carnea* (Henry and Wells 2004). Green lacewings in the genus *Carnea* form a group of cryptic species found in North America, Europe, and Asia that are indistinguishable except through their vibratory songs (Henry et al. 1993, Henry 1994; Henry et al. 1999a, b, 2002; Henry and Wells 2010). Within the 15 cryptic species of *Carnea,* there are several examples of convergent evolution of song traits correlated with substrate preferences (Henry et al. 1999b; Henry and Wells 2004). However, in experimental manipulations comparing transmission of songs of two species, exemplars of which were recorded on a common artificial substrate, no evidence was found for plant-dependent filtering of animal songs (Henry and Wells 2004).

## 12.4 Diversity of Signaling Environments

In the previous section, we described situations where animals spend the majority of their lifetimes on one specific substrate. Under these conditions, signals and/or signaling behavior is predicted to match the average transmission properties of the environment (signal–substrate match). Many animals, however, do not spend their lifetimes confined to one or few substrates. For example, many organisms such as jumping spiders, wolf spiders, and some predatory bug species routinely move through different environments and are found on a diversity of surfaces (Elias et al. 2004; Hebets et al. 2008; Elias et al. 2010a; Elias and Mason 2011). In these situations, the transmission properties of the environment will impose strong constraints as to the reliability of information available. In particular, spectral content will be affected and may dramatically differ from substrate to substrate. In situations such as these, two communication strategies are possible: (1) a specialist strategy, where animals show signal–substrate matching to a specific subset of substrates available in the environment and signal preferentially on selected substrates from among those available. This strategy has a possible cost of lost mating opportunities but a benefit of increased information reliability (signal efficiency). (2) A generalist strategy, where animals produce signals that are effective across the range of environments they typically encounter. Furthermore, a generalist signaling strategy may be achieved in two ways. The first is by producing a "lowest-common-denominator" signal, i.e., one with properties that overlap all available channels as much as possible. This strategy comes with a cost of reduced information reliability (as the match for any given substrate may be less precise) but a benefit of increased mating opportunity (as there is some degree of matching on all available substrates). The second is to produce multiple or complex signals, in which specific signal components are matched to alternative substrates. We discuss these further below in the context of recent work in spiders. Future work is needed to see if the patterns observed in spiders are also present in other taxa.

Percussive signals do not contain specific spectral content. Instead, the signal at the source, to the extent that it approximates a brief impulse, contains all frequencies at equal intensities, and the propagated signal is solely a function of the properties of the substrate (Hebets et al. 2008). In addition, percussion requires no specialized morphology. Percussion is thus a good mechanism to produce signals that will propagate in a variety of habitats (Elias and Mason 2011; Uhl and Elias 2011). In one wolf spider species, *Schizocosa retrorsa,* males court females using percussion of the limbs against the substrate (Hebets et al. 1996, 2008). *Schizocosa retrorsa* can be found on a variety of substrates including pine litter, red clay, and leaf litter, and mating occurs equally in pine litter and red clay substrates, even though they influence signal propagation differently (Hebets et al. 2008). Interestingly, leaf litter propagates percussive signals with the least attenuation, and females were less likely to mate on that substrate. This is in contrast to the congener *S. stridulans*, which show a preference for mating on leaf litter (see below; Hebets et al. 2008). Hebets and colleagues suggested that percussion mechanisms may represent a generalist signaling strategy as it allows effective communication across a wide range of substrates (Hebets et al. 2008; Elias and Mason 2011). This hypothesis has yet to be explicitly tested in other animals that communicate in a variety of substrates. Some anecdotal evidence exists supporting this hypothesis. For example, in the European wolf spider *Hygrolycosa rubrofasciata* that use percussive signals, spiders can be found on several substrates (Kotiaho et al. 1996; Parri et al. 1997a; Kotiaho et al. 1998, 2000). Percussion is a ubiquitous method of substrate-borne signal production, and many animals, including ones that have specialized morphology, include percussive signals in their repertoire. The large variation in the spectra of transmitted information in percussive signals suggests that only temporal properties will be reliable and that generalist species will use only temporal information in mate selection. This hypothesis has yet to be tested. The variability in signal transmission has led some investigators to hypothesize that percussive signals can be used to estimate distance to the source due to the dispersive properties of signals (Michelsen et al. 1982; Dierkes and Barth 1995). This, however, could only be done on specialist species that have an a priori expectation of signal transmission, since variation in the filtering properties of heterogeneous substrates (see above) would confound potential spectral distance cues.

Specialist strategies have been described in several spider species (Rovner 1975; Stratton and Uetz 1981, 1983; Elias et al. 2004, 2006b; Gibson and Uetz 2008; Uetz et al. 2009; Elias et al. 2010a; Elias and Mason 2011). For example, in the jumping spider *Habronattus dossenus*, males and females can be collected on a large variety of surfaces including leaf litter, sandy soil, and rocks (Elias et al. 2004). Each substrate type has very different transmission characteristics, and males court indiscriminately on all substrates (Elias et al. 2004). Females, however, are more likely to mate with males on leaf litter over other available substrates (Elias et al. 2004). Of all the available substrates, leaf litter best transmits the frequencies present in male signals, suggesting that *H. dossenus* signals are specialized to leaf litter environments over other available environments (Elias et al. 2004). In the wolf spider, *Schizocosa stridulans,* substrate-borne signals strongly influence mating success (Hebets 2008). Individuals can be collected on a variety of habitats including pine litter, leaf litter, and sand (Elias et al. 2010a). Substrate-borne signals transmit with the highest efficiency in leaf litter environments, and females are more likely to mate with males on leaf litter over the other available substrates, once again suggesting that spiders can specialize in one available signaling environment over other equally available environments (Elias et al. 2010a). This observation led Elias and Mason (2011) to suggest that specialization in signaling microenvironments may be a mechanism promoting speciation across small invertebrates.

Associated with specialist strategies is morphology and signaling behavior that produce signals with specific spectral properties and may include stridulatory plectrum and file mechanisms, tremulatory movements (vibration of body parts), tymbal mechanisms, and "stick and slip" friction mechanisms (Uetz and Stratton 1982; Chapman 1998; Patek 2001; Gerhardt and Huber 2002; Čokl and Virant-Doberlet 2003; Delaney et al. 2007; Patek and Baio 2007; Uhl and Elias 2011). For example, stridulatory files in spiders are common and allow for the production of substrate-borne vibrations with high-frequency spectral attributes (Rovner 1975; Elias et al. 2003, 2006b; Uhl and Elias 2011). Tremulatory movements are also observed in many spiders and allow for the production of tonal or frequency-modulated signals (Rovner 1980; Uetz and Stratton 1982; Elias et al. 2003, 2006b, 2010b; Sivalinghem et al. 2010; Uhl and Elias 2011). Some spiders use multiple mechanisms to produce substrate-borne signals with a variety of spectral attributes (Elias et al. 2003, 2006b). For specialist species, information in both the spectral and time domain is theoretically available. Temporal content has been demonstrated to be important in a variety of specialist signalers (Elias et al. 2008; Gibson and Uetz 2008; Shamble et al. 2009; Byers et al. 2010; Elias et al. 2010b; Sivalinghem et al. 2010; Cady et al. 2011). Spectral information has been demonstrated to be very important in treehoppers that are host plant specialists and use specialized signaling behaviors and morphology (Cocroft et al. 2008; McNett and Cocroft 2008; Rodriguez et al. 2008). While equivalent experiments have not been conducted in spiders, evidence does suggest that female wolf spiders assess the spectral content in male stridulatory signals (Gibson and Uetz 2008). Future work is needed to test the hypothesis that specialist signalers use spectral information in mate choice. Regardless, in all the experiments on the effects of substrate where substrate mating preferences were observed, males exhibit specialized morphology or behaviors associated with the production of substrate-borne signals with stereotyped spectral and temporal properties (Elias and Mason 2011; Uhl and Elias 2011).

Although the evolution of specialist strategies and signal–substrate matching has been observed in a variety of spider species, recent work has also suggested behavioral adaptations for signaling in non-optimal environments. In these situations, males decrease rates of vibratory signals and increase rates of visual signaling (Taylor et al. 2005). For example, males of the spider *Schizocosa ocreata* are much more likely to court, and females more likely to mate, in leaf litter environments over other available environments (Gordon and Uetz 2011), suggesting that *S. ocreata* are signaling specialists. In non-leaf litter environments, however, male spiders use visual signals at a higher rate, suggesting that they are compensating for signaling in non-optimal environments (Gordon and Uetz 2011). Conditional use of signaling may be a mechanism to maximize signaling opportunity in a variety of environments for specialist signalers.

## 12.5 Practical Issues and Experimental Methods for Studying Vibratory Communication

While wave heterogeneity, habitat heterogeneity, and channel habitat diversity are major reasons why the study of vibratory sensory ecology is so fascinating, these properties present some major experimental challenges for scientists. In the following section, we will discuss issues relating directly to measuring transmission characteristics in substrates and the use of artificial versus natural substrates in behavioral experiments. The complexity of wave propagation, and the large effect of substrates on wave propagation, makes these issues substantial and complex. Necessarily, there are many trade-offs associated with different techniques. It is not our purpose to outline the "perfect" solution but instead to discuss some of the trade-offs and problems inherent with different techniques. It is our hope to stimulate discussion on different techniques in the hopes that common standards for the calibration of substrate effects in experimental paradigms can be implemented in the future.

### *12.5.1 Distance from Sender to Receiver*

Before discussing techniques for measuring transmission and conducting behavioral experiments, it is important to consider the distances over which vibratory signals are transmitted. The greater the distance from source to receiver, the stronger the effect of substrate on information transmission. The further away the source is located, the stronger the effect of attenuation, filtering, and reflections. For some animals such as the jumping spider *Habronattus dossenus*, which are found on a variety of signaling environments, long-distance communication does not include any vibratory signals, and instead, vibratory signaling only occurs when the male is within 1 to 2 body lengths from the female (Elias et al. 2003). In these cases, the effect of substrate on signal transmission will be minimized, although it is still significant (Elias et al. 2004). For animals such as planthoppers and wolf spiders that use vibratory signals to communicate at longer distances, signaling substrates are expected to exert stronger effects, especially if the species can be found on a diversity of substrates (i.e., wolf spiders). For animals signaling on plants (i.e., planthoppers and treehoppers), the effect of reflections or standing waves is expected to depend on distance from the source (and from the plant tips). Considering that signals may comprise multiple wave types and that different waves propagate at different speeds, complex interactions of different wave types will increase with distance from the source. Given this, for biologists interested in understanding senders, it is crucial that signals be recorded as close to the source as possible to minimize substrate effects. For biologists interested in understanding choice (or receiver responses in general), it is crucial to record signals as close to the receiver as possible to capture the full range of signal complexity as it must be evaluated by the receiver. For animals that regularly move around during courtship, this has made understanding signals and mate choice challenging. Nevertheless, the strong effects of substrate on vibratory transmission make it crucial to standardize the positioning of the recording device to the source (signal characterization) or receiver (mate choice) before any inferences can be made regarding the effects of vibratory information—in particular aspects having to do with amplitude, frequency, and/or fine temporal structure. This includes "simple" homogenous natural or artificial substrates because of the complex and unpredictable effects that substrates impose on signal transmission (Elias and Mason unpublished information).

### 12.5.2 Methods for Measuring Information Transmission in Vibratory Environments

There are several things to keep in mind when attempting to measure information transmission through vibratory environments. First is the heterogeneity implicit in vibratory transmission (see above). When conducting experiments, especially on natural substrates, it will be of critical importance to include a broad enough sample of replicates to account for heterogeneity in transmission for a single substrate type (Cocroft et al. 2006). Secondly it is important to consider the methods used to quantify vibratory environments: Below we discuss methods of initiating vibratory energy (actuation) and next we discuss methods to measure vibratory propagation.

#### 12.5.2.1 Actuators

In general, there are three basic techniques that can be used to input energy into a substrate: (1) percussive impacts, (2) moving elements attached directly to substrates, and (3) sympathetic vibrations of substrates elicited by airborne vibrations. The general idea is to use an input signal of known frequency and amplitude and then measure the vibrations on the surface. With this information, at each measurement point, one can calculate a transfer or gain function describing the filtering characteristics of a substrate. By measuring the propagated vibrations at different points relative to the stimulus (in the case of impacts and moving elements), attenuation functions can be computed for different frequencies.

It is of crucial importance to keep one thing in mind when starting these experiments: How does the actuator move at the source? Depending on the input signal and the actuator type, different movements are possible. Typically input signals can be (1) impacts, (2) white (or band-limited) noise, (3) frequency sweeps (or periodic chirps) of constant amplitude, or (4) an animal signal. Impacts, in theory, have all frequencies of equal intensities at the source, so assuming that the stimulus is correctly created and applied, no further calculations are needed and simply analyzing substrate vibrations at each point is sufficient to calculate transfer functions. Note, however, that the transient signals produced by surface impacts are difficult to analyze with typical methods (i.e., fast Fourier transforms), and instead, we suggest other methods such as wavelet analysis (see Casas et al. 2007). White noise stimuli also theoretically include all frequencies at equal intensities, but most actuators have limitations as to the frequencies and amplitudes that can be produced. True white noise is therefore difficult to produce. Close attention must be paid to the properties of the actuator and the way the actuator is coupled to the substrate. We suggest that white noise be carefully evaluated before use, for example, by measuring vibrations directly off the moving element and verifying that the actuator is producing a "proper" (or at least well-characterized) noise spectrum. Pure white noise is probably not a possibility in most commercially available actuators, and we suggest that biologists instead design a band-limited noise stimulus matched to the properties of the actuator and the frequency range of interest, if this is known a priori. Frequency sweeps or periodic chirps of known frequencies and intensities are a practical alternative.

Care must be taken that transfer functions are calculated using the correct mode of vibration. For example, many researchers use piezoelectric devices, which are displacement actuators (i.e., signals at different frequencies will have equal displacement), or mini-shakers, which are acceleration actuators (i.e., signals at different frequencies will have equal acceleration of the actuator and displacement will decrease with increasing frequency). Measurement devices can likewise measure displacement (piezo-devices), velocity (laser vibrometers), or acceleration (accelerometers), and care must be taken to compare displacement to displacement, velocity to velocity, etc., when constructing transfer functions. When the actuator and measurement device are calibrated for different waveform parameters, we suggest that a new input signal be designed that properly matches up with the measurement device or the input (or output signal) be properly corrected (i.e., derive acceleration signal to velocity signal). This will especially be the case with animal signals, and it is essential that the properties of the actuator be accounted for to ensure that animal signals are properly reproduced. In most cases, the most reliable method will be to calculate the transfer function between two measured points, rather than between the actuator and a single measurement point.

Casas and colleagues have used percussive impacts with great success (Casas et al. 1998; Magal et al. 2000; Casas et al. 2007). In particular, percussive stimuli are excellent at producing very high frequencies that are nearly impossible to reproduce using other commonly used actuators (speakers, shakers, piezos). Percussive stimuli are typically produced by dropping an object (usually a ball bearing) from a known height using an electromagnet (Magal et al. 2000; Casas et al. 2007). Measuring vibrations close to where the ball bearing was dropped was difficult, however, because of interference between the measurement device (in this case a laser vibrometer) and the dropped ball (Magal et al. 2000; Casas et al. 2007). Another method for creating impacts is to use a high-amplitude square wave stimulus sent to a moving element actuator that is not initially attached to the substrate. By positioning the actuator at a proper distance from the surface of interest, tapping impacts can be produced, although the frequencies produced will be limited by the speed of the actuator. This will limit the range of frequencies that can be input into the surface.

The most common actuator used by investigators has been devices with moving elements that are attached firmly to the substrate and include mini-shakers, piezo-devices, and speaker cones (Brownell 1977; Michelsen et al. 1982; Miles et al. 2001; Elias et al. 2004; Čokl et al. 2005; Hebets et al. 2008; McNett and Cocroft 2008; Elias et al. 2010a). By firmly attaching the moving element to the substrate, investigators can ensure that the vibrations of interest are properly transmitted into the substrate. Unfortunately, attaching the actuator changes the properties of the substrate by loading the surface. One must take care to minimize these effects by either (1) loading the substrate in a way that approximates that of an animal

(Michelsen et al. 1982; Elias et al. 2004; Čokl et al. 2007) or (2) vibrating the entire surface (as opposed to a point on the surface) by mounting the entire substrate on an actuator. Vibrating the entire surface, however, makes calculating attenuation characteristics problematic. Often, the geometry of the surface and/or the actuator types available makes this type of effect unavoidable. Using sympathetic vibrations induced by a speaker is another technique to characterize substrate properties. Like base stimulations, calculating attenuation characteristics is problematic, as there is no localized position of energy input. A final technique is to vibrate a surface using a magnet glued to the surface and an electromagnet to drive the attached magnet (Michelsen et al. 1982; Čokl et al. 2007). This can be an effective and relatively inexpensive technique, but is notoriously difficult to calibrate and control (Paul De Luca, pers. comm.).

We *strongly* suggest that all input signals be recorded at the source, preferably throughout the course of the entire experiment. For example, an accelerometer attached to the moving element of a shaker throughout the course of the experiment (while a second measurement device records the propagated vibrations at different points). Using this technique, it will be possible to construct accurate transfer functions at every measurement point and compensate for any shortcomings in actuators or the design of the input signal.

#### 12.5.2.2 Measurement Methods

There are currently three major techniques for measuring substrate-borne vibrations: (1) laser Doppler vibrometry, (2) accelerometers, and (3) piezoelectric transducers. Each technique has its own particular nuances, advantages, and disadvantages, discussed below.

The best method available is a laser Doppler vibrometer (LDV). LDVs measure the velocity of a moving object using a laser beam reflected off a point on a surface. LDVs' major advantages include (1) that no contact is needed between the LDV and substrate, hence the properties of the substrate are unaffected by loading, (2) high sensitivity, and (3) high-frequency bandwidth (DC to GHz). Disadvantages include (1) limited capacity for measuring high-amplitude vibrations (see Clark et al. 2011), (2) cost, and (3) limitations in the direction of motion measured (only motions perpendicular to the laser are detected) (McNett et al. 2006). In order to make measurements using the LDV, there must be an unobstructed line of sight between the LDV and substrate. In addition, there must be sufficient reflections from the surface in question. Roughness and heterogeneity of the substrate will increase backscatter and reduce the signal-to-noise ratio of the data. There are several techniques to improve reflection, in particular painting surfaces with reflective white paint and/or attaching reflective materials or objects to the surface. In addition, filtering the laser data can vastly improve the ability of the LDV to measure vibrations of interest and remove spurious transients that can be common when measuring surfaces in the field.

Care must be taken to standardize the position of the laser relative to the surface of the substrate, because different wave types can propagate in different planes and the LDV only detects vibrations perpendicular to the laser beam. It is for the most part unavoidable when using a single LDV (or other devices maximally sensitive to one direction of motion) to underestimate the amplitude of some propagated waves. There are currently 3-dimensional LDVs that use three independent lasers to make measurements, but these are prohibitively costly for most biological laboratories. If two LDVs are available, there are techniques developed using orthogonally positioned points to characterize the major axis of motion (McNett et al. 2006). For some geometries such as thin plates, these types of techniques will be difficult to implement. When possible we suggest that the major axis of motion be identified (McNett et al. 2006) and an experimental paradigm established to keep the angle of the laser relative to the surface constant. In situations where it is not possible to identify the major axis of motion, we suggest that the angle be maintained consistently across experiments and in the direction that makes sense, given the data available (perpendicular to the surface in animals communicating with Rayleigh or Lamb waves, as parallel to the surface as possible in spider webs, etc.). Much more work is needed to model vibrations in objects of different geometries to determine the major axes of motion, and measurements using multiple devices orientated at different angles relative to the substrate. Until then, investigators need to be aware of the possibility of underestimating the amplitudes of propagating waves.

Accelerometers are a less costly alternative to LDVs. Accelerometers measure the acceleration of the surface to which they are attached. Accelerometers' major advantages include (1) relative low cost, (2) ability to measure high-amplitude accelerations, (3) the commercial availability of multi-axis accelerometers, and (4) they are easily transportable. Disadvantages include (1) the measurement device places a load on the surface that may alter the vibrational properties of the substrate, (2) lower bandwidth, and (3) lower sensitivity.

The final category of measurement device is piezoelectric elements. Piezo-devices measure the displacement of a surface and are routinely used as actuators as they can produce large displacements (Brownell 1977; Elias et al. 2004). As a measurement device, the main advantage of piezo-devices is that they are low-cost, come in a variety of shapes and sizes, and are found in a variety of readily available devices (e.g., phonograph cartridges). Piezo-devices have disadvantages in (1) low bandwidth, (2) low sensitivity, and (3) the devices place a load on the surface.

## 12.6 Behavioral Experiments: The Artificial Versus Natural Substrate Conundrum

Given that substrates impose such strong constraints on the propagation of substrate-borne information, when designing experiments on animals that use substrate-borne information, one common concern is what substrate is best to use.

A simple solution would be to use the substrate on which animals are naturally found, but this approach has some substantial problems. First, natural substrates of the same type can vary, often to a great extent (Cocroft et al. 2006). For questions on mate choice, it is often preferable to minimize the effect of substrate, hence artificial substrates can be of great use. In fact‘ in studies of green stinkbugs, mate choice for signals was only observed when experiments were conducted on artificial non-resonant surfaces (Miklas et al. 2001, 2003). Second, if the question of interest focuses on the vibration source (e.g., a sender's signal or cue), then a measurement through a substrate could distort the recorded signal. This is best illustrated again by work on green stinkbugs (Virant-Doberlet and Čokl 2004). Males produce vibrations with their bodies that have certain species-specific temporal and spectral patterns, but the propagated signal includes additional frequencies that the animal does not produce but instead are generated by resonant properties of the plant (Čokl et al. 2005). Questions of how males produced and emitted signals can best be answered by recording behavior on an artificial substrate with no resonant properties in the frequencies of interest (Čokl et al. 2005). Third, many animals may encounter a variety of natural substrates in the field, and experimentally, it may make more sense to conduct choice experiments on controlled artificial substrates.

That is not to say that experiments on natural substrates are not important, in fact, we feel that experiments on natural substrates are *crucial* to understanding the sensory ecology of vibrations. We advocate the use of artificial substrates for questions of signal production and when it makes experimental sense to control for substrate effects. Artificial substrates allow experimenters to design arenas of standardized dimensions and filtering properties. We advocate a combination of artificial and natural substrates for questions on mate choice, habitat choice, and mating behavior. Below we discuss general issues in the selection of artificial substrates.

### 12.6.1 Artificial Substrates

Before selecting an artificial substrate, it is important to know the bandwidth of the vibrational signal in question. Artificial substrates can vary substantially in their properties (Figs. 12.1, 12.2, Table 12.1), and it is important to select substrates that do not differentially distort biologically relevant information. Second, it is important to select a substrate in which an animal has the power to elicit vibrations (or not, if the purpose is to eliminate vibrations). For example, we present data (Fig. 12.1, Table 12.1) showing that in wood substrates, most frequencies are amplified by about 10 dB relative to vibrations produced on a leaf. This assumes that the energy input into the surface is sufficient to elicit substrate-borne vibrations. For small animals, producing vibrations of sufficient amplitude to induce vibrations in a structure such as a wooden board will be a greater challenge than inducing vibrations in filter paper. Once vibrations of sufficient amplitude are

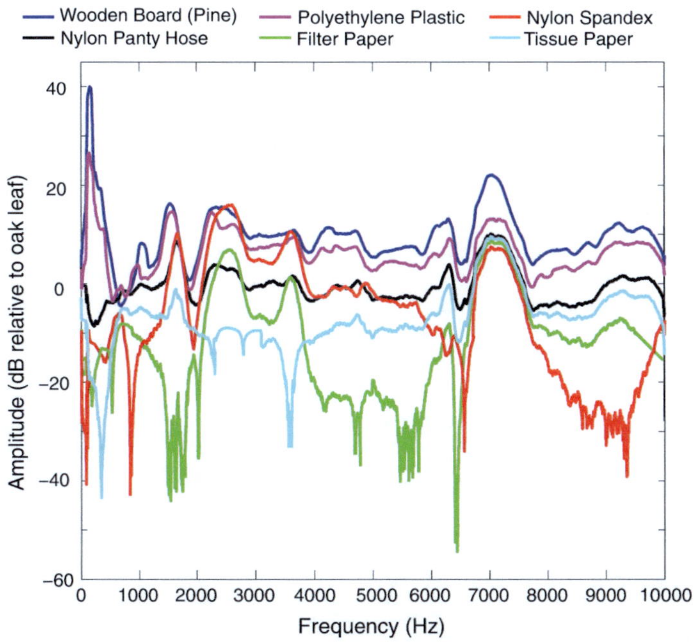


**Fig. 12.1** Vibration amplitude spectra for select artificial materials. Data are presented in dB relative to vibration responses of a natural substrate (oak leaf). *See text* for details of methods. Materials vary in their overall sensitivity and frequency-dependent filtering. These differences could significantly affect the quality and characteristics of communication signals recorded or broadcast in behavioral experiments

induced, our data show that vibrations in wooden boards are greater than in tissue paper ($\sim$10 dB vs. $\sim -12$ dB). The power needed to elicit vibrations in different substrates has not been explicitly studied in the literature, and such information is sorely needed. In one set of experiments where vibrations were produced using a piezo-transducer and recorded using a laser vibrometer, it was demonstrated that at biologically relevant amplitudes, vibrations on granite were on average $-65$ dB lower than the input signal (Elias et al. 2004). At higher amplitudes (or if transfer functions were calculated using base stimulations of the entire rock), resonances occurred at many frequencies (data not shown) and vibrations transmitted through rock with relatively little attenuation (Elias et al. 2004). But vibrations elicited on granite were very small at the input amplitudes tested (Elias et al. 2004). If animals produce signals with enough power, signaling on hard, dense surfaces is possible and can be very efficient (O'Connell-Rodwell 2007). This may explain why in experiments on the drumming wolf spider, *Schizocosa retrorsa*, courting on granite (which was assumed to eliminate all substrate-borne vibrations as was the case in other studies: Hebets 2008; Hebets et al. 2008; Elias et al. 2010a; Rundus et al. 2010) had little effect on mating success (Rundus et al. 2010). It is possible that drumming signals are of sufficient amplitude to propagate on granite.

**Table 12.1** Vibration Characteristics of artificial substrates

| Surface | Amplitude (0.01–100 Hz) relative to leaf (dB) | Amplitude (101–1000 Hz) relative to leaf (dB) | Amplitude (1001–4000 Hz) relative to leaf (dB) | Amplitude (4001–10000 Hz) relative to leaf (dB) |
|---|---|---|---|---|
| Wooden board (pine) | 12.6211 | 9.512 | 9.7436 | 9.6145 |
| Nylon panty hose | −0.6181 | −3.9072 | 0.6126 | −0.5766 |
| Polyethylene plastic | 4.1964 | 5.8872 | 7.536 | 5.6611 |
| Filter paper | −18.8571 | −18.3959 | −11.44 | −15.7159 |
| Cotton twill | −9.6526 | −15.136 | 6.5539 | −7.1778 |
| Nylon spandex | −28.2133 | −17.1691 | 3.3683 | −11.517 |
| Craft foam board | −14.1817 | −17.5717 | −12.5942 | −16.8973 |
| Soft tarlatan | −4.1808 | −9.3812 | −10.3678 | −3.9237 |
| Poster board | −10.7419 | −16.7522 | −12.519 | −14.124 |
| Tissue paper | −7.5828 | −14.6786 | −14.2668 | −4.7292 |

Experiments are needed to determine the power necessary to elicit vibrations of sufficient amplitude that they can be detected by vibratory receptors (Barth 2002). Thirdly, it is preferable to select substrates that maximize the signal-to-noise ratio. Many artificial substrates amplify ambient noise found in laboratory environments (1,000–4,000 Hz), and it is often preferable to select substrates that do not selectively amplify those frequencies but that nevertheless transmit them if they are present in the signals of interest.

For this review, we recorded the filtering properties of a variety of artificial substrates that can be easily obtained. For most materials (excluding the wooden board, polyethylene plastic, and oak leaf), we placed each substrate in a 55-cm-diameter fabric circle and stretched the materials as tautly as possible in the frame. The wooden board and plastic materials were cut approximately 55 cm in diameter. Next, we constructed a 25-cm-high wooden table with a small hole cut out of the center (12 cm). Directly underneath the wooden table, we placed a B&K mini-shaker (Type 4810) with an accelerometer (PCB Piezotronics Model 485B36) glued to the moving element. Each fabric circle was centered on the wooden table top. We fastened the substrate to a small screw mounted on the mini-shaker using super glue. We followed similar procedures for the other substrates and fastened the center of the substrate to the screw. Substrates were stimulated using a periodic chirp stimulus (bandwidth 0.01–10 kHz). We chose periodic chirps to minimize spurious impulses caused by the on- and offsets in frequency sweeps. Substrates were measured using a scanning laser vibrometer (SLDV- Polytec SLV-400, OFV-505 scan head) in a circular array consisting of 40 points. Each point was measured 5 times and averaged. For each point, we calculated a transfer function relative to the input signal recorded using the accelerometer mounted on the mini-shaker (after converting acceleration to velocity). We present average transfer functions for each substrate across all the points. We present amplitude data in decibels (dB) relative to a measurement of a natural substrate (oak leaf) (Fig. 12.1).

We measured 4 types of cloth substrates (nylon pantyhose, cotton twill, nylon spandex, and soft tarlatan), 3 types of paper (filter paper, poster board, tissue paper), and 3 miscellaneous substrates (foam, wood, plastic) and present the average filter functions for a subset of them (Fig. 12.1), as well as a table outlining the average filtering for frequency bands of interest to biologists for all substrates (Table 12.1). Relative to a leaf litter substrate, the least distortion of frequencies between 0.01 Hz and 10 kHz occurs with nylon pantyhose. From 101 to 1000 Hz, there is a slight amplification ($\sim$4 dB), but this amplification is relatively minor compared to other substrates tested. Wooden beam and polyethylene plastic substrates are also relatively flat for most of the frequencies tested (5–12 dB amplification) but show resonances at 300 and 340 Hz, respectively. Paper substrates and some cloth substrates show complex and potentially problematic transmission spectra. Filter paper substrates, for example, show maximal transmission in bands between 200–420, 2100–2800, and 6700–7600 Hz. In other frequency bands, differences of up to 40 dB were observed, suggesting that animal signals may be significantly distorted if transmitted on filter paper. Nylon spandex also severely distorts signals with significant amplifications in bands from 1,000 to 5,000 Hz and reductions from 6000 to 9500 Hz. Note that these measurements show average transmission characteristics across a large area (22 inches) and would be ameliorated if signals are transmitted at close range. If information travels longer distances, extra care must be taken to select the proper artificial substrates for experiments. Also, note that these measurements take into account a single substrate, and often in experiments, multiple substrates are placed on top of each other, for example, filter paper is placed at the bottom of plastic container or table top. In these instances, the transmission spectra will be different than that those reported.

While nylon pantyhose, soft tarlatan, tissue paper, and plastic substrates appear to be the "best" substrates to reduce heterogeneity in frequency transmission, many other factors must be taken into account by biologists (see below). For example, pantyhose and tarlatan substrates represent a mesh-like structure consisting of knotted fabric and open space. Many organisms do not behave normally on these surfaces, and when focusing a laser on the mesh, it is a challenge to stay on the fabric threads and not the floor underneath the arena. Plastic substrates may also be too dense for animals to elicit vibrations.

Regardless of the method used, there are numerous trade-offs to consider when selecting a substrate. We suggest that preliminary observations and recordings on natural substrates be conducted to (1) determine the frequencies of interest, (2) determine the amplitude of vibrations to ensure that animals are transmitting information on the artificial substrate, and (3) ensure that animals behave in a manner consistent with the field. This information will allow investigators to best choose an artificial substrate that will allow proper inferences to natural conditions.

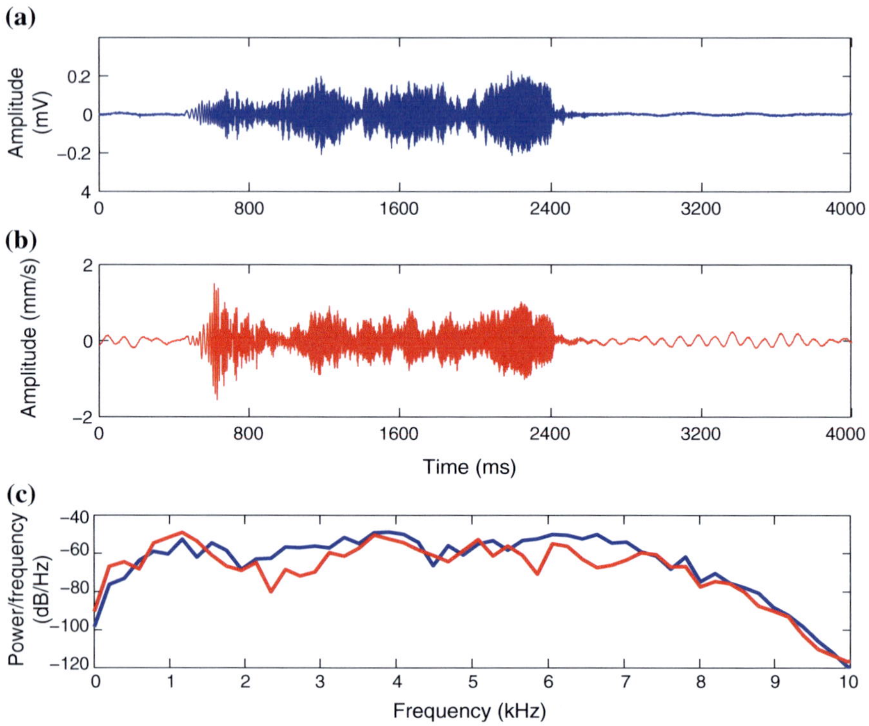


**Fig. 12.2** Sympathetic vibration of tissue paper in response to airborne sound. **a** Microphone signal of frequency sweep broadcast at 90 dB SPL. **b** Surface vibration of tissue paper recorded via laser Doppler vibrometer. **c** Power spectra for microphone and sympathetic surface vibrations. Tissue paper acts as an effective microphone, copying the time–amplitude pattern of airborne sound with high fidelity and sensitivity

We also advocate *always* conducting a follow-up set of experiments on natural substrates. The potential effects of substrate choice are best exemplified in experiments on the parasitoid wasp, *Cotesia marginiventris*. Wasps were raised on a variety of artificial (plastic, glass, chiffon fabric) and natural (maize and bean leaves) substrates (Joyce et al. 2008). *C. marginiventris* males court females using vibratory songs (Joyce et al. 2007), and mating success was highest on the artificial chiffon substrate and lowest on glass and plastic substrates (Joyce et al. 2008). Natural substrates had intermediate levels of mating success (Joyce et al. 2008), demonstrating (1) how important the choice of substrates is for mating trails and (2) the importance of comparing results with natural substrates.

In addition to measurements using directly induced vibrations in these substrates, we also measured their sensitivity to sympathetic vibrations induced by airborne sound, as this is a common potential source of noise contamination in studies of substrate communication signals. Using the same substrates listed above, we broadcast a frequency sweep (0.1–10 kHz) via a stereo speaker, with

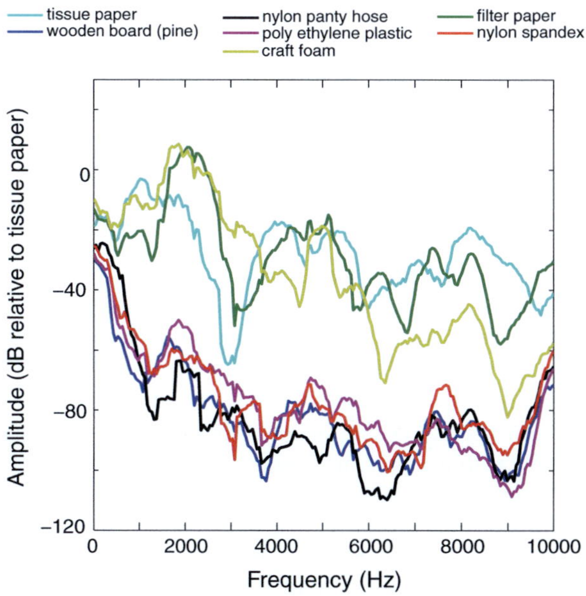


**Fig. 12.3** Sympathetic vibration amplitude spectra for select artificial materials. Data are presented in dB relative to tissue paper (*see text* and Fig. 12.2). Materials vary considerably in their sensitivity to vibration induced by airborne sound. Dense, massive materials (such as wood and plastic) are relatively insensitive, but are likewise difficult for signaling animals to excite. Several paper and fabric substrates are highly sensitive to airborne sound and therefore prone to contamination from background noise. "Stretchy" fabrics, such as nylon, show very low sensitivity to sympathetic vibrations

each substrate resting on a vibration-isolated table. Sound was recorded with a microphone (B&K 4138) positioned within 4 cm of the substrate surface and calibrated to 90 dB SPL (B&K Type 2610) at that location (Fig. 12.2). Sympathetic vibrations of substrates were recorded by LDV (Polytec OFV 3001) at a single, arbitrarily selected location near the center. Substrates varied considerably in both their overall sensitivity to sound-induced vibration and in their filtering properties (Fig. 12.3). The most sensitive substrate under these conditions (i.e., with highest amplitude of induced vibration) was tissue paper (Fig. 12.2), which acted as an effective microphone, capturing the waveform and spectrum of the broadcast sound quite accurately. Other paper types were similar, but with slightly lower amplitudes. For comparison, we present amplitude spectra for transfer functions of representative substrates, calibrated in dB relative to tissue paper (Fig. 12.3).

The data in Fig. 12.3 emphasize the importance of considering the interaction between recording conditions, substrate properties, and the characteristics of the signals of interest in designing studies of substrate communication. Furthermore, these results modify some of the apparent conclusions regarding the suitability of potential artificial substrates derived from measurements of directly induced vibration alone. Specifically, paper substrates appear to have similar transmission characteristics to those of soft "stretchy" fabrics (such as nylon pantyhose): A relatively flat frequency response that should introduce little bias in propagated vibration signals. Clearly, however, paper (and foam) substrates are far more sensitive to contamination by background, airborne sound, and this could introduce two problems in behavioral studies: (1) poorer signal-to-noise ratios in recordings of animal signals and (2) uncontrolled vibrational noise for receivers on these substrates (e.g., in behavioral trials).

Not surprisingly, dense, massive substrates (wood, plastic) were relatively insensitive to sympathetic vibration (60–80 dB below tissue paper over most of the tested range). Nylon fabrics (pantyhose, spandex) were comparable to wood and plastic in showing little or no sympathetic vibration in response to airborne sound. This combined with their flat frequency responses to directly induced vibration and favorable attenuation characteristics suggests that they are the most reliable of the substrates we tested (with the possible limitation that the spacing of the weave may affect animal behavior as mentioned above).

These results are not intended as an exhaustive study of possible substrates, nor as a "certification" of specific substrates. Rather, our purpose is to highlight the range of issues that must be considered in studies of substrate-borne communication signals. Given that such signals have evolved in the context of heterogeneous substrates, with different species adapted to varied substrates and a diversity of possible strategies for substrate–signal matching, the selection of materials and conditions for controlled studies is critical. A simple comparison of signals recorded very close to a signaling animal with recordings of the same types of signals at the outermost edges of the experimental arena (or at the maximum signaling range) is an excellent way to evaluate the extent of distortion imposed by artificial substrates.

## 12.7 Summary and Conclusions

1. Substrate signaling is inherently more complex than other signal modalities, at least in terms of the effects of the medium on propagation. This is largely due to structural (mechanical) heterogeneity of the environment. There are many possible substrates, each with potentially distinct physical properties that may regularly affect transmission between a sender (animal producing substrate-borne signals) and a receiver. In addition, on any given substrate, depending upon its composition, shape, and dimensions, multiple wave types and patterns of propagation are possible. It is not always clear a priori which parameters of a substrate-borne waveform are relevant to a receiver. Since each wave type may be affected differently by variation in substrate properties, there is potentially a combinatorial expansion of signal parameters that must be considered in analyses of substrate communication systems.

2. Signaling adaptations among animals using substrate-borne communication may take several forms. Some species may specialize in specific substrates from among those available and evolve signals that are tuned to match their transmission properties (e.g., frequency filtering, propagation patterns). Others may adopt a generalist strategy and evolve signals with broader characteristics, such that some signal energy will propagate in all, or most, available substrates. This generalist strategy may be realized through increased signal complexity, such as signals comprising multiple components, each of which is suited to different possible signaling channels (substrates). Alternatively, a generalist signaling strategy may be based on a "lowest-common-denominator" signal design.
3. Studies of substrate-borne communication systems must be designed with careful consideration of possible effects of substrates. Such studies require the selection of suitable materials on which to carry out controlled observations and/or manipulations of signaling behavior. In some cases, natural substrates may be available and suitable, but due to the issues of natural substrate heterogeneity and complexity outlined above, these will not necessarily be the best choice for controlled experiments. In those cases, artificial substrates must be selected. In any case, thorough characterization and calibration of a substrate's physical (vibrational) properties is required. This will usually be an iterative process, taking into account the general properties of the signals of interest in order to identify suitable substrates that will allow, in turn, more precise quantification of signal variation.
4. Several aspects of substrate signal transmission have received little study or have been examined only in select species. More research in the following areas would be fruitful: Mathematical models describing transmission characteristics of different substrates combined with empirical verification using behavioral experiments and multiple measurements at different orientations; behavioral studies investigating the role of frequency and fine temporal structure on mate choice using standardized recording positions; comparative studies examining the mechanisms used to produce vibratory signals and habitat usage; and studies examining how vibrational energy is introduced into substrates.

## References

Achenbach JD (1973) Wave propagation in elastic solids. North Holland Pub. Co., Amsterdam

Aicher B, Tautz J (1990) Vibrational communication in the Fiddler Crab, *Uca pugilator*.1. Signal transmission through the substratum. J Comp Physiol A 166:345–353

Aicher B, Markl H, Masters WM, Kirschenlohr HL (1983) Vibration transmission through the walking legs of the Fiddler Crab, *Uca pugilator* (Brachyura, Ocypodidae) as measured by laser doppler vibrometry. J Comp Physiol 150:483–491

Alexander RD (1962) Evolutionary change in cricket acoustical communication. Evolution 16:443–467

Arak A, Eiriksson T (1992) Choice of singing sites by male bush-crickets (*Tettigonia viridissima*) in relation to signal propagation. Behav Ecol Sociobiol 30:365–372

Arak A, Enquist M (1993) Hidden preferences and the evolution of signals. Philos T Roy Soc B 340:207–213

Archer SN, Djamgoz MBA, Loew ER, Partridge JC, Vallerga S (eds) (1999) Adaptive mechanisms in the ecology of vision. Kluwer Academic, Dordrecht

Arnqvist G (2006) Sensory exploitation and sexual conflict. Philos T Roy Soc B 361:375–386

Arnqvist G, Rowe L (2005) Sexual conflict. Princeton University Press, Princeton

Aylor D (1972a) Noise-reduction by vegetation and ground. J Acoust Soc Am 51:197–205

Aylor D (1972b) Sound transmission through vegetation in relation to leaf area density, leaf width, and breadth of canopy. J Acoust Soc Am 51:411–414

Barth FG (1998) The vibrational sense of spiders. In: Hoy RR, Popper AN, Fay RR (eds) Comparative hearing: insects. Springer, New York, pp 228–278

Barth FG (2002) A spider's world: senses and behavior. Springer, Berlin

Barth FG, Schmid A (eds) (2001) Ecology of sensing. Springer, Berlin

Barth FG, Bleckmann H, Bohnenberger J, Seyfarth EA (1988) Spiders of the genus Cupiennius Simon 1891 (Araneae, Ctenidae).2. On the vibratory environment of a wandering spider. Oecologia 77:194–201

Bell WJ, Cardé R (eds) (1984) Chemical ecology of insects. Sinauer Associates, Sunderland

Bennet-Clark HC (1987) The tuned singing burrow of mole crickets. J Exp Biol 128:383–409

Bernays EA, Wcislo WT (1994) Sensory capabilities, information-processing, and resource specialization. Q Rev Biol 69:187–204

Bleckmann H, Barth FG (1984) Sensory ecology of a semi-aquatic spider (*Dolomedes triton*) II. The release of predatory behavior by water surface waves. Behav Ecol Sociobiol 14:303–312

Bleckmann H, Rovner JS (1984) Sensory ecology of a semi-aquatic spider (*Dolomedes triton*). Behav Ecol Sociobiol 14:297–301

Boake CRB (1991) Coevolution of senders and receivers of sexual signals: genetic coupling and genetic correlations. TREE 6:225–227

Boughman JW (2002) How sensory drive can promote speciation. TREE 17:571–577

Bradbury JW, Vehrencamp SL (1998) Principles of animal communication. Sinauer Assoc, Sunderland

Brownell P, Farley RD (1979) Orientation to vibrations in sand by the nocturnal scorpion *Paruroctonus mesaensis*—mechanism of target localization. J Comp Physiol 131:31–38

Brownell PH (1977) Compressional and surface-waves in sand used by desert scorpions to locate prey. Science 197:479–482

Brownell PH, Van Hemmen JL (2000) Vibration sensitivity and prey-localizing behaviour of sand scorpions. Am Zool 40:955–956

Butlin RK, Ritchie MG (1989) Genetic coupling in mate recognition systems: what is the evidence? Biol J Linn Soc 37:237–246

Byers J, Hebets E, Podos J (2010) Female mate choice based upon male motor performance. Anim Behav 79:771–778

Cady AB, Delaney JK, Uetz GW (2011) Contrasting energetic costs of courtship signaling in two wolf spiders having divergent courtship behaviors. J Arachnol 39:161–165

Cardé R, Millar GJ (eds) (2004) Advances in insect chemical ecology. Cambridge University, Cambridge

Casas J, Bacher S, Tautz J, Meyhöfer R, Pierre D (1998) Leaf vibrations and air movements in a leafminer–parasitoid system. Biol Control 11:147–153

Casas J, Magal C, Sueur J (2007) Dispersive and non-dispersive waves through plants: implications for arthropod vibratory communication. P Roy Soc B Bio 274:1087–1092

Chapman RF (1998) The insects: structure and function. Cambridge University, Cambridge

Chittka L, Thomson JD (2001) Cognitive ecology of pollination: animal behavior and floral evolution. Cambridge University, Cambridge

Clark CJ, Elias DO, Prum RO (2011) Aeroelastic flutter produces hummingbird feather songs. Science 333:1430–1433

Cocroft RB (2001) Vibrational communication and the ecology of group-living, herbivorous insects. Am Zool 41:1215–1221

Cocroft RB (2011) The public world of insect vibrational communication. Mol Ecol 20:2041–2043

Cocroft RB, Rodriguez RL (2005) The behavioral ecology of insect vibrational communication. Bioscience 55:323–334

Cocroft RB, Tieu TD, Hoy RR, Miles RN (2000) Directionality in the mechanical response to substrate vibration in a treehopper (Hemiptera: Membracidae: *Umbonia crassicornis*). J Comp Phys A 186:695–705

Cocroft RB, Shugart HJ, Konrad KT, Tibbs K (2006) Variation in plant substrates and its consequences for insect vibrational communication. Ethology 112:779–789

Cocroft RB, Rodriguez RL, Hunt RE (2008) Host shifts, the evolution of communication and speciation in the *Enchenopa binotata* species complex of treehoppers. In: Tilmon KJ (ed) Speciation, specialization and radiation: the evolutionary biology of insect and plant interactions. University of California Press, Berkeley, pp 88–100

Cocroft RB, Rodriguez RL, Hunt RE (2010) Host shifts and signal divergence: mating signals covary with host use in a complex of specialized plant-feeding insects. Biol J Linn Soc 99:60–72

Čokl A (2008) Stink bug interaction with host plants during communication. J Insect Physi 54:1113–1124

Čokl A, Virant-Doberlet M (2003) Communication with substrate-borne signals in small plant-dwelling insects. Annu Rev Entomol 48:29–50

Čokl A, Virant-Doberlet M, Stritih N (2000) The structure and function of songs emitted by southern green stink bugs from Brazil, Florida, Italy and Slovenia. Physiol Entomol 25:196–205

Čokl A, Presern J, Virant-Doberlet M, Bagwell GJ, Millar JG (2004) Vibratory signals of the harlequin bug and their transmission through plants. Physiol Entomol 29:372–380

Čokl A, Zorovic M, Zunic A, Virant-Doberlet M (2005) Tuning of host plants with vibratory songs of *Nezara viridula* L (Heteroptera: Pentatomidae). J Exp Biol 208:1481–1488

Čokl A, Nardi C, Bento JMS, Hirose E, Panizzi AR (2006) Transmission of stridulatory signals of the burrower bugs, *Scaptocoris castanea* and *Scaptocoris carvalhoi* (Heteroptera : Cydnidae) through the soil and soybean. Physiol Entomol 31:371–381

Čokl A, Zorovic M, Millar JG (2007) Vibrational communication along plants by the stink bugs *Nezara viridula* and *Murgantia histrionica*. Behav Process 75:40–54

Čokl A, Zunic A, Millar JG (2009) Transmission of *Podisus maculiventris* tremulatory signals through plants. Centr Eur J Biol 4:585–594

Cremer L, Heckl M, Ungar EE (1973) Structure-borne sound. Springer, Berlin

Dangles O, Irschick D, Chittka L, Casas J (2009) Variability in sensory ecology: expanding the bridge between physiology and evolutionary biology. Q Rev Biol 84:51–74

Daws AG, Bennet-Clark HC, Fletcher NH (1996) The mechanism of tuning of the mole cricket singing burrow. Bioacoustics 7:81–117

Delaney KJ, Roberts JA, Uetz GW (2007) Male signaling behavior and sexual selection in a wolf spider (Araneae: Lycosidae): a test for dual functions. Behav Ecol Sociobiol 62:67–75

Devetak D, Mencinger-Vracko B, Devetak M, Marhl M, Spernjak A (2007) Sand as a medium for transmission of vibratory signals of prey in antlions *Euroleon nostras* (Neuroptera: Myrmeleontidae). Physiol Entomol 32:268–274

Dicke M, Takken W (eds) (2006) Chemical ecology: from gene to ecosystem. Springer, Dordrecht

Dierkes S, Barth FG (1995) Mechanism of signal production in the vibratory communication of the wandering spider *Cupiennius getazi* (Arachnida, Araneae). J Comp Physiol A 176:31–44

Dukas R (ed) (1998) Cognitive ecology: the evolutionary ecology of information processing and decision making. University of Chicago Press, Chicago

Dusenbery DB (1992) Sensory ecology. W. H. Freeman, New York

Elias DO, Mason AC (2011) Signaling in variable environments: substrate-borne signaling mechanisms and communication behavior in spiders. In: O'Connell-Rodwell C (ed) The use of vibrations in communication: properties, mechanisms and function across taxa. Research Signpost, Kerala

Elias DO, Mason AC, Maddison WP, Hoy RR (2003) Seismic signals in a courting male jumping spider (Araneae: Salticidae). J Exp Biol 206:4029–4039

Elias DO, Mason AC, Hoy RR (2004) The effect of substrate on the efficacy of seismic courtship signal transmission in the jumping spider *Habronattus dossenus* (Araneae: Salticidae). J Exp Biol 207:4105–4110

Elias DO, Hebets EA, Hoy RR, Maddison WP, Mason AC (2006a) Regional seismic song differences in sky island populations of the jumping spider *Habronattus pugillis* Griswold (Araneae, Salticidae). J Arachnol 34:545–556

Elias DO, Lee N, Hebets EA, Mason AC (2006b) Seismic signal production in a wolf spider: parallel versus serial multi-component signals. J Exp Biol 209:1074–1084

Elias DO, Kasumovic MM, Punzalan D, Andrade MCB, Mason AC (2008) Assessment during aggressive contests between male jumping spiders. Anim Behav 76:901–910

Elias DO, Mason AC, Hebets EA (2010a) A signal-substrate match in the substrate-borne component of a multimodal courtship display. Curr Zool 56:370–378

Elias DO, Sivalinghem S, Mason AC, Andrade MCB, Kasumovic MM (2010b) Vibratory communication in the jumping spider *Phidippus clarus*: substrate-borne courtship signals are important for male mating success. Ethology 116:990–998

Elias DO, Maddison WP, Peckmezian C, Girard MB, Mason AC (2012) Orchestrating the score: complex multimodal courtship in the *Habronattus coecatus* group of *Habronattus* jumping spiders (Araneae: Salticidae). Biol J Linn Soc 105:522–547

Endler JA (1992) Signals, signal conditions, and the direction of evolution. Am Nat 139:S125–S153

Endler JA (1993) Some general comments on the evolution and design of animal communication systems. Philos T Roy Soc B 340:215–225

Endler JA, Basolo A (1998) Sensory ecology, receiver biases and sexual selection. TREE 13:415–420

Eriksson A, Anfora G, Lucchi A, Virant-Doberlet M, Mazzoni V (2011) Inter-plant vibrational communication in a leafhopper insect. PLoS ONE 6:6

Evans TA, Lai JCS, Toledano E, McDowall L, Rakotonarivo S, Lenz M (2005) Termites assess wood size by using vibration signals. P Natl Acad Sci USA 102:3732–3737

Evans TA, Inta R, Lai JCS, Lenz M (2007) Foraging vibration signals attract foragers and identify food size in the drywood termite, *Cryptotermes secundus*. Insect Soc 54:374–382

Evans TA, Inta R, Lai JCS, Prueger S, Foo NW, Fu EW, Lenz M (2009) Termites eavesdrop to avoid competitors. P Roy Soc B Bio 276:4035–4041

Fertin A, Casas J (2007) Orientation towards prey in antlions: efficient use of wave propagation in sand. J Exp Biol 210:3337–3343

Field LH, Bailey WJ (1997) Sound production in primitive Orthoptera from Western Australia: sounds used in defence and social communication in *Ametrus* sp. and *Hadrogryllacris* sp. (Gryllacrididae: Orthoptera). J Nat Hist 31:1127–1141

Fischer S, Samietz J, Dorn S (2003) Efficiency of vibrational sounding in parasitoid host location depends on substrate density. J Comp Physiol A 189:723–730

Fleishman LJ (1988) Sensory influences on physical design of a visual display. Anim Behav 36:1420–1424

Forrest TG (1982) Acoustic communication and baffling behaviors of crickets. Fla Entomol 65:33–44

Frohlich C, Buskirk RE (1982) Transmission and attenuation of vibration in orb spider webs. J Theor Biol 95:13–36

Gerhardt HC, Huber F (2002) Acoustic communication in insects and anurans. University of Chicago Press, Chicago London

Gibson JS, Uetz GW (2008) Seismic communication and mate choice in wolf spiders: components of male seismic signals and mating success. Anim Behav 75:1253–1262

Gogala M (1985) Vibrational songs of land bugs and their production. In: Kalmring K, Elsner N (eds) Acoustic and vibrational communication in insects. Paul Parey, Berlin Hamburg, pp 143–150
Gordon SD, Uetz GW (2011) Multimodal communication of wolf spiders on different substrates: evidence for behavioural plasticity. Anim Behav 81:367–375
Greenfield M (2002) Signalers and receivers: mechanisms and evolution of arthropod communication. Oxford University Press, Oxford
Hebets EA (2008) Seismic signal dominance in the multimodal courtship display of the wolf spider *Schizocosa stridulans* Stratton 1991. Behav Ecol 19:1250–1257
Hebets EA, Stratton GE, Miller GL (1996) Habitat and courtship behavior of the wolf spider *Schizocosa retrorsa* (Banks) (Araneae, Lycosidae). J Arachnol 24:141–147
Hebets EA, Elias DO, Mason AC, Miller GL, Stratton GE (2008) Substrate-dependent signalling success in the wolf spider, *Schizocosa retrorsa*. Anim Behav 75:605–615
Henry CS (1994) Singing and cryptic speciation in insects. TREE 9:388–392
Henry CS, Wells MLM (2004) Adaptation or random change? The evolutionary response of songs to substrate properties in lacewings (Neuroptera: Chrysopidae: Chrysoperla). Anim Behav 68:879–895
Henry CS, Wells MM (2010) Acoustic niche partitioning in two cryptic sibling species of *Chrysoperla* green lacewings that must duet before mating. Anim Behav 80:991–1003
Henry CS, Wells MM, Pupedis RJ (1993) Hidden taxonomic diversity within *Crysoperla plorabunda* (Neuroptera, Chrysopidae): two new species based on courtship songs. Ann Entomol Soc Am 86:1–13
Henry CS, Brooks SJ, Johnson JB, Duelli P (1999a) Revised concept of *Chrysoperla mediterranea* (Holzel), a green lacewing associated with conifers: courtship songs across 2800 kilometres of Europe (Neuroptera: Chrysopidae). Syst Entomol 24:335–350
Henry CS, Wells MLM, Simon CM (1999b) Convergent evolution of courtship songs among cryptic species of the *Carnea* group of green lacewings (Neuroptera: Chrysopidae: Chrysoperla). Evolution 53:1165–1179
Henry CS, Brooks SJ, Duelli P, Johnson JB (2002) Discovering the true *Chrysoperla carnea* (Insecta: Neuroptera: Chrysopidae) using song analysis, morphology, and ecology. Ann Entomol Soc Am 95:172–191
Herberstein ME, Tso I (2011) Spider webs: evolution, diversity, and plasticity. In: Herberstein ME (ed) Spider behaviour: flexibility and versatility. Cambridge University Press, Cambridge, pp 57–99
Hill PSM (2001) Vibration and animal communication: a review. Am Zool 41:1135–1142
Hill PSM (2008) Vibrational communication in animals. Harvard University Press, Cambridge
Hill PSM, Shadley JR (1997) Substrate vibration as a component of a calling song. Naturwissenschaften 84:460–463
Hoefler CD, Carlascio AL, Persons MH, Rypstra AL (2009) Male courtship repeatability and potential indirect genetic benefits in a wolf spider. Anim Behav 78:183–188
Holland B, Rice WR (1998) Perspective: chase-away sexual selection: antagonistic seduction versus resistance. Evolution 52:1–7
Hunt RE (1994) Vibrational signals associated with mating behavior in the treehopper, *Enchenopa binotata* Say (Hemptera: Homoptera: Membracidae). J New York Entomol S 102:266–270
Joyce AL, Hunt RE, Vinson SB, Bernal JS (2007) Courtship songs of the *Cotesia flavipes* complex. J Insect Sci 7:12
Joyce AL, Hunt RE, Bernal JS, Vinson SB (2008) Substrate influences mating success and transmission of courtship vibrations for the parasitoid *Cotesia marginiventris*. Entomol Exp Appl 127:39–47
Keuper A, Kuhne R (1983) The acoustic behavior of the bushcricket *Tettigonia cantans* 2. Transmission of airborne-sound and vibration signals in the biotope. Behav Process 8:125–145

Klarner D, Barth FG (1982) Vibratory signals and prey capture in orb-weaving spiders (*Zygiella x-notata*, *Nephila clavipes*, Araneidae). J Comp Physiol 148:445–455
Kotiaho J, Alatalo RV, Mappes J, Parri S (1996) Sexual selection in a wolf spider: Male drumming activity, body size, and viability. Evolution 50:1977–1981
Kotiaho JS, Alatalo RV, Mappes J, Nielsen MG, Parri S, Rivero A (1998) Energetic costs of size and sexual signalling in a wolf spider. P Roy Soc B Bio 265:2203–2209
Kotiaho JS, Alatalo RV, Mappes J, Parri S (2000) Microhabitat selection and audible sexual signalling in the wolf spider *Hygrolycosa rubrofasciata* (Araneae, Lycosidae). Acta Etholog 2:123–128
Kroodsma DE, Miller EH (eds) (1996) Ecology and evolution of acoustic communication in birds. Cornell University Press, Ithaca
Lardner B, bin Lakim M (2002) Tree-hole frogs exploit resonance effects. Nature 420:475
Magal C, Scholler M, Tautz J, Casas J (2000) The role of leaf structure in vibration propagation. J Acoust Soc Am 108:2412–2418
Markl H (1983) Vibrational communication. In: Huber F, Markl H (eds) Neuroethology and behavioral physiology. Springer, Berlin, pp 332–354
Mason MJ, Narins PM (2001) Seismic signal use by fossorial mammals. Am Zool 41:1171–1184
Masters WM (1984) Vibrations in the orbwebs of Nuctenea sclopetaria (Araneidae).1. Transmission through the web. Behav Ecol 15:207–215
Masters WM, Markl H (1981) Vibration signal transmission in spider orb webs. Science 213:363–365
Masters WM, Markl HS, Moffatt AJM (1986) Transmission of vibration in a spider's web. In: Shear W (ed) Spiders, web, behaviour and evolution. Stanford University Press, Stanford
McNett GD, Cocroft RB (2008) Host shifts favor vibrational signal divergence in *Enchenopa binotata* treehoppers. Behav Ecol 19:650–656
McNett G, Miles R, Homentcovschi D, Cocroft R (2006) A method for two-dimensional characterization of animal vibrational signals transmitted along plant stems. J Comp Physiol A 192:1245–1251
Michelsen A, Fink F, Gogala M, Traue D (1982) Plants as transmission channels for insect vibrational songs. Behav Ecol Sociobiol 11:269–281
Miklas N, Stritih N, Čokl A, Virant-Doberlet M, Renou M (2001) The influence of substrate on male responsiveness to the female calling song in *Nezara viridula*. J Insect Behav 14:313–332
Miklas N, Čokl A, Renou M, Virant-Doberlet M (2003) Variability of vibratory signals and mate choice selectivity in the southern green stink bug. Behav Process 61:131–142
Miles RN, Cocroft RB, Gibbons C, Batt D (2001) A bending wave simulator for investigating directional vibration sensing in insects. J Acoust Soc Am 110:579–587
Miranda X (2006) Substrate-borne signal repertoire and courtship jamming by adults of *Ennya chrysura* (Hemiptera: Membracidae). Ann Entomol Soc Am 99:374–386
Morris GK (1980) Calling display and mating behavior of *Copiphora rhinoceros* pictet (Orthoptera, Tettigoniidae). Anim Behav 28:42–51
Morton E (1975) Ecological sources of selection on avian sounds. Am Nat 109:855–869
Narins PM (2001) Vibration communication in vertebrates. In: Barth FG, Schmid A (eds) Ecology of sensing. Springer, Berlin, pp 127–149
Nascimento FS, Hrncir M, Tolfiski A, Zucchi R (2005) Scraping sounds produced by a social wasp (*Asteloeca ujhelyii*, Hymenoptera: Vespidae). Ethology 111:1116–1125
O'Connell-Rodwell CE (2007) Keeping an "ear" to the ground: seismic communication in elephants. Physiology 22:287–294
O'Connell-Rodwell CE, Arnason BT, Hart LA (2000) Seismic properties of Asian elephant (*Elephas maximus*) vocalizations and locomotion. J Acoust Soc Am 108:3066–3072
O'Connell-Rodwell CE, Hart LA, Arnason BT (2001) Exploring the potential use of seismic waves as a communication channel by elephants and other large mammals. Am Zool 41:1157–1170

Ota D, Čokl A (1991) Mate location in the southern green stink bug, *Nezara viridula* (Heteroptera: Pentatomidae), mediated through substrate-borne signals on ivy. J Insect Behav 4:441–447
Panizzi AR (1997) Wild hosts of Pentatomids: ecological significance and role in their pest status on crops. Annu Rev Entomol 42:99–122
Parker GA (2006) Sexual conflict over mating and fertilization: an overview. Philos T Roy Soc B 361:235–259
Parri S, Alatalo RV, Kotiaho J, Mappes J (1997a) Female choice for male drumming in the wolf spider *Hygrolycosa rubrofasciata*. Anim Behav 53:305–312
Parri S, Alatalo RV, Kotiaho JS, Mappes J (1997b) Female choice for male drumming in the wolf spider *Hygrolycosa rubrofasciata*. Anim Behav 53:305–312
Parri S, Alatalo RV, Kotiaho JS, Mappes J, Rivero A (2002) Sexual selection in the wolf spider *Hygrolycosa rubrofasciata*: female preference for drum duration and pulse rate. Behav Ecol 13:615–621
Patek SN (2001) Spiny lobsters stick and slip to make sound. Nature 411:153–154
Patek SN, Baio JE (2007) The acoustic mechanics of stick slip friction in the California spiny lobster (*Panulirus interruptus*). J Exp Biol 210:3538–3546
Paul RC, Walker TJ (1979) Arboreal singing in a burrowing cricket, *Anurogryllus arboreus*. J Comp Physiol 132:217–223
Penna M (2004) Amplification and spectral shifts of vocalizations inside burrows of the frog *Eupsophus calcaratus* (Leptodactylidae). J Acoust Soc Am 116:1254–1260
Peretti AV, Aisenberg A (2011) Communication under sexual selection hypotheses: challenging prospects for future studies under extreme sexual conflict. Acta Etholog 14:109–116
Polajnar J, Čokl A (2008) The effect of vibratory disturbance on sexual behaviour of the southern green stink bug *Nezara viridula* (Heteroptera, Pentatomidae). Centr Eur J Biol 3:189–197
Polajnar J, Svenšek D, Čokl A (2012) Resonance in herbaceous plant stems as a factor in vibrational communication of pentatomid bugs (Heteroptera: Pentatomidae). J Roy Soc Interface 9:1898–1907
Proctor HC (1992) Sensory exploitation and the evolution of male mating behaviour: a cladistics test using water mites (Acari: Parasitengona). Anim Behav 44:745–752
Prozesky-Schulze L, Prozesky OPM, Anderson F, Van Der Merwe GJJ (1975) Use of a self-made sound baffle by a tree cricket. Nature 255:142–143
Randall JA (2001) Evolution and function of drumming as communication in mammals. Am Zool 41:1143–1156
Rivero A, Alatalo RV, Kotiaho JS, Mappes J, Parri S (2000) Acoustic signalling in a wolf spider: can signal characteristics predict male quality? Anim Behav 48:188–194
Rodriguez RL, Cocroft RB (2006) Divergence in female duetting signals in the *Enchenopa binotata* species complex of treehoppers (Hemiptera: Membracidae). Ethology 112:1231–1238
Rodriguez RL, Ramaswamy K, Cocroft RB (2006) Evidence that female preferences have shaped male signal evolution in a clade of specialized plant-feeding insects. P Roy Soc B Bio 273:2585–2593
Rodriguez RL, Sullivan LM, Snyder RL, Cocroft RB (2008) Host shifts and the beginning of signal divergence. Evolution 62:12–20
Rohrig A, Kirchner WH, Leuthold RH (1999) Vibrational alarm communication in the African fungus-growing termite genus *Macrotermes* (Isoptera, Termitidae). Insect Soc 46:71–77
Rovner JS (1975) Sound production by Nearctic wolf spiders: a substratum-coupled stridulatory mechanism. Science 190:1309–1310
Rovner JS (1980) Vibration in *Heteropoda venatoria* (Sparassidae): a 3rd method of sound production in spiders. J Arachnol 8:193–200
Rowe L, Arnqvist G (2002) Sexually antagonistic coevolution in a mating system: combining experimental and comparative approaches to address evolutionary processes. Evolution 56:754–767

Rowe L, Cameron E, Day T (2005) Escalation, retreat, and female indifference as alternative outcomes of sexually antagonistic coevolution. Am Nat 165:S5–S18
Rundus AS, Santer RD, Hebets EA (2010) Multimodal courtship efficacy of *Schizocosa retrorsa* wolf spiders: implications of an additional signal modality. Behav Ecol 21:701–707
Ryan MJ (1990) Sexual selection, sensory systems, and sensory exploitation. In: Futuyma D, Antonovics J (eds) Oxford surveys in evolutionary biology. Oxford University Press, Oxford
Ryan MJ, Rand AS (1993) Sexual selection and signal evolution: the ghosts of biases past. Philos T Roy Soc B 340:187–195
Sandeman DC, Tautz J, Lindauer M (1996) Transmission of vibration across honeycombs and its detection by bee leg receptors. J Exp Biol 199:2585–2594
Scheffer SJ, Uetz GW, Stratton GE (1996) Sexual selection, male morphology, and the efficacy of courtship signalling in two wolf spiders (Araneae: Lycosidae). Behav Ecol Sociobiol 38:17–23
Shamble PS, Wilgers DJ, Swoboda KA, Hebets EA (2009) Courtship effort is a better predictor of mating success than ornamentation for male wolf spiders. Behav Ecol 20:1242–1251
Shannon CE (1949) The mathematical theory of communication. University of Illinois Press, Urbana
Sivalinghem S, Kasumovic MM, Mason AC, Andrade MCB, Elias DO (2010) Vibratory communication in the jumping spider *Phidippus clarus*: polyandry, male courtship signals, and mating success. Behav Ecol 21:1308–1314
Stevens M (2012) Sensory ecology, behaviour and evolution. Oxford University Press, Oxford
Stratton GE, Uetz GW (1981) Acoustic communication and reproductive isolation in two species of wolf spiders (Araneae: Lycosidae). Science 214:575–577
Stratton GE, Uetz GW (1983) Communication via substrate-coupled stridulation and reproductive isolation in wolf spiders (Aranea: Lycosidae). Anim Behav 31:164–172
Sullivan-Beckers L, Cocroft RB (2010) The importance of female choice, male-male competition, and signal transmission as causes of selection on male mating signals. Evolution 64:3158–3171
Tautz J (1996) Honeybee waggle dance: recruitment success depends on the dance floor. J Exp Biol 199:1375–1381
Tautz J, Casas J, Sandeman D (2001) Phase reversal of vibratory signals in honeycomb may assist dancing honeybees to attract their audience. J Exp Biol 204:3737–3746
Taylor PW, Roberts JA, Uetz GW (2005) Flexibility in the multi-modal courtship of a wolf spider, *Schizocosa ocreata*. J Ethol 23:71–75
Uetz GW, Roberts JA (2002) Multisensory cues and multimodal communication in spiders: Insights form video/audio playback studies. Brain Behav Evolut 59:222–230
Uetz GW, Stratton GE (1982) Acoustic communication and reproductive isolation in spiders. In: Witt PN, Rovner JS (eds) Spider communication: mechanisms and ecological significance. Princeton University Press, Princeton, pp 123–129
Uetz GW, Roberts JA, Taylor PW (2009) Multimodal communication and mate choice in wolf spiders: female response to multimodal versus unimodal signals. Anim Behav 78:299–305
Uhl G, Elias DO (2011) Communication. In: Herberstein ME (ed) Spider behavior: flexibility and versatility. Cambridge University Press, Cambridge, pp 127–190
Virant-Doberlet M, Čokl A (2004) Vibrational communication in insects. Neotrop Entomol 33:121–134
Walker TJ, Figg DE (1990) Song and acoustic burrow of the prairie mole cricket, *Gryllotalpa major* (Orthoptera, Gryllidae). J Kans Entomol Soc 63:237–242
Wehner R (1987) Matched-filters—neural models of the external world. J Comp Physiol 161:511–531
Wiley C, Shaw KL (2010) Multiple genetic linkages between female preference and male signal in rapidly speciating Hawaiian crickets. Evolution 64:2238–2245
Wood TK (1993) Speciation of the *Enchenopa binotata* complex (Insecta: Homoptera: Membracidae). In: Lees DR, Edwards D (eds) Evolutionary patterns and processes. Academic Press, New York, pp 299–317

Wood TK, Keese MC (1990) Host-plant-induced assortative mating in *Enchenopa* treehoppers. Evolution 44:619–628

Wood TK, Olmstead KL, Guttman SI (1990) Insect phenology mediated by host-plant water relations. Evolution 44:629–636

Zefa E, Martins LD, Szinwelski N (2008) Complex mating behavior in *Adelosgryllus rubricephalus* (Orthoptera, Phalangopsidae, Grylloidea). Iheringia Ser Zool 98:325–328

Zunic A, Čokl A, Doberlet MV, Millar JG (2008) Communication with signals produced by abdominal vibration, tremulation, and percussion in *Podisus maculiventris* (Heteroptera: Pentatomidae). Ann Entomol Soc Am 101:1169–1178

# Chapter 13
# Vibrational Playback Experiments: Challenges and Solutions

**Reginald B. Cocroft, Jennifer Hamel, Quang Su and Jeremy Gibson**

**Abstract** Playbacks are one of the most useful experimental tools in animal communication research. Playbacks of substrate vibrations present special challenges, but conducting high-fidelity vibrational playbacks is not difficult and depends less on the specific equipment used than on avoiding some common pitfalls. We review the major issues, describing both the problems and a range of solutions. Our focus is on playback through living plants, but most of the issues apply to playback through other substrates as well. The major challenge for playback through any substrate is that the vibrational signal is almost always changed by the playback equipment and the substrate, so that the signal received by the focal animal is different from the one intended by the experimenter. The general solution to this problem is to measure the changes imposed by the playback system and to pre-filter the playback signal to compensate for them. A second challenge is to ensure that the focal animal receives a signal at the appropriate amplitude. Achieving the proper amplitude is a straightforward process. However, amplitude is substrate dependent (e.g., on a plant, amplitude is inversely proportional to stem diameter), and the experimenter should choose a realistic amplitude for the substrate. Other issues include choices of playback device, natural versus artificial substrates, single versus multiple substrate exemplars, and playback in laboratory versus field. Our goal in this chapter is to give experimenters, especially those just starting out, the knowledge and confidence needed to conduct high-quality vibrational playbacks.

---

R. B. Cocroft (✉) · J. Gibson
Division of Biological Sciences, University of Missouri, Columbia, MO, USA
e-mail: cocroftr@missouri.edu

J. Hamel
Department of Entomology and Nematology, University of Florida, Gainesville, FL, USA

Q. Su
Watson School of Engineering and Applied Science, Binghamton University, Binghamton, NY, USA

R. B. Cocroft et al. (eds.), *Studying Vibrational Communication*,
Animal Signals and Communication 3, DOI: 10.1007/978-3-662-43607-3_13,

## 13.1 Introduction

The playback experiment is one of the most widely used methods in the study of animal communication. Playbacks are essential for testing hypotheses about receiver responses to signals, including how signals function, what form of selection receivers impose on signals, and how receivers are influenced by signal changes caused by the transmission channel. A substantial literature on the proper design of playback experiments reflects this importance to the field (Kroodsma 1986, 1989; McGregor et al. 1992; McGregor 2000; Wiley 2003). Playback experiments require both proper design and proper execution, i.e., using a statistical design that supports the level of inference desired (Kroodsma et al. 2001) and executing this design so that the signal has the proper characteristics when it reaches the subject (McGregor 2000). The statistical issues cut across communication modalities, and there is some level of agreement about statistical design of playbacks. In contrast, the technical challenges differ greatly between modalities. For example, the technical issues arising in video playbacks (Fleishman et al. 1998; Cuthill et al. 2000) are distinct from those arising in acoustic playbacks (Gerhardt 1995), with multimodal playbacks experiencing both (Uetz and Roberts 2002). Our goal here is to review the technical aspects of playback in yet another modality: substrate vibration.

Communication through substrate vibrations occurs in many vertebrates and invertebrates (Hill 2008). In invertebrates, the use of substrate vibration dwarfs the use of airborne sound. In insects, it has been estimated that of all species communicating via some form of mechanical stimuli transmitted through a medium, over 90 % use substrate vibrations alone or in combination with airborne sound and 70 % use substrate vibration exclusively (Cocroft and Rodríguez 2005). The percentages are undoubtedly higher for spiders (Barth 1982, 2002) and possibly other invertebrate groups. However, although vibrational communication is the most widespread of the mechanical modalities, it is comparatively understudied. This lack of attention is changing, as evidenced by the growth of the literature in recent years (see Chap. 1, this volume), but there has been little discussion in the literature of the technical problems inherent in conducting playback of substrate-borne vibrational signals and of how to solve these problems. Hill and Shadley (2001) provide an excellent discussion of the most important problem (see Sect. 13.2, below), which is to compensate for the frequency response of the playback equipment and transmission channel. Wood and O'Connell-Rodwell (2010) provide a more wide-ranging discussion of playback experiments, addressing issues such as choosing sensors, playback equipment, and recording devices, and how to determine whether the wave type produced by the playback equipment matches that produced by a signaling animal. Here, we provide an overview of these and other issues that arise when conducting vibrational playbacks. Hill and Shadley (2001) and Wood and O'Connell-Rodwell (2010) consider issues arising in studying ground-borne vibrations. Because our research deals with insects that communicate through living plants, we will focus on this class of substrates, but most of the issues we address here apply to other substrates as well.

## 13.2 Frequency Profile of the Played-Back Signal

### *13.2.1 The Problem*

The basic methodological challenge in any playback experiment is to deliver the desired stimulus at the location of the receiver. There are several reasons why accomplishing this goal with vibrational signals is not straightforward and why some common procedures cause the playback subject to receive a signal that differs from what was intended by the experimenter. In this section, we first explore how the frequency spectrum of the played-back signal may be changed in undesirable ways and then show how to correct for this problem.

#### 13.2.1.1 Mismatch Between Recording and Playback Devices

Consider a signal recorded simultaneously with a laser vibrometer and an accelerometer (Fig. 13.1). The waveforms and amplitude spectra of the two recordings are different, although they represent the same physical vibration recorded at the same location. However, the output of a laser vibrometer is proportional to the substrate's velocity, while the output of an accelerometer is proportional to the substrate's acceleration. If the signal were also recorded with a ceramic phonograph cartridge, the waveform would again be different, because the cartridge's output is proportional to displacement. For signals that span a range of frequencies, the displacement, velocity, and acceleration waveforms will differ predictably as a function of frequency (Fig. 13.2).

Just as sensors have their own characteristics, so do playback devices. There are several methods of introducing a signal into a substrate. The most common is an electrodynamic shaker, which uses a coil and magnet to vibrate solid structures. Other common means of vibrating a substrate (Cocroft 2010) include piezoelectric actuators; an electromagnet that drives a magnet attached to the substrate; audio speakers that are modified by removing the diaphragm to reduce airborne sound and driving the substrate by means of a pin attached to the coil; and even airborne sound (Rebar et al. 2012).

One issue to consider is whether the output of the playback device is proportional to the signal's acceleration, velocity, or displacement. In general, the output of shakers is proportional to the stimulus acceleration. Accordingly, a signal recorded with an accelerometer and played back using a shaker should be faithfully reproduced, other things being equal (which they almost never are, as we will see below). On the other hand, if a signal recorded with a laser vibrometer is played back using a shaker, the shaker will reproduce the laser recording as if it reflected the stimulus acceleration, and the higher frequencies will be underrepresented in the playback, with their amplitude dropping by half for every doubling of frequency (see Fig. 13.2). The output of audio speakers (modified to play back vibrations) is proportional to velocity.

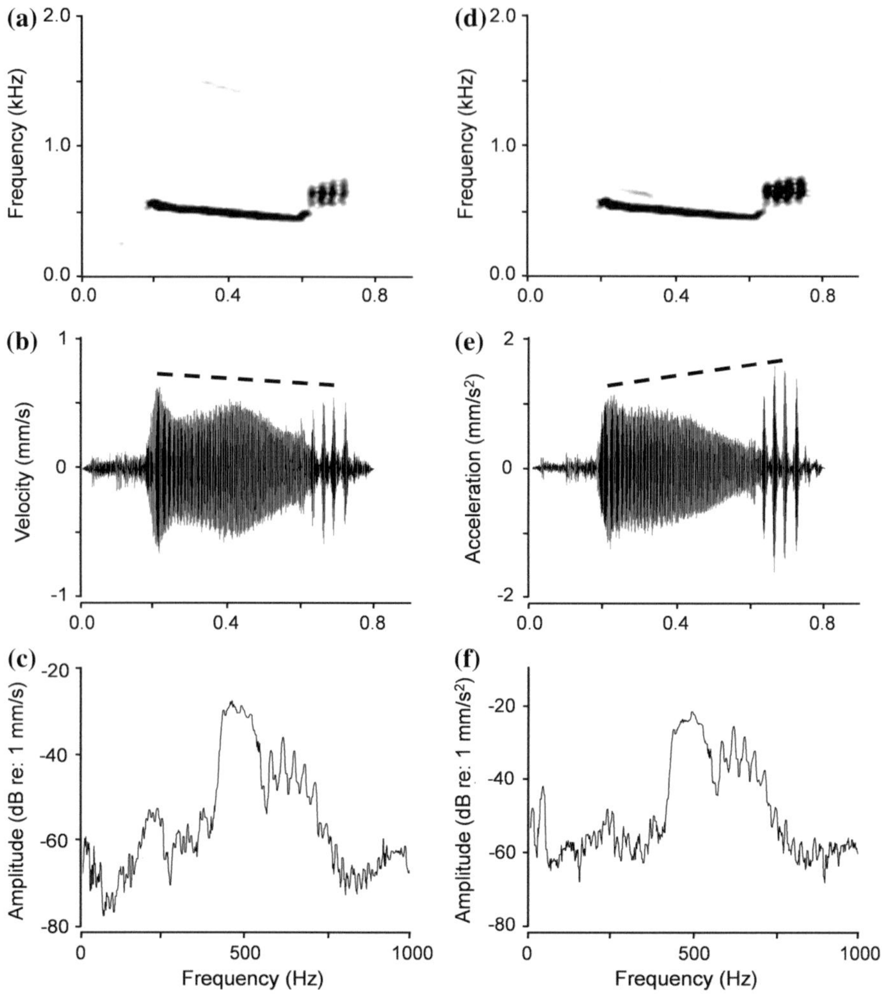


**Fig. 13.1** The velocity and acceleration waveforms of the vibrational signal of a male *Tylopelta gibbera* (Membracidae), recorded with a laser vibrometer and an accelerometer at the same location on the host plant stem. Spectrograms, waveforms, and amplitude spectra are shown for the laser signal (**a**–**c**) and the accelerometer signal (**d**–**f**). Note that the higher-frequency component of the signal (the pulses at the end) has a greater amplitude in the accelerometer recording

Matching the recording and playback devices (e.g., using an accelerometer/shaker combination or a laser/modified speaker combination) is, however, neither sufficient nor necessary. The reason such matching is not sufficient is that most playback devices, whether shaker, electromagnet, speaker, or piezoelectric actuator, do not have a flat frequency response. Some frequencies will have higher amplitudes than others, either because of the device's inherent properties (which vary with the mass being driven) or because of the way the device is coupled to the

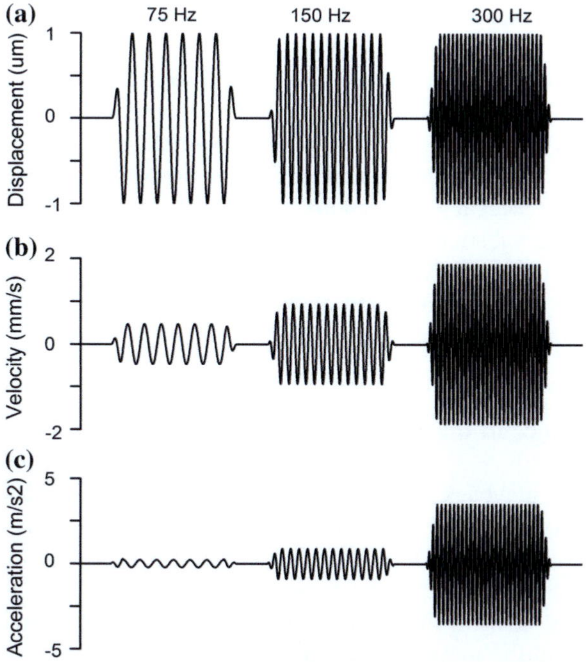


**Fig. 13.2** The amplitude relationship between displacement, velocity, and acceleration. **a** A series of three tones an octave apart, each with a peak displacement of 1 μm. **b** The velocity record of the same three tones. The peak velocity doubles for each doubling of frequency. Consider substrate motion within the first quarter of a cycle: For a 75 Hz sine wave, the change from 0 displacement to 1 mm occurs in 3.3 ms, while for a 150 Hz sine wave, the substrate moves through the same distance in half the time, so the velocity is doubled, while the peak displacement remains the same. **c** The acceleration record of the same three tones; the peak acceleration again doubles (with respect to velocity) with each doubling of frequency. A useful generality: For a signal containing a range of frequencies, the velocity amplitude will increase by 6 dB per octave relative to the displacement amplitude, while the acceleration amplitude will increase by 12 dB per octave

substrate. This problem is analogous to substrate filtering, considered in the next section. In Sect. 13.2.2 (below), we explain how this problem can be solved to conduct high-fidelity vibrational playbacks using virtually any combination of recording and playback devices.

#### 13.2.1.2 The Substrate Filters the Played-Back Signal

Plant stems and leaves, like other substrates, act as filters that attenuate some frequencies more than others. As a consequence, even if one could use idealized, 'perfect' recording and playback devices, the signal at the playback subject will still differ from that intended. This filtering effect can easily be seen by playing back a noise stimulus with equal amplitude across a band of frequencies and recording it some distance away; the amplitude spectrum will typically be far from flat at the second point (Fig. 13.3). The same filtering, of course, will also occur with an animal signal played back through the same substrate. Filtering varies among plant species (McNett and Cocroft 2008), among plants of the same species (Cocroft et al. 2006), among different parts of a single plant (McNett and Cocroft 2008), and even among locations on the same leaf or stem (Michelsen et al. 1982; Cokl et al. 2007; Magal et al. 2000). Given this level of variability, average differences among substrate types can be revealed only with large sample sizes (e.g., McNett and Cocroft 2008). If substrate filtering is not compensated for (see below), then the focal receiver will experience a signal that departs in unknown ways from the intended signal.

#### 13.2.1.3 Resonance in the Playback Setup or the Substrate

In particular, for playback on a larger plant, it may be necessary to mount the playback device on a tripod or other structure to position it next to the substrate. The result will be a structure with its own resonant frequency or frequencies. If the resonant frequency falls within the range of the playback stimulus, then this frequency will be excited in the structure holding the playback device and will be overrepresented in the played-back signal.

Resonance also occurs in natural substrates, but under what conditions it is important for vibrationally communicating insects remains an open question. Polajnar et al. (2012) documented resonance in a grasslike plant with straight, hollow stems: At certain frequencies, there was a pattern of regular increases and decreases in amplitude along the stem. At a given location on the stem, resonance was seen during playback of a frequency sweep, in the form of sharp increases in amplitude at particular frequencies. Evidence that resonance was the cause of these patterns came from a match between the observed patterns and those predicted in an ideal thin elastic rod. The resonant peaks occurred as a consequence of reflected waves that created standing waves in the plant stem. In contrast, in our work with membracid treehoppers on woody host plants, reflected waves are minimal and the striking resonance phenomena observed by Polajnar et al. (2012) are absent. We do not yet have a general framework for predicting when a system will be dominated by reflected waves and resonance and when it will be dominated by transient one-way wave propagation, so researchers will need to approach playback substrates on a case-by-case basis.

#### 13.2.1.4 The Playback Device is not Adequately Coupled to the Substrate

Playback devices must be attached to the substrate. Shakers function like heavy, robust audio speakers, producing a force proportional to the current supplied to the coil that drives the load. Instead of a membrane that produces airborne sound, there is typically a mounting stud to which a bolt or 'stinger' is fitted and then attached to the substrate to be vibrated. For engineering purposes, the stinger is usually a bolt that is coupled to a matching threaded attachment on the substrate or load. For structures such as plant stems, this method of attachment is usually not feasible, and in practice, most investigators position the shaker so that the bolt is in firm contact with the substrate. This method provides less than ideal coupling between shaker and substrate, but the coupling (especially at higher frequencies) can be improved by using an adhesive to attach the bolt to the stem.

Even small shakers are relatively heavy objects (typically a few kg) and can be difficult to position and align to vibrate the substrate at the desired location and along the desired axis. Like any playback device, a shaker requires a fixed base from which it pushes against the substrate. One approach is to place the shaker on a tabletop and position the plant so the desired part contacts the shaker, but this is not feasible for all plants. Another approach, which works even for large branching woody plants, is to use a positioning structure such as a microscope boom stand to securely hold the base of the shaker, allowing the investigator to adjust the shaker position in three dimensions.

The use of audio speakers as playback devices is common in the literature, and usually, the membrane is removed to reduce the production of airborne sound—although in some cases, the insects will respond to playback when placed directly on the speaker membrane (Zunic et al. 2008). To use the speaker more like a shaker, after removing the membrane, a pin or other small rigid device is attached to the moving coil. The speaker is then clamped in position so that the end of the pin contacts the plant substrate. For small plant structures, this approach provides a lighter and more easily positioned device.

Electromagnets have often been used to drive a small magnet glued to the plant stem or leaf (Michelsen et al. 1982). One potential advantage of this method is that it produces a less rigid coupling to the substrate and allows the plant stem to move more freely (though its motion is still constrained by the presence of an attached magnet in the magnetic field of the electromagnet). One disadvantage is that the frequency response of the playback system is highly dependent on the distance between the electromagnet and the magnet, and thus, this distance needs to be maintained. Another disadvantage is that it is more difficult to achieve an alignment that drives the stem or leaf along a single axis, and the magnet may thus produce a more complex whirling motion at the source than does a shaker, modified audio speaker, or piezoelectric actuator.

Piezoelectric actuators are small, light devices that are easy to position and which when properly aligned will produce motion along a single axis. They require more specialized electronics to drive them and in our experience are most useful for signals containing energy above 100 Hz.

One potential disadvantage of all of these methods is that they all require attachment to the substrate, and thus, the coupling of vibrations to the substrate is likely to differ in unknown ways from that produced by a signaling animal. Whether constraining the plant by attaching a playback device alters receiver behavior has been little explored, but two approaches provide a way around the issue that can be used in special circumstances. In one creative study, the investigators glued the insect's back to a stick driven by a speaker coil and allowed the insect to grasp the substrate with its legs. Vibrating the stem via a living insect is the closest any playback experiment has come to replicating signal production by an insect (A. Cokl, personal communication). Another study used broadcast airborne sound (Rebar et al. 2012), which can be perceived by the playback subject either directly (Shaw 1994) or by means of the vibrations produced in the plant by the sound (see Gogala Chap. 3, this volume). This latter approach is especially useful when one needs to expose large numbers of insects on multiple plants to a stimulus, but would not be suitable for more detailed studies of vibration localization or female preferences, because it provides little control over the vibration amplitude and frequency spectrum at the location of the receiver.

### *13.2.2 The Solution*

The changes introduced in a signal by the playback equipment and substrate can be easily diagnosed and in most cases easily solved, by pre-filtering the signal based on measurement of the filter imposed on the signal along the playback path. The solution is conceptually simple and requires only a modest amount of signal processing to carry out. Essentially, one needs to measure the frequency response of the system—i.e., how the signal has been filtered between the computer output and the substrate at the point of the focal receiver—and then compensate for this frequency response (see Hill and Shadley 2001). Figure 13.3 provides an illustration. Once this has been done, then shakers, modified audio speakers, electromagnet/magnet combinations, and piezoelectric actuators can all be used to conduct high-fidelity vibrational playbacks. Furthermore, all will produce bending waves in plant stems, as do signaling insects (Michelsen et al. 1982; Cocroft et al. 2000).

The playback system's frequency response can be compensated for in a number of ways. An approximate, analog method would be to use a graphic equalizer, which uses a series of filters (generally one to three per octave) that allow one to adjust the system's frequency response. This would involve a trial-and-error phase of raising and lowering the amplitude in various bands to approximate a flat frequency response; we are not aware of any published studies that have used this method. Digital signal processing methods are far more flexible. This approach involves playing back a test signal through the system and recording it with a sensor at the site where the playback subject will be placed. Comparing the amplitude spectrum of the test signal between its original and played-back form allows one to characterize the system's frequency response. Essentially, for each frequency bin in the amplitude spectrum, one calculates the ratio of the amplitude of the original test signal to that of the played-back signal. This ratio can be obtained by dividing one amplitude spectrum by the other or by calculating the transfer function between the two signals. The transfer function of the playback

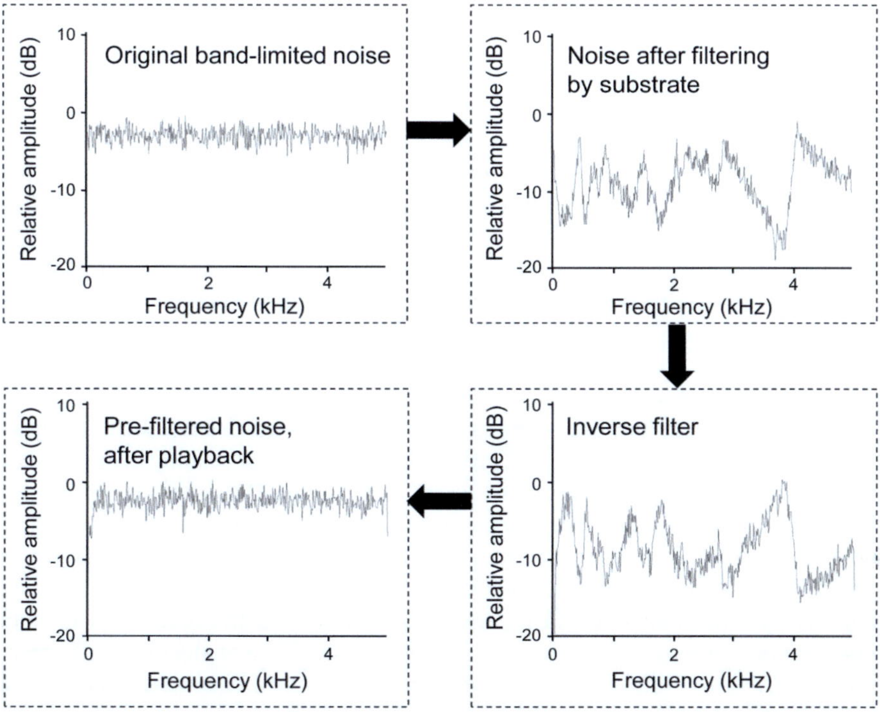


**Fig. 13.3** Compensating for the frequency response of the playback system. A band of noise with equal energy across a range of frequencies is played through a vibration transducer and recorded at some distance along the stem, where it no longer has equal energy in all frequencies. The inverted frequency response of the system is used to design a digital filter (*third panel*), and that filter is applied to the original noise signal. When this compensated signal is played through the system, the signal at the location of the playback subject now has equal energy across the relevant frequency range (*final panel*)

system is the ratio of the FFT of the played-back signal to the FFT of the original signal and can be calculated by taking the cross-spectral density between the two signals, divided by the auto-spectral density of the original signal.

Once the frequency response of the playback system is characterized, it can be compensated for. One approach is to multiply the FFT of the playback signal by the transfer function (original signal/recorded signal) and use an inverse FFT to recreate the playback signal with an adjusted frequency spectrum. An alternative is to use the system's frequency response to design a digital filter; this approach is more flexible in that it allows pre-filtering of playback files of any duration (the previous approach requires the stimulus file to have the same number of sample points as the files used to calculate the transfer function). When the compensated signal is played through the system, its amplitude spectrum should now closely match that of the signal when originally recorded (Fig. 13.3), as should the time-domain waveform. The process often requires two or three iterations, because for frequencies that are severely attenuated, the filter estimation improves after the attenuation has been at least partially compensated for. Note that for playbacks conducted in the field (see below), background noise may alter the estimation of the system's frequency response, and its influence should be minimized to the extent possible (Hill and Shadley 2001).

The frequency response of the playback system must be re-calculated for each new playback location, whether this is a different substrate or a different location on the same substrate. One possible exception is for playback transducers that are driving a negligible load, such as a small leaf. Here, it is possible to calculate the system's frequency response once and to use the resulting compensation filter for all future playbacks using that particular playback transducer. However, once there is an appreciable load on the playback transducer—such as a plant stem—the frequency response will be influenced by the nature of the load.

Depending on the purpose of the experiment, playbacks that do not compensate for substrate filtering may give misleading results. For example, suppose a receiver's actual frequency response curve is flat in the 100–4,000 Hz range, but that the receiver responds more (using some assay of response) at higher amplitudes. If the substrate used for the playback experiment imposed a filter like that in Fig. 13.3, then an experimenter playing back a range of frequencies at an equal amplitude *at the source* would mistakenly conclude that the receiver was much more responsive to some frequencies than to others. This conclusion would be an artifact of the substrate filtering, simply because some frequencies had a much higher amplitude than others at the receiver's location.

Note that if the system filter is calculated using an accelerometer or other sensor that contacts the substrate, removing that sensor will change the filter. If it is necessary to remove the sensor prior to the experiment (e.g., some insects we have worked with will climb onto and court an accelerometer), the influence of this change should be estimated, such as by calculating the filter with and without a second sensor attached.

The above discussion assumes that if one is using recordings of natural animal signals, they were recorded with the same kind of sensor as the one used for calculating the system's frequency response. If different kinds of sensors are used for recording the original signals and for calibrating the playback, then the compensation will be incorrect; for example, if the signal was recorded with a laser vibrometer and the amplitude calibration is done with an accelerometer, the compensation filter or transfer function will be off by 6 dB per octave. In this situation, it is still possible to compensate correctly by adjusting the filter using the transfer function between the two sensors or by using numerical integration/differentiation to convert the waveform from acceleration to velocity or vice versa.

If there is resonance in the playback setup or substrate, the frequency response at the resonant frequencies sometimes cannot be compensated for through signal processing. For example, resonance where the motion corresponds to the motion of the shaker head can be compensated, but other modes where the shaker head is pitching and rocking cannot. How one deals with the problem thus depends on the source of the resonance and the questions being asked. If the resonance arises in the equipment or supporting structures used for conducting the playback, then it may be possible to alter the setup, either by isolating parts of it from the plant with vibration-damping material or by adding mass to lower the resonant frequency out of the frequency range of interest (e.g., at a node of a vibrational mode). If the resonance is in the substrate itself, how one deals with it depends on whether the substrate is typical of those used by the insect in nature and on whether the subject will be stationary. If resonance is present in an artificial substrate, or in a host plant much smaller than those used by the insects in the field, then using a different substrate would be appropriate. If the resonance observed is typical of that occurring in natural substrates used by the insects, then if the subject is stationary, it may be possible to find a location where resonance is not an issue for the frequency range of interest. If the subject is moving, then the resonance is something it would encounter in nature, and the best approach may be to use the 'post hoc' method described below (see Sect. 13.9).

Finally, an exception to the above discussion: There is one situation in which it is more effective to simply calibrate the stimulus amplitude, without first obtaining the system filter. When the stimulus contains only a single frequency, playing the signal through the substrate and adjusting the amplitude at the desired location allows precise adjustment of the amplitude of that frequency; indeed, for single frequencies, this method can be more precise than an FFT-based method. When the playback experiment involves a modest number of single-frequency stimuli (and especially if the amplitude calibration is automated), this approach is quick and straightforward and yields excellent precision. However, for any stimulus containing a band of frequencies (such as a frequency sweep or a broadband click), pre-filtering based on the system filter is required.

A note on software: The authors use MATLAB (MathWorks Inc., Natick, MA, USA) for signal processing. MATLAB scripts for frequency compensation and for conversion of waveforms between displacement, velocity, and acceleration are provided in the Appendix.

## 13.3 Temporal Characteristics of the Played-Back Signal

### *13.3.1 The Problem*

In some circumstances, the temporal features of the playback signal, such as its duration, will be changed between the playback transducer and the focal animal. In particular, resonance in the playback setup or substrate, or distortion from reflected waves can cause signal degradation. Furthermore, if the transmission distance is large between source and receiver, and the signal contains a range of frequencies, then dispersive propagation of bending waves (for which transmission speed is proportional to the square root of frequency) could cause changes in the amplitude envelope of the signal [see Wood and O'Connell-Rodwell (2010) for an example of dispersion-related changes in a signal propagated by Rayleigh waves]. Finally, if there is frequency modulation in the playback signal, and the frequency filtering properties of the playback path are not compensated for, then because some frequencies will have higher amplitudes than others, the overall amplitude envelope of the signal will be changed. For example, the amplitude of a frequency sweep that was constant at the source will now fluctuate in unpredictable ways.

### *13.3.2 The Solution*

If the problem arises from reflected waves in the substrate, their influence can sometimes be reduced by choosing a different substrate or lowering the amplitude of the playback. Signal changes due to dispersive propagation can be avoided by conducting the playback close to the focal animal (on a plant, within a few 10 s of cm). Changes in the amplitude envelope caused by substrate filtering can be eliminated by correcting for the system's frequency response as described above. In our experience, changes in the gross temporal features of playback signals due to resonance or dispersive propagation are a less common issue than changes due to uncorrected frequency filtering, but in some study systems, these changes are much more frequent (Cokl and Virant-Doberlet 2003; Polajnar et al. 2012). The extent of the problem may depend on the nature of the substrate; with small insects on woody host plants, we have encountered little reflected energy, but signals on simple rodlike structures with little damping may create substantial resonance (Polajnar et al. 2012).

## 13.4 Amplitude Calibration

### *13.4.1 The Problem*

For vibrational signals produced on solid structures, the amplitude of the signal depends on properties of the substrate including its density and mass and on the impedance between the signaler and the substrate. On a plant, an insect signaling on a thin leaf petiole will produce a much higher-amplitude signal near the source than the same insect signaling on a thick woody stem (Fig. 13.4). Measurements of signal amplitude alone, then, without reference to the structure on which they were produced, are not very meaningful. Likewise, playback of vibrational signals on plants should take into account the diameter of the stem or petiole on which the signal was recorded or on which amplitude was measured. Whether insects take account of the inverse relationship between mass and velocity to assess signal power has not been investigated, but it seems prudent for investigators to keep this relationship in mind. That is, suppose the peak velocity of the signals an insect produces on a thin leaf petiole is 5 mm/s and the peak velocity of the signals it produces on a thick woody stem is 0.1 mm/s. With a shaker, it would be possible to play back signals to a receiver on a thick woody stem using a peak velocity of

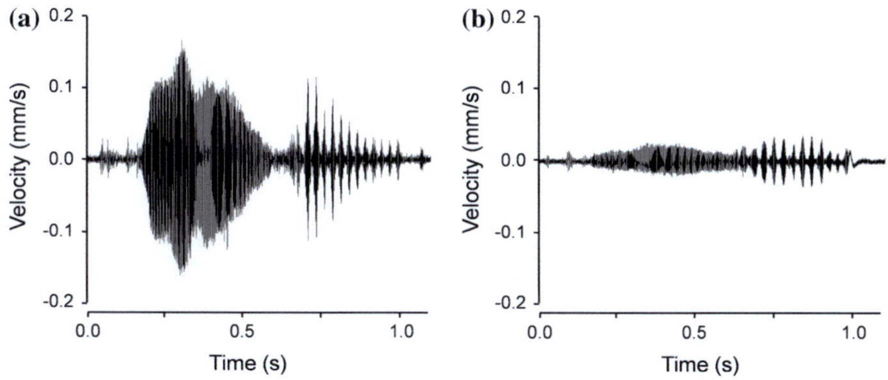


**Fig. 13.4** What is the amplitude of an insect's vibrational signal? A male treehopper (*Tylopelta gibbera*) signaled at two locations on a host plant and at each location was recorded with a laser vibrometer focused on a point 1.5 cm from the male. Signal (**a**) was recorded on a green stem with a diameter of 2.2 mm, while signal (**b**) was recorded on a woody stem with a diameter of 7.2 mm. There is a 13-dB difference between the peak velocity of the signals (0.16 vs. 0.035 mm/s)

5 mm/s, but this would create a very unnatural situation in which the receiver's behavior may be difficult to interpret.

The relationship between amplitude and substrate is a particularly important issue for playbacks in which the insect must localize a vibration source or is otherwise moving around the plant; if the trunk at the base of the plant is shaken at an amplitude typical of that on a leaf petiole, the thinner stems toward the apex of the plant may vibrate at an amplitude outside the range experienced by communicating insects.

### *13.4.2 The Solution*

There are several ways to achieve the appropriate signal amplitude for a playback.

(a) The most straightforward way to ensure the proper signal amplitude is to use a calibrated transducer and data acquisition system and adjust the output of the playback device to match the value of the desired peak or RMS amplitude (at the receiver's location) of the played-back signal. This value might be the amplitude of that same signal when originally recorded, or the average amplitude of a sample of signals, or some other value of experimental relevance. The playback substrate should be similar to the substrate on which the signals were recorded (see Sect. 13.4.1).
(b) The proper playback amplitude can be obtained when using a sensor whose output is not calibrated or when using a recording system with some unknown multiplier of the incoming voltage signal, as is typical for systems with a variable gain (e.g., computer sound card, audio recorder). As long as the

recording system and gain settings are the same as those used to record the original signals, the amplitude can be faithfully reproduced by matching the played-back signal amplitude to that of the original signal.

(c) The correct amplitude can be obtained even with sensors such as phonograph cartridges or piezoelectric film, whose output level varies with the nature of the connection (e.g., the pressure applied, the angle between the stylus and the substrate). If a sensor used to record some insect signals is left at its original position, and the signals are played from the site of the original signaler, then simply matching the amplitude of the original recording is sufficient.

(d) Finally, with any recording system, it is possible to set up the playback, record a sample of individuals on that substrate, and adjust the output of the playback device to achieve the appropriate amplitude.

## 13.5 Substrate Effects on Receiver Behavior

### *13.5.1 The Problem*

For playbacks in the field (see below), the substrate choice will be dictated by the location of the study animals. For playbacks in the laboratory, there are more choices: natural substrate or artificial? Host or non-host? Which plant part? This choice can have important consequences. For example, *Enchenopa binotata* treehoppers are host plant specialists. Although males will signal on a non-host plant, they produce fewer and shorter signals than they do on a host (Sattman and Cocroft 2003). Wolf spiders prefer to produce their vibrational signals on leaf litter, and when signaling on less conductive substrates like soil or rock, the spiders accompany their vibrational signals with more visual signals (Gordon and Uetz 2011).

For studies of vibration localization, the choice of substrate is critical. Artificial substrates may behave completely differently than natural substrates, and this may influence whether or how a focal subject responds to a stimulus. For example, an animal on an artificial substrate that appears unable to localize a signal may not have access to the same information as an animal on a natural substrate. In such cases, we suggest comparing localization cues (e.g., time delays between sensors in different legs, amplitude gradients) between natural and artificial substrates.

### *13.5.2 The Solution*

There is no way to know a priori whether individuals of a given species will behave and respond to playbacks similarly on their usual substrate and on an artificial substrate in the laboratory. For example, in treehoppers, males of a host plant specialist signal differently on hosts and non-hosts (Sattman and Cocroft

2003), while males of a species with a larger host plant range have the same signaling behavior on hosts and non-hosts (Cocroft et al. 2006). The use of a substrate closely resembling the one on which communication usually takes place is the safest bet. However, sometimes artificial substrates are extremely useful, and in that case, one simply needs to compare the behavior of the animals on the natural and artificial substrate.

## 13.6 The Substrate Vibrates Along More Than One Axis

### *13.6.1 The Problem*

The vibrational motion of physical structures occurs in three dimensions. In a plant stem, motion can occur along the long axis of the stem or in a plane perpendicular to that axis. Michelsen (Chap. 11, this volume) points out that no one has yet characterized the extent of motion along the long axis of plant stems, and the importance of longitudinal vibrations along the stem remains an open question. For the remaining two dimensions (Fig. 13.5), there are two issues. First, if the stem is vibrating along only one axis, the focal animal may or may not be aligned with that axis. Rohrseitz and Kilpinen (1997) found that the subgenual organ of honeybees is about 10 dB more sensitive to motion along the long axis of the tibia than to motion perpendicular to that axis. To the extent that the same is true of other species, then this issue needs to be taken into account. For example, suppose a playback device causes the plant stem to move along a single axis. If the most sensitive axis of the study subject's vibration receptors is parallel to the axis of motion, the subject will experience a much higher-amplitude signal than if the most sensitive axis of its sensors is perpendicular to the axis of motion. This alignment issue is important for behavior: Male treehoppers locating a female made more accurate decisions when their dorsoventral axis was aligned with the major axis of stem motion (Gibson and Cocroft, in preparation). Second, the use of two sensors whose axes of sensitivity are perpendicular to each other reveals that although plant stems sometimes vibrate along a single axis, they typically vibrate with a whirling motion (Fig. 13.5), a phenomenon Michelsen et al. (1982) first described by observing plant motion using a strobe light. How the whirling nature of stem motion influences insect vibration perception is still unknown, but two-dimensional motion of the substrate is a ubiquitous feature of the insect's perceptual world.

### *13.6.2 The Solution*

We know of no studies that have controlled the two-dimensional motion of a plant stem at a distance from the playback device. Doing so would require two playback devices placed at right angles to each other, and even this approach

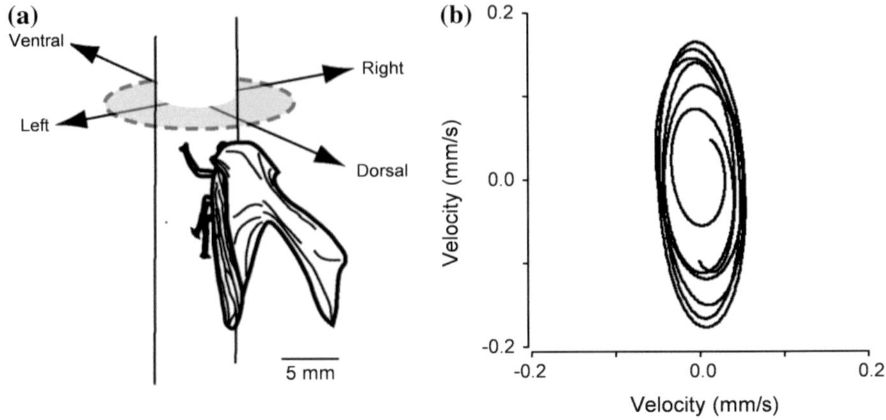


**Fig. 13.5** Vibrational signals produced on plant stems are transmitted as bending waves that cause motion perpendicular to the long axis of the stem. This motion can occur along any axis in a plane perpendicular to the stem (**a**) [Reproduced from McNett et al. (2006), with permission]. Typically, the motion occurs along more than one axis during the transmission of insect signals, so that stems vibrate with a whirling motion (**b**). The motion of a point on the stem can be seen by plotting the output of two transducers whose axes of sensitivity are in the plane shown in (**a**) but are oriented at 90° with respect to each other. The trace in (**b**) shows stem velocity during 3.4 ms of a mating signal from a male *Umbonia crassicornis* [illustrated in (**a**)] recorded close to the signaler using two orthogonally positioned laser vibrometers

would allow one to influence the two-dimensional motion of the stem only at specific locations. Once the signal has travelled beyond the point at which the compensation was done, uncontrolled variation in substrate properties will affect the motion. However, measurements taken at varying distances from a signaling treehopper on a plant stem show that, close to the insect, the vibrations occur with a relatively linear motion aligned with the dorsoventral axis of the insect. Although the nature of the motion imparted to the stem will depend on the method of vibration production, it seems likely that for many signaling species, the results will be similar to the treehopper example. If so, then keeping the playback device within 10 cm or so of the focal insect, with the major axis of motion aligned with the insect's dorsoventral axis, should minimize the variation introduced by changes in the two-dimensional motion of the stem. Note, however, that this assumes that the playback device is imparting uniaxial motion to the stem, which will be largely true for a shaker or piezoelectric stack. For a magnet/electromagnet combination, however, depending on the alignment between magnet and electromagnet and the attachment of the magnet to the stem, the magnet may have a 'fluttering' or side-to-side rocking motion rather than simple motion along one axis (RBC, pers. obs.).

## 13.7 Conducting Vibrational Playback Experiments in the Field

### *13.7.1 The Problem*

Field playbacks present additional challenges over those encountered in the laboratory, including limited choices of playback location, difficulty in maintaining alignment between playback transducer and substrate, the presence of background noise, and variable weather conditions. Only a few researchers have conducted vibrational playback experiments in the field (Hill and Shadley 2001; Morales et al. 2008; O'Connell-Rodwell et al. 2006, 2007; Caldwell et al. 2010). We continue to focus on playbacks of signals through plant substrates.

Only a fraction of potential receivers will be in suitable playback locations. Subjects must be within reach of the playback apparatus and in a location where the influence of environmental noise and of other members of the biological community is minimized.

Using a pre-filtering procedure to assess and compensate for filtering by the substrate and playback apparatus is also more challenging than in a laboratory environment. For example, if wind moves the plant substrate and changes its position relative to the playback transducer, one must re-compensate for frequency filtering. It is thus necessary to stably align transducer and substrate.

In addition to changing the alignment between the playback transducer and plant substrate, wind is also the single greatest source of environmental noise (Cocroft and Rodríguez 2005). Other sources of environmental noise include rain, birdsong, and signaling by other invertebrates on the focal plant. Environmental noise should be minimized because it can alter the estimation of the system's frequency response (Hill and Shadley 2001; O'Connell-Rodwell et al. 2006) and because it can influence receiver behavior. For example, wind inhibits vibrational communication by insects (McNett et al. 2010; Tishechkin 2007), and birdsong inhibits signaling by male wolf spiders (Gordon and Uetz 2012).

Variable weather conditions pose significant challenges for playbacks in the field, given that most electronic gear used in vibrational communication research is fragile and costly. If one is using wax, putty, or similar adhesive to couple a transducer to a substrate, direct sunlight or heat can cause the transducer to shift position.

### *13.7.2 The Solution*

Desirable playback locations are dependent on the playback apparatus, other members of the biological community, environmental and anthropogenic noise sources, and the experimental design. Playback subjects that have fixed locations are advantageous, in that a playback apparatus can be assembled on site and left in place for several treatments (e.g., as with a paired or repeated-measures design).

For playbacks on plants, stationary alignment between transducer and substrate can be achieved by fixing both the transducer and the plant substrate with separate tripods. However, note that metal tripods in particular may introduce resonance into the playback system and that fixing the plant substrate loads it with additional mass and changes it in ways that are likely to affect signal transmission. Therefore, the system filter should be estimated after all of the playback apparatus is in place.

Wind noise in many environments can be largely avoided by conducting playbacks in the early morning. To avoid contact by nearby plants with the focal plant, assessing the focal plant area in advance and clearing or trimming contacting vegetation is recommended. Additionally, the area surrounding the playback subject should be carefully examined, because undetected invertebrates may vibrationally 'chime in' during a playback treatment.

Field playbacks are replete with trade-offs. Part of the benefit of conducting playbacks in the field is to assess not only the effect of the playback stimulus, but also the relative importance of environmental predictors on receiver response. By minimizing the influence of environmental noise, by conducting playbacks during early mornings, when wind is rare and temperatures are coolest, and by modifying the physical environment of the plant substrate, one limits the potential interference. However, with these efforts, field playbacks are possible and add a layer of biological reality that complements laboratory studies. Of course, the communicating animals themselves will only experience these conditions for a limited time during the day, and there are few studies of how vibrationally communicating animals deal with the varied signaling conditions they encounter (McNett et al. 2010; Tishechkin 2007; Lohrey et al. 2009). However, the evidence so far suggests that the animals communicate when the conditions are favorable [as with animals using other modalities, e.g., Brumm and Slabbekoorn (2005)], so conducting the playback under ideal, low-noise conditions is reasonable.

## 13.8 Should Experimental Designs Use Multiple Exemplars of Substrates?

### *13.8.1 The Problem*

Every natural substrate is unique; two leaves from the same plant, two areas of leaf litter, or two square meters of soil will differ from each other in their vibration-transmitting characteristics. For some experimental questions, it is reasonable to use the same substrate exemplar for all replicates of the experiment, while for others, using a single substrate would limit the inferences that are possible.

### *13.8.2 The Solution*

The issue is whether the focal animal's response is dependent on variation in the nature of the substrate. For a female preference test, for example, as long as the substrate is appropriate and the system's frequency response is taken into account, using one substrate exemplar may be sufficient. In contrast, for any question for which the nature of the substrate is relevant, using more than one substrate exemplar is important. In particular, localization abilities are highly substrate dependent (depending on propagation velocity, reflected waves, angle of motion relative to the subject's body, etc.), so for making general statements about localization ability, it is critical to use more than one substrate. One approach is to use 'substrate' as an additional experimental treatment. For example, if localization is occurring on a plant and there are to be 15 experimental subjects, one might use three plants, with 5 subjects within each plant. 'Plant' could then be treated as a random effect in a statistical model to evaluate the results. While this would not necessarily provide a representative sample of variation in plant substrates, it would allow one to assess the extent to which differences among plants influence localization.

## 13.9 Mobile Playback Subjects

### *13.9.1 The Problem*

We have argued that the most basic and widespread problem in vibrational playbacks is the unpredictable frequency response of substrates and playback devices. This problem is easily solved, but only for a single location on a plant or other substrate. If the focal animal remains at or very near the location where the system frequency response was calculated, the animal will experience the desired signal or one very close to it. This method works well for playbacks to animals whose position can be predicted and/or controlled by the experimenter. The picture changes if the playback subject will be moving around during the playback. Because the frequency response of solid substrates varies so much from place to place, the animal will experience different signal properties at different locations.

### *13.9.2 The Solution*

There are at least three solutions to the moving-subject problem. One approach is to determine the frequency response at multiple points on the substrate and generate a series of playback stimuli, each adjusted for the frequency response at a different location. When the animal is at a given point, the playback stimulus appropriate for that location can be played back. This approach is time-consuming but feasible (Gibson et al., in preparation) and is appropriate for some experiments with moving subjects. For example, suppose one wanted to characterize female preference functions with respect to some signal feature such as frequency, but females made substantial movements between responses to signals. If the signal properties have only been controlled at one location, then using female responses at other locations where the signal differs in unknown ways from that intended will introduce a large amount of noise (and possibly bias) into the statistical results. In this case, it would be necessary to play back signals that were previously adjusted for the female's potential locations (or to use the 'post hoc' approach, below). It is not possible to say a priori how close is 'close enough'; for example, would the system response need to be calculated at 1-cm intervals, 5-cm intervals, or 10-cm intervals? It is difficult to predict the signal properties at a given distance from the source because of unknown material properties and complicated branch geometry. To estimate the appropriate spatial scale, one should first characterize substrate filtering at different locations. For example, if the amplitude of a tone changes by less than 3 dB within 10-cm, then adjusting for the system filtering every 10-cm may be sufficient.

A second approach to the mobile-subject problem is to compensate the signals for the system's frequency response very close to the location of the playback device and to assume that the distance-dependent changes in the signal are similar to those that would occur with a real signaling animal at the same location on the substrate. This assumption may not be warranted if the playback substrate is not representative of natural substrates and ideally would be evaluated before conducting the playbacks. This approach may not work well for some experiments, such as tests that relate signal properties to receiver responses, because those signal properties will vary between locations. But animals in nature always have to deal with substrate filtering, so this approach is fine for questions such as those dealing with the timing of movement or signaling relative to the playback stimulus (Cocroft 2005; Legendre et al. 2012).

A third approach to dealing with the mobile-subject problem is to determine 'post hoc' what signal properties the animal experienced at each location (Gibson and Cocroft, in preparation). This first requires tracking the locations at which the animal responded to the signal, then after the experiment is finished, playing back the signals again and recording them at each of the relevant locations. This method allows the signal to be influenced in unpredictable ways at different locations on the substrate, but the measurements taken at each of those locations provide high precision in relating signal properties to behavioral responses. This post hoc approach requires the use of a vibration transducing method that does not influence the substrate's properties, such as laser vibrometry for small structures or accelerometers for large structures or the soil. Signal characteristics (e.g., amplitude) at each location can then be used as predictors in a multiple regression, with the behavior of the animal (e.g., move forward or reverse direction) as the dependent variable.

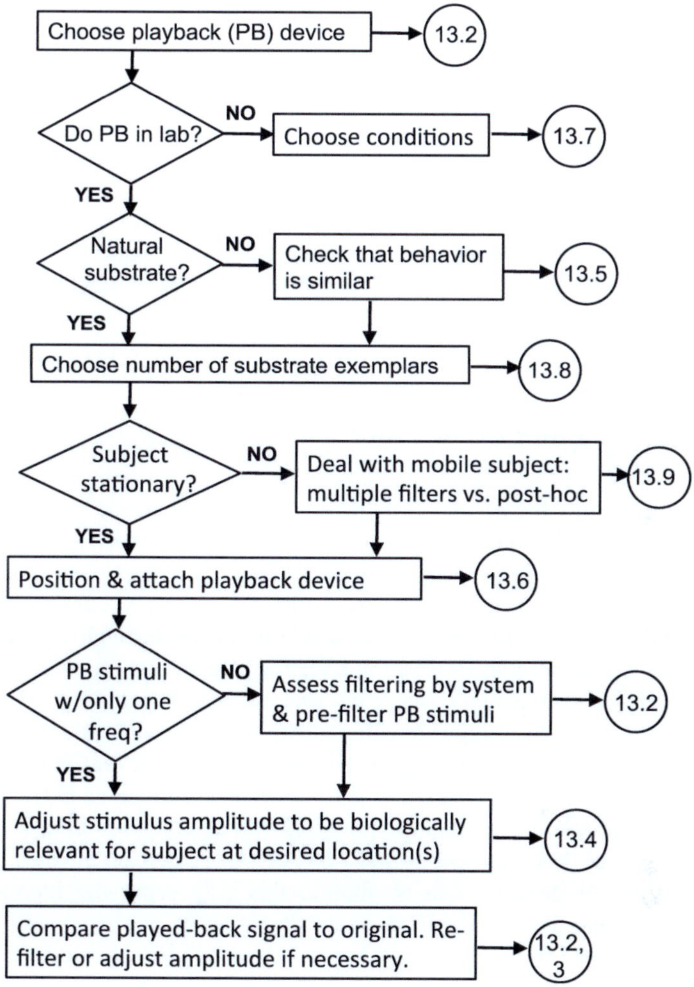


**Fig. 13.6** A decision tree for conducting vibrational playbacks. *Numbers in the circles* indicate the section(s) of the text relevant to the decision

## 13.10 A Decision Tree for Conducting Vibrational Playbacks

Here, we provide a flowchart (Fig. 13.6) to assist in the design and execution of vibrational playbacks and to address the issues discussed in this chapter. Although our discussion has largely focused on playbacks on living plants, most of the issues are relevant to vibrational playbacks through other substrates such as soil or leaf litter.

**Acknowledgments** RBC acknowledges support from NSF IOB 0820533. The manuscript benefitted from conversations with Ron Miles, Carol Miles, Peggy Hill, Matija Gogala, and Andreas Wessel.

# A.1 Appendix

The following MATLAB programs use the Signal Processing Toolbox and the Control System Toolbox and have been confirmed to work in MATLAB 6.5 (R13) through to MATLAB 8.1 (R2013a).

## *A.1.1 Digital Equalization Filter*

This program obtains the system equalization filter from stored measurements. A typical system would include the digital-to-analog converter, amplifier and vibration exciter, vibration medium, measurement transducer, anti-aliasing filter, and analog-to-digital converter. Prior to running this code, a continuous random signal (stored in WAVE file 'Playback1.wav') is played through the system, and the response is measured and saved (WAVE file 'Recording1.wav').

The power spectral density functions are estimated and used to obtain the magnitude of the input-to-output transfer function. The useful data range is taken between the specified lower and upper frequencies in hertz (variables 'f_lo' and 'f_hi'), and the digital filter coefficients are estimated and saved (MATLAB data file 'FilterCoefs.mat'). For evaluation purposes, the equalization filter is applied to the original playback signal and stored (WAVE file 'Playback2.wav'). Arbitrary signals of different duration can be filtered this way using the identified filter coefficients, as long as the sample rates are the same.

MATLAB script for acquiring and implementing digital equalization filter:

```
close all, clear all


[out,fs,NBITS]=wavread('Playback1.wav'); %WAVE file with
original playback
[in,fs,NBITS]=wavread('Recording1.wav'); %WAVE file with
recorded signal dt=1/fs;
t_out = [0:dt:(length(out)-1)*dt];
t_in = [0:dt:(length(in)-1)*dt];


fftLength=4096;
[PSDout,Freq]=pwelch(out,ones(fftLength,1),
[],fftLength,fs);
```

```
[PSDin,Freq]=pwelch(in,ones(fftLength,1),
[],fftLength,fs);
Hcmp=sqrt(PSDout./PSDin); %Amplitude compensation filter


f_lo=40; f_hi=10000; %lower and upper cutoff frequencies in
Hz.
lo=round(f_lo/(fs/fftLength))+1;
hi=round(f_hi/(fs/fftLength))+1;
Hcmp(1:lo)=0; Hcmp(hi:length(Hcmp))=0;


wn=Freq/max(Freq);
B=fir2(fftLength,wn,Hcmp); %this calculates the digital
filter coefficients
A=1;
save FilterCoefs.mat B A


outcmp=filter(B,A,out); %this applies the digital filter to
the signal
outcmp=outcmp*.9/max(abs(outcmp));
wavwrite(outcmp,fs,16,'Playback2.wav');
```

## *A.1.2 Differentiation and Integration of Playback Signal*

This MATLAB script numerically differentiates and integrates the time signal stored in a WAVE file ('ArbPlayback.wav'). Differentiation of the signal can be approximated using the finite difference method (with 'diff.m'), while integration of the signal can be approximated using trapezoidal integration (with 'cumtrapz.m'). These methods work well if the time step is sufficiently small and if there is no noise in the signal.

When the signal has additional noise, the higher-frequency noise is increased by the differentiation process, while the lower-frequency noise is increased by integration. This noise can be reduced by using a first-order band-pass filter to perform the differentiation or integration. The band-pass center frequency is set to a high frequency for differentiation (variable 'f_hi'), so the frequencies below the center frequency approximate a differentiation filter, while frequencies above are attenuated. For integration, the center frequency is set to a low frequency (variable 'f_lo'), so frequencies below the center frequency are attenuated, while frequencies above approximate an integration filter. The appropriate center frequency also depends on the frequency content of the signal.

MATLAB script for differentiation and integration of playback signal:

```
wavfile='ArbPlayback.wav'; %WAVE file name with playback
signal


[out,fs,NBITS]=wavread(wavfile);
dt=1/fs;
nt=length(out);
t_out =[0:dt:(nt-1)*dt];


%numerical differentiation by finite difference:
outdiff=diff(out)/dt;
outdiff(nt)=outdiff(nt-1);


%numerical integration by trapezoidal rule:
outint=cumtrapz(t_out,out);


% Differentiation filter: Band pass filter with high corner
frequency
f_hi=10000; %upper cutoff frequency in Hz.
SYSc=tf((2*pi*f_hi)^2*[1    0],conv([1    f_hi*2*pi],[1
f_hi*2*pi])); SYSd=c2d(SYSc,1/fs,'foh');
[Bcmp,Acmp]=tfdata(SYSd);
outfiltdiff=filter(Bcmp{1},Acmp{1},out);


% Integration filter: Band pass filter with low corner
frequency
f_lo=10; %lower cutoff frequency in Hz.
SYSc=tf([1 0],conv([1 f_lo*2*pi],[1 f_lo*2*pi]));
SYSd=c2d(SYSc,1/fs,'foh');
[Bcmp,Acmp]=tfdata(SYSd);
outfiltint=filter(Bcmp{1},Acmp{1},out);
```

## References

Barth FG (1982) Spiders and vibratory signals: sensory reception and behavioral significance. In: Witt PW, Rovner JS (eds) Spider communication: mechanisms and ecological significance. Princeton University Press, Princeton, pp 67–120

Barth FG (2002) A spider's world: senses and behavior. Springer, Heidelberg
Brumm H, Slabbekoorn H (2005) Acoustic communication in noise. Adv Stud Behav 35:151–209
Caldwell MS, Johnston GR, McDaniel JG, Warkentin KM (2010) Vibrational signaling in the agonistic interactions of red-eyed treefrogs. Curr Biol 20:1012–1017
Cocroft RB (2005) Vibrational communication facilitates cooperative foraging in a phloem-feeding insect. Proc R Soc B Biol 272:1023–1029
Cocroft RB (2010) Vibrational communication. In: Breed MD, Moore J (eds) Encyclopedia of animal behavior, vol 3. Academic Press, Oxford, pp 498–505
Cocroft RB, Rodríguez RL (2005) The behavioral ecology of insect vibrational communication. Bioscience 55:323–334
Cocroft RB, Tieu T, Hoy RR, Miles R (2000) Mechanical directionality in the response to substrate vibration in a treehopper. J Comp Physiol 186:695–705
Cocroft RB, Shugart HJ, Konrad KT, Tibbs K (2006) Variation in plant substrates and its consequences for insect vibrational communication. Ethology 112:779–789
Cokl A, Virant-Doberlet M (2003) Communication with substrate-borne signals in small plant-dwelling insects. Annu Rev Entomol 48:29–50
Cokl A, Zorovic M, Millar JG (2007) Vibrational communication along plants by the stink bugs *Nezara viridula* and *Murgantia histrionica*. Behav Process 75:40–54
Cuthill IC, Hart NS, Partridge JC, Bennett ATD, Hunt S, Church SC (2000) Avian colour vision and avian video playback experiments. Acta Ethol 3:29–37
Fleishman LJ, McClintock WJ, D'Eath RB, Brainard DH, Endler JA (1998) Colour perception and the use of video playback experiments in animal behavior. Anim Behav 56:1035–1040
Gerhardt HC (1995) Phonotaxis in female frogs and toads: execution and design of experiments. In: Klump GM, Dooling RR, Fay RR, Stebbins WC (eds) Animal psychophysics: design and conduct of sensory experiments. Birkhäuser Verlag, Basel, pp 209–220
Gordon SD, Uetz GW (2011) Multimodal communication of wolf spiders on different substrates: evidence for behavioural plasticity. Anim Behav 81:367–375
Gordon SD, Uetz GW (2012) Environmental interference: impact of acoustic noise on seismic communication and mating success. Behav Ecol 23:700–714
Hill PSM (2008) Vibrational communication in animals. Harvard University Press, Cambridge
Hill PSM, Shadley JR (2001) Talking back: sending soil vibration signals to lekking prairie mole cricket males. Am Zool 41:1200–1214
Kroodsma DE, Byers BE, Goodale E, Johnson S, Liu W-C (2001) Pseudoreplication in playback experiments, revisited a decade later. Anim Behav 61:1029–1033
Kroodsma DE (1986) Design of playback experiments. Auk 103:640–642
Kroodsma DE (1989) Suggested experimental designs for song playbacks. Anim Behav 37:600–609
Legendre F, Marting PR, Cocroft RB (2012) Competitive masking of vibrational signals during mate searching in a treehopper. Anim Behav 83:361–368
Lohrey AK, Clark DL, Gordon SD, Uetz GW (2009) Antipredator responses of wolf spiders (Araneae: Lycosidae) to sensory cues representing an avian predator. Anim Behav 77:813–821
Magal C, Scholler M, Tautz J, Casas J (2000) The role of leaf structure in vibration propagation. J Acoust Soc Am 108:2412–2418
McGregor PK, Catchpole CK, Dabelsteen T, Falls JB, Fusani L, Gerhardt HC, Gilbert F, Horn AG, Klump GM, Kroodsma DE, Lambrechts MM, McComb KE, Nelson DA, Pepperberg IM, Ratcliffe L, Searcy WA, Weary DM (1992) Design and interpretation of playback: the Thornbridge Hall NATO ARW consensus. In: McGregor PK (ed) Playback and studies of animal communication. Plenum Press, New York, pp 1–9
McGregor PK (2000) Playback experiments: design and analysis. Acta Ethol 3:3–8
McNett GD, Miles RN, Homentcovschi D, Cocroft RB (2006) A method for two-dimensional characterization of animal vibrational signals transmitted along plant stems. J Comp Physiol A 192:1245–1251

McNett GD, Cocroft RB (2008) Host shifts favor vibrational signal divergence in *Enchenopa binotata* treehoppers. Behav Ecol 19:650–656

McNett GD, Luan L, Cocroft RB (2010) Wind-induced noise alters signaler and receiver behavior in vibrational communication. Behav Ecol Sociobiol 64:2043–2051

Michelsen A, Fink F, Gogala M, Traue D (1982) Plants as transmission channels for insect vibrational songs. Behav Ecol Sociobiol 11:269–281

Morales MA, Barone JL, Henry CS (2008) Acoustic alarm signalling facilitates predator protection of treehoppers by mutualist ant bodyguards. Proc R Soc B Biol 275:1935–1941

O'Connell-Rodwell CE, Wood JD, Rodwell TC, Puria S, Partan SR, Keefe R, Shriver D, Arnason BT, Hart LA (2006) Wild elephant (*Loxodonta africana*) breeding herds respond to artificially transmitted seismic stimuli. Behav Ecol Sociobiol 59:842–850

O'Connell-Rodwell CE, Wood JD, Kinzley C, Rodwell RC, Poole JH, Puria S (2007) Wild African elephants (*Loxodonta africana*) discriminate between familiar and unfamiliar conspecific seismic alarm calls. J Acoust Soc Am 122:823–830

Polajnar J, Svensek D, Cokl A (2012) Resonance in herbaceous plant stems as a factor in vibrational communication of pentatomid bugs (Heteroptera: Pentatomidae). J R Soc Interface 9:1898–1907

Rebar D, Höbel G, Rodríguez RL (2012) Vibrational playback by means of airborne stimuli. J Exp Biol 215:3513–3518

Rohrseitz K, Kilpinen O (1997) Vibration transmission characteristics of the legs of freely standing honeybees. Zoology 100:80–84

Sattman DA, Cocroft RB (2003) Phenotypic plasticity and repeatability in the mating signals of *Enchenopa* treehoppers, with implications for reduced gene flow among host-shifted populations. Ethology 109:981–994

Shaw S (1994) Detection of airborne sound by a cockroach 'vibration detector': a possible missing link in insect auditory evolution. J Exp Biol 193:13–47

Tishechkin DY (2007) Background noises in vibratory communication channels of Homoptera (Cicadinea and Psyllinea). Russ Entomol J 16:39–46

Uetz GW, Roberts JA (2002) Multisensory cues and multimodal communication in spiders: insights from video/audio playback studies. Brain Behav Evol 59:222–230

Wiley RH (2003) Is there an ideal behavioural experiment? Anim Behav 66:585–588

Wood JD, O'Connell-Rodwell CE (2010) Studying vibrational communication: equipment options, recording, playback and analysis techniques. In: O'Connell-Rodwell CE (ed) The use of vibrations in communication: properties, mechanisms and function across taxa. Transworld, Kerala, pp 163–182

Zunic A, Virant Doberlet M, Cokl A (2008) Preference of the southern green stink bug (*Nezara viridula*) males for female calling song parameters. B Insectol 61:183–184

# Part IV
# Vibration Detection and Orientation

# Chapter 14
# Functional Morphology and Evolutionary Diversity of Vibration Receptors in Insects

**Reinhard Lakes-Harlan and Johannes Strauß**

**Abstract** Vibratory signals of biotic and abiotic origin occur commonly in the environment of all living organisms. Many species deliberately produce such signals for communication purposes. Thus, it is not only useful but also advantageous and/or necessary to be able to detect and process vibratory signals with appropriate receptor organs. Mechanoreception is suggested to be evolutionarily ancient among animals (Kung 2005; Thurm 2001). Given the long evolutionary history, such receptors have very different anatomical structures and corresponding physiological properties. Responding to mechanical stress is a basic property of cells, even outside the nervous system. In the nervous system, specialized sensory cells and organs register mechanosensory signals and impart the information to higher centers. Structural and molecular adaptations in various mechanoreceptors can push these systems to a sensitivity at or near to the physical limits, e.g., with respect to the noise–stimuli relation. Here, we will deal with the vibratory receptor systems of insects, with a focus on the specialized scolopidial sensory organs from molecular mechanisms to systems analysis.

## 14.1 Anatomical Diversity of Sensilla

In insects, some vibration receptor types are located at the external surface or embedded in the cuticle, like campaniform sensilla, hair sensilla, or hairplates (Fig. 14.1). Other receptors are internal sensilla, like scolopidial sensilla or

R. Lakes-Harlan (✉) · J. Strauß
Justus-Liebig University Gießen, Institute for Animal Physiology,
Heinrich-Buff-Ring 26, 35392 Giessen, Germany
e-mail: reinhard.lakes-harlan@uni-giessen.de

J. Strauß
e-mail: Johannes.Strauss@physzool.bio.uni-giessen.de

R. B. Cocroft et al. (eds.), *Studying Vibrational Communication*,
Animal Signals and Communication 3, DOI: 10.1007/978-3-662-43607-3_14,

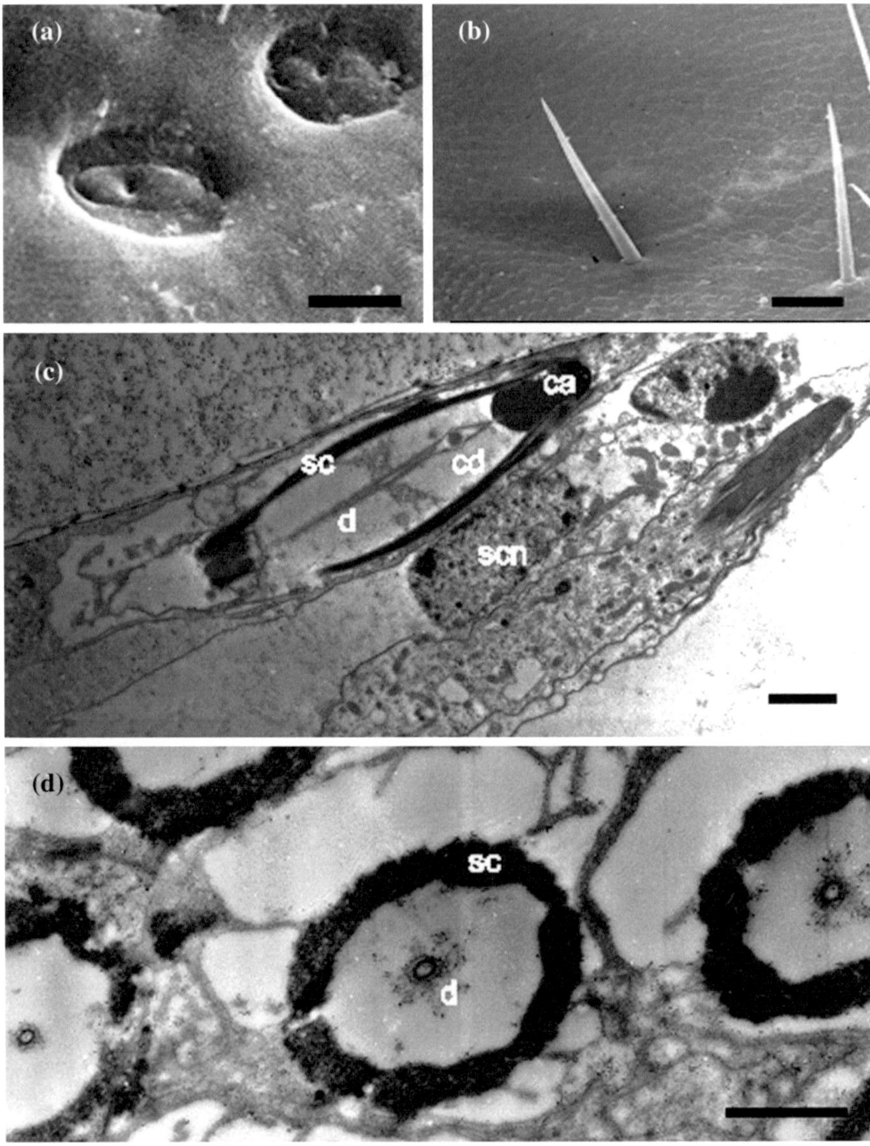


**Fig. 14.1** Different mechanoreceptive sense organs of insects. **a**, **b** Scanning electron microscopic photos of campaniform sensilla **a** and hair sensilla, **b** of the appendages of *Decticus verrucivorus* (Orthoptera). **c**, **d** Transmission electron microscopic photos of scolopidial sensilla. **c** Longitudinal section of a scolopidium in the prosternal organ of *Homotrixa alleni* (Diptera). Note the dendrite (*d*) running in the lumen of the scolopale (*sc*). *ca* cap, *cd* ciliary dilation, *scn* nucleus of the scolopale cell. **d** Cross section of scolopidia in *Phormia regina* (Diptera). *Scales* **a** 10 μm, **b** 50 μm, **c** 2 μm, and **d** 1 μm (Unpublished data)

multipolar/multidendritic sensilla. Among these types, the scolopidial sensilla are particularly sensitive to vibrational stimuli. However, a classification and discrimination between the different mechanical processes, like cell expansion, stretching of organs, touch, substrate vibration, and airborne sound, is sometimes difficult, and adequate stimuli may overlap in a given type of sensilla. In the following, different types of sensilla are briefly reviewed with respect to their properties for mechanoreception of vibration.

### *14.1.1 Campaniform Sensilla*

Campaniform sensilla (CS) are a type of external sensilla that is embedded in the cuticle. From the outside, the sensilla are characterized by a shallow dome (Fig. 14.1). Often, the anatomical structure is also indicative of the direction of stimulus perception. The campaniform sensilla react to mechanical stress in the cuticle—the stimuli can originate either from the animal's own movements (proprioception) or from external sources (substrate vibration). CS are located at different positions on the body segments and on appendages, such as the wings or halteres, where mechanical deformations can be detected (Keil 1997). CS can occur as single sensilla, or they can be arranged in groups. CS of the legs have been found to possess minimum acceleration thresholds at frequencies of 50 Hz and lower (Kühne 1982). Stimuli with such frequencies were responded to with a phase-lock of the action potentials. The upper cutoff frequency is some hundred hertz. The CS of legs typically have an axonal projection that is ramified in the lateral part of the corresponding ganglion (Schmitz et al. 1991; Merritt and Murphey 1992; Mücke and Lakes-Harlan 1995) (see also Fig. 14.6). CS occurring dorsally at the proximal tibia are involved in feedback loops of leg movement in many insects (Burrows and Pflüger 1988; Zill et al. 2011). The spatial location of the CS relative to vibration-receptive organs, like the subgenual organ, raises the question of whether CS responses might also be integrated in the neuronal vibratory network for filtering or modulation of information.

### *14.1.2 Hair Sensilla*

Hair sensilla (HS) come in various anatomical shapes of the hair shaft (and were consequently differently named) and with different physiological functions (Fig. 14.1). Mechanosensory bristles can be found on all body parts. They are, just as CS, usually constructed by four cells during development: a sensory neuron, a sheath cell, a socket cell, and a hair shaft cell. The mechanosensory neuron has ultrastructural specializations in its dendrite, like the tubular body (Keil 1997). The tubular body is a massive complex of microtubules in the dendritic tip that is involved in mechanotransduction. Destruction of the tubular body leads to a decrease in the receptor potential (Erler 1983). The interaction between stimuli, structural elements, and ion channels is the subject of an ongoing analysis of transduction mechanisms (Liang et al. 2011) (see also Sect. 14.2.2). Hair sensilla typically react to touch stimuli. However, filiform hairs might also react to vibrational stimuli, either direct or indirect. An example of indirect measurements is a vibrating leaf that moves air particles above its surface, which in turn deflect the filiform hairs of the insect. Parasitoid wasps might locate their vibration-producing prey within the leaves by such a mechanism (Meyhöfer and Casas 1999). Hair sensilla (trichobotria) of water striders also serve as vibration receptors (Goodwyn et al. 2009). Neuronal networks involving HS have been studied with respect to touch perception (Burrows and Newland 1993). The central projections of hair sensilla are somatotopically ordered in the lateral neuropile of the respective ganglion (Mücke and Lakes-Harlan 1995) (see also Fig. 14.6) and correlate with the receptive field of interneurons (Burrows and Newland 1993).

### *14.1.3 Multipolar Sensilla*

In comparison with other sensilla in insects, least is known about function and physiology of multipolar sensilla. In contrast to bipolar neurons, these sensilla have multiple processes, which can cover much of the body surface (Grueber et al. 2002); in *Drosophila,* they are known as multidendritic (md) neurons. Multidendritic neurons might react to a variety of stimuli: Subsets of md neurons might be involved in proprioception (Grueber et al. 2002), whereas other subsets might respond to temperature or nociception (Tracey et al. 2003). Multipolar sensilla of locusts occur on different body positions and as joint receptors. They can react to vibrational stimuli, but are relatively insensitive (Kühne 1982).

### *14.1.4 Scolopidial Sensilla*

Scolopidial sensilla or chordotonal organs in general have already been subject of a detailed treatise (Field and Matheson 1998). Therefore, their basic features are briefly summarized in the present review, and the focus here is on recent findings of the transduction mechanisms and a discussion of the complexity of sensory organs, their position, and evolution. Scolopidial sensilla are internal sensilla homologous to external sensilla and are characterized by an electron-dense structure, the "scolopale." The scolopale is formed by actin filaments inside the scolopale cell and surrounds an extracellular space into which the dendrite of the sensory cells extends (Fig. 14.1). The dendrite terminates at an electron-dense cap that is formed by an attachment cell (Fig. 14.1). Further characters of the dendrite are a ciliary dilation and rootlets at its base in the cell body. The sensory cell is surrounded by a glial cell. Such a unit with basically a sensory cell, a glial cell, a scolopale cell, and an attachment cell is termed a scolopidium. Based on ultrastructural data, two basic types are distinguished. Type 1 scolopidia have cilia with constant diameter and an extracellular cap. The proposed stimulus is an axial extension, and these sensilla are probably most important for vibration perception. Type 2 scolopidia have an enlarged distal segment in the cilium and an elongated tube. They occur often in crustaceans, but have also been shown in the Johnston's organ of insects (for review see Moulins 1976). The cellular composition of scolopidia varies between organs, as some scolopidia possess two sensory cells or additional ligament cells. Within the insect body, the scolopidia are arranged as single units, in small groups or in large, complex sensory organs. In the latter cases, distinct groups of scolopidia, named scoloparia, can occur within the organ complex and might have distinct physiological properties.

## 14.2 Scolopidial Organs

Scolopidial organs are located at different positions in the insect body. They are part of the sensory complement in the basic bauplan of an insect's segment (Fig. 14.2). For example, in *Drosophila* larvae, most segments contain three scolopidial organs, with a specific number of sensory units (Campos-Ortega and Hartenstein 1997; Hertweck 1931). Besides *Drosophila*, the scolopidial organs have been mapped in orthopteran body segments (Field and Matheson 1998; Meier et al. 1991), whereas in many other insects, only exemplary scolopidial organs have been described. The scolopidial organs are located within a specific segment or are connecting two segments. The position determines the physiological function and sensitivity, but the function may not be obvious on first sight. For example, it is possible that substrate vibrations are perceived with body scolopidial organs, although the body does not have direct contact with the substrate (see Sect. 14.2.3.4 on the prosternal organ). Among the head appendages, the Johnston's organ is well known as a receiver of antennal motion in many insects. Within the insect legs, scolopidial organs have typical positions (Fig. 14.2): in the proximal femur (femoral chordotonal organ), in the proximal tibia (subgenual organ, complex tibial organ), and in the distal tibia and tarsae (chordotonal organs with only a few scolopidia). In the following sections, the development, mechanisms of transduction, as well as functional morphology and physiology of chordotonal organs are reviewed in more detail, with a focus on the scolopidial organs in the proximal tibia.

### 14.2.1 Development

For different types of sensilla, an underlying developmental sequence has been identified for the cellular differentiation of chordotonal cells, CS, and HS (microchaetes) of *Drosophila* (Lai and Orgogozo 2004). During that sequence, a

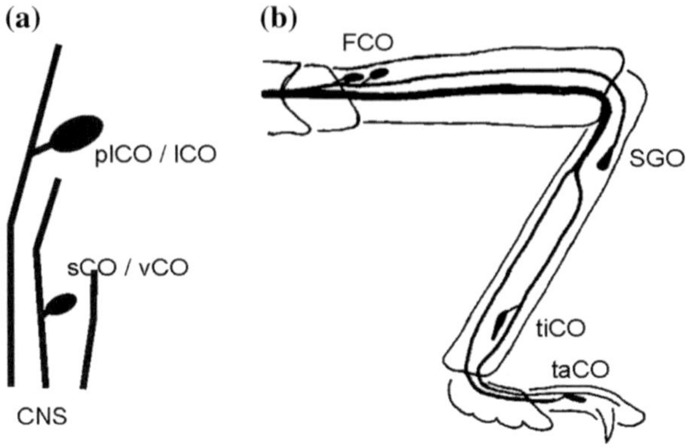


**Fig. 14.2** Schematic overview of the location of scolopidial organs in the body segments and legs. **a** In a typical segment, two to three nerves innervate scolopidial organs at two positions (in Orthoptera, the pleural chordotonal organ (plCO) and the sternal chordotonal organ (sCO); in *Drosophila*, the lateral chordotonal organ (lCO) and the ventral chordotonal organ (vCO). **b** In a typical insect leg, the femoral chordotonal organ (FCO) might have different subparts; the subgenual organ (SGO) is often accompanied by other scolopidial organs (see text), and chordotonal organs are located in the distal tibia (tiCO) and tarsae (taCO). (Modified from Campos-Ortega and Hartenstein 1997; Meier et al. 1991; Mücke 1991)

sensory mother cell divides asymmetrically, and after further cell divisions, four to five cells are generated. Thereby, two accessory cells (e.g., socket cell and shaft cell) differentiate and one sensory neuron with glial or sheath cells. By differential gene expression, the different types of receptor organs are generated. For example, mutations in the *cut* gene transform a hair sensillum into a scolopidial sensillum (Bodmer et al. 1987).

By contrast, the development of complex scolopidial organs is only partially understood. For formation of sensory units in the relatively large femoral chordotonal organ of *Drosophila,* the epidermal growth factor receptor promotes continuous generation of sensory mother cells (zur Lage and Jarman 1999). Complexity is not only a question of cell numbers, but also a question of distinct subparts of an organ, as these subparts might have different physiological properties. The complex tibial organ (CTO) of some Ensifera consists of three parts with different functions (see Sect. 14.2.3). The CTO develops during embryonic development (Klose 1996; Lakes-Harlan and Strauß 2006; Meier and Reichert 1990), but at least in some species of Tettigoniidae, the sensory cells of the parts of the CTO have different developmental origins (Fig. 14.3, unpublished results). The cells of the subgenual organ (SGO) and the *crista acustica* homolog (CAH) arise from sensory mother cells that proliferate at separate positions in the epidermal cell layer. Later, during embryogenesis, the cells form the CTO together and establish their specific central projections. Scolopidial sensory neurons grow their axons during embryogenesis or metamorphosis to the CNS (Lakes-Harlan and Pollack 1993; Lakes and Pollack 1990). These axons orientate along preformed

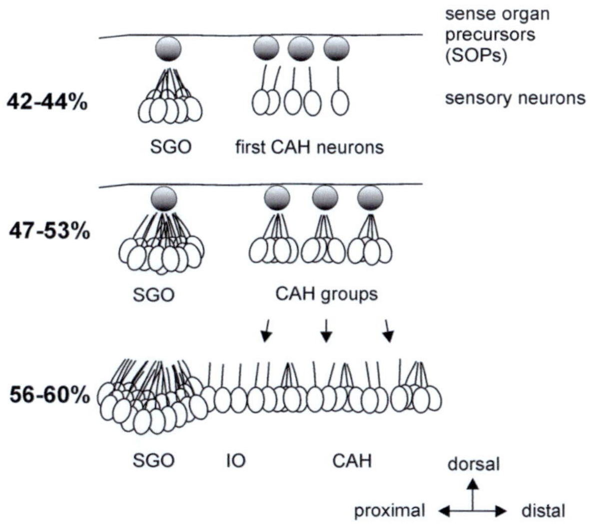


**Fig. 14.3** Model of the developmental sequence of the complex tibial organ of *M. elongata* (Ensifera). The model is based on sequential preparations of embryonic stages (indicated by percentages of developmental time to larval hatching) with immunolabeling of the sensory cells. The cells of the subgenual organ (SGO) and the crista acustica homolog (CAH) have different developmental origins (sense organ precursor cells) and form together the complex sensory organ, including the intermediate organ (IO). (Based on unpublished data)

pathways in the embryo; in cases where the pathway is missing, the axons arrest their growth (Klose and Bentley 1989). For axonal guidance, the cell surface molecule Fasciclin I, which is expressed by the sensory cells and the target neuropiles in the CNS, could be important (Schäffer and Lakes-Harlan 2001). In the holometabolic Diptera, complex scolopidial organs like the femoral chordotonal organ (FCO) develop during metamorphosis. Interestingly, the axons of the FCO extend to the CNS, but after mid-metamorphosis, the organ is retracted within the femur and the axons are apparently shortened during this process (Lakes and Pollack 1990). By contrast, in the moth *Manduca sexta*, sensory units of the FCO are already present during larval stages (Kent et al. 1996). Vibration receptors are fully functional after hatching, and the sensitivity to vibration stimuli does not change during postembryonic development in tettigoniids and locusts (Rössler et al. 2006).

## *14.2.2 Mechanotransduction*

Transduction of mechanical stimuli is a multistep process ranging from mechanical force acting on the receptors to neuronal activity of sensory neurons (French 1992). The first step is a mechanical coupling between the stimulus, e.g., a substrate vibration, and the receptor structure. Such a structure could be the cuticle transmitting the external force. Consequently, embedded sensilla, the campaniform sensilla, can measure compression or strain of the cuticle that leads to activation of the receptor. In the case of the internal scolopidial sensilla, a longitudinal stretch of the dendrite is likely to be the adequate stimulus. The micromechanics of vibration inside a leg or other body parts are rarely investigated (see Sect. 14.2.3 for some discussion), although they might be decisive for the physiological function of the scolopidial organs. Subcellular structures in the scolopidial units, like a ciliary dilation in the dendrite of the sensory neuron, may also be important as suggested by mutant studies (reviewed in Field and Matheson 1998). The next steps in mechanotransduction are the opening of ion channels and changes in the membrane potential. In recent years, the molecular physiology of sensory transduction was pushed forward especially in the model system of the antennal Johnston's organ of *Drosophila* (Kernan 2007; Lu et al. 2009). The Johnston's organ consists of several hundred scolopidial units and is mainly an auditory and gravitation receptor.

Behavioral screens for flies defective in mechanotransduction identified candidate molecules for transduction gating, including ion channels of the transient receptor potential (TRP) ion channel family. TRP channels are cation channels with six transmembrane domains and cytoplasmic N- and C-terminals found in different sense organs; TRP channels can be classified into seven distinct groups based upon sequence similarities and structural characteristics (Christensen and Corey 2007; Matsuura et al. 2009).

Three TRP channels have been identified as candidates for the gating channel in mechanosensory cells: "No mechanoreceptor potential C" (NompC), "Inactive" (Iav), and "Nanchung" (Nan). These channels are specifically located in subcellular structures important for mechanotransduction: Iav and Nan are localized in the proximal cilia of chordotonal neurons (Gong et al. 2004); NompC is localized in the distal cilia of scolopidial neurons and the tubular body of campaniform sensilla (Lee et al. 2010; Liang et al. 2011; Keil 2012). Nan and Iav are expressed in most sensory neurons in the *Drosophila* Johnston's organ. These channels form heterodimers, and Nan is dependent on Iav for proper localization within the scolopidia (Gong et al. 2004). If Iav is deleted, the neuronal response to acoustic stimuli is abolished. It was discussed whether the Nan-Iav dimers may be mechanically activated by ciliary deflections or have a function in signal control downstream of the transducing channel, as disruption of Nan-Iav increases the active amplification in the hearing organ (Göpfert et al. 2006). While the majority of the Johnston's organ sensory neurons express NompC (Lee et al. 2010), only the neurons mediating hearing require it functionally for generating a regular neural response (Effertz et al. 2011). NompC includes an ankyrin spring of 29 ankyrin repeats (Howard and Bechstein 2004) and has been shown to be directly opened by mechanical stimuli (Yan et al. 2013). It may form the gating spring in campaniform sensilla (Liang et al. 2013) though not the gating channel in auditory neurons (Lehnert et al. 2013). The definite contribution of the respective molecules to mechanotransduction in *Drosophila* hearing is so far unresolved (Lehnert et al. 2013; Gong et al. 2013; Matsuo and Kamikouchi 2013). Given this functional specialization between Johnston's organ neurons, the functional conservation or similarity of TRP channels in sensory organs for substrate vibrations like the subgenual organ or femoral chordotonal organ will be relevant to explore.

Related to the identity of the gating, channel is the question for the mechanism of channel opening ultimately causing mechanotransduction. According to a membrane force model, direct opening could result from transfer of forces on the neuron's cell membrane and a pull on the ion channel, opening the pore (Chalfie 2009; Christensen and Corey 2007). Indirect tether models in general assume that force is transferred to specific molecules/proteins which are linked to the ion channel and activate it by induction of a conformational change. Such molecules could be linked to the intracellular cytoskeleton or extracellular matrix molecules or a combination thereof (Chalfie 2009; Christensen and Corey 2007). In scolopidia of Johnston's organ of *Drosophila*, opening of the gating channel is supposedly direct and via gating springs (Albert et al. 2007). The short latencies in chordotonal organs in the submillisecond range support the hypothesis of a direct activation process (Albert et al. 2007).

In mechanoreceptors, the sheath cells produce a high concentration of potassium in the surrounding of the dendritic cilium compared to hemolymph (French 1988; McIver 1985). In scolopidial units, the scolopale space is filled by an endolymph with high potassium concentration, but low calcium concentration (Todi et al. 2004). The receptor potential of mechanoreceptors is proportionally graded to the stimulus intensity (French 1988). The primary sensory neurons are supposed to generate action potentials at the beginning of the axon. In the scolopidia of an orthopteran hearing organ, small non-propagating spikes are passively conducted back into the soma and the dendrites (Hill 1983).

## *14.2.3 Functional Morphology and Physiology*

Scolopidial organs differ largely in their functional morphology. Here, we focus on those found within the legs, and especially in the proximal tibia, that have been investigated for several insect taxa, with another example of a vibration-sensitive organ in the thorax of an insect.

### 14.2.3.1 Femoral Chordotonal Organ

The femoral chordotonal organ (FCO) is probably present in all insects, and in many species, it is a large mechanosensory organ with up to hundreds of sensory units (Debaisieux 1938). It is located dorso-proximally in the femur and attaches via a cuticular apodeme at the femur–tibia joint (Fig. 14.2). While the FCO is in general a proprioceptive organ involved in sensory feedback loops required for limb coordination, it is also a vibration receptor in some insects. The functional specialization may be associated with structural subdivision. For example, in orthoptera and stick insects, two distinct scoloparia are found, whereby the locust FCO and *Carausius* FCO contain heterodynal type I scolopidia with two sensory neurons each (Füller and Ernst 1973; Matheson and Field 1990). In the locust, the FCO of the middle leg contains 42 sensory neurons in the distal scoloparium and several hundred relatively small sensory neurons in the proximal scoloparium (Field and Pflüger 1989). Physiological investigations in the locust showed a functional distinction between the two groups: The distal scoloparium mediates the postural resistance reflex, while the proximal group does not affect this reflex (Field and Pflüger 1989). The latter group was suggested to be a functional receiver of vibrational stimuli between 50 and 300 Hz, as it responds to vibrations with displacements of 4-μm amplitude (Field and Pflüger 1989). Similarly, in the stick insect *Carausius morosus,* a proprioceptive function could be ascribed to the ventral scoloparium in the resistance reflex, leaving open the role of the larger dorsal scoloparium (Kittmann and Schmitz 1992). In the stick insect *Cuniculina impigra*, a high number of sensory neurons in the FCO were proven to be vibrosensitive (Sauer and Stein 1999; Stein and Sauer 1999), but a localization of the structural correlate was not given. A distinction into two scoloparia is also clearly present in crickets (Nishino 2000) and bush crickets (Theophilidis 1986), although in some species, the separation into two scoloparia may not be complete (Matheson and Field 1990).

In the green lacewing *Chrysoperla* (Neuroptera), the femoral chordotonal organ contains up to 26 scolopidia and it is vibroceptive with a maximum sensitivity at about 1 kHz with a threshold between 0.1 and 1 $ms^{-2}$ (Devetak and Amon 1997). In some insects, the number of scolopidia can be as low as 12, e.g., in the stink bug, *Nezara viridula* (Michel et al. 1982). The number of sensory cells, however, is not size related, as in the small *Drosophila melanogaster,* the FCO has three subunits with 14–32 scolopidia each (Shanbhag et al. 1992). The FCO of *D. melanogaster* also reacts to low-frequency vibrational stimuli (Lakes-Harlan and Lefevre 2012). Together with a rather insensitive tibial scolopidial organ (Schneider 1950), the FCO might be an important source for vibration perception, as Diptera do not possess a subgenual organ (see below). The vibratory function of the FCO is further indicated by the central projection because parts of the dipteran FCO axons project in the mVAC (Merritt and Murphey 1992; see Sect. 14.2.4).

#### 14.2.3.2 Scolopidial Organs in the Proximal Tibia

In many insect taxa, scolopidial organs are found in the proximal tibia; known exceptions are the Diptera and Coleoptera. The most widely distributed organ is the so-called subgenual organ (SGO), by which scolopidial sensilla are named according to their location just distally of the femur–tibia joint (Fig. 14.4). However, in a number of species, additional scolopidial organs in the proximal tibia are documented: a distal organ (in Caelifera and Blattodea: Friedrich 1929), an intermediate organ and a crista acustica or its atympanate homolog (in Ensifera), a Nebenorgan (Blattodea and Mantophasmatodea), and an accessory organ (in

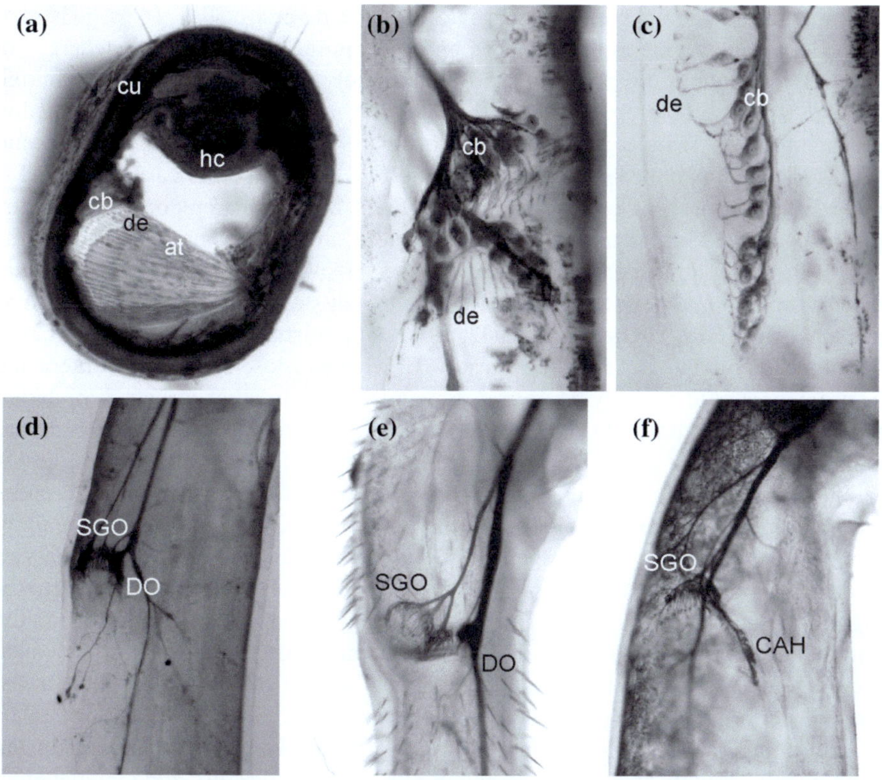


**Fig. 14.4** Photoplate of scolopidial sense organs of in the proximal tibia of different insects. **a–c**, Midleg of *M. elongata* (Ensifera; photos courtesy of Jan Häusler, unpublished). **a** View from proximal end into an opened tibia to visualize the localization of the subgenual organ (SGO). Anterior is to the *left*; staining with Janus Green. **b** Frontal view of a nerve backfill of the SGO with distinguishable cell bodies and dendrites. **c** Frontal view of the crista acustica homolog (CAH). **d–f**, Lateral view of the SGO and further scolopidial sense organs in different orthopteran species (unpublished). **d** *Schistocerca gregaria* (Caelifera), **e** *Hierodula membranacea* (Mantodea), and **f** *Stenopelmatus spec*. (Ensifera). *at* attachment of the SGO; *cb* cell bodies of sensory neurons; *cu* cuticle; *de* dendrites; *DO* distal organ; *hc* hemolymph channel; *SGO* subgenual organ. *Scale* 100 μm

Mantodea, Blattodea, and others: Strauß and Lakes-Harlan 2013). In contrast to the SGO, the specific functions and the evolutionary relations of the other organs are so far largely unknown.

The best-known vibration receiver is the subgenual organ, which has been studied in detail in a number of insect taxa and was documented in most Pterygota. It is not connected to leg joints and detects external stimuli transferred to the leg. The SGO is considered to be the most sensitive vibration receptor in insects (Cokl and Virant-Doberlet 2003). All subgenual organs contain type 1 scolopidia with a single sensory neuron per scolopidium (Field and Matheson 1998), but the organization of the subgenual organ is of notable diversity (Fig. 14.5): it is club shaped in some Hymenoptera species and in termites (Howse 1964), whereas in other Hymenoptera, the organ hangs in the hemolymph cavity (Vilhelmsen et al. 2001, 2008). In orthopteroid insects, the SGO is often sail shaped and suspended in the hemolymph (Lin et al. 1995; Schnorbus 1971). In this case, the cap and accessory cells span the hemolymph channel (Fig. 14.4); in other cases, the SGO is more like a "mass" of cells (Nishino and Field 2003). In Lepidoptera, it is diffusely organized and apparently the scolopidia are distally unattached (Howse 1968). In Hemiptera, like *N. viridula* or a species of Membracidae, *Stictocephala bisonia,* scolopidial cells are attached to a ligament stretching along the tibia from proximal to distal (Michel et al. 1982, Roye unpublished). This anatomy of scolopidial cells might even raise the question of whether those cells in Hemiptera and Neuroptera should actually be named subgenual cells. In other insects, scolopidial units adjacent to the SGO that extend into the longitudinal axis of the tibia are named intermediate organ or distal organ (see below). Further anatomical, physiological, and developmental studies are needed to clarify whether the differences are purely nomenclatorial or based on the possibly convergent evolution of tibial sense organs. The subgenual organ can be innervated by two different leg nerves. In orthopteroid insects, the sensory nerve 5B1 innervates one part of the subgenual organ and another part is innervated by a branch of the mixed leg nerve 5B2. In the tettigoniid *Mecopoda elongata*, the nerve 5B2 innervates the posterior portion of the subgenual organ and contains more axons of the subgenual scolopidia than the sensory nerve innervating the complex tibial organ.

The numbers of scolopidia in the SGO vary largely between species. In the stink bug *N. viridula* (Heteroptera), it consists of just two scolopidia (Michel et al. 1982), and in the neuropteran *Chrysoperla carnea*, it consists of three scolopidia (Devetak and Amon 1997), and many orthopteroid species have 20–80 scolopidia (Rössler 1992; Schnorbus 1971). For adult tobacco hornworm moths, *M. sexta*, about 30 neurons are found in the SGO (Kent and Griffin 1990). A large variance occurs within Hymenoptera: The SGO of ants contains 10–40 scolopidia (Howse 1964; Menzel and Tautz 1994), that of the honeybee *Apis mellifera* around 40 scolopidia (Kilpinen and Storm 1997), and in parasitoid wasps, females of certain species can have 300–400 SGO scolopidia (Vilhelmsen et al. 2008). The functional relevance of the SGO structure and cell numbers has only been discussed for the parasitoid wasps (see Sect. 14.2.4).

The subgenual organ reacts to substrate vibration, but in cockroaches is so sensitive to mechanical stimuli that it might react to airborne stimuli despite lack of tympana (Shaw 1994). For physiological characterization of substrate vibration, an important parameter is the threshold: the lowest displacement or acceleration stimulus that elicits a neuronal response. For the cockroach *Periplaneta americana,* displacements of 0.22–5 nm at a frequency of 1.57 kHz can be detected by subgenual receptors (Shaw 1994). Similar values have been found in the green lacewing, which corresponds to an acceleration threshold of 0.02 m/s$^2$ (Devetak and Amon 1997). The vibroceptors in *N. viridula* are tuned to different frequencies but share minimal thresholds around 0.01 m/s$^2$ (Cokl 1983). The cricket *Gryllus*

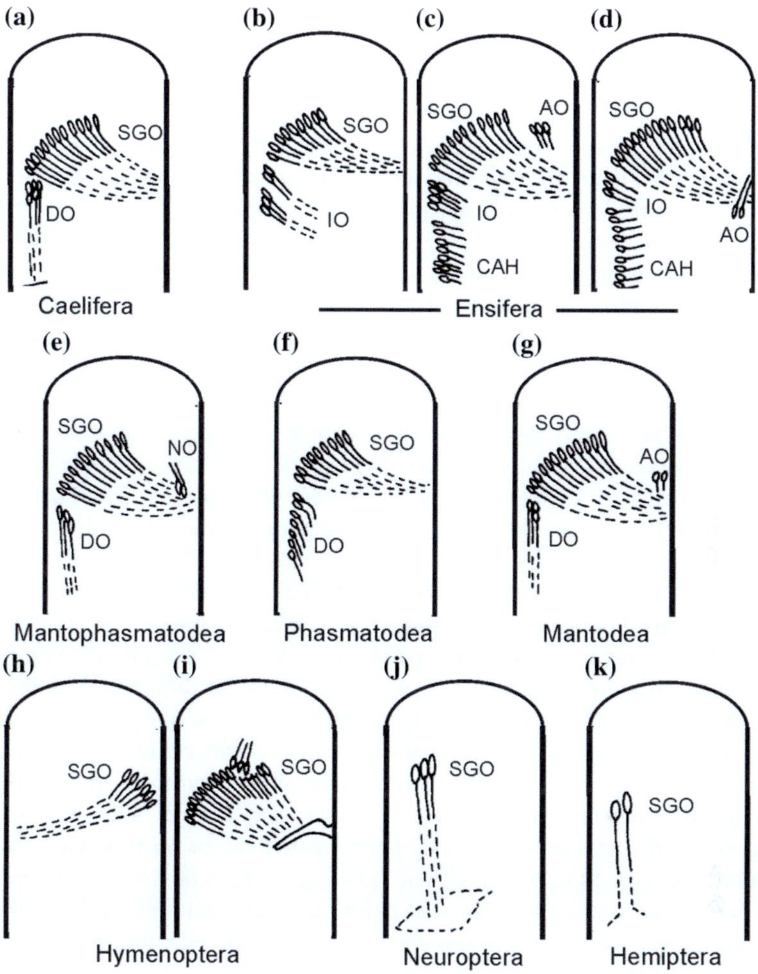


**Fig. 14.5** Schematic overview of the sensory arrangement of scolopidial organs in the proximal tibia of different taxa. Drawn are sensory cell bodies with their dendrites and their attachment (stippled). Compare the overall similarities in Orthoptera (**a–d**), and in orthoptera-related groups (**e–f**), to more distantly related taxa (**h–k**). Note the large differences within the Hymenoptera (**h**: ant, **i**: parasitic wasp) that are probably caused by specialized functional adaptations. Drawings are based on the following: **a** *Schistocerca gregaria* (Lin et al. 1995), **b** *Troglophilus neglectus* (Jeram et al. 1995), **c** *Comicus calcaris* (Strauß and Lakes-Harlan 2010), **d** *Hemideina femorata* (Nishino and Field 2003), **e** *Karoophasma biedouwense* (Eberhard et al. 2010), **f** *Sipyloidea sipylus* (Strauß and Lakes-Harlan 2013), **g** *Hierodula membranacea* (unpublished), **h** *Camponotus ligniperda* (Menzel and Tautz 1994) and *Apis mellifica* (Schön 1911), **i** *Orussus abietinus* (Vilhelmsen et al. 2001), **j** *C. carnea* (Devetak and Pabst 1994), and **k** *N. viridula* (Michel et al. 1982) and *S. bisonia* (Roye, unpublished). *AO* accessory organ, *CAH* crista acustica homolog, *DO* distal organ, *IO* intermediate organ, *NO* Nebenorgan, and *SGO* subgenual organ. Anterior is to the *left*, proximal to the *top*. Drawn not to scale; cell numbers are represented, but not exactly depicted

*bimaculatus* has been found to be highly sensitive: The threshold of midleg subgenual receptors was at only 0.0018 $m/s^2$ at frequencies from 700 to 1,000 Hz (Dambach 1972). This sensitivity is not seen throughout the Ensifera, as the tree weta has a threshold of 0.015 $m/s^2$ for vibrational signals at 1 kHz (McVean and Field 1996). Subgenual thresholds in both locusts and bush crickets range between 0.01 and 1 $m/s^2$ with little species-specific variation (Kühne 1982).Whereas in crickets, the midleg reacts most sensitively to vibrational stimuli (Dambach 1972), in *C. carnea,* the metathoracic hindlegs were most sensitive (Devetak and Amon 1997). In Mantodea, the vibration sensitivity has been tested with respect to functional adaptations of the leg pairs. The foreleg is adapted for prey capture and typically held in the position known for mantis without contact to the substrate; the other legs are used for standing and foraging and are in contact with the substrate. However, no major functional adaptation could be found: The sensory organs in the three legs are highly similar, and furthermore, the legs do not have different physiological sensitivities (unpublished results). The characteristic frequency (CF), defined as the frequency with the lowest threshold, is in crickets around 700–1,000 Hz in the midlegs and 400–500 Hz in the hindlegs (Dambach 1972). In the midlegs of various species of bush crickets, the CFs range from 500 to 1,500 Hz (Kühne 1982). Tuning curves as well as intensity response curves vary between recorded neurons, suggesting that vibratory stimuli can be precisely fractioned and coded in the SGO. Even the few SGO neurons in *Nezara* may discriminate different frequencies (Cokl 1983). A major difficulty with the interpretation of some physiological data is that they cannot be unequivocally ascribed to the subgenual organ. Extracellular recordings could contain responses from unidentified elements; in intracellular recordings, SGO neurons were identified via (an assumed) physiology and by their probable central projection. Only in a very few cases has the peripheral cell body been labeled and its position unequivocally identified (see Fig. 14.6, recording and labeling by A. Stumpner, Göttingen).

An important, but rarely addressed question is what are the mechanical forces and parameters driving the physiological reaction of the vibration receptors. It has been suggested that the SGO acts like an accelerometer (Schnorbus 1971). In the honeybee, it has been possible to document the vibrations of the SGO itself. Substrate vibrations are transferred to the hemolymph, and the sensory organ is actually oscillating with the hemolymph rather than in the hemolymph (Kilpinen and Storm 1997). Thus, the subgenual organ's oscillations are matched with the hemolymph oscillations and model calculations show that it behaves as an overdamped system (Storm and Kilpinen 1998).The model suggests that the sensory cells of the SGO are displacement sensitive. Velocity threshold curves of SGO neurons from *Nezara* run in parallel with equal acceleration values below best frequency and in parallel with equal displacement lines above the best frequency (Cokl 1983). In addition, the SGO of honeybees was shown to have stimulus-direction-specific responses (Rohrseitz and Kilpinen 1997). These results indicate that careful control of the stimulus application is important and that further research is needed to understand the micromechanics of vibration perception in the SGO and other scolopidial sense organs. Interestingly, the structure of the SGO of

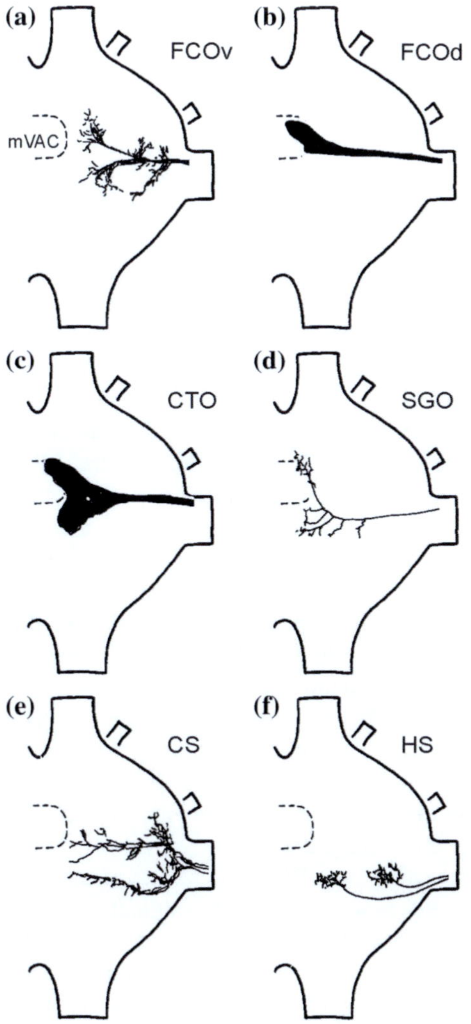


**Fig. 14.6** Schematic overview of typical central axonal projections of mechanoceptive sensory fibers. The projections are outlined in a generalized segmental hemiganglion. The median ventral association center (mVAC) is outlined, to indicate which fibers project into this internal neuropile. **a**, **b** The femoral chordotonal (FCO) can have different central projections: **a** Fibers of the dorsal scoloparium of crickets (Nishino 2000) or the proximal part in locusts (Field and Pflüger 1989) project into the mVAC; **b** Other fibers of the FCO have a more lateral projection. **c** Projection of the midleg CTO of *Stenopelmatus* (Strauß and Lakes-Harlan 2008b). **d** Projection of single fiber originating in the SGO. This neuron has been recorded from and it has been stained completely with neurobiotin, thereby confirming unequivocally its origin in the SGO of *Ancistrura nigrovittata,* Ensifera (courtesy of Dr. Stumpner, Göttingen, unpublished). **e** Central projection of campaniform sensilla from the leg of locusts. **f** Central projection of hair sensilla on leg parts, which are somatotopically ordered within the neuropile (two positions are indicated by the two arborization areas). Drawings generalized after (Pflüger et al. 1988 for the ganglion outline) [**a**, **b** after (Nishino 2000), **c** after (Strauß and Lakes-Harlan 2008b), **e** after (Hustert et al. 1981) and **f** after (Mücke and Lakes-Harlan 1995)]

aqueous larvae and land-living adults of Plecoptera is rather similar (Wittig 1955), although different mechanical forces might act on the legs.

In many taxa, the subgenual organ is not the only scolopidial organ in the proximal tibia, although vibration perception can obviously be well achieved with the SGO alone (Fig. 14.5). Thus, questions arise about the function and the evolution of the other sensory organs. As an example, this can be studied in the Ensifera, which possess the so-called complex tibial organ (CTO). Within the foreleg of Tettigoniidae, this organ complex is associated with sound-propagating structures and can perceive vibratory stimuli as well as acoustic stimuli. The CTO comprises the subgenual organ, an intermediate organ, and a crista acustica (in tympanate legs). The *crista acustica* is a conspicuous feature of the CTO due to its more or less linear arrangement of sensory cells, which correlates with physiological response properties of the auditory receptors (review: Stumpner and von Helversen 2001). Recently, such a tripartite CTO has also been described in atympanate Ensifera (Strauß and Lakes-Harlan 2008a, b), including a structure homologous to the *crista acustica*. The distinct parts of the CTO have commonalities, but also differences: The parts have different attachments, partly different adequate stimuli and different developmental origin. The scolopidia of the SGO are orientated circularly within the leg and clearly separated from the other two parts. The scolopidia of the intermediate organ point toward an attachment fixed at the dorsal tibia. The scolopidia of the *crista acustica* homolog (CAH) point toward a supporting structure that extends in the longitudinal axis of the leg. The intermediate organ and the CAH are probably vibration receivers, but physiological details are so far unknown. The different dendritic attachment in comparison with that of the SGO might indicate perception of other physiological parameter. For example, the CAH might not vibrate with the hemolymph as the SGO (see above). Thereby, other parameters, like different waveforms or different oscillation planes, might be perceived by the CAH. Biophysical measurements of the oscillations of the CAH will hopefully resolve the physiological properties. Interestingly, the CAH is present not only in deaf ensiferans, but also in the atympanate legs of hearing Tettigoniidae. Thus, it is likely that the organ complex has an important function in the sensory world of Ensifera. Generally, it might be more than a coincidence that the CTO has been formed in the proximal tibia. The position just distally of the femur–tibial joint seems to be well suited for perception of vibratory signals, due to filter properties of the leg (Cokl personal communication). Consequently, also other taxa have distinct scolopidial organs in the proximal tibia besides the SGO (Fig. 14.5): In Caelifera and Mantophasmatodea, a distal organ has been described (Eberhard et al. 2010; Lin et al. 1995). The Mantophasmatodea have a pronounced vibrational communication with species-specific signals (Eberhard and Picker 2008). The vibrational sensitivity of the leg nerves reaches thresholds of 0.01–0.001 $m/s^2$ within a frequency range from 600 to 1,200 Hz (Eberhard et al. 2010). The precise origin of this physiological response is not known, although the SGO is likely to be the most sensitive vibration receptor. Phasmatodea possess an elaborated distal organ besides the SGO (Strauß and Lakes-Harlan 2013).

Another organ in the proximal tibia is the accessory organ, which has only a few sensory units attached to the dorsal cuticle. It is present in mantids as well as in some orthopterans in close association with the other parts of the sensory complex. The function of the organ is unknown; it might contribute to the information processing of relevant vibratory signals.

#### 14.2.3.3 Tarsal Chordotonal Organs

Scolopidial organs can be found in different parts of tarsi of many insects (Mücke 1991; Goodwyn et al. 2009; Wiese 1972). The function of these scolopidial organs has been addressed in water striders, which detect water surface waves for prey location. The respective sensors of water vibrations are located in the tarsi, as cutting of the entire tarsi abolished the orientation of the animal (Murphey 1971). In *Aquarius paludum*, three scoloparia occur in the tarsi. Each scoloparium contains 2–3 scolopidia and is orientated in different directions, which might be seen as sensory adaptation to perception of complex water wave vibrations (Goodwyn et al. 2009). In the backswimmer *Notonecta glauca*, the tarsal chordotonal organ enables localization of the wave-producing prey (Wiese 1972). The tarsal chordotonal organ is located in the distal tarsomere and consists of two scoloparia (proximal and distal) with three and five sensory neurons, respectively. The sensory units respond in the behaviorally relevant frequency intensity range (Wiese 1972). However, not all tarsal chordotonal organs might serve as vibration receivers: In *C. carnea*, a sensitive vibration response was lacking (Devetak and Amon 1997), and in Mantophasmatodea, the destruction of the tarsal chordotonal organ did not change the vibration sensitivity (Eberhard et al. 2010).

#### 14.2.3.4 Prosternal Chordotonal Organ

The prosternal chordotonal organ of Diptera is interesting for several aspects. This organ presents an example of a sense organ that is *not* immediately obvious for substrate vibration perception. The prosternal chordotonal organ is located in the prothorax, directly behind the head. It attaches to a neck membrane and has therefore been proposed to monitor head movements (Preuss and Hengstenberg 1992). However, it could be shown that the organ reacts sensitively to high-frequency vibrations not found in movements of the head or during flight (Lakes-Harlan et al. 1999; Stölting et al. 2007). The organ might therefore be able to pick up substrate vibrations. This finding is furthermore in accordance with the hypothesis that the prosternal organ was modified during evolution into a hearing organ (Lakes-Harlan and Heller 1992; Lakes-Harlan et al. 1999; Robert et al. 1992). The evolutionary scenario implicates that the animals first perceived vibratory signals and that modifications in the sound-propagating structures (enlargement of a tracheal chamber and thinning of tympanal membranes) resulted in the capacity to perceive airborne sound.

The physiology of the prosternal organ is also interesting with respect to unknown vibratory sense organs in other insects. For example, Membracidae have been shown to communicate with vibratory signals and to react very sensitively to vibratory stimuli (Cocroft 1996; Cocroft and McNett 2006). The vibratory stimuli might be sensed by the SGO, but perhaps also with scolopidial organs at other body locations. Many Membracidae have a spectacular morphology with protuberances and extensions of the body surface. These structures have been shown to vibrate in response to substrate vibration, whereby the maxima and minima of vibrations follow a complex system (Cocroft et al. 2000). If an internal scolopidial organ were located near the maximum of such vibrations, this organ could pick up the substrate vibrations, similar to the prosternal organ. Future experiments will certainly unravel the behaviorally relevant sensory organs.

### *14.2.4 Neural Networks and Neuroethology*

The central projections of the sensory fibers are the first level in the neural networks processing vibrational information. Probably in order to facilitate efficient neuronal processing, these central projections are usually ordered within the central nervous system by type of the sensory neuron and/or by position of the sensory cell body (Fig. 14.6). Consequently, the central projection of the sensory neuron may be indicative of its function. In each segmental ganglion of insects, a neuropile area, the median ventral association center (mVAC), is known for processing of vibratory, auditory, and proprioceptive information (Pflüger et al. 1988). The projection into this neuropile may also be indicative of a vibration-receptive function of its sensory cells.

As described above, the FCO often contains two distinct parts (scoloparia) and these parts also have different projections in the corresponding ganglion. In locusts, the proximal part has a dense projection close to the midline within the mVAC and the distal part has a rather loose projection, not merging with that of the proximal scoloparium (Field and Pflüger 1989; Mücke and Lakes-Harlan 1995). In crickets and in wetas, neurons of dorsal parts of the FCO project into the mVAC (Fig. 14.6; Nishino 2000, 2003). In these species, even a more detailed order in the central projection of small groups of neurons could be shown. Furthermore, distinct central projections also correlate with different physiological properties (Matheson 1992). In the Diptera, the FCO projects into the mVAC as well (Merritt and Murphey 1992). Given that the mVAC is often the first-order neuropile involved in vibration processing, the projection in Diptera could indicate that at least parts of the FCO play a role in sensing vibrations. The sensory neurons of the prosternal chordotonal organ of Diptera project among other areas into the mVAC of all three thoracic neuromeres, in both atympanate and tympanate Diptera (Stölting et al. 2007; Stumpner et al. 2006).

On the other hand, the sensory cells of the vibrosensitive SGO do not project completely in the mVAC. The central projection of the SGO has unequivocally been documented in orthopterans either by labeling a complete single cell during recording (Stumpner 1996) or by careful anterograde backfills (Nishino and Field 2003). The fibers of the tettigoniid SGO might project into the mVAC, but not close to the midline (Fig. 14.6; Stumpner 1996). Axons with a bifurcating morphology project anteriorly in the mVAC, whereas subparts of the SGO also have posterior projecting fibers that do not reach the mVAC, but establish an ordered projection outside the mVAC (Nishino and Field 2003). The complex tibial organs of the atympanate Ensifera also have a projection within the mVAC, but so far it has not been unequivocally resolved whether the cells of the CAH project into the neuropile (Fig. 14.6; Strauß and Lakes-Harlan 2008a, b). Nevertheless, in the different atympanate Ensifera analyzed so far, the distinct overall projections in the taxon-specific anatomies of the mVAC are likely to originate at least partly in the CAH (Strauß and Lakes-Harlan 2008a, b, 2010). The accessory chordotonal organ in wetas also projects into the mVAC (Nishino and Field 2003). A CO that does not project in the mVAC is the proprioceptive tarsal CO in locusts (Mücke and Lakes-Harlan 1995). It will be interesting to study the central projection of the tarsal CO in the water strider, which is supposed to register water vibrations (Goodwyn et al. 2009).

In contrast to the scolopidial sensilla, mechanosensitive hair sensilla have somatotopically ordered, often tufted-like projections mainly in the lateral neuropiles of the respective ganglion (Fig. 14.6; Burrows and Newland 1993; Mücke and Lakes-Harlan 1995). Campaniform sensilla often have a widely arborized projection in lateral neuropiles, as shown in locusts and flies (Fig. 14.6; Hustert et al. 1981; Merritt and Murphey 1992). Some of the CS may reach the median ventral association center with single axonal branches. As mentioned above, external sense organs (es) like sensory hairs can be transformed into chordotonal sensilla in *Drosophila* embryos. Such transformed sensilla exhibit a variety of central projection anatomies ranging from those of es neurons to those of chordotonal neurons (Merritt et al. 1993). Thus, the formation of a central projection is probably controlled by a number of genes.

The mechanosensory neurons synapse onto vibratory interneurons that distribute and compute the information in the CNS (Rössler et al. 2006). One of the first features of such networks is probably the localization of the source of the stimuli. Such directional discrimination is possible by calculation of the input from the receptors of different legs (Virant-Doberlet et al. 2006). However, it has to be kept in mind that the sensitivity of the different legs can be different (see above) and that additional input from scolopidial organs in various parts of the body could contribute to the networks. Biologically relevant answers of discrimination are given by the behavior of the animal: ant lions that locate their prey in sandy pits (Devetak 1998); toktok beetles in the Namib Desert tap on the sand surface with their abdomen and attract each other (unpublished observations). In other species, vibrotaxis might supplement phonotaxis. In crickets or tettigoniid, species-specific vibratory signals can facilitate orientation in a complex 3D habitat, like bushes, toward conspecifics (Latimer and Schatral 1983; Stiedl and Kalmring 1989; Weidemann and Keuper 1987). Holometabolous caterpillars can detect and discriminate vibratory signals occurring on leaves with so far unidentified sense organs (Guedes et al. 2012).These exemplary observations show the ability of various species to locate vibrational stimuli.

### 14.2.5 Evolution

Vibration receptor organs have evolved in relation to selective pressures on vibration perception, and some organs have later been further modified for perception of airborne sound. The receptor organs have to match the distinct parameters of vibratory signals relevant for reproduction and offspring. But can receptor organ complexity be related to vibratory signaling or parameters of vibratory signals, like frequency, displacement, and others?

Insects that possess a subgenual organ as a sensitive vibration receiver have between two and several hundred sensory units. Orthoptera with elaborated acoustic and vibratory communication signals have about 20–80 scolopidia in each of the subgenual organs of their legs. So far, no correlation between cell number and signal has been worked out for Orthoptera. However, it has been shown that subgenual organs with few neurons are apparently sufficient to serve vibration communication and may even discriminate different frequencies (Cokl 1983). The subgenual organs of the neuropteran *Chrysoperla* contain only three scolopidia (Devetak and Pabst 1994), and only two scolopidia are in the SGO of the bug *Nezara* (Michel et al. 1982). Thus, from a strictly numerical perspective, a low number of receptor units might be functionally sufficient. It can be asked what the receptors in insects with more sensory neurons are used for, or whether they are functionally redundant.

On the other hand, an example for sensory adaptation has been proposed in some groups of parasitoid wasps. These female wasps tap on the substrate with their antenna to evoke "echoes" of vibration by which they locate their hosts. This behavior is termed vibrational sounding. Apparently, the receptor organ for the echoes is the subgenual organ, which contains 300–400 scolopidia in 55 species of Orussidae (Vilhelmsen et al. 2001) and five of 39 subfamilies of Ichneumonidae (Broad and Quicke 2000). This enlargement of the subgenual organ correlates with the vibrational behavior and ecological factors, like host size and substrate (Broad and Quicke 2000). The co-organization of sounding and signal detection indicates a coevolution between signal evocation and signal detection. The increase in number of receptors may functionally improve the ability to detect the hidden hosts. For specific taxa, phylogeny is helpful to infer the ancestral situation of a sensory system. However, the Orussidae are a basal group (Vilhelmsen et al. 2001), and therefore, the enlarged subgenual organ may not be apomorphic in this lineage.

The evolution of the tibial scolopidial organ in relation to vibrational signals has also been discussed for Ensifera. While crickets and bush crickets mainly use acoustic signaling (Bailey 1990), several taxa lack hearing organs but instead use vibrational signals by substrate drumming with hindlegs or abdomen (Field and Bailey 1997; Gwynne 2004; Weissman 2001). In the ensiferan phylogeny, the plesiomorphic situation seems to be the possession of a subgenual organ together with an intermediate organ (which might be related to the distal organ or Nebenorgan in other taxa). Such neuroanatomical organization is present in a cave cricket (Raphidophoridae: Jeram et al. 1995). Other taxa of atympanate Ensifera (Stenopelmatidae, Schizodactylidae, and Gryllacrididae) possess a tripartite CTO (Strauß and Lakes-Harlan 2008a, b, 2009, 2010) with receptor structures homologous to that in tettigoniids, but without adaptations to sound reception. Since the serial organization in all three leg pairs is similar (e.g., the numbers of scolopidia), it was presumed that the sensory organ structure represents the ancestral organization of hearing organs and therefore might have a vibroceptive function (Strauß and Lakes-Harlan 2009). In this case, it can be argued that the detection of vibration was the ancestral function of the subgenual organ, to which further receptor structures were added with a presumptive function in analyzing intraspecific signaling by substrate vibration. The additional receptor structures might be necessary for detecting vibration parameters independent from hemolymph oscillations that can be perceived by the SGO (see above for details).

The question of evolution of vibration receptors has rarely been analyzed in detail. Hopefully, future studies will show which selective pressures acted on the formation of scolopidial organs and what constraints influence the evolution of these interesting sense organs.

## References

Albert J, Nadrowski B, Göpfert M (2007) Mechanical signatures of transducer gating in the *Drosophila* ear. Curr Biol 17:1000–1006

Bailey WJ (1990) Acoustic behaviour of insects. An evolutionary perspective. Chapman and Hall, London, New York, Tokyo

Bodmer R, Barbel S, Sheperd S, Jack J, Jan L, Jan Y (1987) Transformation of sensory organs by mutations of the cut locus of *D. melanogaster*. Cell 51:293–307

Broad G, Quicke D (2000) The adaptive significance of host location by vibrational sounding in parasitoid wasps. Proc R Soc Lond B 267:2403–2409

Burrows M, Newland PI (1993) Correlation between the receptive fields of locust interneurons, their dendritic morphology, and the central projections of mechanosensory neurons. J Comp Neurol 329:412–426

Burrows M, Pflüger H-J (1988) Positive feedback loops from proprioceptors involved in leg movements of the locust. J Comp Physiol A 163:425–440

Campos-Ortega J, Hartenstein V (1997) The embryonic development of *Drosophila melanogaster*. Springer, Heidelberg

Chalfie M (2009) Neurosensory mechanotransduction. Nature Rev Mol Cell Biol 10:44–52

Christensen A, Corey D (2007) TRP channels in mechanosensation: direct or indirect activation? Nature Rev Neurosci 8:510–521

Cocroft RB (1996) Insect vibrational defence signals. Nature 382:679–680

Cocroft RB, McNett GD (2006) Vibratory communication in treehoppers (Hemiptera: Membracidae). In: Drosopoulos S, Claridge MF (eds) Insect sound and communication. CRC Press, Boca Raton, pp 305–317

Cocroft RB, Tieu TD, Hoy RR, Miles RN (2000) Directionality in the mechanical response to substrate vibration in a treehopper (Hemiptera: Membracidae: *Umbonia crassicornis*). J Comp Physiol A 186:695–705

Cokl A (1983) Functional properties of vibroreceptors in the legs of *Nezara viridula* (L.) (Heteroptera, Pentatomidae). J Comp Physiol A 150:261–269

Cokl A, Virant-Doberlet M (2003) Communication with substrate-borne signals in small plant-dwelling insects. Ann Rev Entomol 48:29–50

Dambach M (1972) Der Vibrationssinn der Grillen. I. Schwellenmessungen an Beinen frei beweglicher Tiere. J Comp Physiol 79:281–304

Debaisieux P (1938) Organes scolopidiaux des pattes d'ìnsectes. La Cellule 47:77–202

Devetak D (1998) Detection of substrate vibration in Neuropteroidea: a review. Acta Zool Fennica 209:87–94

Devetak D, Amon T (1997) Substrate vibration sensitivity of the leg scolopidial organs in the green lacewing *Chrysoperla carnea*. J Insect Physiol 43:433–437

Devetak D, Pabst M (1994) Structure of the subgenual organ in the green lacewing, *Chrysoperna carnea*. Tiss Cell 26:249–257

Eberhard M, Picker M (2008) Vibrational communication in two sympatric species of Mantophasmatodea (Heelwalkers). J Insect Behav 21:240–257

Eberhard M, Lang D, Metscher B, Pass G, Picker M, Wolf H (2010) Structure and sensory physiology of the leg scolopidial organs in Mantophasmatodea and their role in vibrational communication. Arthrop Struct Devel 39:230–241

Effertz T, Wiek R, Göpfert MC (2011) NompC TRP channel is essential for *Drosophila* sound receptor function. Curr Biol 21:592–597

Erler G (1983) Reduction of mechanical sensitivity in an insect mechanoreceptor correlated with destruction of its tubular body. Cell Tiss Res 234:451–461

Field LH, Bailey WJ (1997) Sound production in primitive Orthoptera from Western Australia: sounds used in defence and social communication in *Ametrus* sp. and *Hadrogryllacris* sp. (Gryllacrididae: Orthoptera). J Nat Hist 31:1127–1141

Field LH, Matheson T (1998) Chordotonal organs of insects. Adv Ins Physiol 27:1–228

Field L, Pflüger H-J (1989) The femoral chordotonal organ: a bifunctional orthopteran (*Locusta migratoria*) sense organ? Comp Biochem Physiol A 93:729–743

French AS (1988) Transduction mechanism of mechanosensilla. Ann Rev Entomol 33:39–58

French AS (1992) Mechanotransduction. Ann Rev Physiol 54:135–152

Friedrich H (1929) Vergleichende Untersuchungen über die tibialen Scolopalorgane einiger Orthopteren. Z wiss Zool 134:84–148

Füller H, Ernst A (1973) Die Ultrastruktur der femoralen Chordotonalorgane von *Carausius morosus* Br. Zool Jb Anat 91:574–601

Gong Z, Son W, Chung YD, Kim J, Shin DW, McClung CA, Lee Y, Lee HW, Chang DJ, Kaang BK, Cho H, Oh U, Hirsh J, Kernan MJ, Kim C (2004) Two interdependent TRPV channel subunits, inactive and Nanchung, mediate hearing in *Drosophila*. J Neurosci 24:9059–9066

Gong J, Wang Q, Wang Z (2013) NOMPC is likely a key component of *Drosophila* mechanotransduction channels. Eur J Neurosci 38:2057–2064

Göpfert M, Albert J, Nadrowski B, Kamikouchi A (2006) Specification of auditory sensitivity by *Drosophila* TRP channels. Nat Neurosci 9:999–1000

Grueber WB, Jan LY, Jan YN (2002) Tiling of the *Drosophila* epidermis by multidendritic sensory neurons. Development 129:2867–2878

Guedes R, Matheson S, Frei B, Smith M, Yack J (2012) Vibration detection and discrimination in the marked birch caterpillar (*Drepana arcuata*). J Comp Physiol A 198:325–335

Gwynne D (2004) Reproductive behavior of ground weta (Orthoptera: Anostostomatidae): drumming behavior, nuptial feeding, post-copulatory guarding and maternal care. J Kansas Entomol Soc 77:414–428

Hertweck H (1931) Anatomie und Variabilität des Nervensystems und der Sinnesorgane von *Drosophila melanogaster* (Meigen). Z Wiss Zool 139:560–664

Hill KG (1983) The physiology of locust auditory receptors. II. Membrane potentials associated with the response of the receptor cell. J Comp Physiol 152:483–493

Howard J, Bechstein S (2004) Hypothesis: a helix of ankyrin repeats of the NOMPC-TRP ion channel is the gating spring of mechanoreceptors. Curr Biol 14:R224–226

Howse P (1964) An investigation into the mode of action of the subgenual organ in the termite, *Zootermopsis angusticollis* Emerson, and the cockroach, *Periplaneta americana* L. J Insect Physiol 10:409–424

Howse P (1968) The fine structure and functional organization of chordotonal organs. Symp Zool Soc Lond 23:167–198

Hustert R, Pflüger H-J, Bräunig P (1981) Distribution and specific central projections of mechanoreceptors in the thorax and proximal leg joints of locusts. III. The external mechanoreceptors: the campaniform sensilla. Cell Tiss Res 216:97–111

Jeram S, Rössler W, Cokl A, Kalmring K (1995) Structure of atympanate tibial organs in legs of the cave-living Ensifera, *Troglophilus neglectus* (Gryllacrididae, Raphidophoridae). J Morphol 223:109–118

Keil TA (1997) Functional morphology of insect mechanoreceptors. Microsc Res Tech 39:506–531

Keil TA (2012) Sensory cilia in arthropods. Arthrop Struct Dev 41:515–534

Kent KS, Griffin LM (1990) Sensory organs of the thoracic legs of the moth *Manduca sexta*. Cell tiss Res 259:209–223

Kent KS, Fjeld CC, Anderson R (1996) Leg proprioceptors of the tobacco hornworm, *Manduca sexta*: organisation of central projections at larval and adult stages. Micr Res Tech 35:265–284

Kernan M (2007) Mechanotransduction and auditory transduction in *Drosophila*. Pflügers Arch 454:703–720

Kilpinen O, Storm J (1997) Biophysics of the subgenual organ of the honeybee, *Apis mellifera*. J Comp Physiol A 181:309–318

Kittmann R, Schmitz J (1992) Functional specialisation of the scoloparia of the femoral chordotonal organ in stick insects. J Exp Biol 173:91–108

Klose M (1996) Development of leg chordotonal sensory organs in normal and heat shocked embryos of the cricket *Teleogryllus commodus* (Walker). Roux's Arch Dev Biol 205:344–355

Klose M, Bentley D (1989) Transient pioneer neurons are essential for formation of an embryonic peripheral nerve. Science 245:982–984

Kühne R (1982) Neurophysiology of the vibration sense in locusts and bushcrickets: Response characteristics of single receptor units. J Insect Physiol 28:155–163

Kung C (2005) A possible unifying principle for mechanosensation. Nature 436:647–654

Lai E, Orgogozo V (2004) A hidden program in *Drosophila* peripheral neurogenesis revealed: fundamental principles underlying sensory organ diversity. Dev Biol 269:1–17

Lakes R, Pollack GS (1990) The development of the sensory organs of the legs in the blowfly, *Phormia regina*. Cell Tiss Res 259:93–104

Lakes-Harlan R, Heller K-G (1992) Ultrasound-sensitive ears in a parasitoid fly. Naturwissenschaften 79:224–226

Lakes-Harlan R, Lefevre C (2012) The femoral chordotonal organ of adult *Drosophila melanogaster* Meigen 1830. Mitt Dtsch Ges Allg Angew Ent 18:71–74

Lakes-Harlan R, Pollack GS (1993) Pathfinding of peripheral neurons in the central nervous system of an embryonic grasshopper (*Chorthippus biguttulus*). Cell Tiss Res 273:97–106

Lakes-Harlan R, Strauß J (2006) Developmental constraint of insect audition. Front Zool 3:27

Lakes-Harlan R, Stölting H, Stumpner A (1999) Convergent evolution of insect hearing organs from a preadaptive structure. Proc R Soc B 266:1161–1167

Latimer W, Schatral A (1983) The acoustic behaviour of the bushcricket *Tettigonia cantans*. I. Behavioural responses to sound and vibration. Behav Proc 8:113–124

Lee J, Moon S, Cha Y, Chung YD (2010) Drosophila TRPN(= NOMPC) channel localizes to the distal end of mechanosensory cilia. PLoS ONE 5:e11012

Lehnert BP, Baker AE, Gaudry Q, Chiang A-S, Wilson RI (2013) Distinct roles of TRP channels in auditory transduction and amplification of *Drosophila*. Neuron 77:115–128
Liang X, Madrid J, Saleh H, Howard J (2011) NOMPC, a member of the TRP channel family, localizes to the tubular body and distal cilium of *Drosophila* campaniform and chordotonal receptor cells. Cytoskeleton 68:1–7
Liang X, Madrid J, Gärtner R, Verbavatz J-M, Schiklenk C, Wilsch-Bräuninger M, Bogdanova A, Stenger F, Voigt A, Howard J (2013) A NOMPC-dependent membrane-microtubule connector is a candidate for the gating spring in fly mechanoreceptors. Curr Biol 23:755–763
Lin Y, Rössler W, Kalmring K (1995) Morphology of the tibial organs of Acrididae: comparison of the subgenual organ and distal organs in fore-, mid-, and hindlegs of *Schistocerca gregaria* (Acrididae, Catantopidae) and *Locusta migratoria* (Acrididae, Oedipodinae). J Morphol 226:351–360
Lu Q, Senthilan P, Effertz T, Nadrowski B, Göpfert M (2009) Using *Drosophila* for studying fundamental processes in hearing. Integr Comp Biol 49:674–680
Matheson T (1992) Morphology of the central projections of physiologically characterised neurones from the locust methathoracic femoral chordotonal organ. J Comp Physiol A 170:101–120
Matheson T, Field LH (1990) Innervation of the metathoracic femoral chordotonal organ of *Locusta migratoria*. Cell Tissue Res 259:551–560
Matsuo E, Kamikouchi A (2013) Neuronal encoding of sound, gravity, and wind in the fruit fly. J Comp Physiol A 199:253–262
Matsuura H, Sokabe T, Kohno K, Tominaga M, Kadowaki T (2009) Evolutionary conservation and changes in insect TRP channels. BMC Evol Biol 9:228
McIver S (1985) Mechanoreception. In: Kerkut G, Gilbert L (eds) Comprehensive insect physiology, biochemistry and pharmacology. Pergamon Press, Oxford, pp 71–132
McVean A, Field LH (1996) Communication by substratum vibration in the New Zealand tree weta, *Hemideina femorata* (Stenopelmatidae: Orthoptera). J Zool 239:101–122
Meier T, Reichert H (1990) Embryonic development and evolutionary origin of the orthopteran auditory organs. J Neurobiol 21:592–610
Meier T, Chabaud F, Reichert H (1991) Homologous patterns in the embryonic development of the peripheral nervous system in the grasshopper *Schistocerca gregaria* and the fly *Drosophila melanogaster*. Development 112:241–253
Menzel J, Tautz J (1994) Functional morphology of the subgenual organ of the carpenter ant. Tiss Cell 26:735–746
Merritt DJ, Murphey RK (1992) Projections of leg proprioceptors within the CNS of the fly *Phormia regina* in relation to the generalized insect ganglion. J Comp Neurol 322:16–34
Merritt DJ, Hawken A, Whitington PM (1993) The role of the *cut* gene in the specification of central projections by sensory axons in *Drosophila*. Neuron 10:741–752
Meyhöfer R, Casas J (1999) Vibratory stimuli in host location by parasitic wasps. J Insect Physiol 45:967–971
Michel K, Amon T, Cokl A (1982) The morphology of the leg scolopidial organs in *Nezara viridula* (L.)(Heteroptera, Pentatomidae). Rev Can Biol Exp 42:139–150
Moulins M (1976) Ultrastructure of chordotonal organs. In: Mill PJ (ed) Structure and function of proprioceptors in the invertebrates. Chapman and Hall, London, pp 387–426
Mücke A (1991) Innervation pattern and sensory supply of the midleg of *Schistocerca gregaria*. Zoomorphology 110:175–187
Mücke A, Lakes-Harlan R (1995) Central projections of sensory cells of the midleg of the locust, *Schistocerca gregaria*. Cell Tissue Res 280:391–400
Murphey R (1971) Motor control of orientation to prey by the waterstrider *Gerris remigis*. Z vergl Physiol 72:150–167
Nishino H (2000) Topographic mapping of the axons of the femoral chordotonal organ neurons in the cricket *Gryllus bimaculatus*. Cell Tiss Res 299:145–157

Nishino H (2003) Somatotopic mapping of chordotonal organ neurons in a primitive ensiferan, the New Zealand tree weta *Hemideina femorata*: I. Femoral chordotonal organ. J Comp Neurol 464:312–326

Nishino H, Field LH (2003) Somatotopic mapping of chordotonal organ neurons in a primitive ensiferan, the New Zealand tree weta *Hemideina femorata*: II. Complex tibial organ. J Comp Neurol 464:327–342

Goodwyn PP, Katsumata-Wada A, Okada K (2009) Morphology and neurophysiology of tarsal vibration receptors in the water strider *Aquarius paludum* (Heteroptera: Gerridae). J Insect Physiol 55:855–861

Pflüger H-J, Bräunig P, Hustert R (1988) The organization of mechanosensory neuropils in locust thoracic ganglia. Phil Trans R Soc Lond B 321:1–26

Preuss T, Hengstenberg R (1992) Structure and kinematics of the prosternal organs and their influence on head position in the blowfly *Calliphora erythrocephala* Meig. J Comp Physiol A 171:483–493

Robert D, Amoroso J, Hoy RR (1992) The evolutionary convergence of hearing in a parasitoid fly and its cricket host. Science 258:1135–1137

Rohrseitz K, Kilpinen O (1997) Vibration transmission characteristics of the legs of freely standing honeybees. Zoology 100:80–84

Rössler W (1992) Funcional morphology and development of tibial organs in the legs I, II and III of the bushcricket *Ephippiger ephippiger* (Insecta, Ensifera). Zoomorphology 112:181–188

Rössler W, Jatho M, Kalmring K (2006) The auditory-vibratory sensory system in bushcrickets. In: Drosopoulos S, Claridge MF (eds) Insect sound and communication. CRC Press, Boca Raton, pp 35–69

Sauer AE, Stein W (1999) Sensorimotor pathways processing vibratory signals from the femoral chordotonal organ of the stick insect. J Comp Physiol A 185:21–31

Schäffer S, Lakes-Harlan R (2001) Embryonic development of the central projection of auditory afferents (*Schistocerca gregaria*, Orthoptera, Insecta). J Neurobiol 46:97–112

Schmitz J, Dean J, Kittmann R (1991) Central projections of leg sense organs in *Carausius morosus* (Insecta, Phasmida). Zoomorphology 111:19–33

Schneider W (1950) Über den Erschütterungssinn von Käfern und Fliegen. Z vergl Physiol 32:287–302

Schnorbus H (1971) Die subgenualen Sinnesorgane von *Periplaneta americana*: Histologie und Vibrationsschwellen. Z vergl Physiol 71:14–48

Schön A (1911) Bau und Entwicklung des tibialen Chordotonalorgans bei der Honigbiene und bei Ameisen. Zool Jb Anat 31:439–472

Shanbhag SR, Singh K, Singh RN (1992) Ultrastructure of the femoral chordotonal organs and their novel synaptic organization in the legs of *Drosophila melanogaster* Melgen (Diptera: Drosophilidae). Int J Insect Morphol Embryol 21:311–322

Shaw SR (1994) Re-evaluation of the absolute threshold and response mode of the most sensitive known "vibration" detector, the cockroach's subgenual organ: a cochlea-like displacement threshold and a direct response to sound. J Neurobiol 25:1167–1185

Stein W, Sauer A (1999) Physiology of vibration-sensitive afferents in the femoral chordotonal organ of the stick insect. J Comp Physiol A 184:253–263

Stiedl O, Kalmring K (1989) The importance of song and vibratory signals in the behaviour of the bushcricket *Ephippiger ephippiger* Fiebig (Orthoptera, Tettigoniidae): taxis by females. Oecologica 80:142–144

Stölting H, Stumpner A, Lakes-Harlan R (2007) Morphology and physiology of the prosternal chordotonal organ of the sarcophagid fly *Sarcophaga bullata* (Parker). J Insect Physiol 53:444–454

Storm J, Kilpinen O (1998) Modelling the subgenual organ of the honeybee, *Apis mellifera*. Biol Cybern 78:175–182

Strauß J, Lakes-Harlan R (2008a) Neuroanatomy and physiology of the complex tibial organ of an atympanate Ensiferan, *Ametrus tibialis* (Brunner von Wattenwyl, 1888) (Gryllacrididae, Orthoptera) and evolutionary implications. Brain Behav Evol 71:167–180

Strauß J, Lakes-Harlan R (2008b) Neuroanatomy of the complex tibial organ of *Stenopelmatus* (Orthoptera: Ensifera: Stenopelmatidae). J Comp Neurol 511:81–91

Strauß J, Lakes-Harlan R (2009) The evolutionary origin of auditory receptors in Tettigonioidea: the complex tibial organ of Schizodactylidae. Naturwissenschaften 96:143–146

Strauß J, Lakes-Harlan R (2010) Neuroanatomy of the complex tibial organ in the splay-footed cricket *Comicus calcaris* IRISH 1986 (Orthoptera: Ensifera: Schizodactylidae). J Comp Neurol 518:4567–4580

Strauß J, Lakes-Harlan R (2013) Sensory neuroanatomy of stick insects highlights the evolutionary diversity of the orthopteroid subgenual organ complex. J Comp Neurol (in press). doi:10.1002/cne.23378

Stumpner A (1996) Tonotopic organization of the hearing organ in an bushcricket. Naturwissenschaften 83:81–84

Stumpner A, von Helversen D (2001) Evolution and function of auditory system in insects. Naturwissenschaften 88:159–170

Stumpner A, Allen G, Lakes-Harlan R (2006) Hearing and frequency dependence of auditory interneurons in the parasitoid fly *Homotrixa alleni* (Tachinidae: Ormiini). J Comp Physiol A 193:113–125

Theophilidis G (1986) The femoral chordotonal organ of *Decticus albifrons* (Orthoptera: Tettigoniidae)—I. Structure. Comp Biochem Physiol A 84:529–536

Thurm U (2001) Evolutionary aspects of mechanoreception: from ciliates to man. In: Backhaus W (ed) Neuronal coding of perceptual systems. World Scientific Publishing, Singapore, pp 237–248

Todi S, Sharma Y, Eberl DF (2004) Anatomical and molecular design of the *Drosophila* antenna as a flagellar auditory organ. Microsc Res Tech 63:388–399

Tracey WJ, Wilson R, Laurent G, Benzer S (2003) *Painless*, a *Drosophila* gene essential for nociception. Cell 113:261–273

Vilhelmsen L, Nunzio I, Romani R, Basibuyuk H, Quicke D (2001) Host location and oviposition in a basal group of parasitic wasps: the subgenual organ, ovipositor apparatus and associated structures in the Orussidae (Hymenoptera, Insecta). Zoomorphology 121:63–84

Vilhelmsen L, Turrisi GF, Beutel RG (2008) Distal leg morphology, subgenual organs and host detection in Stephanidae (Insecta, Hymenoptera). J Nat Hist 42:1649–1663

Virant-Doberlet M, Cokl A, Zorovic M (2006) Use of substrate vibrations for orientation: from behaviour to physiology. In: Drosopoulos S, Claridge MF (eds) Insect sound and communication. CRC Press, Boca Raton, pp 81–97

Weidemann S, Keuper A (1987) Influence of vibratory signals on the phonotaxis of the gryllid *Gryllus bimaculatus* DeGeer (Ensifera: Gryllidae). Oecologia 74:316–318

Weissman D (2001) Communication and reproductive behaviour in North American Jerusalem crickets (*Stenopelmatus*) (Orthoptera: Stenopelmatidae). In: Field LH (ed) The biology of wetas, king crickets and their allies. CABI Publishing, Wallingford, pp 351–373

Wiese K (1972) Das mechanorezeptive Beuteortungssystem von *Notonecta*. I. Die Funktion des tarsalen Scolopidialorgans. J Comp Physiol 78:83–102

Wittig G (1955) Untersuchungen am Thorax von *Perla abdominalis* Burm. (Larve und Imago). Zool Jhrb Anat Ontog 74:491–570

Yan Z, Zhang W, He Y, Gorcuyca D, Xiang Y, Cheng LE, Meltzer S, Jan LY, Jan YN (2013) *Drosophila* NOMPC is a mechanotransduction channel subunit for gentle-touch sensation. Nature 493:221–225

Zill S, Büschges A, Schmitz J (2011) Encoding of force increases and decreases by tibial campaniform sensilla in the stick insect, *Carausius morosus*. J Comp Physiol A 197:851–867

zur Lage P, Jarman AP (1999) Antagonism of EGFR and Notch signalling in the reiterative recruitment of Drosophila adult chordotonal sense organ precursors. Development 126:3149–3157

# Chapter 15
# Echolocation in Whirligig Beetles Using Surface Waves: An Unsubstantiated Conjecture

**Jonathan Voise and Jérôme Casas**

**Abstract** Understanding the physics of signal transmission in insect vibratory communication is nearly impossible for most natural substrates on which insect communicates, except for the water surface, on which the propagation of linear capillary surface waves is reasonably well mastered. The water surface is an environment where several insect species live and exploit its static and dynamic deformations. Whirligig beetles use surface waves in a wide range of situations, and it has been hypothesized that they could use the echo of their own waves to scan the water surface. These insects were also suspected to perceive menisci, i.e., static deformations of the water surface. However, no manipulative experimental studies, based on quantitative predictions of the surface deformation, have been conducted to determine whether echolocation or meniscus perception is used. The aim of this work was to test the hypothesis that whirligig beetles detect the meniscus around immobile objects. If the perception of an immobile object on the water surface can be explained by resorting to the meniscus only, then the hypothesis of echolocation should be reconsidered. We used cylinders of varying diameter and wettability crossing the water surface to experimentally modify the meniscus profiles. Contacts between beetles and cylinders were recorded using a high-speed camera. Loops in trajectories before or after a contact, as well as unfolding of forelegs before a contact, were used as criteria for the distance at which cylinders were perceived. Based on a theoretical modeling of the meniscus profiles, we predicted the distance at which one type of cylinder was detected based on the meniscus amplitude corresponding to the distance of perception of another type of cylinder. Both diameter size and wettability affected cylinder perception. Our predictions were unfortunately contradictory, and the unfolding of forelegs could not be explained by meniscus perception only. Complementary experiments are needed to eliminate any influence of vision. We obtained

J. Voise · J. Casas (✉)
University of Tours, IRBI UMR CNRS 6035, Av. Monge, 37200 Tours, France
e-mail: casas@univ-tours.fr

R. B. Cocroft et al. (eds.), *Studying Vibrational Communication*,
Animal Signals and Communication 3, DOI: 10.1007/978-3-662-43607-3_15,

conflicting evidence about meniscus perception, with the statistically most powerful of our tests being strongly in favor of it. Thus, echolocation by whirligig beetles should still be considered as a conjecture, at best.

## 15.1 Introduction

Understanding signal propagation during insect vibratory communication, here used in the sense of being able to predict the amplitude and phase of a signal at the location of the receiver, is a task nearly impossible to achieve in most situations. Leaves, trunks, rocks, or soil have complex structural and material properties that defy straightforward modeling. Even the simplest prediction, such as a decline of amplitude with distance on a plant, cannot be made with certainty, as waves may form complex patterns of varying amplitude through interference or wave mode changes. This explains partly why the science of insect vibratory communication still lacks, as of today, rigorous experiments on natural substrates about the exact properties of the signals which matter most.

The water surface is an interface for which the prospects for quantifying and understanding the physics of vibratory communication are better. The number of wave types relevant to insect vibratory communication is much more limited than on any solid surface and is basically restricted to gravity– capillary waves. The theory of *linear* capillary wave propagation has been worked out in detail, and many experiments have confirmed the theory (see e.g., Lighthill 1978). The theory of propagation of surface waves cannot be considered as complete, however, in particular regarding the nonlinear aspects as well as the different drag contributions around an immersed body (see Le Merrer et al. 2011 for the latest developments). It is, however, much more mature and applicable than any theory for surface waves on the solids encountered by insects in the wild. Deformation of the water surface can be dynamic, in the case of surface waves, or static, in the case of a meniscus. Animals living at the water surface have evolved the ability to use such deformations in a wide range of situations. On the one hand, surface waves are used for prey detection by fishes and amphibians preying on flying insects trapped on the water surface (Bleckmann and Toop 1981; Claas et al. 1993). Arthropods use surface waves to detect their prey, for example, water striders (Heteroptera, Gerridae), backswimmers (Heteroptera, Notonectidae), and fishing spiders (Araneae, Pisauridae) (Bleckmann et al. 1994; Lang 1980; Murphey 1971; Weise 1974). Waves propagating on the water surface are also used to detect potential predators and cause escape or fright behavior in whirligig beetles (Coleoptera, Gyrinidae), e.g., (Eggers 1926). Vibratory communication in some insects depends on surface waves , as shown in water striders (Wilcox 1972). On the other hand, static deformations of the water surface, resulting from an equilibrium between capillary forces and gravity, are also used by some insects to keep group cohesion or to move on water (Hu and Bush 2005; Voise et al. 2011).

These two kinds of surface deformations could be used by whirligig beetles. Whirligig beetles are semi-gregarious insects that are very sensitive to surface waves and can detect surface deformations of a few micrometers in amplitude using their antennae (Rudolph 1967). Several authors have also suggested perception of menisci by these beetles (Eggers 1926, 1927; Rudolph 1967; Wilde 1941). The fact that they are able to avoid immobile obstacles or capture prey in the absence of light has also raised the hypothesis that they use the echo of their own waves to scan their environment (Eggers 1927; Wilde 1941). This hypothesis was championed by Tucker (1969), who published his purely observational reasoning in the journal *Science*. This venue assured a wide spreading of the hypothesis. Since then, the unexpected and exotic, but still purely hypothetical, use of echolocation by whirligig beetles has penetrated most textbooks (Denny 1993; Dusenbery 1992; Hill 2008; Vogel 2003). No experimental study, however, has shown conclusive results in favor of echolocation, and a comprehensive work on the use of surface waves is missing. The main experimental difficulty is to dissociate echolocation and the detection of the meniscus around objects. It is indeed very difficult, if at all possible, to totally eliminate the meniscus around objects or to stop surface wave reflection.

The aim of this work was to study the detection of immobile objects on the water surface by whirligig beetles and to determine whether this detection can be interpreted solely on the basis of the extent of the meniscus. In contrast to echolocation using surface waves , there is a sufficiently well-founded theoretical basis for making inferences about the use of the meniscus in object detection. Previous studies have shown that whirligig beetles are attracted by small wires vertically crossing the water surface (Bott 1928). We therefore decided to use cylinders crossing the water surface for this work. We manipulated the meniscus in two ways. First, we used cylinders with different diameters; the diameter affecting the size of the meniscus around cylinders for radii less than the capillary length (de Gennes et al. 2003). Secondly, we used two different wettabilities, hydrophobic and hydrophilic, to influence the polarity of the meniscus. We thus produced concave and convex menisci, respectively. Considering the hypothesis that the meniscus is used to detect immobile objects, we predicted the distance at which one type of cylinder (i.e., with a specific diameter and wettability) is detected based on the meniscus amplitude corresponding to the distance of perception of another type of cylinder. The meniscus profile around each type of cylinder was thus used to estimate the relationship of the meniscus amplitude at which whirligig beetles reacted with the size and hydrophobicity of cylinders. If the perception of immobile objects on the water surface by whirligig beetles can be explained by resorting to the meniscus only, then the hypothesis of echolocation should be reconsidered. Indeed, one may assume that the perception of objects using reflected surface waves occur at much larger distances, tens of centimeters.

## 15.2 Methods

### 15.2.1 Beetles

Whirligig beetles of the species *Gyrinus substriatus* Steph. were collected from temporary ponds in Indre et Loire and Vendée, France. They were kept in ten-l aquariums, in groups of eight to ten, and were fed daily with *Drosophila* fruit flies. Aquariums were filled with tap water, which was changed weekly.

### 15.2.2 Experimental Setup

For contact and loop analysis described below, beetles were placed individually in a white plastic tank 0.9 m in diameter and 0.6 m high, filled with tap water. We placed a square wave absorber device inside the tank. The wave absorber was 50 cm × 50 cm, and its borders were made with strips of extruded polystyrene (5 × 50 cm) inclined with an angle of 20°. This apparatus did not totally eliminate surface waves reflecting from the walls but attenuated them sufficiently to decrease reflections. Beetles were placed inside the arena formed by the wave absorber (Fig. 15.1). A water depth of 10 cm is complying with the deep water condition for capillarity waves (Dean and Dalrymple 1991).

A cylinder was placed in the tank, in the center of the arena. It crossed the water surface and was fixed from above (Fig. 15.1). Cylinders of four different diameters (0.2, 2.5, 6, and 19 mm) and two different surfaces (hydrophilic and hydrophobic) were used. All the cylinders were painted white, providing a hydrophilic surface. A hydrophobic surface was obtained by soaking the cylinders in hot wax. The wax layer was thin enough not to modify the cylinder diameter.

Whirligig beetles were filmed with a Fastec Imaging TroubleShooter high-speed camera equipped with a Computar zoom lens (18-108/2.5). Video resolution was 640 × 480 pixels. Light was provided by four 500-W projectors. The camera and lights were fixed on a 1 m × 1 m × 2 m aluminum structure placed above the tank. The lights were at a height of 1 m and the camera was at a height of 1.2 or 2 m.

When beetles seize prey such as *Drosophila* struggling on the water surface, they do this with the first pair of legs. This seizing behavior is identical when approaching objects such as a cylinder. In this study, perception distances were based on loops in trajectories made before or after a physical contact with cylinders and on unfolding of forelegs. Thus, two kinds of video sequence were recorded. The first type of sequence was used to analyze *contacts* between whirligig beetles and the cylinder. The camera was positioned 1.2 m above the plastic tank, and the focal length was set to have a visual field of approximately 7 cm × 5 cm. This setting was chosen to see the widest area possible around the cylinder while keeping a good resolution to see beetle legs. All four projectors were switched on. Video sequences were recorded at a speed of 250 frames $s^{-1}$.

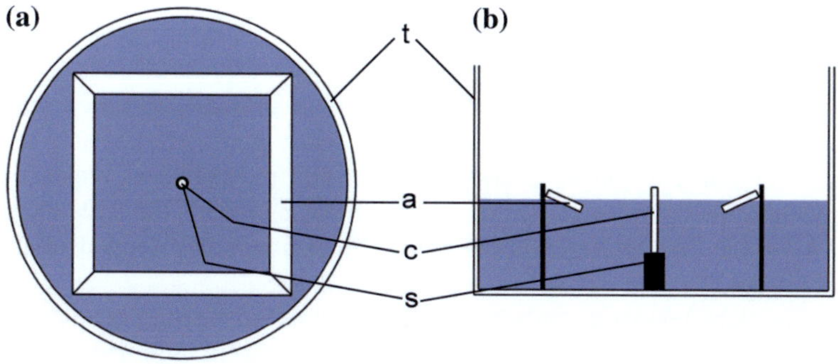


**Fig. 15.1** Experimental arena used for testing detection of immobile objects by whirligig beetles on the water surface. **a** *top view* and **b** *cross section* of the water tank. Beetles were placed inside the arena formed by the wave absorber. *a* wave absorber; *c* cylinder; *s* support of the cylinder; *t* plastic tank. All these components were painted in *white*

The purpose of the second kind of video sequence was to record trajectories, and more precisely *loops* made by insects before or after a contact with the cylinder. A large visual field was necessary to record trajectories in the whole arena, so the camera was fixed at a height of 2 m, on top of the aluminum structure. Videos were recorded at a rate of 50 frames $s^{-1}$, and the focal length of the zoom lens was set at its maximum value. Only two projectors on opposite sides of the tank were switched on.

The camera was focused on a dead beetle, and a floating ruler was used for scale. The water was maintained at ambient temperature (approximately 20 °C). We avoided leaving the lights on for too long, to prevent large increases in the temperature of the water in the tank. The water was kept as clean as possible, and floating dust was removed before each video recording.

### *15.2.3 Excluding Visual Perception of the Cylinders*

To study perception of surface deformations, we needed first to exclude visual perception of the cylinders. To this end, all the component of the experimental arena, i.e., water tank, wave absorber, cylinder and the support of the cylinder, were painted in white using the same paint. This dramatically decreased the contrast produced by the cylinder. In addition, the brightness produced by the 500-W projectors was very important into the water tank. Whirligig beetles needed several minutes of adaptation when the light brightness changes drastically (Carthy and Goodman 1964; Bennet 1967). In order to not allow beetles to adapt their vision, they were put in the experimental arena just before recording video sequences and removed just after. They stayed inside the experimental area no longer than 5 min. If no contact was recorded during 5 min, the insect was removed and replaced several minutes after.

### 15.2.4 Contact and Loop Analyses

For each type of cylinder (four diameters and two surfaces), 30 video sequences were recorded using ten beetles (three sequences per beetle). Each recorded sequence contained a contact between beetles and the cylinder. We recorded a *contact* when beetles passed within 1 mm of the cylinder. The video sequence started when the insect entered in the field of the camera and stopped when it left the camera field.

Several parameters were analyzed. First, we observed whether beetles unfolded their forelegs before or during the contact. If forelegs were unfolded, the distance between the head of the insect and the border of the cylinder was calculated using Image J (http://rsb.info.nih.gov/ij/). The type of leg kinematics before the contact was also noticed. These leg kinematics type I, II, and III—the type being characterized by the stroke pattern—produce mean speeds of 15.05, 18.34, and 48. 41 $cm.s^{-1}$, respectively (Voise and Casas 2010).

To analyze contacts at a larger scale, we recorded video sequences when beetles made a *loop* in their trajectories, in a time range of 3 s before and 3 s after the contact. When several loops appeared in the trajectory, we selected the loop with the closest crossing from the cylinder (crossing being defined as the location of loop closure). For each type of cylinder, 30 video sequences were recorded using ten beetles (three sequences per beetle). We also recorded loops on 30 video sequences without cylinders crossing the water surface as a control.

The trajectories were reconstructed from videos using Image J, using the extremity of the beetle head, which is the center of perception. Four parameters of the loops were measured. We calculated the minimum distance between the loop and the border of the cylinder, and the distance between the crossing of the loop and the border of the cylinder. The perimeter of the loops was also calculated and used to obtain the mean speed of whirligig beetles during the loop. For control sequences, distances were calculated using the center of the cylinder support.

Statistical analyses were done using R software (http://www.R-project.org). Multiple comparisons were made using a Kruskal–Wallis rank sum test followed by post hoc tests of multiple comparisons. Comparison between two samples was made using a Wilcoxon test. Comparison between predicated and observed values was made using a one-sample t-test (*df* is the degree of freedom). Linear correlations were tested using simple linear models (*lm*). *N* represents the number of individuals in the dataset, *p* is the *p*-value of the test, and we considered a significant difference when *p* was less than 0.05.

### 15.2.5 Meniscus Profiled Around Cylinders

The meniscus profile produced by each type of cylinder was calculated numerically. For the meniscus, the contact angle $\theta_0$ of the hydrophilic and hydrophobic surfaces of the cylinders was first measured. These angles were measured on 10-μl drops deposited on small horizontal plates covered with the same paint and wax as cylinders, using a digital camera CCD (696 × 512 pixels) and a stereo microscope. We found angles of 45° and 110° for hydrophilic and hydrophobic surfaces, respectively. Then, we computed the meniscus around each type of cylinder. For the small diameters of 0.2 and 2.5 mm, we used the general solution provided by James (1974), where the meniscus profile *z(x)* is given by

$$z(x) = r_0 \sin\theta \left[ \ln 2x - \ln\left\{ x + \left(x^2 - r_0^2 \sin^2\theta\right)^{1/2} \right\} + K_0 \left\{ \left(\rho g/\sigma\right)^{1/2} x \right\} \right] \quad (15.1)$$

where $r_O$, the radius of the cylinder; $\theta = 90° - \theta_O$; $K_O$, a modified Bessel function of order zero; $\rho$, the water density; $g$, the acceleration due to gravity, and $\sigma$, the surface tension. For the large cylinders, i.e., with their radius greater than the capillary length $\kappa^{-1} = (\sigma/\rho g)^{1/2}$, corresponding to cylinders of diameters of 6 and 19 mm, the meniscus profile was calculated using the formula given by de Gennes et al. (2003):

$$x - x_0 = k^{-1} \arg\cosh\left(2k^{-1}/z\right) - 2k^{-1}\left\{1 - \left(z^2/4k^{-2}\right)\right\}^{1/2} \quad (15.2)$$

where $x_O$ corresponds to the point on the surface were $x = 0$ (the border of the cylinder) and $z = h$. The parameter $h$ is the maximum or minimum height of the meniscus (i.e., the maximal deformation of the water surface) and is calculated as

$$h = 2^{1/2} k^{-1} (1 - \sin\theta_0)^{1/2} \quad (15.3)$$

The meniscus profile obtained for each cylinder is shown in Fig. 15.2.

## 15.3 Results

### *15.3.1 Contacts with the Cylinder*

Forelegs were never unfolded when beetles swam at high speed (i.e., using type III leg kinematics), except in one case where the insect unfolded its leg due to the shock created by the contact. The distance at which forelegs are unfolded was greater with type II leg kinematics than with type I (Fig. 15.3a) (Wilcoxon, $p < 0.01$, $N = 42$).

The number of contacts where whirligig beetles unfolded their forelegs decreased as the diameter of the cylinder increased (Table 15.1, Fig. 15.3b). The distance between the beetle head and the cylinder when insects unfolded their forelegs was lower for hydrophobic cylinders than hydrophilic cylinders (Wilcoxon, $p < 0.01$, $N = 43$) (Fig. 15.3c).

Considering the 0.2 diameter cylinder, for which many data were available (Table 15.1), a significant difference was found between hydrophilic and

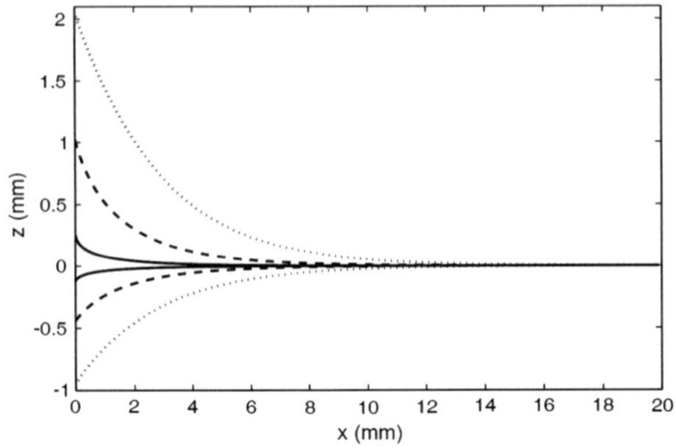


**Fig. 15.2** Theoretical meniscus profiles *z(x)* around cylinders obtained numerically. *Black line* corresponds to the 0.2 mm diameter, *dashed line* corresponds to the 2.5 mm diameter, and *dotted line* corresponds to the 6 and 19 mm diameter. Hydrophobic and hydrophilic surfaces created a concave meniscus ($z < 0$) and a convex meniscus ($z > 0$), respectively. An important asymmetry can be observed between the different surfaces

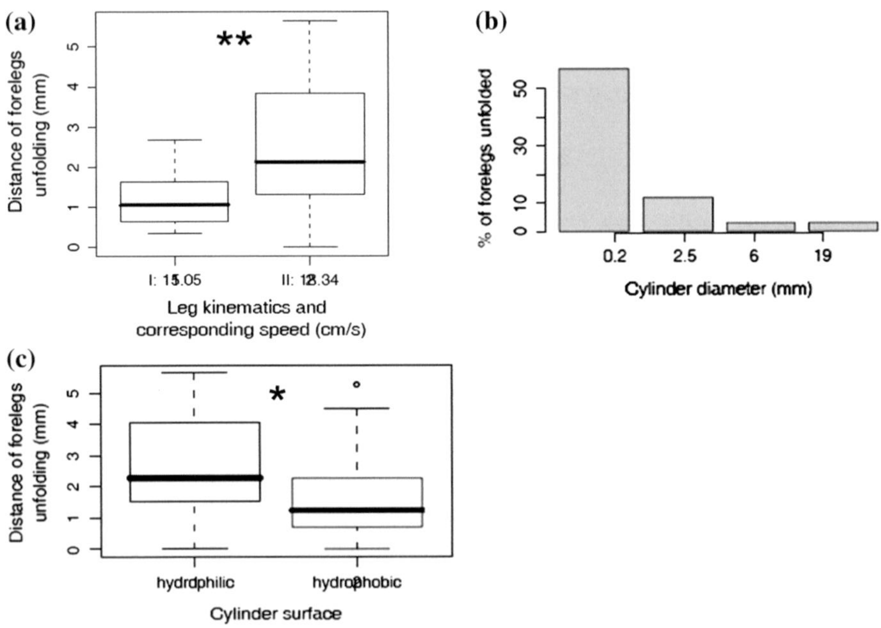


**Fig. 15.3** Foreleg unfolding by whirligig beetles. **a** Boxplot of the distance at which whirligig beetles unfolded their forelegs before contact with the cylinders for the two types of low-speed leg kinematics I and II (speed values are obtained from Voise and Casas 2010). **b** Percentage of foreleg unfolded before contact for each cylinder diameter. **c** Boxplot of the distance at which whirligig beetles unfolded their forelegs before contact for hydrophilic and hydrophobic cylinders. The *horizontal line* within the box represents the median, the *upper* and *lower borders* of the *box* are the 25th and 75th percentiles, and the *error bars* show the 10th and 90th percentiles. Points represent outliers. $*p < 0.05$; $**p < 0.01$

**Table 15.1** Number of observations recorded for each type of cylinders during experiments

| Number of observations | Cylinders | | | | | | | |
|---|---|---|---|---|---|---|---|---|
| | ∅ 0.2 mm | | ∅ 2.5 mm | | ∅ 6 mm | | ∅ 19 mm | |
| | hphilic | hphobic | hphilic | hphobic | hphilic | hphobic | hphilic | hphobic |
| Loops | 30 | 30 | 30 | 30 | 30 | 30 | 30 | 30 |
| Contacts | 30 | 30 | 30 | 30 | 30 | 30 | 30 | 30 |
| Legs unfolding | 18 | 16 | 0 | 6 | 0 | 1 | 0 | 2 |

*Hphilic* hydrophilic; *hphobic* hydrophobic

hydrophobic surfaces for the unfolding distance (Wilcoxon, $p < 0.05$, $N = 34$). For this diameter, we could calculate the amplitude of the surface deformation corresponding to the mean distance at which beetles unfolded their legs in the case of hydrophilic cylinders. We found an amplitude of 26.67 μm at that distance of 2.84 ± 1.46 mm ($N = 18$). We predicted the distance of unfolding legs corresponding to the same amplitude of surface deformation to be 1.53 mm for hydrophobic cylinders. The observed distance was 1.75 ± 1.40 mm ($N = 16$), which is not significantly different (t-test, $df = 15$, $p = 0.6204$). Thus, we were able to predict the seizing behavior toward a hydrophobic surface on the basis of the behavior displayed toward a hydrophilic surface for the smallest cylinder.

Another prediction based on the meniscus amplitude was made using the unfolding distance for the 0.2-mm hydrophobic cylinder to predict the unfolding distance for the 2.5-mm hydrophobic cylinder. This is the only other prediction we can make, given the low number of observations, and its statistical power is less than the previous prediction. The mean unfolding distance measured for the 0.2-mm hydrophobic cylinder (1.75 ± 1.40 mm; $N = 16$) corresponds to a meniscus amplitude of 23.53 μm. We predicted a distance of 5.72 mm for the 2.5-mm hydrophobic cylinder, and we observed a distance of 1.53 ± 1.34 mm ($N = 6$) corresponding to a meniscus amplitude of 177.7 μm. Predicted and observed distances were significantly different (t-test, $df = 6$, $p < 0.001$). Moreover, observed distances at which forelegs were unfolded were not significantly different between 0.2 and 2.5-mm hydrophobic cylinders (t-test, $df = 9.38$, $p = 0.7475$). Thus, despite the low number of observations, we conclude that the behavior displayed toward small-diameter cylinders cannot be used to predict the behavior toward cylinders of larger diameter.

### *15.3.2 Loops in Trajectories*

Parameters extracted from the loops displayed during the biotests (Fig. 15.4a–d) showed values significantly different from the control situation. This is true for the minimum distance between the loop and the cylinder, for the distance between the crossing of the loop and the cylinder, and for the perimeter of the loop

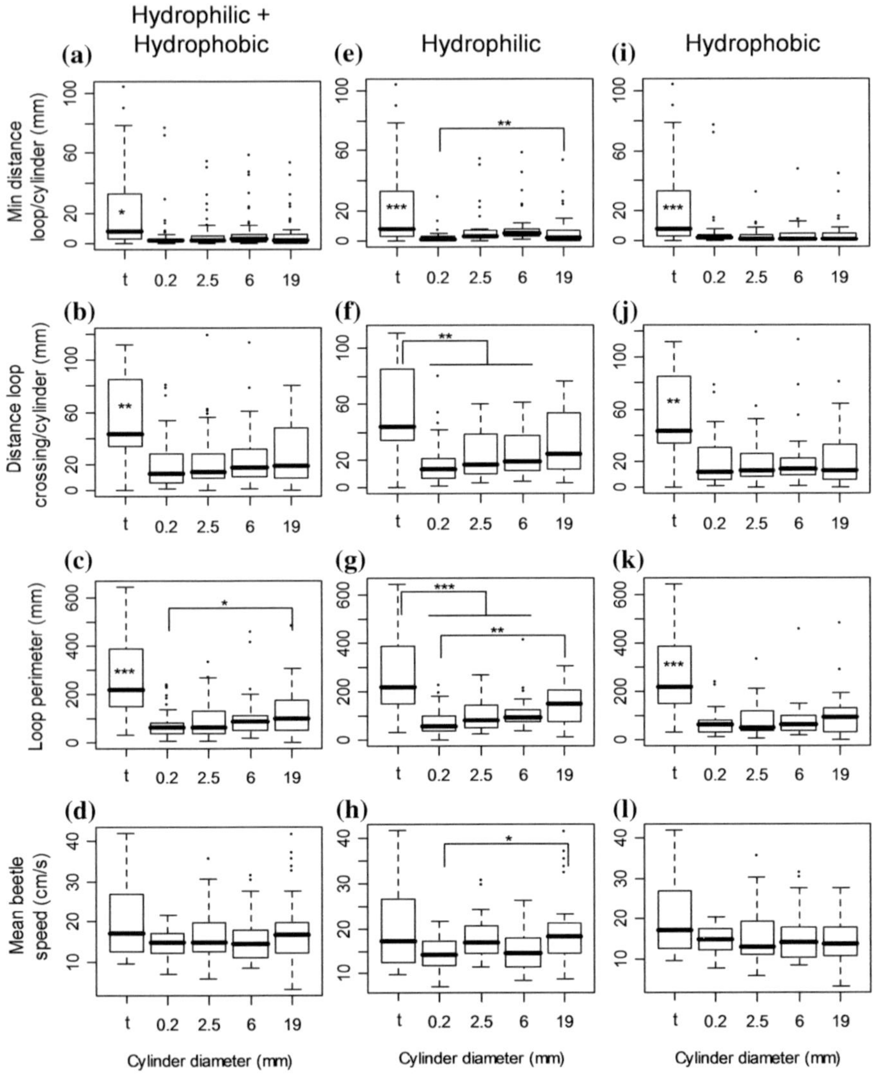


**Fig. 15.4** Effect of the diameter of the cylinders on loop parameters. **a–d** boxplots of the four loop parameters measured for each diameter and for the control group without cylinder (t). Loop parameters are **a** the minimum distance between the loop and the cylinder, **b** the distance between the crossing of the loop and the cylinder, **c** the loop perimeter, **d** the mean beetles speed during the loop. The *horizontal line* within the *box* represents the median, the *upper* and *lower borders* of the *box* are the 25th and 75th percentiles, and the *error bars* show the 10th and 90th percentiles. Points represent outliers. $*p < 0.05$; $**p < 0.01$; $***p < 0.001$. **e–l** Linear regressions of the four loop parameters as a function of the cylinder diameters for hydrophilic cylinders (**e–h**) and hydrophobic cylinders (**i–l**)

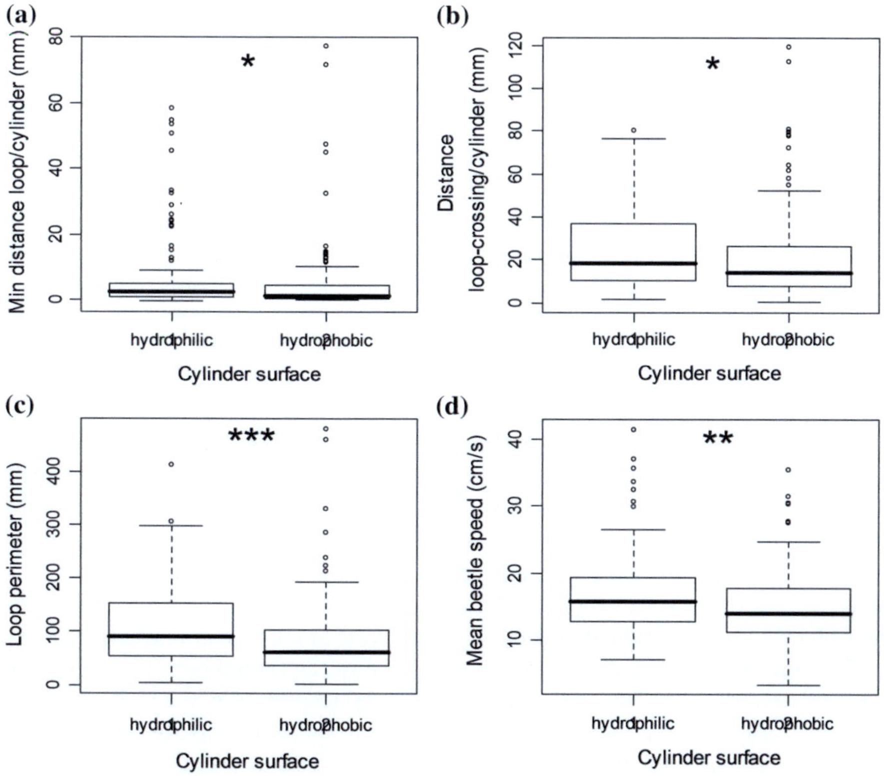


**Fig. 15.5** Effect of the surface of the cylinders on the loop parameters. Boxplots of the four parameters measured for loops in trajectories for each surface. Loop parameters are **a** the minimum distance between the loop and the cylinder, **b** the distance between the crossing of the loop and the cylinder, **c** the loop perimeter, **d** the mean beetles speed during the loop. The *horizontal line* within the *box* represents the median, the *upper* and *lower borders* of the *box* are the 25th and 75th percentiles, and the *error bars* show the 10th and 90th percentiles. Points represent outliers. *$p < 0.05$; **$p < 0.01$; ***$p < 0.001$

(Kruskal–Wallis, $p < 0.05$, $N = 270$). No difference was observed for the mean speed of insect during the loop (Kruskal–Wallis, $p > 0.05$, $N = 270$).

When hydrophilic cylinders were analyzed separately, a positive correlation was found between the diameter and loop perimeter, the minimum distance loop/cylinder and the distance loop-crossing/cylinder (lm, $df = 118$, $p < 0.01$) (Fig. 15.4e–h). For hydrophobic cylinders, no correlation was found between the diameter and the loop parameters (lm, $df = 118$, $p > 0.5$) (Fig. 15.4i–l).

Regarding the effect of the surface of cylinders on the four loop parameters, we found significant difference between hydrophobic and hydrophilic surfaces (Wilcoxon, $p < 0.05$, $N = 240$), with the hydrophilic cylinders corresponding to larger loops (Fig. 15.5a–d).

## 15.4 Discussion

Experiments carried out in this study were not fully conclusive about meniscus perception, and more generally about the mechanism of perception of immobile objects in whirligig beetles. Cylinder seizing (i.e., unfolding of forelegs before contact) and the size of the loop around cylinders were the only two characteristics that showed systematic variations as a function of cylinder diameter, wettability, and beetle speed. We discuss these aspects in turn. The main limitation of our study is that the insect's behavior was affected in an unexpected way: the amount of cylinder seizing (i.e., when forelegs were unfolded) decreased as the diameter increased and became close to zero for diameters of 6 and 19 mm. Thus, the insects did not show the same interest in cylinders of different sizes. The seizing behavior, very similar to the behavior involved in prey capture (Kolmes 1983), seems to indicate that small cylinders (∅ 0.2 and 2.5 mm) are perceived as prey, in contrast to big cylinders (∅ 6 and 19 mm). Interestingly, Wilde (1941) encountered the same problem in a different way, when he investigated the "avoiding-power" in whirligig beetles. He concluded that whirligig beetles are able to avoid relatively large objects, but not small objects, such as *Lemna* leaves or dead beetles. We can thus hypothesize that whirligig beetles are attracted by objects on the water surface and identify them as potential prey if their size is smaller than a given threshold ($\leq$2.5 mm) but not if they are too large ($\geq$6 mm). Differentiation between small and large objects could be done only at a short distance. Indeed, the beetles seemed to be still attracted, but they did not unfold their forelegs for capture for the largest cylinder used in this study (∅19 mm). The present study did not allow us to determine how the estimation of the cylinder diameter is carried out.

Trajectories around cylinders, and more specifically loops, were affected by the wettability of the cylinder. Loops were smaller and closer to the cylinder for hydrophobic cylinders than for hydrophilic cylinders. These results are consistent with a detection of the cylinder by the meniscus as it is always smaller around hydrophobic cylinders than around hydrophilic cylinders. If the meniscus is small, the cylinder is detected at a smaller distance and maneuvers are tight, generating small loops.

We found that the size of the loop was correlated linearly with the cylinder diameter. However, the difference in loop size among cylinders of varying diameter was relatively low compared to the difference among cylinders of varying wettability. By contrast, meniscus profiles showed that the effect of the diameter on the meniscus' absolute amplitude is more important than the effect of the surface wettability (Fig. 15.2). Thus, if the observed differences were due to the meniscus only, a larger difference should be found between cylinders of different sizes than between hydrophilic and hydrophobic cylinders. Moreover, we cannot exclude the possibility that the correlation between cylinder diameter and loop perimeter was simply due to the fact that the probability of random contact by beetles was higher for larger cylinders. Finally, close circles are not possible around large cylinders, increasing the mean value of the loop perimeter. Thus, loops are unfortunately poorly discriminating criteria for our purposes. Our interest in loops arose from the fact that the animal revisits the same site twice, indicating a potential interest in that region.

We found that the forelegs were unfolded at different distances from the cylinder for hydrophilic and hydrophobic surfaces. The two menisci (concave and convex) not being mirror images, we used the distance at which 0.2-mm hydrophilic cylinders, were producing the same amplitude of the surface deformation. This amplitude was then used to predict the distance at which the forelegs should be unfolded for hydrophobic cylinders of the same diameter. The results validated our prediction. On the contrary, the prediction for the 2.5-mm hydrophobic cylinder based on the 0.2-mm hydrophobic cylinder was wrong, and the observed amplitude of the meniscus at which forelegs are unfolded was one order of magnitude higher for the 2.5-mm cylinders. This could indicate that something else intervened, such as vision, for example, which we discuss below.

The distance at which forelegs are unfolded increases with increasing beetle speed. This result concords with perception of the meniscus as it is in agreement with the prediction made by Rudolph (1967). He predicted that, if whirligig beetles are able to detect menisci, they should detect them at a longer distance if they move faster. This assumes that beetles do not perceive the absolute surface deformation but the gradient of the deformation. Moreover, forelegs were never unfolded when whirligig beetles swam at high speed, even for the smallest cylinder diameter, suggesting that they are not able to detect cylinders at that speed. The bow wave formed just ahead of insects swimming at such speeds should complicate, if not prevent, the detection of menisci and the perception of surface waves. The fact that any contact with the cylinder looks than more like a crash (i.e., when beetles enter in contact with cylinder at high speed) seems to confirm this hypothesis. Rudolph (1967) also concluded that, considering detection of large obstacles using the meniscus, whirligig beetles could not detect these objects in time to avoid them when swimming too fast. This problem of perception could explain the typical intermittent locomotion behavior observed at high-speed swimming (Voise and Casas 2010) as it might allow perception of surface deformations (static or dynamic) during the deceleration phases.

Complementary experiments are needed. First, we need to make sure that vision is not used by these beetles. In fact, differences between hydrophobic and hydrophilic cylinders could lead to differences in UV reflection between paint and wax applied on cylinders. It is also possible that the meniscus produces a kind of optical lens, or some other optical effect, which could be perceived. Another experimental setup should be therefore used to eliminate vision altogether. If vision is not involved, we should then find better ways to discriminate between the use of the meniscus and reflected waves, the prediction being that the use of reflected waves enables an earlier detection of an object, hence at a larger distance. A quantitative model testing this hypothesis is, however, very difficult to make: the speed of the beetle, the type of leg kinematics, and the wave amplitude of incoming and reflected waves all matter. In terms of experimental setup, testing this model would furthermore imply taking measurements at very high speed over an extended period of time, a real conundrum. In any case, we need to define other behavioral criteria reflecting the distance of perception. Here, we used the distance of forelegs unfolding, but whirligig beetles could detect objects at longer distances and unfold their forelegs only at short distance.

Conflicting evidence was obtained for meniscus perception by whirligig beetles. It is sufficient to fully explain the seizing behavior of the smallest cylinder diameter, and nearly irrelevant for larger cylinders. In the first case, echolocation seems thus an unnecessary explanation, while it is inconclusive in the second instance. This study being the only one based on manipulative experiments designed on the basis of predictions; its inconclusiveness casts serious doubts on sweeping statements about the use of echolocation obtained from simple observational studies (Eggers 1926; Tucker1969; Wilde 1941; and following textbooks). As of today, the use of echolocation in whirligig beetles is nothing but an appealing conjecture.

## References

Bennett RR (1967) Spectral sensitivity studies on whirligig beetles, *Dineutes ciliatus*. J Insect Physiol 13:621–626, IN9, 627–633

Bleckmann H, Borchardt M, Horn P, Gorner P (1994) Stimulus discrimination and wave source localization in fishing spiders (*Dolomedes triton* and *D. okefinokensis*). J Comp Physiol A 174:305–316

Bleckmann H, Topp G (1981) Surface wave sensitivity of the lateral line organs of the topminnow *Aplocheilus lineatus*. Naturwissenschaften 68:624–625

Bott RH (1928) Beitrage zur kenntnis von *Gyrinus natator substriatus* Steph. Z Morphol Oekol Tiere 10:207–306

Carthy JD, Goodman LJ (1964) An electrophysiological investigation of the divided eye of *Gyrinus bicolor*. J Insect Physiol 10:431–436

Claas B, Munz H, Gorner P (1993) Reaction to surface waves by *Xenopus laevis* Daudin. Are sensory systems other than the lateral line involved? J Comp Physiol A 172:759–765

de Gennes PG, Brochard-Wyart F, Quéré D (2003) Capillary and wetting phenomena: drops, bubbles, pearls and waves. Springer, Berlin

Dean RG, Dalrymple RA (1991) Water wave mechanics for engineers and scientists. World Scientific, Singapore

Denny MW (1993) Air & water: the biology and physics of life's media. Princeton University Press, Princeton

Dusenbery DB (1992) Sensory ecology: how organisms acquire and respond to information. Freeman, New York

Eggers F (1926) The presumed function of Johnston's sensory organ in *Gyrinus*. Zool Anz 68:184–192

Eggers F (1927) Further information on the Johnston's sensory organ and avoidance ability of whirligig beetles. Zool Anz 71:136–156

Hill PSM (2008) Vibrational communication in animals. Harvard University Press, Cambridge

Hu DL, Bush JWM (2005) Meniscus-climbing insects. Nature 437:733–736

James DF (1974) The meniscus on the outside of a small circular cylinder. J Fluid Mech 63:657–664

Kolmes SA (1983) Ecological and sensory aspects of prey capture by the whirligig beetle *Dineutus discolor* (Coleoptera, Gyrinidae). J New York Entomol S 91:405–412

Lang HH (1980) Surface-wave discrimination between prey and nonprey by the back swimmer *Notonecta glauca* L (Hemiptera, Heteroptera). Behav Ecol Sociobiol 6:233–246

Le Merrer M, Clanet C, Quéré D, Raphaël E, Chevy F (2011) Wave drag on floating bodies. P Natl Acad Sci USA 108:15064–15068

Lighthill J (1978) Waves in fluids. Cambridge University Press, Cambridge

Murphey RK (1971) Sensory aspects of control of orientation to prey by the waterstrider, *Gerris reminis*. Z vergl Physiol 72:168–185

Rudolph P (1967) Zum Ortungsverfahren von *Gyrinus substriatus* Steph. Z vergl Physiol 56:341–375

Tucker VA (1969) Wave-making by whirligig beetles (Gyrinidae). Science 166:897–899

Vogel S (2003) Comparative biomechanics: life's physical world. Princeton University Press, Princeton

Voise J, Casas J (2010) The management of fluid and wave resistances by whirligig beetles. J Roy Soc Interface 7:343–352

Voise J, Schindler M, Casas J, Raphaël E (2011) Capillary-based static self-assembly in higher organisms. J Roy Soc Interface 8:1357–1366

Wiese K (1974) The mechanoreceptive system of prey localization in *Notonecta* II. The principle of prey localization. J Comp Physiol 92:317–325

Wilcox RS (1972) Communication by surface waves: mating behavior of a water strider (Gerridae). J Comp Physiol 80:255–266

Wilde JD (1941) Contribution to the physiology of the Johnston organ and its part in the behavior of the *Gyrinus*. Arch Neer de Physiol 25:381–400

# Chapter 16
# Sand-Borne Vibrations in Prey Detection and Orientation of Antlions

**Dušan Devetak**

**Abstract** Pit-building antlions capture their prey by digging funnel-shaped pits in loose sand and then laying in wait for prey to fall inside the trap. Behavioral experiments studying predator–prey interactions and measurements of vibrations propagated in sandy substrates revealed that antlions are extremely sensitive to substrate vibrations produced by prey crawling on the sand surface. Prey produce low-frequency sand-borne vibrations, and to locate a source of vibration, antlions rely on time differences of waveforms arriving at their receptors—tufts of hairs positioned on lateral parts of the mesothorax and metathorax. In this chapter, the role of physical properties of sand in substrate-borne vibration transmission is discussed.

## 16.1 Introduction

It is well known that some insect predators detect their prey according to vibrations produced by the prey during crawling on solid surfaces (Čokl and Virant-Doberlet 2003b; Cocroft and Rodríguez 2005). This chapter describes the role of vibrations in predator–prey interactions in antlion larvae.

Antlions (Myrmeleontidae) are holometabolous insects whose larvae are known to dig conical pitfall traps in sand or loose soil to catch prey at the bottom of the trap. However, only one-tenth of antlion species are pit-builders. The other sand-dwelling antlions lurk, buried in the substrate, without constructing pits. Often, only the jaws and antennae of the buried pit-building larva are visible. This sit-and-wait predator feeds on small arthropods that slide into the trap.

Well documented is the catching behavior of pit-building antlion species of the genera *Myrmeleon* and *Euroleon* (for a review see Griffiths 1980; Napolitano 1998; Scharf and Ovadia 2006; Gepp 2010; Scharf et al. 2011). The most intensively

D. Devetak (✉)
Department of Biology, Faculty of Natural Sciences and Mathematics, University of Maribor, Koroška 160, 2000 Maribor, Slovenia
e-mail: dusan.devetak@guest.arnes.si

R. B. Cocroft et al. (eds.), *Studying Vibrational Communication*,
Animal Signals and Communication 3, DOI: 10.1007/978-3-662-43607-3_16,

studied species is a common European species, *Euroleon nostras*. An antlion can wait motionless at the bottom of the conical pit for hours and then suddenly moves like a flash to capture its prey, usually an ant or another small soft-bodied arthropod. An antlion waiting for its prey never moves from the center of the pit, but by opening its mandibles, it shows us that it has detected the ant, crawling on the sand surface, at a distance of a few decimeters from the pit. The larva with violent jerks of the head and thorax tosses sand upon the prey in order to disorient it.

If the prey falls into the conical pitfall trap, things go very fast. The antlion immediately tries to grasp the ant sliding into the pit. If the larva does not succeed in capturing the ant during its first attempt, and/or if the prey evades the larva and starts to climb out of the pit, the larva tosses sand with violent movements of its head and the prothorax, thereby causing miniature landslides carrying the prey back to the antlion's jaws. The subsequent behavioral pattern is prey grasping. Should the ant prove difficult, the predator will start with prey beating (Napolitano 1998), i.e., it will often flick the ant back and forth against the sides of the pit. Prey beating is a behavior that tends to disorient the prey and thereby gives the predator valuable time to insert its mandibles. In so-called submergence, the larva holding the prey moves down and back into the substrate until the entire antlion and at least part of the prey are not visible. Then, poison is injected through the jaws. When the prey becomes motionless, emergence follows, i.e., the antlion with its prey moves up and forward until the entire prey and at least part of the antlion's head is visible. After the ant is dead, enzymes are introduced and the body contents are digested and extracted. Prey clearing follows, when the jaws are used to position the dead prey on the dorsal head surface of the larva and then the head is flicked rapidly back, expelling the empty carcass from the pit. Finally, to repair the shape of the pit, pit clearing occurs when surplus sand accumulated on the bottom of the pit is expelled by violent flicks of the head.

This complex predatory behavior of antlions is obviously based on a considerable amount of sensory information. A number of questions arose:

1. What signals are important in prey detection?
2. How are the vibrations transmitted from prey to predator?
3. What kinds of receptors are involved in the vibration detection?
4. Over what range can the antlion detect its prey?
5. Does the antlion detect the direction from which the stimulus is coming? Does the predator orient itself toward the prey?

## 16.2 Substrate Vibrations and Topography of the Receptors

Sensory receptors on the body surface of antlion larvae (Fig. 16.1) are involved in predator–prey interactions. The first study of the morphology and histology of antlion hair sensilla and eyes was done one hundred years ago when Doflein

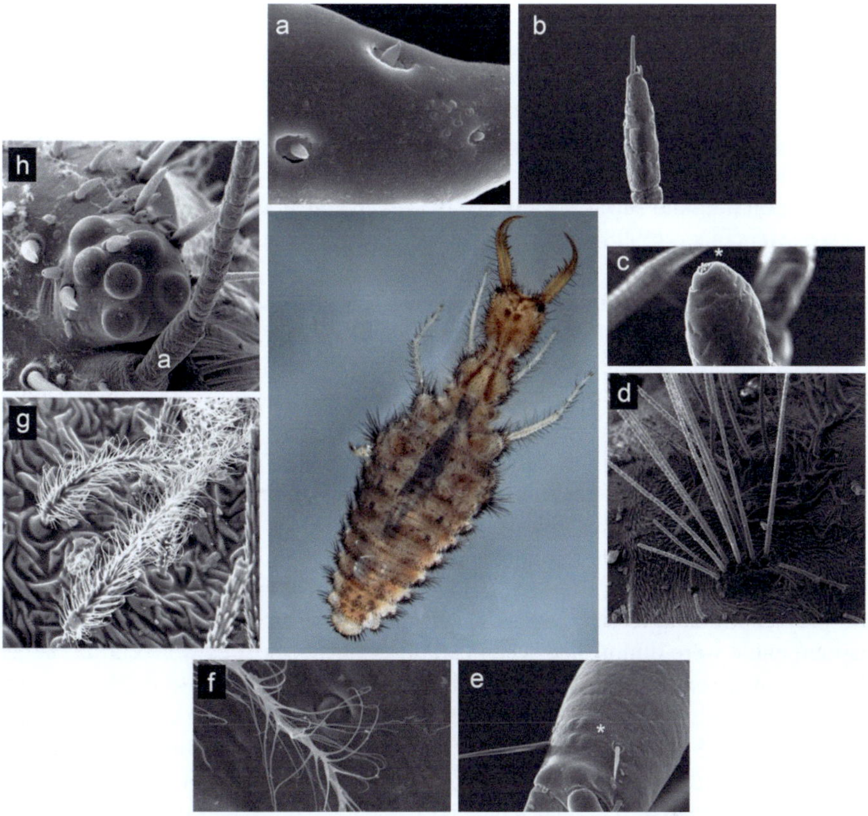


**Fig. 16.1** Sensory equipment of a pit-building antlion larva (*Euroleon nostras*). **a** Sensilla coeloconica on the mandibles. **b** Sensilla basiconica on the tip of the antenna. **c** Sensilla basiconica (*asterisk*) on the tip of the labial palp. **d** Tuft of the mechanoreceptive bristles (sensilla chaetica). **e** Two campaniform sensilla (*asterisk*) on the tarsus of hind leg. **f** Campaniform sensillum and plumose hair on head. **g** Plumose hairs on mesothorax. **h** Larval eyes (consisting of six stemmata in each eye) positioned on eye tubercle, close to the antenna (*a*). Electron micrographs courtesy of M. A. Pabst

(1916) published his fundamental book on the biology of antlion larvae. Existing information on the presence and topography of sensilla using scanning electron microscopy is available for only a few antlion species (Nicoli Aldini 2007; Eisenbeis and Wichard 1987; Lipovšek Delakorda et al. 2009; Cesaroni et al. 2010; Devetak et al. 2010a, b, 2013).

Chemoreceptors, involved in tasting prey, are sensilla coeloconica on the jaws and sensilla basiconica on the labial palps (Fig. 16.1). Any role of the larval eyes in predatory behavior has not yet been clarified sufficiently, although their structure is well known (Jockusch 1967). Antlions catch their prey even (or especially) during the night when their prey, small arthropods, is active. So vision does not play a major role. Campaniform sensilla (Fig. 16.1) serve to detect deformation of the cuticle, produced during movement of the legs, locomotion of the antlion and very probably also during direct contact with prey, for example, when the antlion holds prey or during prey beating.

The antlion detects its prey by sensing the vibrations that prey generate during crawling on the sand surface (Devetak 1985; Devetak et al. 2007; Fertin and Casas 2007). Mencinger (1998) demonstrated that the predator detects its prey even when vision is excluded. A larva with eyes covered with opaque paint still detects the source of vibration. Intact antlions react with sand tossing even if an observer with gentle movements of a twig or a pencil elicits substrate vibrations on the sand surface. Furthermore, antlion larva responded with violent sand tossing behavior to play back of vibrations recorded during locomotion of prey (Mencinger 2003; Fertin and Casas 2007).

Which receptors are the candidates for detecting substrate vibrations? Le Faucheux (1972) demonstrated that tufts of the mechanoreceptive bristles, sensilla chaetica, positioned on the thorax, play a certain role in detection of substrate vibrations. On the mesothorax and the metathorax, tufts of bristles occur in pairs, one tuft pair on each lateral side of the body segment (Fig. 16.1d). In intact larvae, the prey capture angle was 280°–290°. Both the ability to catch an ant and prey capture angle were diminished when certain groups of bristles were cut off. When the tufts of the sensilla were excluded unilaterally, the ability to catch prey was then limited only to the prey approaching the antlion from the contralateral side (Le Faucheux 1972). When one pair of the mesothoracic tufts was suppressed, the prey capture angle was reduced to 240°–280°. When two pairs of the mesothoracic tufts were eliminated, the angle changed to 200°–210°. When both mesothoracic pairs and one metathoracic pair of bristles were cut off, the prey capture angle was then reduced to 50°–60°. When all thoracic tufts were eliminated, larvae did not react to the presence of prey at all.

## 16.3 Vibrations Produced by the Prey

Prey animals—ants and similar small arthropods—produce low-frequency disturbances during walking or crawling on the sand surface. The frequency range of the vibrations produced during locomotion of four arthropod species (mealworm *Tenebrio molitor,* firebug *Pyrrhocoris apterus,* ant *Formica* sp. and woodlouse *Trachelipus rathkii*) is 0.1–4.5 kHz (Fig. 16.2), and acceleration values (peak level) of the vibrations, recorded at a distance 2–5 cm from the accelerometer, vary from 400 to 600 μm $s^{-2}$ (ant, firebug) to 1–1.5 mm $s^{-2}$ (mealworm, woodlouse) (Devetak et al. 2007) (Fig. 16.2).

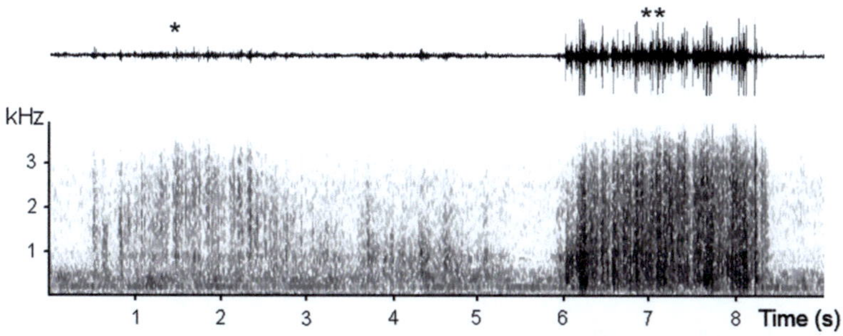


**Fig. 16.2** Oscillogram and sonogram of a firebug (*Pyrrhocoris apterus*) crawling on the sloping sand surface (*asterisk*) and falling on its back (*double asterisk*)

### *16.3.1 How are Vibrations Transmitted from Prey to Predator?*

The biological significance of substrate vibrations and mechanisms of signal transmission in plant-dwelling insects is well explored (Michelsen et al. 1982; Gogala 1985; Čokl and Virant-Doberlet 2003a, b). Vibration transmission in a sandy medium, however, has been studied only in desert scorpions (*Paruroctonus*), fiddler crabs (*Uca*), and antlions (*Euroleon*) (Brownell 1977; Brownell and Farley 1979a, b, c; Aicher and Tautz 1990; Devetak et al. 2007).

Transmission depends on a variety of factors, especially on frequency characteristics of the vibrations and physical properties of the sand. Natural sand occupied by antlions is a non-homogenous medium, containing small particles with different sizes and usually with larger stones or plant particles, like twigs and leaves, on the sand surface (Devetak 2000). The vibrations traveling from prey to predator are attenuated by twigs or small stones and reflected from solid objects (larger stones, rocks) (Fig. 16.3).

In the field, antlions are capable of discriminating between areas of sand differing in particle size, constructing pits in fine sand and avoiding coarser sand. This was confirmed in laboratory conditions when pit-building decisions and pit diameters depended on sand particle size (Botz et al. 2003; Devetak et al. 2005). When four fractions of sands differing according to particle size were offered to antlions, *Euroleon nostras*, the larvae preferred to build pits in the finer sand fraction, with a particle size of 0.23–0.54 mm (Devetak et al. 2005).

Sand strongly attenuates vibrations. The frequency spectra of vibrations of crawling prey differed when propagated in sands with different particle sizes (Devetak et al. 2007). The greater the sand particle size, the broader was the frequency range of the vibrations. The same was confirmed for artificial vibrations. Finer sand (particle size $\leq$0.23 mm) attenuated vibrations highly, and those recognized by antlion larvae traveled only a short distance. Five sands differing in particle size were tested. The damping coefficient ($\alpha_{10}$) of vibrations at a frequency

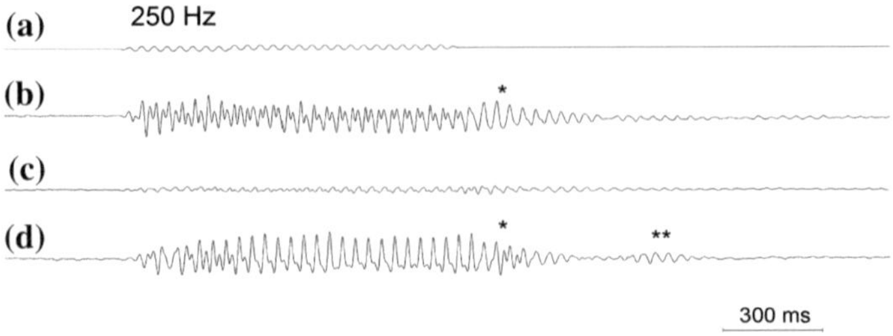


**Fig. 16.3** Reflections and attenuation of an artificial vibrational signal (250 Hz) produced by a Brüel & Kjaer 4810 vibration exciter and recorded with B and K 4381 accelerometer placed on the sand surface 20 cm apart; fine dry sand (particle size $\leq$0.5 mm) in a layer of 5-cm depth was placed in a plastic container. **a** Pure artificial signal with 250 Hz; **b** signal distortion by the reflection (*asterisk*) from the bottom of the container; **c** signal attenuation by an obstacle (a stone 5 × 4 × 0.5 cm) placed between the sender and the receiver; and **d** signal distortion by the reflection from the bottom of the container (*asterisk*) and by the reflection from a stone positioned 4 cm behind the receiver (*double asterisk*)

of 300 Hz for the finest sand fraction ($\leq$0.23 mm) was 2.61 dB cm$^{-1}$; for two median sand fractions (0.23–0.54 mm, 0.54–1 mm), it was 0.74 and 0.45 dB cm$^{-1}$, and for two coarser sand fractions (1–1.54 mm, 1.54–2.2 mm), it was 0.29 and 0.26 dB cm$^{-1}$ (Devetak et al. 2007). In natural habitats, antlions usually occupy fine sands or sands with median particle size, so in those substrates moderate damping occurs, and due to low propagation velocities, the sand enables prey localization.

### *16.3.2 Over What Range Can the Antlion Detect Its Prey?*

Antlions react to prey at maximal distances of a few decimeters (20–30 cm). Reaction distance is correlated positively with mean particle size (Devetak et al. 2007). In the finest sand fraction, the mean reaction distance was 3.3 cm; in two medium sand fractions, mean reaction distances were 5.5–9.1 cm, and in coarser sand, it was 12 cm.

## 16.4 Vibrations Produced by the Antlion

During pit construction, pit clearing and sand tossing toward the prey, jerking movements of the head and prothorax, both parts of the body serving as a shovel, are involved, thereby generating substrate vibrations (Fig. 16.4a). The head is moved left and right to collect sand, also with the help of foreleg movement, and then, the sand is tossed with a violent jerk of the head and prothorax in a dorsal direction. One vibrational "pulse" is composed of the head movements (left and

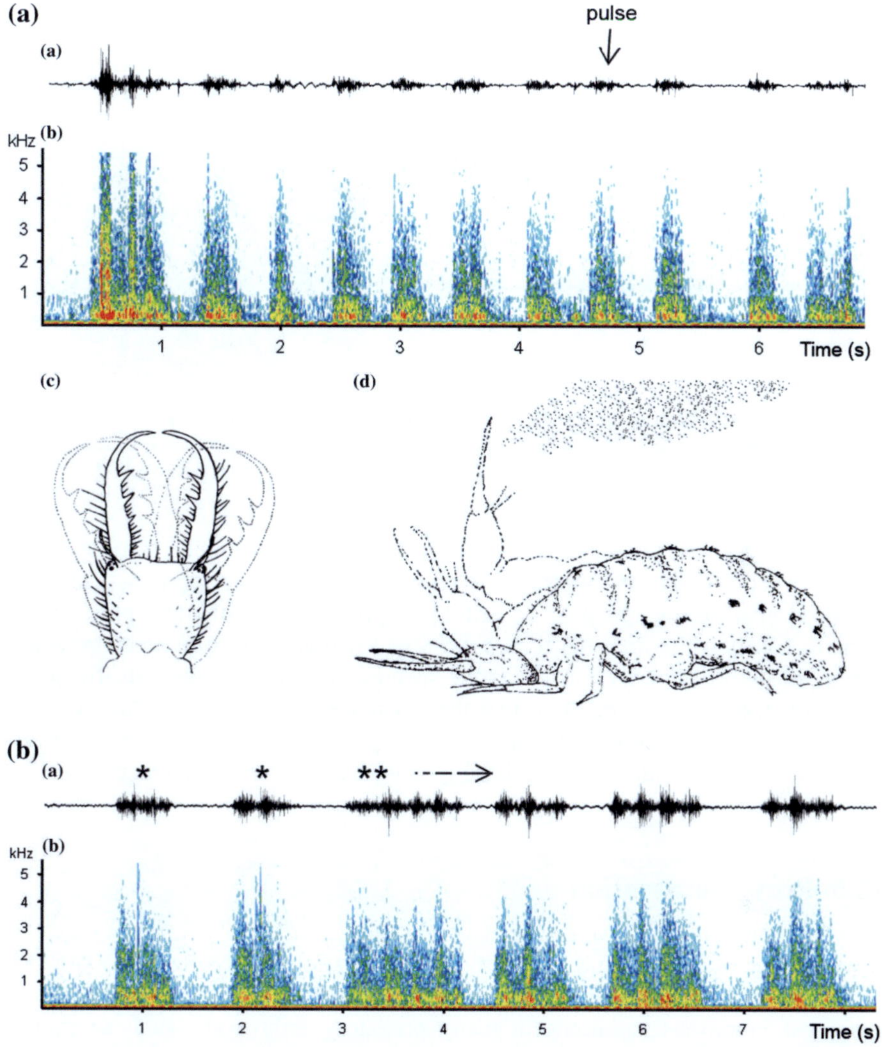


**Fig. 16.4** **a** Signals of sand tossing in *Euroleon*: *a* oscillogram and *b* sonogram of jerking the head *left* and *right* *c* and throwing the sand toward the prey (*d*). **b** Oscillogram (*a*) and sonogram (*b*) of the signals produced during grasping the prey (*asterisk*) and during rhythmic movements of the larva burying the prey in the sand (submergence) (*double asterisk*). For details see text

right) and the dorsally oriented jerk. Pulse duration is about 150–300 ms, and the pause between two consecutive pulses lasts for about 200–400 ms. Vibrations produced thereby are very probably important also as an alerting mechanism for conspecifics to maintain a certain distance from the nearest neighbor.

It is presumed that the abdominal campaniform sensilla have a role in the control of backward digging in sand. During digging, rhythmic jerking movements of the

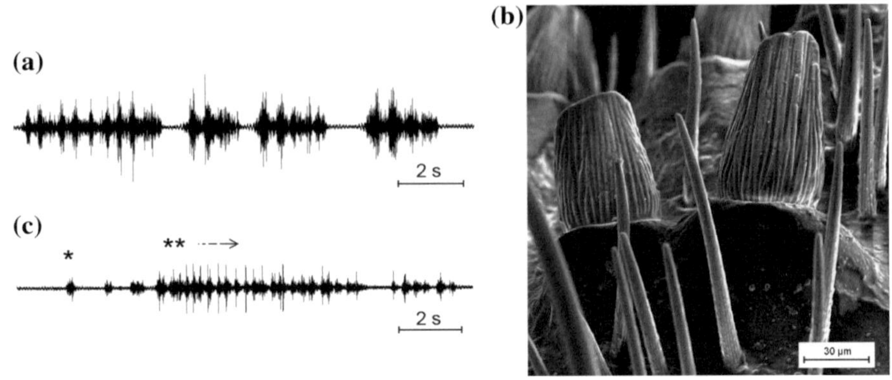


**Fig. 16.5** **a** Signals of digging into the substrate without prey consist of the pulses produced by three to nine jerks of the abdomen. **b** Digging bristles on the tip of the abdomen. **c** Signals of grasping an ant (*asterisk*) and prey beating (*double asterisk*). *Photo* B courtesy of M. A. Pabst

abdomen occur, and the digging bristles on the tip of the abdomen are involved in these jerks (Fig. 16.5a, b). Vibrations with similar frequencies (up to 4–5 kHz) are produced also during grasping an arthropod prey (Fig. 16.4b) and prey beating (Fig. 16.5c). It seems that the time pattern of a vibrational pulse series produced during predatory activity carries more specific information than its frequency spectrum structure. It is possible that these vibrations are important as signals in conspecific communication. As Barkae et al. (2010) supposed, sand tossing may play a role in disturbance of neighboring conspecifics or even heterospecifics.

## 16.5 Prey Localization

A wide range of insects can locate a source of vibrations (Cocroft et al. 2000; Čokl and Virant-Doberlet 2003b; Virant-Doberlet et al. 2006). The exploitation of prey-generated vibrations is known in many predatory arthropods, such as spiders, scorpions, stinkbugs, and parasitoid wasps (Brownell and Farley 1979a, b, c; Pfannenstiel et al. 1995; Casas et al. 1998; Barth 2002).

### *16.5.1 Does the Antlion Detect the Direction from Which the Stimulus is Coming?*

To answer this question, the accuracy of the sand-tossing behavior of *Euroleon* as a response to the presence of prey was measured using a video recording method (Mencinger 1998; Mencinger-Vračko and Devetak 2008). Sand tossing was elicited most frequently by prey behind the antlion; in contrast, there was no response

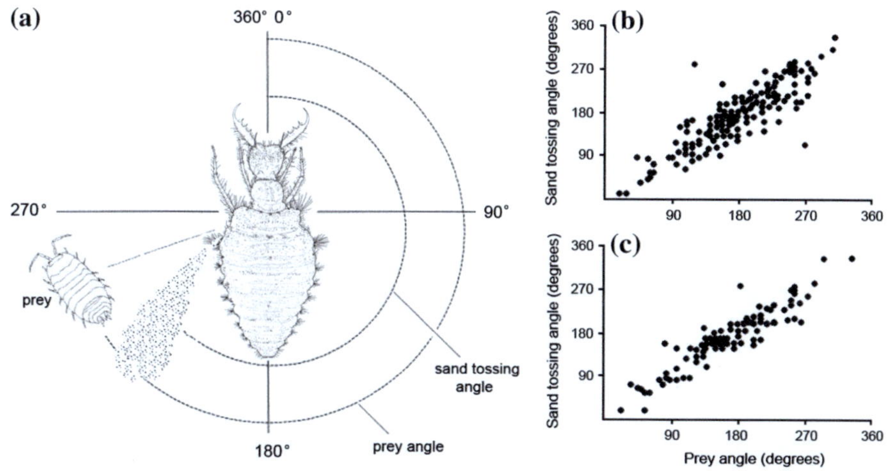


**Fig. 16.6** Sand-tossing angle as a function of prey angle. **a** The antlion is positioned in the center of the pit and its prey—the woodlouse *Trachelipus*—crawls outside the pit. The predator and the prey are not shown to scale. **b** and **c** Accuracy of sand-tossing response of antlions in the presence of the woodlouse: **b** unimpeded antlions; **c** antlions with eyes covered with opaque paint. Each dot represents a single response

when prey was in front of the antlion, in the so-called dead angle zone. The sand-tossing angle was highly positively correlated with the prey angle. The response was unaffected even when vision was excluded. When the antlion's eyes were covered, the sand-tossing angle was still highly positively correlated with the prey angle, in response to mealworm beetles (Mencinger-Vračko and Devetak 2008) and woodlice (Fig. 16.6).

Propagation velocities of surface vibrations (R-waves) in dry loose sand with particles $\leq$0.5 mm amount to 25–35 m/s and depend on the frequency (Mencinger-Vračko and Devetak 2008). Due to the low propagation velocities, the time and phase differences of the vibrations at the receptors—tufts of hairs on both lateral sides of the mesothorax and metathorax—may be expected to determine the prey angle. Time differences between vibrations originating in the lateral side of the sand were in the range of 0.2–0.5 ms.

## 16.6 Vibrations and Associative Learning in the Antlion

The learning ability of antlions in a context of detecting substrate vibrations has been proven recently (Guillette et al. 2009; Guillette and Hollis 2010; Hollis et al. 2011). In experimental conditions, individual antlions (*Myrmeleon crudelis*) received, once per day, either a vibrational cue presented immediately before feeding (the learning group) or the same cue presented independently of feeding (the control group). Vibrations simulating prey arrival not only produced an anticipatory learned response but also conferred a fitness advantage: Associative learning enabled antlions to extract food more efficiently, construct larger pits, and decrease the amount of time spent in the larval stage. The finding is important because the antlions do not fit the "learning profile" of active approach and avoidance behavior, and thus, they are unlike all other insect species studied to date (Guilette and Hollis 2010).

**Acknowledgments** I thank Professor A. Čokl for helpful suggestions on an early version of the manuscript. This project was supported by the Slovene Ministry of Higher Education, Science and Technology within the Biodiversity Research Programme (Grant No. P1-0078).

## References

Aicher B, Tautz J (1990) Vibrational communication in the fiddler crab, *Uca pugilator*. I. Signal transmission through the substratum. J Comp Physiol A 166:345–353

Barkae ED, Scharf I, Subach A, Ovadia O (2010) The involvement of sand disturbance, cannibalism and intra-guild predation in competitive interactions among pit-building antlion larvae. Zoology 113:308–315

Barth FG (2002) A spider's world: senses and behavior. Springer, New York

Botz JT, Loudon C, Barger JB, Olafsen JS, Steeples DW (2003) Effects of slope and particle size on ant locomotion: implications for choice of substrate by antlions. J Kans Entomol Soc 76:426–435

Brownell P, Farley RD (1979a) Detection of vibrations in sand by tarsal sense organs of the nocturnal scorpion, *Paruroctonus mesaensis*. J Comp Physiol 131:23–30

Brownell P, Farley RD (1979b) Orientation to vibrations in sand by the nocturnal scorpion *Paruroctonus mesaensis*: mechanism of target localization. J Comp Physiol 131:31–38

Brownell P, Farley RD (1979c) Prey-localizing behaviour of the nocturnal desert scorpion, *Paruroctonus mesaensis*: orientation to substrate vibrations. Animal Behav 27:185–193

Brownell PH (1977) Compressional and surface waves in sand: used by desert scorpions to locate prey. Science 197:479–482

Casas J, Bacher S, Tautz J, Meyhofer R, Pierre D (1998) Leaf vibrations and air movements in a leafminer—parasitoid system. Biol Control 11:147–153

Cesaroni C, Nicoli Aldini R, Pantaleoni RA (2010) The larvae of *Gymnocnemia variegata* (Schneider, 1845) and *Megistopus flavicornis* (Rossi, 1790) (Neuroptera: Myrmeleontidae): a comparative description. In: Devetak D, Lipovšek S, Arnett AE (eds) Proceedings of the tenth international symposium on neuropterology, Piran, Slovenia, 2008. FNM, Maribor, pp 135–144

Cocroft RB, Rodríguez RL (2005) The behavioral ecology of insect vibrational communication. Bioscience 55:323–334

Cocroft RB, Tieu T, Hoy RR, Miles R (2000) Mechanical directionality in the response to substrate vibration in a treehopper. J Comp Physiol A 186:695–705

Čokl A, Virant-Doberlet M (2003a) Communication with substrate-borne signals in small plant-dwelling insects. Ann Rev Entomol 48:29–50

Čokl A, Virant-Doberlet M (2003b) Vibrational communication. In: Resh VH, Carde RT (eds) Encyclopedia of insects. Academic Press, San Diego, pp 1167–1171

Devetak D (1985) Detection of substrate vibrations in the antlion larva, *Myrmeleon formicarius* (Neuroptera: Myrmeleonidae). Biol Vestn 33(2):11–22

Devetak D (2000) Competition in larvae of two European ant-lion species (Neuroptera: Myrmeleontidae). J Neuropterol 3:51–60

Devetak D, Špernjak A, Janžekovič F (2005) Substrate particle size affects pit building decision and pit size in the antlion larvae *Euroleon nostras* (Neuroptera: Myrmeleontidae). Physiol Entomol 30:158–163
Devetak D, Mencinger-Vračko B, Devetak M, Marhl M, Špernjak A (2007) Sand as a medium for transmission of vibratory signals of prey in antlions *Euroleon nostras* (Neuroptera: Myrmeleontidae). Physiol Entomol 32:268–274
Devetak D, Lipovšek S, Pabst MA (2010a) Larval morphology of the antlion *Neuroleon microstenus* (McLachlan, 1898) (Neuroptera, Myrmeleontidae), with notes on larval biology. Zootaxa 2428:55–63
Devetak D, Lipovšek S, Pabst MA (2010b) Morphology and biology of the antlion *Myrmeleon yemenicus* Hölzel, 2002 (Neuroptera, Myrmeleontidae). Zootaxa 2531:48–56
Devetak D, Klokočovnik V, Lipovšek S, Bock E, Leitinger G (2013) Larval morphology of the antlion *Myrmecalurus trigrammus* (Pallas, 1771) (Neuroptera, Myrmeleontidae), with notes on larval biology. Zootaxa 3641:491–500
Doflein F (1916) Der Ameisenlöwe. Eine biologische, tierpsychologische und reflexbiologische Untersuchung. Fischer, Jena
Eisenbeis G, Wichard R (1987) Ant lions (Myrmeleonidae). In: Eisenbeis G, Wichard (eds) Atlas on the biology of soil arthropods. Springer, Berlin, pp 278–283
Fertin A, Casas J (2007) Orientation towards prey in antlions: efficient use of wave propagation in sand. J Exp Biol 210:3337–3343
Gepp J (2010) Ameisenlöwen und Ameisenjungfern. Myrmeleontidae, Westarp Wissenschaften Hohenwarsleben
Gogala M (1985) Vibrational communication in insects (biophysical and behavioural aspects). In: Kalmring K, Elsner N (eds) Acoustic and vibrational communication in insects. Parey, Hamburg Berlin, pp 117–126
Griffiths D (1980) The feeding biology of ant-lion larvae: prey capture, handling and utilization. J Anim Ecol 49:99–125
Guillette LM, Hollis KL (2010) Learning in insects, with special emphasis on pit-digging larval antlions (Neuroptera: Myrmeleontidae). In: Devetak D, Lipovšek S, Arnett AE (eds) Proceedings of the Tenth International Symposium on Neuropterology, Piran, Slovenia, 2008. FNM, Maribor, pp 159–170
Guillette LM, Hollis KL, Markarian A (2009) Learning in a sedentary insect predator: antlions (Neuroptera: Myrmeleontidae) anticipate a long wait. Behav Processes 80:224–232
Hollis KL, Cogswell H, Snyder K, Guillette LM, Nowbahari E (2011) Specialized learning in antlions (Neuroptera: Myrmeleontidae), pit-digging predators, shortens vulnerable larval stage. PLoS One 6:1–7
Jockusch B (1967) Bau und Funktion eines larvalen Insektenauges. Untersuchungen am Ameisenlöwen (*Euroleon nostras* Fourcroy, Planip., Myrmel.). Z vergl Physiol 56:171–198
Le Faucheux M (1972) Le role des soies thoraciques dans la capture des proies par la larve d'*Euroleon nostras* Fourcroy (Névroptère). Rev Comp Animal 6:217–221
Lipovšek Delakorda S, Pabst M-A, Devetak D (2009) Morphology of the eyes and sensilla in the antlion larvae (Neuroptera: Myrmeleontidae). In: Pabst MA, Zellnig G (eds) MC 2009—microscopy conference, Graz, Austria, vol 2, Life Sciences. Verlag der Technischen Universität Graz, pp 403–404
Mencinger B (1998) Prey recognition in larvae of the antlion *Euroleon nostras* (Neuroptera: Myrmeleontidae). Acta Zool Fenn 209:157–161
Mencinger Vračko B (2003) Substrate vibrations as important stimulus for the prey recognition by the antlion larva Euroleon nostras (Geoffroy in Fourcroy, 1785). University of Ljubljana, Biotechnical Faculty, Ljubljana. M.Sc. Thesis. /in Slovene with English summary/
Mencinger-Vračko B, Devetak D (2008) Orientation of the pit-building antlion larva *Euroleon* (Neuroptera, Myrmeleontidae) to the direction of substrate vibrations caused by prey. Zoology 111:2–8
Michelsen A, Fink F, Gogala M, Traue D (1982) Plants as transmission channels for insect vibrational songs. Behav Ecol Sociobiol 11:269–281

Napolitano JF (1998) Predatory behavior of a pit-making antlion, *Myrmeleon mobilis* (Neuroptera, Myrmeleontidae). Fla Entomol 81:562–566
Nicoli Aldini R (2007) Observations on the larval morphology of the antlion *Myrmeleon bore* (Tjeder, 1941) (Neuroptera Myrmeleontidae) and its life cycle in the Po Valley (northern Italy). Ann Mus civ Storia nat Ferrara 8:59–66
Pfannenstiel RS, Hunt RE, Yeargan KV (1995) Orientation of a hemipteran predator to vibrations produced by feeding caterpillars. J Insect Behav 8:1–9
Scharf I, Ovadia O (2006) Factors influencing site abandonment and site selection in a sit-and-wait predator: a review of pit-building antlion larvae. J Insect Behav 19:197–218
Scharf I, Lubin Y, Ovadia O (2011) Foraging decisions and behavioural flexibility in trap-building predators: a review. Biol Rev 86:626–639
Virant-Doberlet M, Čokl A, Zorović M (2006) Use of substrate vibrations for orientation: from behaviour to physiology. In: Drosopoulos S, Claridge MF (eds) Insect sounds and communication: physiology, behaviour, ecology, and evolution. Taylor and Francis, New York, pp 81–97

# Part V
# Biology and Evolution of Vibrational Communication in Some Well-Studied Taxa

# Chapter 17
# Mechanical Signals in Honeybee Communication

**Axel Michelsen**

**Abstract** Honeybees make use of several kinds of mechanical signals in their communication. Touching other bees is very common in connection with food exchange. Vibrations transmitted through the combs or during physical contact between bees play a role in communication between workers and also between queens and workers, but their function is debated. This is also the case for the vibrations associated with various vibratory dances. Bees cannot hear the pressure fluctuations of sounds, but they are sensitive to airflows associated with sounds. Most of these airflows are bidirectional and caused by vibrating wings or bodies, but unidirectional jet airflows are produced by dancing bees. By mimicking the movements and airflows with a robot, it is possible to tell the bees where to fly.

## 17.1 Communication by Touch and Vibrations

Honeybees (*Apis mellifera*) are social insects that live in colonies with up to about 70,000 individuals, who carry out a large number of jobs. The jobs include such diverse activities as collection of food, defense of the colony, building the nest, feeding larvae, cleaning cells, and keeping a constant temperature in the nest. Obviously, much communication among the individuals is needed in order to coordinate the work that needs to be done, but each bee decides itself what to do and for how long. However, the probability that an individual bee does a particular job depends on its age and genetic disposition. In-nest jobs are performed by young individuals, whereas outside tasks (like the collection of food) are performed by older bees. Vision is probably poor or impossible inside most of the nest, and the bees have to rely on chemical and mechanical cues and signals.

---

A. Michelsen (✉)
Department of Biology, University of Southern Denmark, 5230 Odense M, Denmark
e-mail: a.michelsen@biology.sdu.dk

R. B. Cocroft et al. (eds.), *Studying Vibrational Communication*,
Animal Signals and Communication 3, DOI: 10.1007/978-3-662-43607-3_17,

**Fig. 17.1** Two bees (*left*) begging food from another bee (*center*) (*Photo* Axel Michelsen)

By observing the behavior of bees inside the hive in dark-red light, which is invisible to the bees, one finds that most bees pass each other without touching. Somehow the bees must be able to sense the presence of other bees and keep a short distance from them. Many cases of communication involve nest mates touching each other. An exchange of food (Fig. 17.1), where the bees beg by touching other bees with their antennae, is probably the most common type of communication in the bee hive. An exchange of food occurs, not only between bees with food and bees lacking food, but very often also when bees with sufficient food meet each other. During the wagging dances, bees following the dance may emit 320-Hz begging vibrations (also known as stop signals) as a request to the dancer to stop and offer food samples. Artificial vibration signals between 300 Hz and 3 kHz and with a velocity amplitude of at least 1 mm/s also cause bees to freeze (Michelsen et al. 1986a). The dorsoventral abdominal vibration (DAV) signal (also known as the shaking signal) is very common. Here, a worker swings its abdomen up and down, frequently while grasping another worker or a queen. The occurrence of the DAVs is temporally correlated with swarming and foraging activities.

New colonies are established by swarming, which may occur when the hive is too small to hold the adult bees (Huber 1792). The process begins with the construction of queen cells, in which the queen deposits eggs. The queen receives an increasing number of DAVs until the queen cells are sealed and then a diminishing number until she leaves the nest with about half of the bees (Fletcher 1975). The swarm settles not far from the nest, and the shaking workers now perform DAVs on some of the workers, while other workers engage in house hunting. Later, swarms may occur when the young queens emerge. Piping sounds may be heard by human listeners 7–11 days after the first swarm, and this closely precedes the emergence of another swarm (Butler 1609).

The piping is composed of two signals: a tooting signal produced by a queen who has emerged from her cell and walks around and quacking signals made by queens who are confined to their cells by worker bees. Both signals can be heard by us and

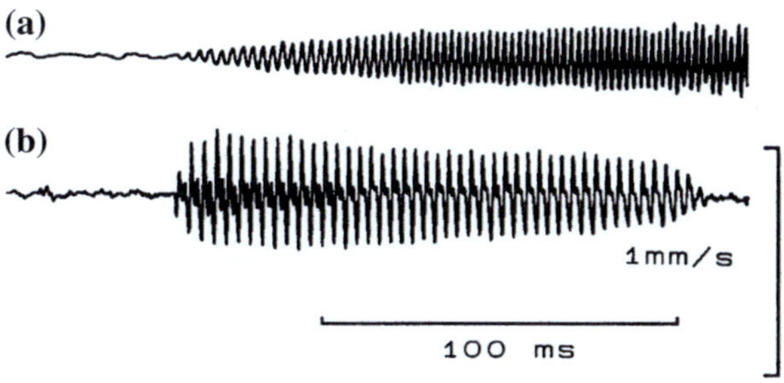


**Fig. 17.2** Fine structure of the tooting (**a**) and quacking (**b**) vibration signals. Note the frequency sweep in **a**, and the different rise times of the two signals (from Michelsen et al. 1986b)

measured as airborne sounds, but the signals propagate and are perceived by the bees as substrate-borne vibrations (Hansson 1945; Michelsen et al. 1986b). We used a laser vibrometer for measuring the vibrations (Fig. 17.2) and found the behavioral threshold of the bees to be about 1 mm/s (the same threshold as with the stop signals) when the tooting signals released quacking. We tried to find out how the bees distinguish toots from quacks and found that the difference in rise time (Fig. 17.2) is a likely cue. Artificial toots with long (100 ms) rise time are much better in releasing quacks in confined queens than are toots with short (10 ms) rise time. Although two of my students discovered how the vibration receptor organ (the subgenual organ) works (Kilpinen and Storm 1997, see also Chap. 6, this volume), almost nothing is known about the processing of the vibrational information in the central nervous system. It thus remains to be learned whether the bees perceive the differences between the frequency spectra of the two signals and/or the fine structures like the initial frequency sweep in the toots (Fig. 17.2a).

Although the piping behavior has been studied for four centuries, the exact messages carried by these signals are still not known. The occurrence of toots causes a delay in the emergence of confined queens, and the worker bees probably act as intermediaries by performing DAVs on the queen cells (Bruinsma et al. 1981). The workers may also be involved in the tooting, because the free queen has often received a DAV just before tooting. The amplitude of the piping signals decreases with distance, and the range of communication is only about 10 cm (Michelsen et al. 1986b). The free queen therefore spends much time walking briskly on the combs, making a stop, and tooting for each 5–10 cm.

Several vigorous vibratory "dances" have been described. Buzzing runs (*Schwirrlauf*) are made by workers during the last 2 h before a swarm leaves the hive (Lindauer 1955). A shaking dance made by a dirty bee may cause neighbor bees to groom the dancer. Other dances include round dances and waggle dances (discussed in the next section) as well as jostling, spasmodic, and trembling dances. We can distinguish these dances visually, but it is not known how the bees distinguish them in the darkness of the hive.

## 17.2 The Dance Language

Honeybees can recruit other bees to visit a food source by performing a dance, in which information on the distance and direction from the hive to the food is codified (Frisch 1967). The communication about remote events in the dance language is an exception to the general rule that animals tend to communicate about immediate events connected with the actor and its surrounds. Furthermore, in the dance language, a symbolic code is used to transmit an impressive amount of information. Most dances are performed on the vertical wax combs in the hive. In the waggle dance, the dancer moves in a straight line (the waggle run) and circles back, alternating between a left and a right return path so that the entire dance path takes on a figure-eight shape. During the waggle run, the dancer moves her body from side to side 13–15 times per second and emits 280 Hz sounds by vibrating her wings. The angle between the direction of the waggle run and the vertical indicates the angle which the recruits should later maintain between their flight path to the food and the sun's azimuth (the direction to the projection of the sun on the horizon). The distance to the food is indicated by the speed of the dancer.

A few bees (called follower bees) keep close contact with the dancer during most of the dance, and these bees may be recruited to search for the food at the position indicated in the dance. The dancer carries the odor of the flowers visited, and this information makes it easier for the followers to find the food. This is understandable since the dance does not contain any information about the shape or color of the flowers. The dance language is also used by the bees for indicating the position of resin, water, and suitable new nests (during swarming). The dance language is important for the collection of food when the food is scarce, but not when there is plenty of food (Seeley 2011).

Most of our work has been focused on two interconnected problems: How can the follower bees obtain the specific information about distance and direction from dances performed in the darkness of the hive? And which of the many components of the dance do the bees perceive as signals? The studies were done over a period of about 15 years in close cooperation between our research group in Odense and that of the late professor Martin Lindauer in Würzburg. Most of the laboratory work was done in Odense, whereas the field experiments were carried out near Würzburg (where the weather is more predictable than in Denmark).

Karl von Frisch suggested two possible strategies for how the dance followers detect the dancer's movements in the darkness of the hive: The dancer might generate vibrations in the wax comb that could be detected by the follower bees, or the followers might touch the dancer. We now know that dance-generated vibrations occur in the comb near the dancer and that most follower bees have contact with the dancer with one or both antennae during each waggle run (Nieh and Tautz 2000; Rohrseitz and Tautz 1999). The vibrations seem to attract follower bees to attend the dances, but there is no evidence that the follower bees can obtain the specific information about direction or distance from these sources.

In 1987, we proposed a third possible mechanism. It was based on the observation of two three-dimensional near fields of intense oscillating airflows around the dancer, caused by the wagging motion and wing vibration, respectively. This suggested that the follower bees might keep a small distance from the dancer and estimate her dance path by means of receptor organs sensitive to such airflows (Michelsen et al. 1987). The forces generated by the vibrating wings and causing the airflows could be measured by means of a pair of probe microphones. The probe tubes were about 1 mm in diameter, and the microphones were mounted next to each other such that the distance between the tips could be varied. With the tips 2 mm apart and one probe 1 mm above and the other 1 mm below the edge of a vibrating wing, the sound pressures were close to 100 dB (re. 20 μPa) and totally out of phase. The velocity of the airflow caused by the pressure difference was calculated to about 0.5 m/s. With this technique and various distances between the probe tips, we could map the airflows around the dancing bee. In agreement with the theoretical expectations for the wing as a dipole emitter, the airflows decreased with the third power of distance to a few percent at the distance of one bee length. Bees often dance a few bee lengths from each other, and the wing-generated airflows thus seemed well suited as carriers of "private" dance-specific information. Unfortunately, the wagging motion has a much lower Reynold's number than the vibrating wings and hardly generates sounds, so these airflows could not be studied in this manner.

On the basis of these findings, we proposed that the follower bees obtain information about the distance and direction to the target by measuring the airflows generated by the dancer. In order to test this hypothesis, we built a mechanical robot model of a dancing honeybee, which generated low- and high-frequency oscillating airflows similar to those generated by the body and wings of live dancers. The robot was made of brass, and the wings were simulated by a single piece of razor blade, which was vibrated by an electromagnet. A step motor rotated the model and caused it to waggle during the wagging runs. Two motors moved the model in a figure-eight path, and a fourth motor pumped "food samples" (sugar water) through a plastic tube that terminated near the "head." All motors were controlled by a computer. This made it possible to mimic natural waggle dances and to create dances that differed from the natural ones (Michelsen et al. 1992). The robot and sugar water were given a faint floral scent. The robot was deployed on the dance floor of an observation hive (Fig. 17.3). Faintly scented baits were placed at various locations in the field, and human observers noted the number of visits to the baits. The distribution of the visits to the baits varied with the specific information provided by the robot (Fig. 17.4), but the robot did not recruit so many bees to the target as live dancers.

Several parameters of the dance vary with the distance and direction to the food, so the information is redundant. Therefore, one cannot be sure which part of the dance is perceived by the follower bees. The robot allowed us to solve this problem. With the computer, we could make manipulated dances, which provided the bees with conflicting information about distance or direction (Fig. 17.5). For example, at short distances to the target, the dancers run fast during the entire dance, whereas

**Fig. 17.3** A mechanical robot model of a dancing honeybee photographed during its performance of a waggle dance among the bees on the dance floor of an opened observation hive. A number of bees are facing the robot and running along with it, and some of these bees later appear at the location indicated by the robot's dance (from Michelsen et al. 1992)

they run more slowly when they signal longer distances. With the computer, we could program the robot dancer to run slowly during the wagging run and fast during the return run, and vice versa. The experiments showed that in both cases the bees followed the instructions given by the wagging run and ignored the information given by the return run. This was also the result of experiments, in which the dance figure was changed so that the wagging run and the entire dance path pointed at opposite directions (Fig. 17.5c). The wagging run is thus the "master parameter" for the transfer of information about both direction and distance.

The robot experiments received much attention, also from the general public. Some colleagues also tried to build improved versions of robot dancers, but so far without success. As so often in science, our approach has been criticized. For example, Harald Esch recently reported that he has built a number of robots over the years, but—like with our robot—their recruitment success was consistently smaller than that of real dancers (Esch 2011). A probable explanation for the small number of recruits in our robot experiments is that the robot did not touch the wax comb and therefore did not generate vibrations in the comb. Dancers on combs with empty cells recruit three times as many nest mates to feeding sites as those that dance on a much heavier comb with capped brood cells (Tautz 1996), and vibrations spread much better in light combs (Sandeman et al. 1996). Esch argued further that von Frisch had insisted on not using odors on robots or feeding sites, because odors alone can attract bees to feeders. On the other hand, information about both location and odors are needed when dance followers try to find the new food source (Seeley 2011). In the two experiments in Fig. 17.4, the same small amount of odor was present at all 8 observation positions, and only the positions indicated by the robot dancer had significantly more visits than the other 7 positions. Randolf Menzel and his staff have recently tried to build an improved version of a robot dancer, but unfortunately, its activity only seemed to increase the motivation of the bees to search for food (Menzel 2011). We hope that their robot will work soon, because this research group has the equipment for following the flight path of single bees with radar.

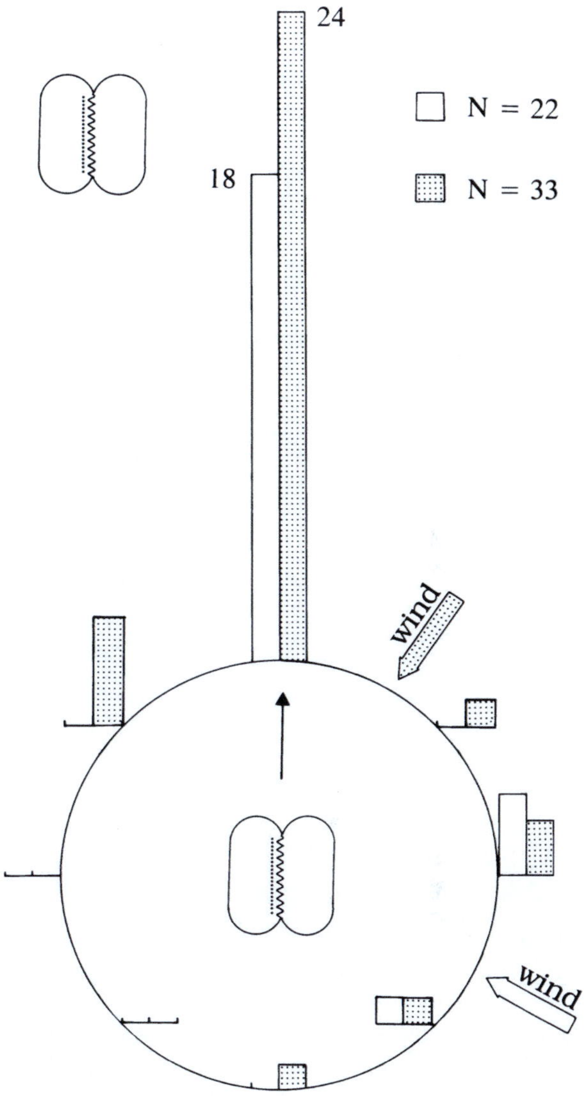


**Fig. 17.4** Two experiments (*open* and *dotted bars*, respectively) with normal wagging dances testing the transfer of directional information. The direction indicated by the robot is indicated with an *arrow*. The number of bees observed at each of eight directions is indicated. The direction to the target differed by 90° in the two experiments (the direction of the wind was constant) (From Michelsen et al. 1992)

## 17.3 Revision of a Hypothesis

Simultaneously with the experiments with the robot, the behavioral thresholds for the perception of oscillatory airflows had been determined (Kirchner et al. 1991). The results seemed to fit the amplitudes of the airflows generated by the live dancers, and for a few years, we had the impression that the problem had been solved (see e.g., Kirchner and Towne 1994). However, it gradually became apparent that there had been flaws in the methods used by Kirchner et al. for determining the behavioral threshold. They had connected a tube to a loudspeaker

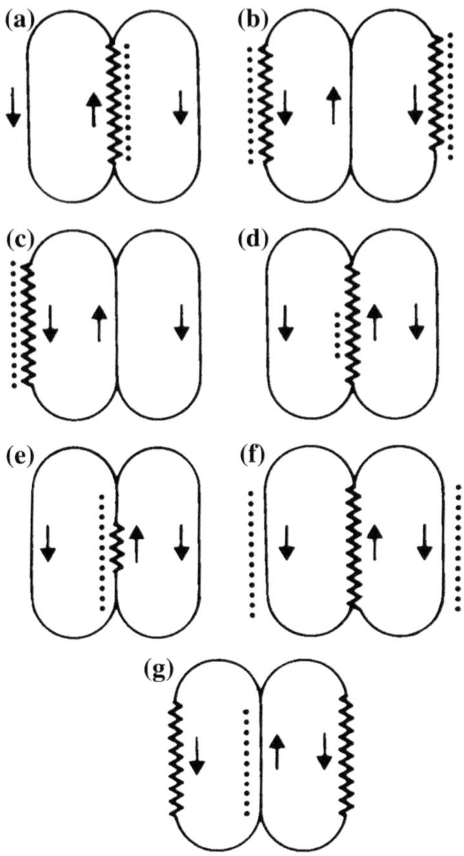


**Fig. 17.5** Seven dance patterns tested in experiments with the robot. Wagging and sound emission by the vibrating wing are indicated by a *zigzag line* and a series of *dots*, respectively, **a** is the normal dance, **b** was not understood. Dance patterns **c**–**g** were understood, but the scatter of the data was larger than with **a** (From Michelsen et al. 1992)

and assumed the oscillating airflows at some distance from the open end to be laminar and homogeneous. In fact, two ring-shaped vortices with opposite directions of rotation are formed during each cycle when air is forced to flow in and out of a circular tube, and the relationship between the voltage to the loudspeaker and the air velocity can be highly nonlinear (Nitsche and Krasny 1994). A quantitative disagreement thus seemed to exist between the reported behavioral threshold and the amplitudes of the oscillating airflows that might be used by the follower bees for perceiving the specific information from the dances (Michelsen 1999). This disappointing conclusion was strengthened when we began to visualize the airflows of model dancers by means of particle image velocimetry (PIV).

PIV exploits a standard trick in fluid mechanics: A laser beam is transformed to a thin sheet of light by passing through a cylindrical lens, and the sheet (thickness 0.6 mm) is made visible by smoke and photographed by means of a camera mounted normal to the sheet. Airflows in the plane of the sheet can be detected as different positions of the smoke particles on two photographs taken with a very short interval using a digital camera with about a million light-sensitive cells and a

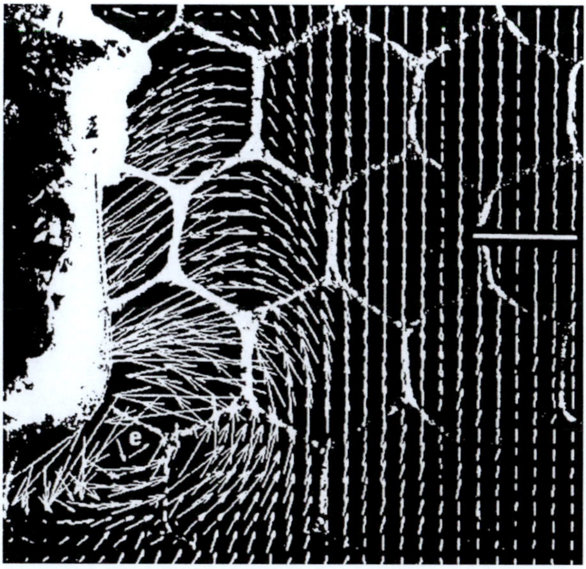


**Fig. 17.6** Airflow caused by simulated wagging of a stuffed bee. The body is moving like a pendulum and passing the vertical position on its way toward left. Air is flowing from the right to fill up the space left by the body. Simultaneously, displaced air from the left side of the bee collides with air from the right, thus creating an eddy (e). The airflow is visualized by means of the PIV technique, and the air velocities are indicated as vectors (*arrows*). The *bar* to the right corresponds to 100 mm/s

similar number of storage cells. The processing of the data involves cross-correlation of the positions of smoke particles within about 1,000 "interrogation areas" and results in an average particle displacement vector for each area.

Obviously, PIV is a powerful method for studying flows around physical models of animals in a laboratory, but it cannot be used on live bees in a beehive. In the observation hive, we therefore used hot-wire anemometry. The principle of this method is that a very hot, thin wire is part of a balanced electrical bridge, which sends current through the wire when it is cooled by an airflow. Normally, the temperature of the wire is maintained at some 100 °C. The output is then a linear function of the airflow, but rather noisy. The presence of such a hot wire would not be tolerated by live bees. However, the bees accepted the presence of a few mm of wire operated at about 50 °C, where there was much less noise. We had to accept that the output voltage was no longer a linear function of the airflow velocity.

High-speed films had shown that the wagging movement of dancing bees is close to that of a pendulum with the rotation point located 1 mm in front of the head. PIV measurements of dead bees stuffed with plastic and moved by a machine with the amplitude and speed of normal wagging showed that the airflows were much more complicated than we had imagined (Fig. 17.6). The main reason is that moving a body with a certain volume displaces air in the direction of the movement and leaves an empty space at its previous position. The movement is

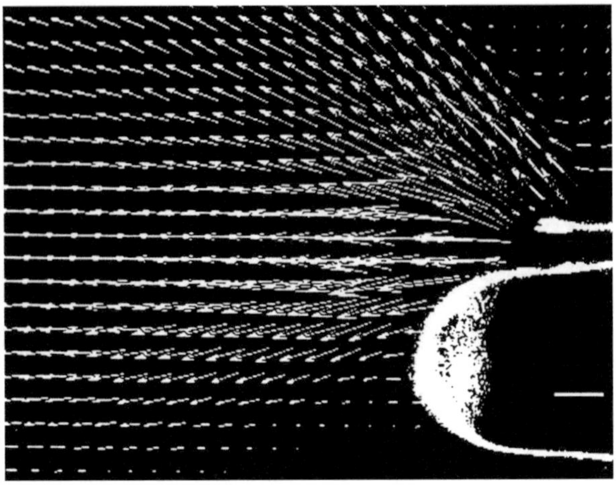

**Fig. 17.7** A jet airflow caused by a 250-Hz wing vibration in a metal model with only one wing. The tip of the abdomen and tip of the wing are photographed when the wing was moving up. The bar to the lower right corresponds to 100 mm/s. The largest vectors correspond to about 300 mm/s. Perpendicular to the plane shown here, the width of the jet is about 1 mm when it leaves the tip of the wing, but the jet recruits air and becomes broader when it travels away from its origin

therefore not only pushing air, but also sucking air to fill up the empty space. A large part of the air arriving to fill the empty space comes from the other side of the animal and flows over the back of the animal and around the head and the tip of the abdomen. These airflows tend to collide, creating eddies (e in Fig. 17.6). Such airflows are therefore not suited for providing the follower bees with precise information about the movements of the dancer. A similar conclusion was reached for the oscillating airflows generated by the wings.

The PIV studies thus falsified the hypothesis that had guided our work since 1987, but they also led to a discovery and a new hypothesis. The oscillation of the wings cause not only oscillatory airflows, but also a jet airflow, which is generated at the tip of the wings and move backward. The jet is fan shaped and much broader in the dorsoventral direction (Fig. 17.7) than in the lateral directions. In contrast to the oscillatory airflows, where the masses of air are flowing to and from, the air flows in only one direction in the jet (away from the dancer). Measurements with the small hot-wire anemometers showed that the passing of the tip of the abdomen of the wagging dancer is associated with a distinct maximum of airflow (Fig. 17.8a).

How do the follower bees determine their position behind the dancer and thus the direction to the target announced in the dance? A single follower bee located behind the dancer may be able to position herself at the middle of the dancer's wagging, but two or three follower bees cannot all be at the middle. One possibility is that follower bees take advantage of the dancer's very regular rhythm of wagging and determine their angular position relative to the midline from a time code. As illustrated in Fig. 17.9, the temporal pattern of the maxima of airflows experienced by the follower bees behind the dancer depends on the position of the follower bee

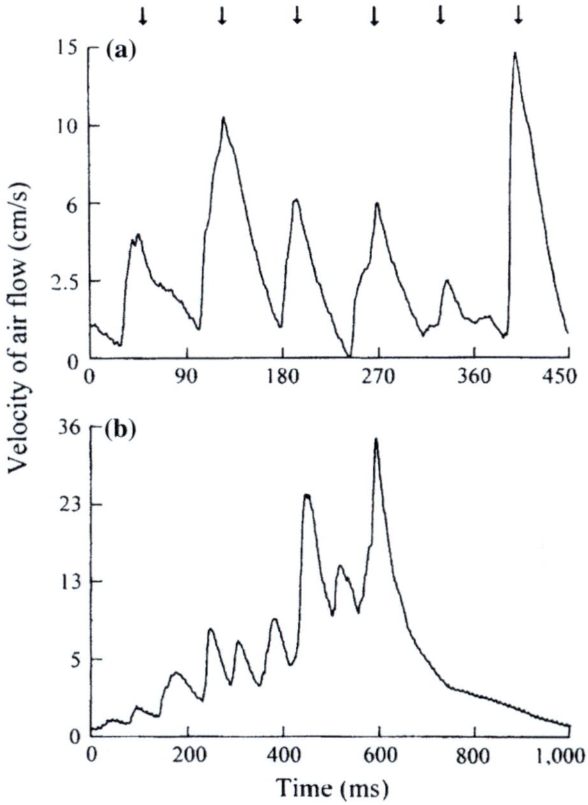


**Fig. 17.8** Airflows measured with a hot-wire anemometer behind dancing foragers. **a** The anemometer is at an extreme lateral position within the angle of wagging behind the dancer. The *arrows* indicate the time when the dancer's abdomen pointed toward the hot wire. **b** Airflows measured with an anemometer behind a dancer, which produces both narrow jets and a broad airflow

relative to the direction of the waggle run. One may therefore speculate that the temporal pattern is used as a cue for calculating the direction signaled by the dancer. That the position behind the dancer is well suited for determining the direction announced in the dance is supported by studies in which the positions of the follower bees around the dancers were correlated with their later success in finding the food. The specific information about the direction to the food is available only (Judd 1995) or mainly (Tanner and Visscher 2009) to follower bees, which for some time have been located behind the dancer within the angle of the wagging.

High-speed video (at least 250 frames/s) is needed in order to avoid blurred pictures of the wagging motion and the mechanisms responsible for the generation of the jet airflows. During the recordings, one should be able to measure sound from the vibrating wings and local airflows. Since the directions of the dances vary, the instruments should be fixed on a microscope head, which can easily be rotated. Our solution (Fig. 17.10) includes a prism, which allows us to observe and film the dancer and one of her follower bees, simultaneously, both from above and from the side. This allows us to observe how fast the follower bees move their antennae when the body of the wagging dancer approaches (with less time resolution, one gets the impression of a collision between the antenna and the body of

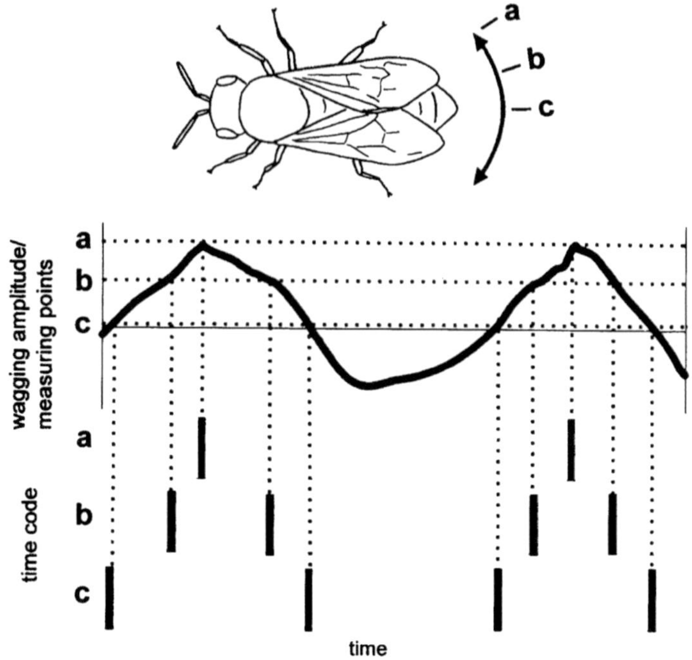


**Fig. 17.9** Predicted time patterns for reception of a jet emitted by a wagging dancer by receivers at positions *a*, *b*, and *c*. The time patterns have been observed in recordings with hot-wire anemometers behind live dancers

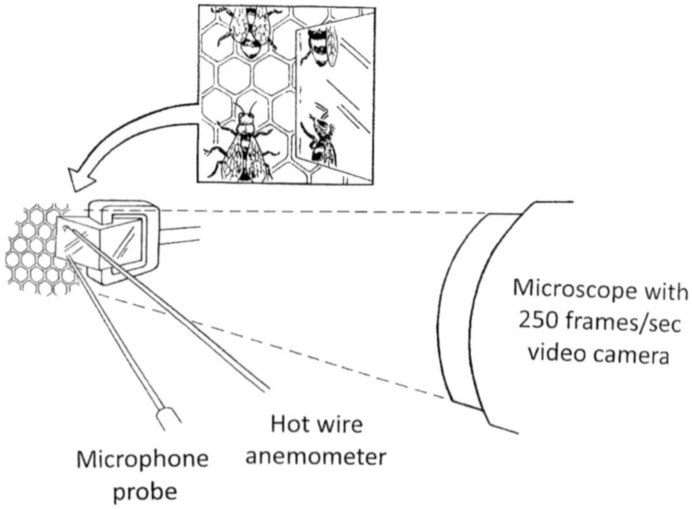


**Fig. 17.10** Setup for simultaneous recordings of dance sounds, airflows, and high-speed video. One half of each video frame shows the bees from *above*, while the other half shows the same bees from the *right*

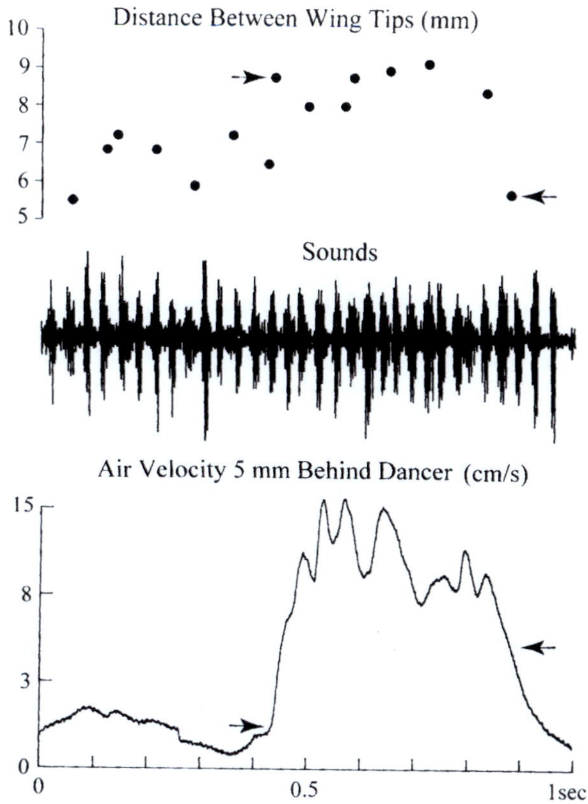


**Fig. 17.11** Simultaneous recordings of wing position, dance sounds, and air velocity about 5 mm behind a dancer during the last second of a waggle run of 2.2-s duration. The dancer emitted sounds (that is, vibrated its wings) during the entire record, but it only produced jet airflows (a broad jet and narrow jets) during the last second when the distance between the tips of the wings had increased by about 2.5 mm. *Arrows* indicate the times of wing opening and closure (From Michelsen 2003)

the dancer). The only major problem with this arrangement is that the experimenters have to spend hours cleaning prisms. Bees have very dirty feet, and they love making frequent inspection trips on the surface of the prism, which often requires a good bath in chemically clean acetone.

The studies of the generation of jet airflows by live dancers led to yet another surprise: The dancers are able to generate two kinds of jet airflows, the spatially narrow jets and broad jets (Fig. 17.8a, b), and both may be switched on and off by an adjustment of the position of the wings (Fig. 17.11). One may speculate that the broad jets (a displacement of about 10 ml of air per second) serve to mark the zone behind the dancer, where the follower bees can obtain specific information, and/or to mark periods of particularly stable dancing (perhaps by means of an odor?). It remains to be learned how and why the dancers control the generation of the two kinds of jet airflows and how the jets are perceived by the follower bees. It seems reasonable to conclude that the dance language will keep the investigators busy for many years to come.

**Acknowledgments** I am grateful to my colleague Ole Naesbye Larsen for comments on the manuscript.

## References

Bruinsma O, Kruijt JP, van Dusseldorp W (1981) Delay of emergence of honey bee queens in response to tooting sounds. Proc Kon Ned C 84:381–387

Butler C (1609) The feminine monarchie. Barnes, Oxford

Esch H (2011) Foraging honey bees: How foragers determine and transmit information about feeding site locations. In: Eisenhardt D, Galizia CG, Giurfa M (eds) Honeybee neurobiology and behaviour—a tribute for Randolf Menzel. Springer, Berlin, pp 101–121

Fletcher DCJ (1975) Significance of dorsoventral abdominal vibration among honey bees (*A. mellifera* L.). Nature 256:721–723

Frisch K von (1967) Tanzsprache und Orientierung der Bienen. Springer, Berlin (The dance language and orientation of bees. Harvard University Press, Cambridge 1993, is a paperback edition in which the main progress from 1967 to 1993 is reviewed.)

Hansson Å (1945) Lauterzeugung und Lautauffassungsvermögen der Bienen. Opusc Entomol Suppl VI:1–124

Huber F (1792) Nouvelles observations sur les abeilles. Paschoud, Genève

Judd TM (1995) The waggle dance of the honey bee: which bees following a dancer successfully acquire the information? J Insect Behav 8:343–354

Kilpinen O, Storm J (1997) Biophysics of the subgenual organ of the honeybee, *Apis mellifera*. J Comp Physiol A 181:309–318

Kirchner WH, Towne WF (1994) The sensory basis of the honeybee's dance language. Sci Am 270:74–80

Kirchner WH, Dreller C, Towne WF (1991) Hearing in honeybees: operant conditioning and spontaneous reactions to air-borne sound. J Comp Physiol A 168:85–89

Lindauer M (1955) Schwarmbienen auf Wohnungssuche. Z Vergl Physiol 37:263–324

Menzel R (2011) Commentary to section 2. In: Eisenhardt D, Galizia CG, Giurfa M (eds) Honeybee neurobiology and behaviour—a tribute for Randolf Menzel. Springer, Berlin, pp 101–121

Michelsen A (1999) The dance language of honeybees: recent findings and problems. In: Hauser MD, Konishi M (eds) The design of animal communication. MIT Press, Cambridge, pp 111–131

Michelsen A (2003) Signals and flexibility in the dance communication of honeybees. J Comp Physiol A 189:165–174

Michelsen A, Kirchner WH, Lindauer M (1986a) Sound and vibration signals in the dance language of the honeybee, *A. mellifera*. Behav Ecol Sociobiol 18:207–212

Michelsen A, Kirchner WH, Andersen BB, Lindauer M (1986b) The tooting and quacking vibration signals of honeybee queens: a quantitative analysis. J Comp Physiol A 158:606–611

Michelsen A, Towne WF, Kirchner WH, Kryger P (1987) The acoustic near field of a dancing honeybee. J Comp Physiol A 161:633–643

Michelsen A, Andersen BB, Storm J, Kirchner WH, Lindauer M (1992) How honeybees perceive communication dances, studied by means of a mechanical model. Behav Ecol Sociobiol 30:143–150

Nieh JC, Tautz J (2000) Behaviour-locked signal analysis reveals weak 200–300 Hz comb vibrations during the honeybee waggle dance. J Exp Biol 203:1573–1579

Nitsche M, Krasny R (1994) A numerical study of vortex ring formation at the edge of a circular tube. J Fluid Mech 276:139–161

Rohrseitz K, Tautz J (1999) Honey bee dance communication: waggle run direction coded in antennal contacts? J Comp Physiol A 184:463–470

Sandeman DC, Tautz J, Lindauer M (1996) Transmission of vibration across honeycombs and its detection by bee leg receptors. J Exp Biol 199:2585–2594

Seeley TD (2011) Progress in understanding how the waggle dance improves the foraging efficiency of honey bee colonies. In: Eisenhardt D, Galizia CG, Giurfa M (eds) Honeybee neurobiology and behaviour—a tribute for Randolf Menzel. Springer, Berlin, pp 101–121

Tanner D, Visscher K (2009) Does the body orientation of waggle dance followers affect the accuracy of recruitment? Apidologie 40:55–62

Tautz J (1996) Honeybee waggle dance: recruitment success depends on the dance floor. J Exp Biol 199:1375–1381

# Chapter 18
# Vibratory Communication in Stingless Bees (Meliponini): The Challenge of Interpreting the Signals

**Michael Hrncir and Friedrich G. Barth**

**Abstract** Foragers of several species of stingless bees (Apidae; Meliponini), a group of eusocial bees comprising more than 400 mainly tropical species, produce pulsed thoracic vibrations inside the nest when returning from a successful foraging trip. These vibrations do not provide navigational information on the direction and distance of a food source. Instead, both their occurrence and their temporal pattern correlate with the net gain during a foraging trip. The vibrations are therefore considered important information for potential foragers about the profitability of a food patch. Their repeated presentation lowers the foraging threshold of potential food collectors. The vibrations are considered as an alerting signal, which increases the colony's foraging activity. So far, nothing is known about how foragers of stingless bees perceive the pulsed thoracic vibrations of the recruiters. Yet, consideration of the corresponding receptors and their thresholds in honeybees suggests three possible pathways for their transmission to the nestmates: (1) the substrate (vibrations), (2) the air (air particle movements), and (3) direct physical contact (tactile stimuli). The corresponding differ significantly. Whereas substrate vibrations will reach receivers up to ten bee lengths away (medium-range transmission), air particle oscillations and direct vibrations can be detected only by bees very close to, or in contact with, the forager (short-range transmission). Thus, depending on the transmission pathway and the recipient's sensory capacity, the signal generated by thoracic vibrations will have different meanings. Indeed, substrate vibrations attract both food processors and potential foragers to the vibrating bee, whereas air particle oscillations and direct contact vibrations, in addition to important olfactory and gustatory information, may well

---

M. Hrncir (✉)
Departamento de Ciências Animais, Universidade Federal Rural do Semi-Árido,
Av. Francisco Mota 572, Mossoró 59625-900, Brazil
e-mail: michael@ufersa.edu.br

F. G. Barth
Department for Neurobiology, Faculty of Life Sciences, University of Vienna,
Althanstrasse 14, 1090 Vienna, Austria
e-mail: friedrich.g.barth@univie.ac.at

R. B. Cocroft et al. (eds.), *Studying Vibrational Communication*,
Animal Signals and Communication 3, DOI: 10.1007/978-3-662-43607-3_18,

be used by prospective recruits to evaluate the profitability of the advertised food source. In contrast to the honeybee waggle dance vibrations, there is no indication in stingless bees of an air jet potentially providing directional information.

## 18.1 Introduction

There are more than 18,000 described species of bees worldwide, and an estimate of the total number of species is near or even above 20,000 (Michener 2000). Thoracic vibrations not related to flight and generated by contractions of the indirect flight muscles (Simpson 1964; Esch and Wilson 1967) are widespread among bees (Michener 2000). They have been reported in a variety of behavioral contexts, such as nest construction (Michener 1974, 2000), nest defense (Vicidomini 1998; Hrncir et al. 2006a), and the detection of females by males (Larsen et al. 1986). There are male "sounds" during mating (Eickwort and Ginsberg 1980; Larsen et al. 1986; Roubik 1989; Conrad et al. 2010), vibrations used for pollen collection (Michener 1962; Wille 1963; Buchmann 1983; Harter et al. 2002; Nunes-Silva et al. 2010), and, in social bees, vibrations for the communication among nestmates (Hrncir et al. 2006a).

The term "bee communication" is most frequently associated with the honey bee's famous waggle dance, the stereotyped figure-eight movements performed by successful food collectors on their return to the nest. Ever since the pioneering discovery by Karl von Frisch (1946) that these dances convey information about both the distance and the direction of the visited food source, scientists have been searching for similar forms of symbolic communication, that is, an abstract code providing information about an object without causal relation or similarity between signal and object (Menzel 2012), in closely related bee groups. Outstanding among these are the stingless bees (Apidae; Meliponini), which represent a group of highly eusocial bees with more than 400 species mainly found in the tropics (Michener 2000; Camargo and Pedro 2007). The degree of social organization of stingless bees is similar to that of the honeybees (Michener 1974), and above all, their impressive capacity to recruit nestmates to food sources (Lindauer 1956; Lindauer and Kerr 1958, 1960; Nieh and Roubik 1995; Jarau et al. 2000) furthered speculations about intranidal signals providing prospective recruits with navigational information about the position of a food patch (Esch et al. 1965; Esch 1967; Nieh and Roubik 1998).

Lindauer and Kerr (1958, 1960) were the first to investigate in detail the behavior of stingless bees within their nest during food exploitation processes. These authors observed three conspicuous behaviors shown by the foragers upon their return from a profitable food source: zigzag runs, jostling of nestmates, and buzzing sounds (Fig. 18.1). Of these, the best-studied displays related to recruitment communication are the buzzing sounds, which originate from thoracic

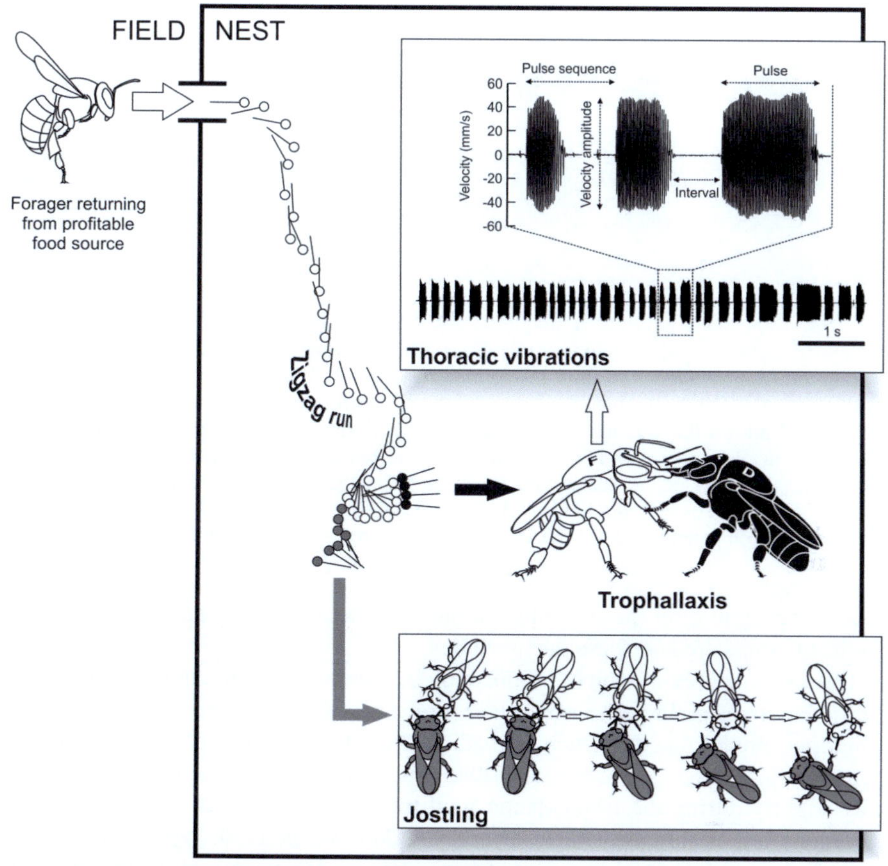


**Fig. 18.1** Intranidal behaviors of stingless bee foragers. When foragers (see *empty/white bees* and *symbols*) return from a profitable food source, they excitedly run through the colony (zigzag run), thereby jostling their nestmates (*gray symbols* and bees in *inset*). While running, but predominantly during trophallaxis (see *filled/black bee* and *symbols*), the foragers generate pulsed thoracic vibrations. *F* forager; *D* food receiver. *Inset* shows parameters of the temporal pattern of the vibratory signals recorded with a laser vibrometer: pulse duration, interval duration, pulse sequence, velocity magnitude. Symbols (*circle* head; *line* long axis of body) indicate change of position of the bees, video-taped at 25 frames per second [Adapted from Hrncir (2009)]

vibrations generated by foragers collecting at a highly profitable food source (Hrncir 2009) (Fig. 18.1). The pulsed structure of these vibrations, reminiscent of a Morse code, promoted the idea that information about the food source may be encoded within the temporal pattern of the sounds. The first attempts to decode the message and meaning of the thoracic vibrations suggested that the duration of the pulses provides a measure of the distance to a food source (*Melipona quadrifasciata*, *Melipona seminigra*: Esch et al. 1965; Esch 1967; *Melipona panamica*: Nieh and Roubik 1998) or even its height (*M. panamica*: Nieh and Roubik 1998).

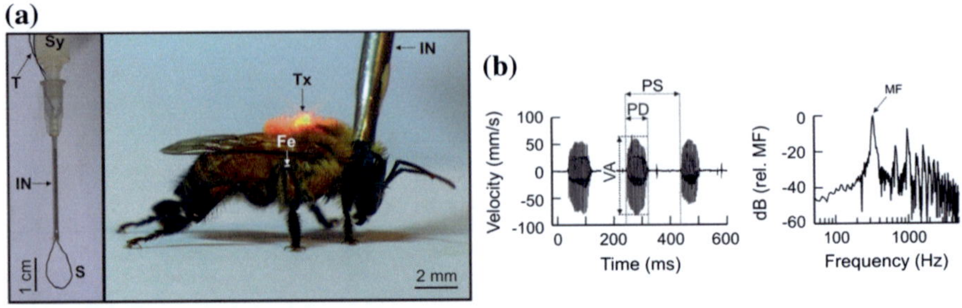


**Fig. 18.2** Annoyance buzzing in stingless bees. Stingless bees (shown: worker of *Melipona rufiventris*) generate pulsed thoracic vibrations when tethered by a sling around their neck. **a** Sling-tethering method: the sling (*S*) formed by a nylon thread (*T*) and guided through an injection needle (*IN*). *Sy*, syringe for fixing the thread. Using one or even two laser vibrometers, this method allows the detailed measurement of the vibrations at various body parts such as thorax (*Tx*) or distal mesothoracic femur (*Fe*) and the calculation of signal transmission. **b** The following parameters of the pulsed vibrations can be analyzed for a comparison with those of forager vibrations: velocity amplitude (*VA*), duration of single pulses (*PD*), pulse sequence (*PS*), and the main component (*MF*) of the frequency spectrum [Adapted from Hrncir et al. (2006b)]

Below, we reinterpret these results, taking into account additional factors that had not been considered in these early studies.

Unlike honeybees, which produce their communication signals exclusively during locomotion through the nest, the foragers of stingless bees generate thoracic vibrations predominantly when unloading their food to nestmates (trophallaxis) (Hrncir et al. 2006a, b; Barth et al. 2008; Hrncir 2009). Since the individuals move only slightly during the trophallactic food transfers, the recording of the thoracic vibrations is possible with high accuracy. Likewise, methodological innovations over the past decade, like using laser vibrometry instead of sound pressure microphones (Hrncir et al. 2004a, b, 2006b; Schmidt et al. 2006, 2008; Morawetz 2007) and the investigation of the vibrations generated by sling-tethered bees (Hrncir et al. 2008a, b) (Fig. 18.2), provided new insights about stingless bee vibratory signals. The present chapter outlines our current knowledge of the message and meaning of forager-produced thoracic vibrations, the mechanisms of their generation, and the possible pathways of transmission during recruitment communication.

## 18.2 Message of Thoracic Vibrations of Stingless Bee Foragers

The key to decoding the message of a putative signal is the unequivocal identification of all the factors that influence and shape the respective behavioral display (Seeley 1992). The first attempts to decode the message of meliponine vibratory signals (Esch et al. 1965; Esch 1967; Nieh and Roubik 1998) suffered from premature conclusions regarding the existence in stingless bees of a referential communication of a food source's location. Probably biased by the expectation of finding a precursor of the honeybee's "dance language," these early studies searched for correlations between the temporal pattern of the vibratory pulses and the spatial parameters of a food patch. However, they did not consider numerous additional criteria potentially influencing the temporal pattern of the foragers' sounds. In four *Melipona* species (*M. quadrifasciata, M. seminigra*: Esch et al. 1965; Esch 1967, *Melipona bicolor, and Melipona mandacaia*: Nieh et al. 2003), the duration of the vibratory pulses (pulse duration) was found to increase with increasing distance of the food source (Fig. 18.3). In *M. panamica* (Nieh and Roubik 1998), pulse duration during food unloading was longer when bees collected food at ground level than when collecting at the canopy top, whereas after unloading, pulse duration increased with increasing foraging distance (Fig. 18.3). In all these studies, pulse duration varied by up to 60 % (Esch 1967; Nieh et al. 2003) or even by more than 200 % (Nieh and Roubik 1998) at each investigated distance/height (Variation = Standard Deviation × 100/mean value; values obtained from the respective publication). This variability raises the question whether potential recruits could extract reliable information about food source position from the temporal pattern of the foragers' vibrations. Furthermore, these results, which were interpreted to support the referential communication hypothesis, could never be replicated by other researchers studying the same (*M. quadrifasciata*: Hrncir et al. 2000, *M. seminigra*: Samwald 2000) or closely related bee species (*M. costaricensis*: Aguilar and Briceño 2002; just as *M. panamica*, *M. costaricensis* had formerly been classified as subspecies of *M. fasciata*: Camargo and Pedro 2007) (Fig. 18.3). The hypothesis that the thoracic vibrations of *Melipona* code the distance to a food source was also greatly weakened by the later finding that the visual flow (lateral image motion experienced by the bees during flight) used by foragers to estimate the distance to a food source (shown for *M. seminigra*: Hrncir et al. 2003, following the establishment of the "visual flow hypothesis" for the honeybee: Esch and Burns 1995; Srinivasan et al. 2000) does not affect the temporal pattern of the thoracic vibrations (Hrncir et al. 2004a).

According to recent studies, both the occurrence and the temporal pattern of the vibrations are related to the profitability of the food source experienced by the forager (Hrncir 2009) (Figs. 18.3, 18.4) rather than encoding spatial information about the food patch visited. The most obvious evidence supporting this conclusion is that foragers do not generate thoracic vibrations at all as long as the value of a food source is below a certain threshold (Esch 1967; Hrncir et al. 2000; Schmidt et al. 2006, 2008). Because in all experimental studies so far sugar solution had been offered ad libitum, this "excitement threshold" was determined by the sugar concentration of the collected food. However, the food profitability experienced by a forager and, consequently, her disposition to generate thoracic vibrations, may as well be determined by parameters different from sugar concentration, such as solution flow, handling time, and even the presence of competitors (Hrncir 2009; Hrncir et al. 2011) (Fig. 18.4).

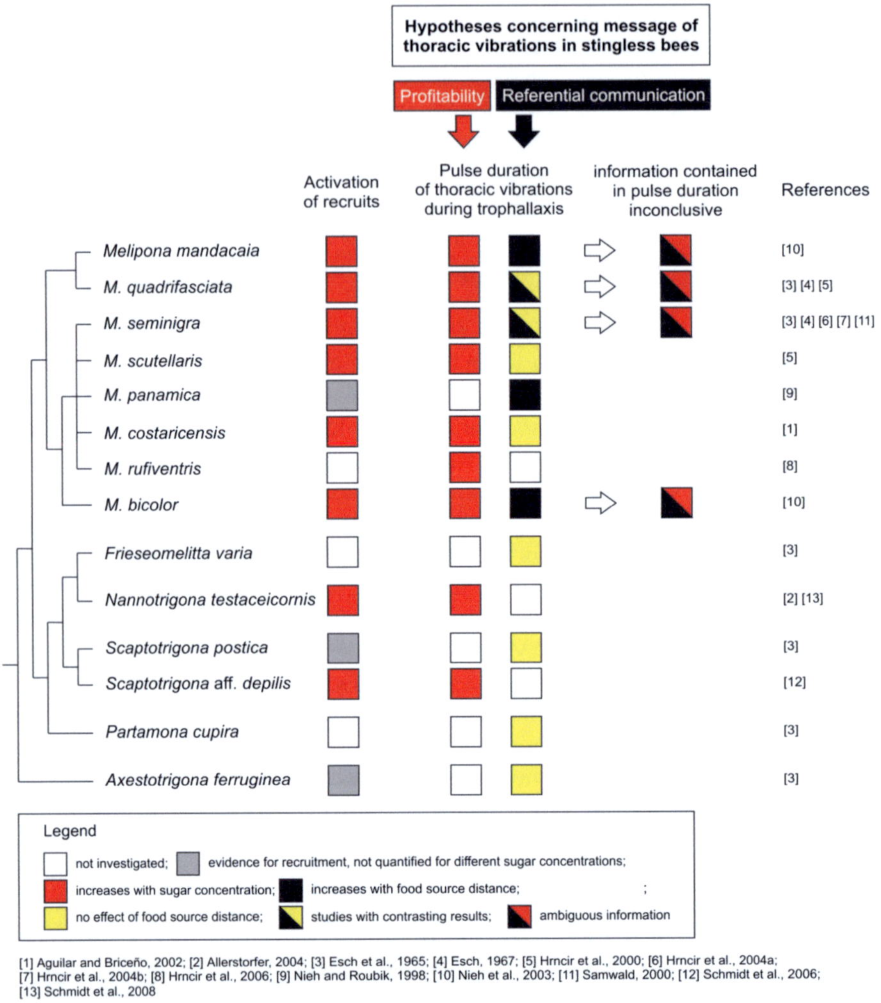


**Fig. 18.3** Hypotheses concerning the message of vibratory signals. Shown is the phylogenetic relationship of stingless bee species [adapted from Ramírez et al. (2010) and Rasmussen and Cameron (2010)] studied in regard to recruitment activity and/or the message of their thoracic vibrations. Recruitment success (number of activated recruits) usually increased with increasing food profitability (*red squares*). Studies corroborating the profitability hypothesis found an increase in pulse duration with increasing sugar concentration (*red squares*). Studies corroborating the referential communication hypothesis found an increase in pulse duration with increasing foraging distance (*black squares*). If both hypotheses were true, pulse duration would not provide conclusive information for potential recruits. According to the profitability hypothesis, pulse duration should decrease with foraging distance (*see text*), whereas, according to the referential communication hypothesis, it increases with distance of the food source

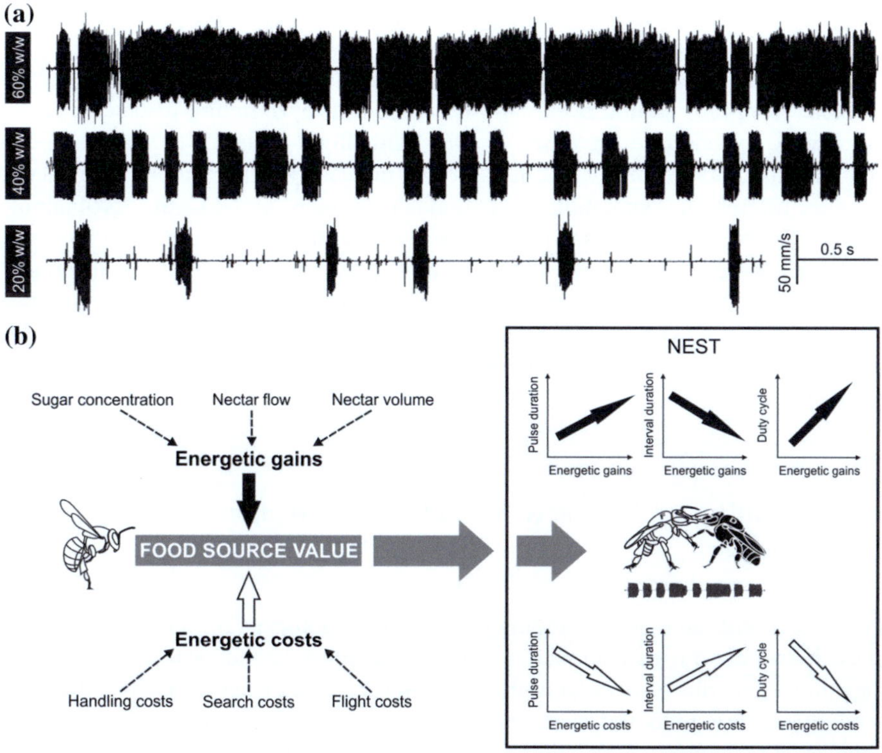


**Fig. 18.4** Message of vibratory signals. **a** The thoracic vibrations generated by foragers of many stingless bee species correlate with the concentration of the collected sugar solution or nectar. To show this, an example is given of vibrations generated by a forager of *Melipona seminigra* collecting an aqueous solution containing 60, 40, or 20 % sugar weight on weight (w/w). **b** In addition to sugar concentration, other parameters determine the value of a food source for collecting bees and, consequently, influence the temporal pattern of the foragers' thoracic vibrations: Increasing energetic gains at the food patch result in longer pulses, shorter intervals, and consequently, an increasing duty cycle (duty cycle = pulse duration/[pulse duration + interval duration]). Increasing energetic costs, by contrast, result in shorter pulses, longer intervals, and a decreasing duty cycle. *F* forager; *D* food receiver [Adapted from Hrncir (2009)]

Once the food profitability exceeds the "excitement threshold" of a forager, the temporal pattern of her vibrations is strongly influenced by the energy intake (sugar concentration). Pulse duration increases and the interval between pulses decreases with increasing profitability of the food source, which implies an increasing duty cycle as well (*M. costaricensis*: Aguilar and Briceño 2002; *M. bicolor*, *M. mandacaia*: Nieh et al. 2003; *M. rufiventris*: Hrncir et al. 2006a; *M. seminigra*: Hrncir et al. 2004a, b; *N. testaceicornis*: Allerstorfer 2004; Schmidt et al. 2008) (Figs. 18.3, 18.4). Along this line of thought, increased energetic expenses experienced during a collecting trip should reduce the "excitement" of a forager. And indeed, in *M. seminigra*, the effect of increased flight costs on the temporal pattern of the foragers' thoracic vibrations was exactly the opposite of that of increased energetic gains (Hrncir et al. 2004a; Hrncir 2009).

In accordance with the profitability hypothesis, the temporal pattern of the thoracic vibrations should eventually be influenced by foraging distance because the energy expenditure increases linearly with flight distance (Hanauer-Thieser and Nachtigall 1995). Yet, in contrast to the increase in pulse duration postulated by the referential communication hypothesis (see above), the profitability hypothesis predicts a decrease in pulse duration (Hrncir et al. 2004a). In any case, the large differences in energy uptake at a food source among individual foragers (Hrncir et al. 2004b) would strongly disguise differences in energy consumption due to different food source distances.[1] It seems, therefore, unlikely that thoracic vibrations of *Melipona* bees contain reliable information about the distance of a food source.

## 18.3 Meaning of Thoracic Vibrations of Stingless Bee Foragers

For a comprehensive understanding of the vibratory signals produced by stingless bees, it is essential to decipher not only their message but also their potential meaning in recruitment communication (*message*: information provided by the sender, *meaning*: influence on the behavior of the receiver, Seeley 1992). Since the behavioral response to a signal depends both on the behavioral context and on the recipient's motivation, revealing the signal's meaning often is an even greater challenge than revealing its message.

Observations suggest that in meliponine bees, the foragers' thoracic vibrations have a modulatory function, raising the activity level of nestmates and increasing their propensity to forage (Hrncir 2009). According to a detailed study of the intranidal case histories of individually marked recruits in *M. seminigra* (Kronberger 2000), the agitation of inactive foragers, measured as jostling contacts, abruptly increased after the first contact with an active collector (Hrncir 2009). The sudden increase of their locomotor activity is taken to indicate the increased motivation to forage in response to the interactions with the food collectors (Hrncir 2009).

[1] The individual variation in sugar intake of *M. seminigra* foragers collecting at an artificial food source was 3.32 mg (Hrncir et al. 2004b). Taking measurements in honeybees, which are of similar body size as *M. seminigra*, as reference, the bees spend 0.70 mg sugar for each 1,000 m of flight (Hanauer-Thieser and Nachtigall 1995). Nestmates receiving the thoracic vibrations of a forager would have to decide whether the forager loaded 3.32 mg less sugar at the food source (less energy intake) or spent more energy due to an additional 4,740 m of flight (consumption of additional 3.32 mg sugar). The energy budget, and thus thoracic vibrations reflecting it, would be the same under both conditions provided that thoracic vibrations are influenced to the same degree by energy intake and energy consumption.

Further evidence for the effect of the forager's agitation on the nestmates' motivation to forage (both experienced, inactive foragers and novice foragers) comes from the observation that no newcomers arrive at the food source as long as the value of the food is below the foragers' "excitement threshold" (Jarau et al. 2000). As soon as the profitability of the food source exceeds this threshold, however, the recruitment success increases with increasing sugar concentration of the collected food. This could be shown for several *Melipona* species already (*M. bicolor*, *M. mandacaia*: Nieh et al. 2003; *M. panamica*: Nieh and Sanchez 2005) and for *Nannotrigona testaceicornis* (Schmidt et al. 2008). Since in these species, the recruiters' excitement correlates with their energetic gains at the food source (see above), it cannot be decided whether the recruitment success depends on either the sugar concentration of the collected and distributed food, or on the foragers' "excitement," or both. *Scaptotrigona* aff. *depilis* is the only meliponine species so far known where recruitment success does not directly depend on the concentration of the sugar water collected by the foragers (Schmidt et al. 2006). In this species, the recruiter's thoracic vibrations depended on past foraging experiences rather than the current food profitability. A steadily increasing sugar concentration did not change the temporal pattern of the vibrations, nor the recruitment success (Schmidt et al. 2006). Hence, in this case, the quality of the received food samples did not influence the foraging motivation of the hive bees. Yet, when the profitability of the food source continuously decreased, both the recruiters' agitation and their recruitment success decreased (Schmidt et al. 2006). From these findings it follows that (at least in *S.* aff. *depilis*) the foraging motivation of inexperienced bees does not depend on the quality of the food brought in by the foragers but, indeed, on the degree of "excitement" of the recruiters.

So far, no studies have been performed to specify whether the foragers' vibrations cause a general increase in foraging activity, where individual recruits search for their own food source, or whether recruits use odor cues to find the same source as the one advertised by the vibrating bee. Recent studies indeed provide strong evidence that meliponine foragers use olfactory information received within the nest for their search for food (Jarau 2009; Roselino and Hrncir 2012). Therefore, as also proposed for honeybees (Grüter and Farina 2009), the combination of vibratory information about a profitable food source with olfactory/gustatory information appears to serve the coordination of foraging processes in two ways. First, it may alert experienced but inactive foragers and inform them that a known food source, identified through the scent, has become profitable, as indicated by the vibratory signals. Provided a sufficiently lowered foraging threshold, these experienced bees will resume their collecting activity at the known food patch (Biesmeijer et al. 1998; Biesmeijer and Slaa 2004). Second, the vibratory signals may lower the foraging threshold of new, inexperienced foragers. In this case, the olfactory information provided by the vibrating bee will bias the search of the naive foragers toward the advertised food source in the field (Jarau 2009; Roselino and Hrncir 2012).

## 18.4 The Generation of Thoracic Vibrations

Many groups of insects use airborne sounds and substrate vibrations to communicate by periodically oscillating specialized organs at their resonant frequency (Bennet-Clark 1999). Bees are not equipped with such structures (Snodgrass 1956; Schneider 1975), their thorax being the only body part capable of generating adequate rhythmic oscillations. As in many other insects, the most prominent purpose of rhythmic thoracic oscillations is to move the wings. The periodic up- and down-strokes of the wings are maintained through stretch activation of the antagonistic indirect flight muscles at the resonant frequency of the oscillating system (Snodgrass 1956; Nachtigall 2003).

Thoracic vibrations associated with nestmate communication or buzz pollination are characterized by fundamental frequencies significantly higher than that of flight vibrations (King 1993; King et al. 1996; Nachtigall 2003; Hrncir et al. 2008a; Burkart et al. 2011) (Table 18.1). According to a study on the thoracic flight and non-flight vibrations generated by *M. seminigra,* the average fundamental frequency of annoyance buzzing (produced by tethered individuals) was 305 Hz, whereas that of forager vibrations was 487 Hz, and 182 Hz was the value found during tethered flight (Hrncir et al. 2008a).[2] The cycle frequency of flight vibrations did not change significantly during the entire oscillation period. In both types of non-flight vibrations, by contrast, the cycle frequency dropped to 215 Hz (annoyance buzzing) and 225 Hz (forager vibrations), respectively, within the last four to six oscillation cycles (Fig. 18.5). This frequency change is explained by the fact that an oscillating system driven by a periodic force at a frequency higher than its natural frequency will vibrate at the excitation frequency as long as the force is applied. As soon as the force stops, however, the vibration magnitude will decay and the frequency drop to the system's resonant frequency (Nocke 1971; Bennet-Clark 1999).

## 18.5 Transmission Pathways of Vibratory Signals

In order to justify the terms "signal" and "communication," a crucial question has to be answered: Who understands these signals? The identification of potential recipients requires knowledge of the exact physical nature of the signal and of the

---

[2] Wasps and bees produce thoracic vibrations when trying to escape from any form of confinement, such as when pushing through narrow nest entrances (Michener 2000), or when trying to escape from the grasp of predators or researchers (Esch and Wilson 1967; Schneider 1975; Larsen et al. 1986; Hrncir et al. 2008a). This form of thoracic vibrations (termed "disturbance buzzes": Larsen et al. 1986; "annoyance buzzing": Hrncir et al. 2008a) are known from both solitary bees (*Colletes cunicularius*: Larsen et al. 1986) and social bees (Bombini; *Bombus terrestris*: Schneider 1975; Meliponini; *Melipona* spp.: Esch and Wilson 1967; Hrncir et al. 2008a, b; Nunes-Silva 2011).

**Table 18.1** Thoracic vibrations by stingless bees

| Species | Thoracic vibrations | | | |
|---|---|---|---|---|
| | BC | MF (Hz) | VA (mm/s) | References |
| *Melipona bicolor* | RC | 538 | n.i. | Nieh et al. (2003) |
| *M. costaricensis* | RC | 493 | n.i. | Aguilar and Briceño (2002) |
| *M. fasciculata* | BP | 245–249 | 392–398 | Nunes-Silva (2011) |
| | AB | 275–332 | 376–492 | Nunes-Silva (2011) |
| *M. flavolineata* | AB | 282–294 | 255–318 | Nunes-Silva (2011) |
| *M. fuliginosa* | AB | 316 | 224 | Nunes-Silva (2011) |
| *M. mandacaia* | RC | 551 | n.i. | Nieh et al. (2003) |
| *M. marginata* | AB | 321–351 | 125–167 | Nunes-Silva (2011) |
| *M. melanoventer* | BP | 266–270 | 409–486 | Nunes-Silva (2011) |
| | AB | 294–307 | 395–435 | Nunes-Silva (2011) |
| *M. panamica* | RC | 433 | n.i. | Nieh (1998) |
| *M. quadrifasciata* | RC | 464–600 | n.i. | Hrncir et al. (2000), Lindauer and Kerr (1958) |
| | AB | 241–263 | 176–205 | Nunes-Silva (2011) |
| *M. rufiventris* | RC | 389–517 | 51–91 | Hrncir unpublished |
| | AB | 226 | 155 | Hrncir and Nunes-Silva unpublished |
| *M. scutellaris* | RC | 350–520 | n.i. | Hrncir et al. (2000) |
| | AB | 264–349 | 198–290 | Burkart et al. (2011), Hrncir and Nunes-Silva unpublished, Nunes-Silva (2011) |
| | FL[e] | 196–229 | n.i. | Burkart et al. (2011), Hrncir unpublished |
| *M. seminigra*[a] | RC | 259–525 | 57–115 | Hrncir et al. (2004a, b), Hrncir et al. (2006b), Hrncir et al. (2008a) |
| | AB | 268–305 | 196–209 | Hrncir et al. (2008a), Hrncir and Nunes-Silva unpublished |
| | FL[f] | 182 | 39 | Hrncir et al. (2008a) |
| *M. seminigra*[b] | AB | 295–301 | 216–374 | Nunes-Silva (2011) |
| *Nannotrigona testaceicornis* | RC | 400 | 90 | Schmidt et al. (2008) |
| | AB | 238 | 99 | Hrncir and Nunes-Silva unpublished |
| *Scaptotrigona* aff. *depilis* | RC | 311–365 | 44–142 | Schmidt et al. (2006) |
| | AB | 258 | 122 | Hrncir and Nunes-Silva unpublished |
| *Scaptotrigona postica*[c] | RC | 391 | n.i. | Lindauer and Kerr (1958) |
| *Schwarziana bipunctata* | AB | 249 | 189 | Hrncir and Nunes-Silva unpublished |
| *Tetragonisca angustula*[d] | RC | 246–326 | n.i. | Lindauer and Kerr (1958) |

Given are species name, the behavioral context (BC) in which the vibrations were measured (RC, recruitment communication; BP, buzz pollination; AB, annoyance buzzing; FL, flight), the average main frequency content (MF), the average velocity amplitude (VA), and the bibliographic reference of the underlying study; n.i., no information given in the respective study

[a] not-identified subspecies from Mato Grosso

[b] not-identified subspecies from Pará

[c] identified as *Trigona rustica*

[d] identified as *Trigona jaty*

[e] free flight

[f] tethered flight

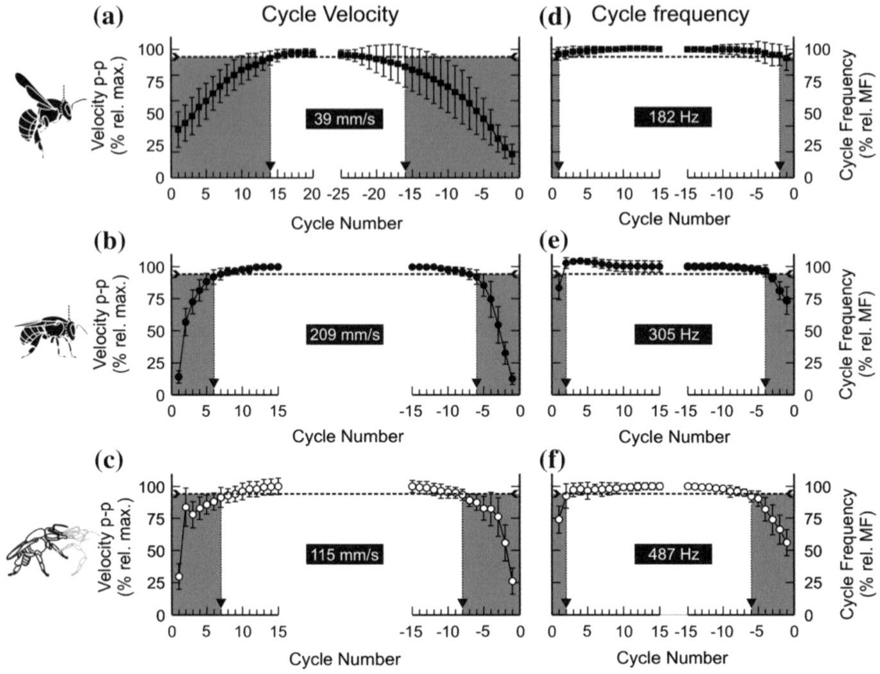


**Fig. 18.5** Comparison of flight and non-flight thoracic vibrations of stingless bees (*Melipona seminigra*). The first and the last 15–20 oscillation cycles of thoracic vibrations (measured with a laser vibrometer) during stationary flight (**a**, **d** *filled squares*, N = 15 individuals), annoyance buzzing (**b**, **e** *filled circles*, N = 15), and forager vibrations (**c**, **f** *open circles*, N = 15) were analyzed regarding velocity amplitude (**a–c**) and cycle frequency (**d–f**). Graphs show the means ± s.d. of relative values (percent of the maximum velocity or of the main frequency, MF). Shaded area indicates the buildup and decay of thoracic oscillations. Broken lines indicate 95 % of maximum. Medium values of velocity amplitude and main frequency are given in the respective plot [Adapted from Hrncir et al. (2008a)]

mechanisms underlying both its transmission to and perception by the receiver. In stingless bee recruitment communication, three transmission pathways of the vibratory signals have been suggested and analyzed: (1) the substrate (substrate vibrations), (2) the air (air particle movements), and (3) direct physical contact (tactile stimuli) (Fig. 18.6). The degree of signal attenuation and, therefore, the range of signal transmission differ greatly between these pathways (Hrncir et al. 2006a, b, 2008b; Morawetz 2007). Whereas substrate-borne vibrations will reach receivers at a distance of up to ten bee lengths from the signaler (medium-range transmission), air particle oscillations and direct vibrations are only detected by bees very close to or in actual contact with the vibrating forager (short-range transmission). Thus, the meaning of the original signal may well differ depending on the type of transmission considered. Like in honeybees, *Apis mellifera*, substrate vibrations are believed to attract hive bees to the forager unloading the collected food (Tautz and Rohrseitz 1998). Air particle oscillations and direct

**Fig. 18.6** Possible pathways of vibratory signal transmission. A forager of *Melipona scutellaris* distributing food to nestmates. During trophallactic contacts, the vibratory signals generated by the forager (*F*) may be transmitted to nestmates as substrate vibrations (medium-range transmission pathway), air particle movements, or directly during trophallaxis as contacts between forager and receivers (short-range transmission pathways). Considering physiological thresholds of vibration receptors of honeybees (*see text*), *S*-bees should perceive only substrate vibrations, whereas *A*-bees perceive air particle oscillations and substrate vibrations, and *D*-bees direct vibrations, air particle oscillations and substrate vibrations

vibrations, on the other hand, in combination with olfactory and gustatory information originating from the food collector, may serve the prospective recruits to evaluate the advertised food source (Michelsen 2003; Grüter and Farina 2009). In the following, arguments supporting these conjectures are given.

### *18.5.1 Substrate Vibrations: Medium-Range Transmission*

When vibrating their thorax, meliponine foragers generate substrate vibrations that can be measured (Hrncir et al. 2000, 2006b), their legs representing the mechanical link between thorax and substrate (Rohrseitz 1998; Tautz et al. 2001; Hrncir et al. 2006a, b). The vibrations are transmitted from the forager's thorax to her leg without loss in velocity amplitude, but are strongly attenuated when passing from the leg to the substrate (Fig. 18.7). In *M. seminigra,* an attenuation of about 50 dB was found between the signal amplitude on the forager's femur and the substrate halfway between forager and food receiver, respectively (Hrncir et al. 2006b) (Fig. 18.7). However, albeit strongly reduced in amplitude, the signal's temporal

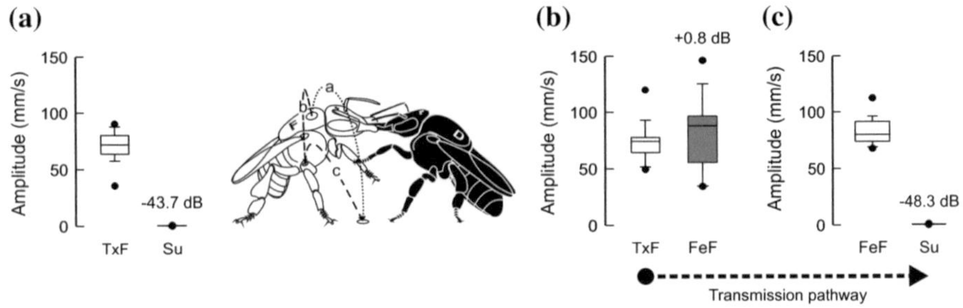


**Fig. 18.7** Vibration transmission to the substrate (*Melipona seminigra*). **a** Comparison of the velocity amplitudes (*boxplot*) of the forager's thorax (*TxF*) and of the substrate (*Su*) close to the forager's leg. Average signal attenuation on its way to the substrate was 43.7 dB in the given example. **b–c** Details of the vibration transmission from the forager's thorax (*TxF*) to its femur (*FeF*) and from there to the substrate (*Su*). *F* forager; *D* food receiver. Data from simultaneous recordings with two laser vibrometers are presented as *boxplots*. Differences between vibration amplitudes picked up at the same body parts (compare *TxF* in **a** and **b**, and *FeF* in **b** and **c**) are due to differences between vibrating individuals [Adapted from Hrncir et al. (2006b)]

pattern and, thus, the information about the forager's degree of "excitement" were well preserved in the substrate vibrations (Hrncir et al. 2006b).

The propagation of substrate vibrations depends on the transmission properties of the respective substrate (Michelsen and Nocke 1974; Barth et al. 1988; Rohrseitz 1998; Sandeman et al. 1996; Barth 1998; Morawetz 2007). In stingless bees, trophallactic interactions and the generation of thoracic vibrations by foragers predominantly occur inside the nest's entrance tunnel (Hrncir et al. 2006b; Morawetz 2007; Hrncir 2009). This is a narrow, tubular structure built from batumen, a mixture of mud, wax, and floral materials (Schwarz 1948; Wille and Michener 1973; Roubik 2006). Analysis of the transmission properties of diverse nest structures in *M. scutellaris* and *M. bicolor* showed that bee generated non-flight vibrations (tethered bees used as vibration generators) are propagated with an attenuation of between 1.5 and 2 dB/cm through the batumen of the entrance tube (Morawetz 2007). Given a velocity amplitude of bee-produced substrate vibrations of 0.37 mm/s right next to the vibrating individual (Fig. 18.7), the vibratory output at a distance of 1 cm from the forager would be at least 0.29 mm/s, at 4 cm 0.15 mm/s, and at 8 cm 0.06 mm/s (output calculated for an attenuation of 2 dB/cm).

To date, the reception of substrate vibrations has not been studied in stingless bees. The only way to get a preliminary idea about their detection by hive bees is through a comparison with the well-studied honey bee. In *A. mellifera*, the reception of substrate vibrations has been predominantly attributed to the subgenual organ, a chordotonal organ found in the proximal part of the tibia of each leg (Schön 1911; Autrum and Schneider 1948). This sensory organ responds to vibrations in the axial direction of the tibia. When the leg is accelerated by substrate vibrations, inertia causes the hemolymph and the subgenual organ suspended in it, to lag behind the movement of the leg, which mechanically stimulates the receptor cells (Autrum and Schneider 1948; Kilpinen and Storm 1997; Storm and Kilpinen 1998). When studied electrophysiologically, its sensory cells were most sensitive to vertical vibrations of the leg at frequencies between 150 and 900 Hz, with an average response threshold between 0.06 and 0.15 mm/s peak–peak (Kilpinen and Storm 1997; Rohrseitz and Kilpinen 1997). Assuming the threshold of the meliponine subgenual organ to be similar to that of the honeybee, the range of just noticeable vibrations would be between 4 and 8 cm from the forager generating them (Morawetz 2007). In case of *Melipona* bees with a body length of 0.8–1.4 cm (Schwarz 1948), this corresponds to between three and ten bee lengths.

### 18.5.2 Airborne Sound: Short-Range Transmission

Non-flight thoracic vibrations of stingless bees are transformed into airborne sound well audible for the human ear (Hrncir et al. 2004a, 2008b). Since, different from us, bees do not have sound pressure receivers (Snodgrass 1956; Hrncir et al. 2006a), the physical parameter most relevant for the perception of airborne sound is air particle movement. In dancing honeybees (*A. mellifera*), two different forms of air particle movement have been described. First, the oscillating wings create intense air particle oscillations close to their edges (Michelsen et al. 1987). Second, air that moves out from the space between the wings and the abdomen during wing vibrations creates an air jet moving away from the bee's abdomen (Michelsen 2003). In the honeybee, both these forms of air particle movement depend on the wing oscillations that go along with the thoracic vibrations. In stingless bees, however, wings play a minor role for the transformation of thoracic vibrations into airborne sounds and medium flow, respectively. According to a detailed investigation in sling-tethered stingless bees (*Melipona scutellaris*), the sound field (particle movement) around a vibrating bee is predominantly generated by the oscillations of the thorax itself (Hrncir et al. 2008b). Although the wings vibrate with velocity amplitudes of close to 700 mm/s along with the thorax (measured in *M. seminigra*; Hrncir et al. 2008a), they significantly affect the vertically oriented particle velocity close to the abdomen only (Fig. 18.8). The different impact of the wings on the generation of air particle movement in *A. mellifera* and *M. scutellaris*, respectively, is believed to be due to a difference in their position when the bees are vibrating. Whereas stingless bees vibrate their thorax with their wings closely folded over the abdomen (Lindauer and Kerr 1958; Hrncir et al. 2006a, b, 2008a), honeybees do it with their wings splayed (wing tips 5–9 mm apart) when dancing (Michelsen 2003). This spreading of the wings increases the effective wing area (Schneider 1975). Consequently, the volume of air between the wings and the abdomen that is moved by every wing stroke is increased, as well, and most likely responsible for the air jet found in honeybees by Michelsen (2003).

In stingless bees, airborne sounds going along with the thoracic vibrations repeatedly have been assumed to transmit information (Esch 1967; Nieh et al.

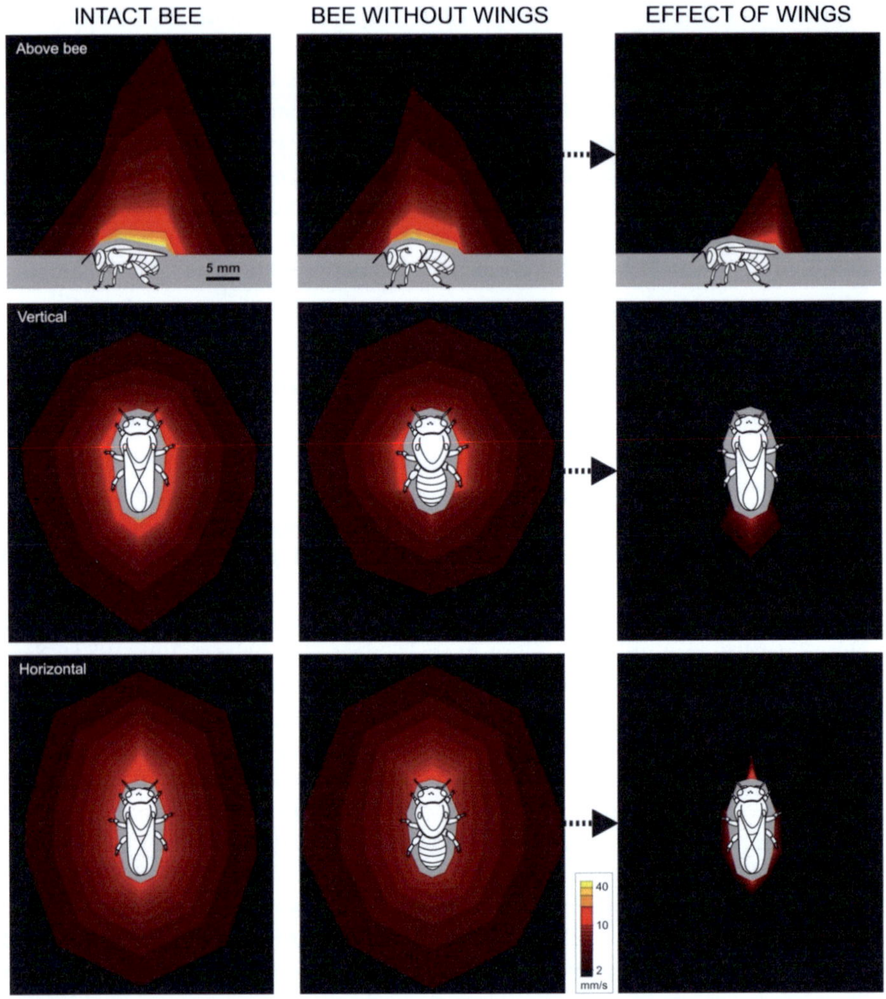


**Fig. 18.8** Air particle oscillations generated by vibrating bees (*Melipona scutellaris*) and measured with airflow sensors. Ranges above and around (vertically or horizontally oriented) vibrating bees in which air particle velocities have the same mean amplitudes. *Different colors* indicate mean velocity amplitudes between 2 and 40 mm/s as explained by the logarithmic color scale. *Left panels* intact individuals; *middle panels* wingless individuals; and *right panels* fraction of particle velocity generated by wings only. Air particle oscillations cannot be accurately measured or estimated at distances below 1 mm from the vibrating bee (*shaded area*). For measurements of the air particle movement above bees, the airflow sensors were positioned at least 5 mm above the substrate. Therefore, no values are given for the region below 5 mm (*shaded area*) [Adapted from Hrncir et al. (2008b)]

2003). Whereas the temporal pattern of the thorax vibrations (pulse duration, pulse sequence, and main frequency component) is indeed well preserved in the air particle oscillations (Hrncir et al. 2004a, 2008b), the crucial question of whether the air particle velocity close to a vibrating bee is strong enough to be detected by the hive bees still awaits an answer.

The candidate mechanosensory organ detecting air particle velocity is Johnston's organ in the antennal pedicel, which is stimulated when the flagellum is deflected by air movement (Snodgrass 1956; Heran 1959). Up to now, neither the physiological nor the mechanical properties of this mechanoreceptor are known in stingless bees. Again, a comparison with data available for *A. mellifera* may be helpful. Heran (1959) found that Johnston's organ of the honeybee had physiological thresholds of 0.37 mm/s (oscillation velocity measured at the tip of the antenna) at a stimulation frequency of 200 Hz, 0.75 mm/s at 300 Hz, and 4.5 mm/s at 400 Hz. However, particle velocity around the antenna has to be about 100 times stronger (i.e., 37–75 mm/s) in order to generate such oscillation velocities of its tip (Kirchner 1994).

When adopting these physiological and mechanical properties for stingless bees, velocities of at least 37 mm/s are needed to effectively stimulate their Johnston's organs. Vibrating stingless bees (*M. scutellaris*) indeed produce air particle velocities sufficiently strong close to their body surface (1 mm above the thorax; estimated particle velocity 43 mm/s) and to the wings (estimated particle velocity: 61 mm/s) (Hrncir et al. 2008b). Hive bees attending trophallactic events stay within less than 5 mm from the forager (distance between head of receiver and body of forager) with their splayed antennae close to or even touching the vibrating forager (Hrncir et al. 2008b). Similarly, in *M. panamica,* the antennal tips of hive bees were found to be only up to 2 mm away from the vibrating forager's body during trophallaxis, and in about 30 % of the cases, the antennal tips were above the wings or the thorax of the forager (Nieh 1998). These behavioral observations taken together with the available measurements of air particle velocity and of the response thresholds of Johnston's organ of the honeybee (Heran 1959) suggest that in stingless bees, hive bees can detect the air particle velocity induced by the forager's thoracic vibrations within a range of 5 mm.

### *18.5.3 Direct Transmission During Physical Contacts*

Unlike honeybees, nectar-collecting foragers of stingless bees generate their vibratory signals predominantly during their trophallactic interactions with food receiving bees (Hrncir et al. 2006a, b; Hrncir 2009). By these mouth-to-mouth contacts, hive bees learn about the sugar concentration, the secretion rate, and the odor of a nectar source (Farina and Grüter 2009; Jarau 2009). In addition, the food receivers are vibrated by the foragers during trophallaxis (Fig. 18.9), thereby receiving information about the profitability of a food patch. The vibratory input received during direct contact with the forager by far exceeds the vibratory stimulation through the substrate (Fig. 18.9). Bees in the immediate vicinity of the vibrating bee but not touching it will detect these substrate vibrations despite their small amplitude (see above). However, it will be difficult for receiver bees to

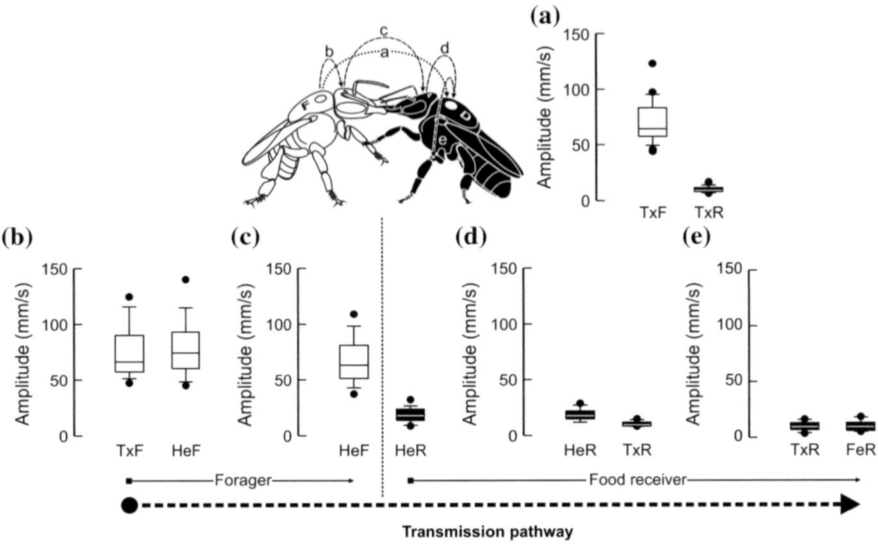


**Fig. 18.9** Vibration transmission during trophallaxis (*Melipona seminigra*). **a** Comparison of velocity amplitudes (*boxplot*) of the vibrations recorded from both the forager's thorax (*TxF*) and the receiver's thorax (*TxR*) using laser vibrometers. **b–e** Transmission pathway in more detail: *Boxplots* of velocity amplitudes simultaneously measured on the forager's thorax (*TxF*) and its head (*HeF*), on the forager's head and the food receiver's head (*HeR*), on the food receiver's head and its thorax (*TxR*), and on the food receiver's thorax and its femur (*FeR*). *F* forager; *D* food receiver. Differences between vibration amplitudes picked up at the same body parts (compare *TxF* in **a** and **b**, *HeF* in **b** and **c**, *HeR* in **c** and **d**, and *TxR* in **d** and **e**) are due to differences between vibrating individuals [Adapted from Hrncir et al. (2006b)]

extract information from these substrate vibrations as soon as two or more foragers returning from different food sources are within their perceptive range. As soon as a hive bee has direct trophallactic contact with the forager, its vibratory input will drastically exceed stimulation by way of the substrate. Information about the profitability of a single food source will then be easy to recognize by the vibration's magnitude.

Assuming similar properties for the subgenual organ of stingless bees and honeybees, the vibratory stimulation of the food receivers during trophallaxis ($\sim$10 mm/s) is well above the sensory threshold in stingless bees (average response threshold between 0.06 and 0.15 mm/s peak–peak at frequencies between 150 and 900 Hz; Kilpinen and Storm 1997; Rohrseitz and Kilpinen 1997). Yet, the subgenual organs are not the only vibration receptors in bees (Sandeman et al. 1996). An additional receptor had its highest sensitivity at low vibration frequencies between 20 and 100 Hz, with a displacement threshold of about 2 μm (corresponding to a velocity threshold between 0.5 and 1.5 mm/s at these frequencies; calculated from Sandeman et al. 1996). The unidentified receptor organ was suggested to be one of the other three chordotonal organs found in the femur, tibia, and tarsus of each leg (Snodgrass 1956). Additionally, a pair of small fusiform chordotonal organs in the head of honeybees and campaniform sensilla in the legs and the head potentially serve as vibration detectors (Snodgrass 1956).

A crucial question is whether potential recruits do actually have trophallactic contacts with the foragers or, alternatively, trophallaxis is restricted to hive bees unloading and storing the incoming food. According to studies of intranidal case histories of individually marked recruits of *M. quadrifasciata* and *M. seminigra*, prospective food collectors indeed do have trophallactic contacts with the foragers before they leave the nest to collect at an advertised food source (Hrncir et al. 2000; Kronberger 2000). The number of trophallactic food transfers and contacts even increases shortly before the prospective recruits leave the nest (*M. quadrifasciata*: Hrncir et al. 2000; *M. seminigra*: Kronberger 2000).

## 18.6 Conclusions and Outlook

Thoracic vibrations generated by foragers on their return from a profitable food source are a feature common among eusocial bees, that is the stingless bees (Meliponini), the honeybees (Apini), and the bumblebees (Bombini) (Hrncir et al. 2006a, 2011). To this day, few species have been studied in some detail. Yet, the available data all show that both the occurrence and the temporal pattern of the pulsed vibrations correlate with the profitability of the exploited food source (Meliponini: see above; Apini: *A. mellifera*; Esch 1962; Hrncir et al. 2011; Bombini: *Bombus terrestris*; Oeynhausen and Kirchner 2001). So far, it remains an open question whether these similarities in vibratory recruitment communication among eusocial bees derive from a common evolutionary origin or whether they have developed independently in the different bee groups. However, the dependence of the vibrational signals on the foragers' motivation as well as their correlation with recruitment success (stingless bees: see above; honeybees: Esch 1962; Dyer 2002; Hrncir et al. 2011) suggests a similar function of the thoracic vibrations for the coordination of foraging processes in eusocial bees. At least in stingless bees, this function is not the transfer of navigational information but of information on the profitability of the food source. Similar interpretations exist for the honeybee (Tautz 1996; Hrncir et al. 2011). It may come as a surprise, however, that even in the well-studied honeybee the question of how exactly the recruits perceive the dance information is far from being fully answered (Esch 2012; Michelsen 2012).

Although foraging strategies differ significantly among social bees, a principal function of intranidal recruitment mechanisms like the generation of vibratory signals is the rapid mobilization of a colony's foraging force. Among the Meliponini, a highly successful strategy is aggressive group foraging, described for many species of the genera *Trigona* and *Oxytrigona* (Hubbell and Johnson 1978; Johnson 1983; Biesmeijer and Slaa 2004). Here, large groups of aggressive foragers dislodge less aggressive species from a specific food patch and monopolize clumped and rich resources (Johnson 1983; Biesmeijer and Slaa 2004). The success of these aggressive species relies on the guidance of the entire group toward a specific goal. This is accomplished by the use of pheromone marks at and near the food patch (Lindauer and Kerr 1958; Schmidt et al. 2003; Jarau et al. 2004, 2006; Schorkopf et al. 2007, 2011; Barth et al. 2008; Jarau 2009). In addition, a quick activation of large numbers of individuals is fundamental to successfully chasing other species away from a food patch and to defending this patch against other aggressive colonies. The trade-off for this increased competitive ability is a reduced capacity to discover new food sources or even neighboring food patches independently (Hubbell and Johnson 1978; Biesmeijer and Slaa 2004). The foraging success of little or non-aggressive species, such as *Melipona* or *Nannotrigona* (Hubbell and Johnson 1978; Johnson 1983; Biesmeijer and Slaa 2004), relies on the quick detection of many food patches and a rapid activation of all available foragers. Thus, when dislodged from a food location by aggressive groups, these species are able to switch the colony's foraging focus to another food patch. Hence, although aggressive and non-aggressive species employ fundamentally different foraging strategies, a quick mobilization of unemployed foragers is required in both cases.

Based on our current knowledge of both the message and the potentially relevant transmission pathways of the vibratory signals of stingless bee foragers, we attribute three behavioral functions to the thoracic vibrations in recruitment communication. (1) *Medium-range transmission—attraction of hive bees to the forager*. Nectar-uptaking bees and food processors wait close to the nest entrance (Sommeijer and De Bruijn 1994; Hart and Ratnieks 2002). An increased "excitement" of a forager returning from a high-profit food source, and the resulting increase in pulse duration and duty cycle of her vibratory signals (Fig. 18.4), increasingly attracts food receivers to the forager (honeybee: Tautz and Rohrseitz 1998; Hasegawa and Ikeno 2011; stingless bees: Hart and Ratnieks 2002). Thereby, the resulting nectar transfer will accelerate the colony's food intake because foragers can resume their collecting activity faster. On the other hand, the gustatory and olfactory information about a profitable food source will spread more quickly through the colony, thereby arousing experienced but inactive foragers (Biesmeijer et al. 1998). (2) *Short-range transmission—reactivation of temporarily inactive foragers*. Just like the food receivers and nectar processors, unemployed experienced foragers stay close to the nest entrance (Nieh 1998; Hrncir 2009) and may be attracted toward the vibrating forager by the substrate vibrations received. Even without participating directly in the nectar transfer, these foragers receive confirming information about a known food source through the scents clinging to the forager's body (honeybee: Grüter and Farina 2009). In addition, they will receive information about the current state of profitability of the resource through the temporal pattern of the vibratory signals transmitted through air particle movement close to the vibrator's body. This latter information is thought important for the temporarily inactive individuals when deciding whether to resume their collecting activity or not (Biesmeijer et al. 1998; Biesmeijer and Slaa 2004). (3) *Direct transmission during trophallaxis–activation*. The quick activation of foragers to a particular food source helps to efficiently exploit ephemeral, high-profit food sources and necessitates the recruitment of collectors inexperienced regarding a particular food source. The novice foragers (Biesmeijer and de Vries 2001) receive multiple categories of information about a particular resource during trophallactic interactions: Once attracted to the forager, during the mouth-to-mouth food transfer, novices receive information about sugar concentration, nectar secretion rate, and the odor of a food source (Farina and Grüter 2009; Grüter and Farina 2009). In addition, they learn about the current profitability of the nectar source through the forager's vibrations. The sum of the information received lowers the foraging threshold of the novice bees (Biesmeijer et al. 1998; Biesmeijer and Slaa 2004; Hrncir 2009), which then leave the nest and search for the advertised food source.

Although knowledge about the vibratory signals in stingless bees has advanced considerably during the past decade, we are still far from a complete understanding of this intriguing communication system, which so efficiently coordinates the foraging processes. So far, some of the conclusions drawn are based on knowledge derived from studies on honeybees (*A. mellifera*). Future research will have to investigate the sensory mechanisms underlying the perception and processing of vibratory signals in the Meliponini, themselves. Only after having determined the physiological thresholds of the sensory organs involved, we will be able to determine the actual range of signal transmission and, subsequently, focus behavioral observations on hive bees within this range. The fact that stingless bees generate thoracic vibrations when tethered (Hrncir et al. 2008a) (Fig. 18.2) will help considerably in designing key experiments. Using annoyance-buzzing bees, thoracic vibrations can be generated under controlled laboratory conditions. This in turn permits the detailed investigation of both the pathways and respective attenuation of the signals on their way to the receivers (Hrncir et al. 2008b) and of the mechanical and physiological responses of receptors to genuine bee-produced vibrations instead of synthetic airborne sounds or substrate vibrations.

**Acknowledgments** Our sincere thanks go to Ronaldo Zucchi, Stefan Jarau, Dirk-Louis P. Schorkopf, and Veronika Schmidt whose participation, help, and discussion were decisive for many of the studies described in this chapter. Financial support for the research came from grants FAPESP (2006/50809-7) and CNPq (304722/2010-3) to MH, and FWF (P 14328, P 17530) to FGB.

## References

Aguilar I, Briceño D (2002) Sounds in *Melipona costaricensis* (Apidae: Meliponini): effect of sugar concentration and nectar source distance. Apidologie 33:375–388

Allerstorfer S (2004) Rekrutierung und Kommunikation bei Nannotrigona testaceicornis Lep. (1836) (Hymenoptera; Apidae; Meliponini). Diploma thesis, University of Vienna, Austria

Autrum H, Schneider W (1948) Vergleichende Untersuchungen über den Erschütterungssinn der Insekten. Z vergl Physiol 31:77–88

Barth FG (1998) The vibrational sense of spiders. In: Hoy RR, Popper AN, Fay RR (eds) Comparative hearing: insects. Springer, Berlin, pp 228–278

Barth FG, Bleckmann H, Bohnenberger J, Seyfarth EA (1988) Spiders of the genus *Cupiennius* Simon 1891 (Araneae, Ctenidae) II. On the vibratory environment of a wandering spider. Oecologia 77:194–201
Barth FG, Hrncir M, Jarau S (2008) Signals and cues in the recruitment behavior of stingless bees (Meliponini). J Comp Physiol A 194:313–327
Bennet-Clark HC (1999) Resonators in insect sound production: how insects produce loud pure-tone songs. J Exp Biol 202:3347–3357
Biesmeijer JC, de Vries H (2001) Exploration and exploitation of food sources by social insect colonies: a revision of the scout-recruit concept. Behav Ecol Sociobiol 49:89–99
Biesmeijer JC, Slaa EJ (2004) Information flow and organization of stingless bee foraging. Apidologie 35:143–157
Biesmeijer JC, van Nieuwstadt MGL, Lukács S, Sommeijer MJ (1998) The role of internal and external information in foraging decisions of *Melipona* workers (Hymenoptera: Meliponinae). Behav Ecol Sociobiol 42:107–116
Buchmann SL (1983) Buzz pollination in angiosperms. In: Jones CE, Little RJ (eds) Handbook of experimental pollination biology. Van Nostrand Reinhold, New York, pp 73–113
Burkart A, Lunau K, Schlindwein C (2011) Comparative bioacoustical studies on flight and buzzing of neotropical bees. J Pollination Ecol 6:118–124
Camargo JMF, Pedro SRM (2007) Meliponini Lepeletier, 1836. In: Moure JS, Urban D, Melo GAR (eds) Catalogue of bees (Hymenoptera, Apoidea) in the neotropical region. Sociedade Brasileira de Entomologia, Curitiba, pp 272–578
Conrad T, Paxton RJ, Barth FG, Francke W, Ayasse M (2010) Female choice in the red mason bee, *Osmia rufa* (L.) (Megachilidae). J Exp Biol 213:4065–4073
Dyer FC (2002) The biology of the dance language. Annu Rev Entomol 47:917–949
Eickwort GC, Ginsberg HS (1980) Foraging and mating behavior in Apoidea. Annu Rev Entomol 25:421–446
Esch H (1962) Über die Auswirkung der Futterplatzqualität auf die Schallerzeugung im Werbetanz der Honigbiene. Verh Deut Z 26:302–309
Esch H (1967) Die Bedeutung der Lauterzeugung für die Verständigung der stachellosen Bienen. Z vergl Physiol 56:199–220
Esch H (2012) Foraging honey bees: how foragers determine and transmit information about feeding site locations. In: Galizia CG, Eisenhardt D, Giurfa M (eds) Honeybee neurobiology and behavior. Springer, Dordrecht, pp 53–64
Esch HE, Burns JE (1995) Honeybees use optic flow to measure the distance of a food source. Naturwissenschaften 82:38–40
Esch H, Wilson D (1967) The sounds produced by flies and bees. Z vergl Physiol 54: 256–267
Esch H, Esch I, Kerr WE (1965) Sound: An element common to communication of stingless bees and to dances of the honey bee. Science 149:320–321
Farina WM, Grüter C (2009) Trophallaxis—a mechanism of information transfer. In: Jarau S, Hrncir M (eds) Food exploitation by social insects—ecological, behavioral, and theoretical approaches. CRC Press, Taylor & Francis Group, Boca Raton, pp 183–197
Frisch K von (1946) Die Tänze der Bienen. Österr Zool Z 1:1–48
Grüter C, Farina W (2009) The honeybee waggle dance: can we follow the steps? TREE 24:242–247
Hanauer-Thieser U, Nachtigall W (1995) Flight of the honey bee. VI. energetics of wind tunnel exhaustion flights at defined fuel content, speed adaptation and aerodynamics. J Comp Physiol B 165:471–483
Hart AG, Ratnieks FLW (2002) Task-partitioned nectar transfer in stingless bees: work organisation in a phylogenetic context. Ecol Entomol 27:163–168
Harter B, Leistikow C, Wilms W, Truylio B, Engels W (2002) Bees collecting pollen from flowers with poricidal anthers in a south Brazilian *Araucaria* forest: a community study. J Apicult Res 40:9–16
Hasegawa Y, Ikeno H (2011) How do honeybees attract nestmates using waggle dances in dark and noisy hives? PLoS ONE 6:e19619

Heran H (1959) Wahrnehmung und Regelung der Flugeigengeschwindigkeit bei *Apis mellifica* L. Z vergl Physiol 42:103–163

Hrncir M (2009) Mobilizing the foraging force—mechanical signals in stingless bee recruitment. In: Jarau S, Hrncir M (eds) Food exploitation by social insects—ecological, behavioral, and theoretical approaches. CRC Press, Taylor & Francis Group, Boca Raton, pp 199–221

Hrncir M, Jarau S, Zucchi R, Barth FG (2000) Recruitment behavior in stingless bees, *Melipona scutellaris* and *M. quadrifasciata* II. Possible mechanisms of communication. Apidologie 31:93–113

Hrncir M, Jarau S, Zucchi R, Barth FG (2003) A stingless bee (*Melipona seminigra*) uses optic flow to estimate flight distances. J Comp Physiol A 189:761–768

Hrncir M, Jarau S, Zucchi R, Barth FG (2004a) Thorax vibrations of a stingless bee (*Melipona seminigra*). I. No influence of visual flow. J Comp Physiol A 190:539–548

Hrncir M, Jarau S, Zucchi R, Barth FG (2004b) Thorax vibrations of a stingless bee (*Melipona seminigra*). II. Dependence on sugar concentration. J Comp Physiol A 190:549–560

Hrncir M, Barth FG, Tautz J (2006a) Vibratory and airborne-sound signals in bee communication (Hymenoptera). In: Drosopoulos S, Claridge MF (eds) Insect sound and communication – physiology, behaviour, ecology and evolution. CRC Press, Taylor & Francis Group, Boca Raton, pp 421–436

Hrncir M, Schmidt VM, Schorkopf DLP, Jarau S, Zucchi R, Barth FG (2006b) Vibrating the food receivers: a direct way of signal transmission in stingless bees (*Melipona seminigra*). J Comp Physiol A 192:879–887

Hrncir M, Gravel AI, Schorkopf DLP, Schmidt VM, Zucchi R, Barth FG (2008a) Thoracic vibrations in stingless bees (*Melipona seminigra*): Resonances of the thorax influence vibrations associated with flight but not those associated with sound production. J Exp Biol 211:678–685

Hrncir M, Schorkopf DLP, Schmidt VM, Zucchi R, Barth FG (2008b) The sound field generated by tethered stingless bees (*Melipona scutellaris*): inferences on its potential as a recruitment mechanism inside the hive. J Exp Biol 211:686–698

Hrncir M, Maia-Silva C, McCabe SI, Farina WM (2011) The recruiter's excitement—features of thoracic vibrations during the honey bee's waggle dance related to food source profitability. J Exp Biol 214:4055–4064

Hubbel SP, Johnson LK (1978) Comparative foraging behavior of six stingless bee species exploiting a standardized resource. Ecology 59:1123–1136

Jarau S (2009) Chemical communication during food exploitation in stingless bees. In: Jarau S, Hrncir M (eds) Food exploitation by social insects—ecological, behavioral, and theoretical approaches. CRC Press, Taylor & Francis Group, Boca Raton, pp 223–249

Jarau S, Hrncir M, Zucchi R, Barth FG (2000) Recruitment behavior in stingless bees, *Melipona scutellaris* and *M. quadrifasciata*. I. Foraging at food sources differing in direction and distance. Apidologie 31:81–91

Jarau S, Hrncir M, Zucchi R, Barth FG (2004) A stingless bee uses labial gland secretions for scent trail communication (*Trigona recursa* SMITH 1863). J Comp Physiol A 190:233–239

Jarau S, Schulz CM, Hrncir M, Francke W, Zucchi R, Barth FG, Ayasse M (2006) Hexyl decanoate, the first trail pheromone compound identified in a stingless bee (*Trigona recursa*). J Chem Ecol 32:1555–1564

Johnson LK (1983) Foraging strategies and the structure of stingless bee communities in Costa Rica. In: Jaisson P (ed) Social insects in the tropics 2. Université Paris-Nord, Paris, pp 31–58

Kilpinen O, Storm J (1997) Biophysics of the subgenual organ of the honeybee, *Apis mellifera*. J Comp Physiol A 181:309–318

King MJ (1993) Buzz foraging mechanism of bumble bees. J Apicult Res 32:41–49

King MJ, Buchmann SL, Spangler H (1996) Activity of asynchronous flight muscle from two bee families during sonication (buzzing). J Exp Biol 199:2317–2321

Kirchner W (1994) Hearing in honeybees: the mechanical response of the bee's antenna to near field sound. J Comp Physiol A 175:261–265

Kronberger E (2000) Futterplatzrekrutierung bei Melipona seminigra merillae. Diploma thesis, University of Vienna, Austria
Larsen O, Gleffe G, Tengö J (1986) Vibration and sound communication in solitary bees and wasps. Physiol Entomol 11:287–296
Lindauer M (1956) Über die Verständigung bei indischen Bienen. Z vergl Physiol 38:521–557
Lindauer M, Kerr WE (1958) Die gegenseitige Verständigung bei den stachellosen Bienen. Z vergl Physiol 41:405–434
Lindauer M, Kerr WE (1960) Communication between workers of stingless bees. Bee World 41(29–41):65–71
Menzel R (2012) Navigation and communication: commentary. In: Galizia CG, Eisenhardt D, Giurfa M (eds) Honeybee neurobiology and behavior. Springer, Dordrecht, pp 117–122
Michelsen A (2003) Signals and flexibility in the dance communication of honeybees. J Comp Physiol A 189:165–174
Michelsen A (2012) How do honey bees obtain information about direction by following dances? In: Galizia CG, Eisenhardt D, Giurfa M (eds) Honeybee neurobiology and behavior. Springer, Dordrecht, pp 65–76
Michelsen A, Nocke H (1974) Biophysical aspects of sound communication in insects. Adv Insect Physiol 10:247–296
Michelsen A, Towne WF, Kirchner WH, Kryger P (1987) The acoustic near field of a dancing honeybee. J Comp Physiol A 161:633–643
Michener CD (1962) An interesting method of pollen collecting by bees from flowers with tubular anthers. Rev Biol Trop 10:167–175
Michener CD (1974) The social behavior of the bees: a comparative study. Harvard University Press, Cambridge
Michener CD (2000) The bees of the world. Johns Hopkins University Press, Baltimore
Morawetz L (2007) Reichweite und Übertragung vibratorischer Signale bei der Kommunikation stachelloser Bienen. Diploma thesis, University of Vienna, Austria
Nachtigall W (2003) Insektenflug. Springer, Berlin
Nieh JC (1998) The food recruitment dance of the stingless bee, *Melipona panamica*. Behav Ecol Sociobiol 43:133–145
Nieh JC, Roubik DW (1995) A stingless bee (*Melipona panamica*) indicates food location without using a scent trail. Behav Ecol Sociobiol 37:63–70
Nieh JC, Roubik DW (1998) Potential mechanisms for the communication of height and distance by a stingless bee, *Melipona panamica*. Behav Ecol Sociobiol 43:387–399
Nieh JC, Sanchez D (2005) Effect of food quality, distance and height on thoracic temperature in the stingless bee *Melipona panamica*. J Exp Biol 208:3933–3943
Nieh JC, Contrera FAL, Rangel J, Imperatriz-Fonseca VL (2003) Effect of food location and quality on recruitment sounds and success in two stingless bees, *Melipona mandacaia* and *Melipona bicolor*. Behav Ecol Sociobiol 55:87–94
Nocke H (1971) Biophysik der Schallerzeugung durch die Vorderflügel der Grillen. Z vergl Physiol 74:272–314
Nunes-Silva P (2011) Capacidade vibratória e polinização por vibração nas abelhas do gênero Melipona (Apidae, Meliponini) e Bombus (Apidae, Bombini). PhD thesis, University of São Paulo-Ribeirão Preto, Brazil
Nunes-Silva P, Hrncir M, Imperatriz-Fonseca VL (2010) A polinização por vibração. Oecolog Aust 14:140–151
Oeynhausen A, Kirchner WH (2001) Vibrational signals of foraging bumblebees (Bombus terrestris) in the nest. In: Proceedings of the meeting of the European sections of IUSSI, Berlin, Germany, p 31
Ramírez SR, Nieh JC, Quental TB, Roubik DW, Imperatriz-Fonseca VL, Pierce NE (2010) A molecular phylogeny of the stingless bee genus *Melipona* (Hymenoptera: Apidae). Mol Phylogenet Evol 56:519–525
Rasmussen C, Cameron SA (2010) Global stingless bee phylogeny supports ancient divergence, vicariance, and long distance dispersal. Biol J Linn Soc 99:206–232

Rohrseitz K (1998) Biophysikalische und ethologische Aspekte der Tanzkommunikation der Honigbienen (Apis mellifera carnica Pollm.). Doctoral thesis, Julius Maximilian University Würzburg, Germany

Rohrseitz K, Kilpinen O (1997) Vibration transmission characteristics of the legs of freely standing honeybees. Zoology 100:80–84

Roselino AC, Hrncir M (2012) Repeated unrewarded scent exposure influences the food choice of stingless bee foragers, *Melipona scutellaris*. Anim Behav 83:755–762

Roubik DW (1989) The ecology and natural history of tropical bees. Cambridge University Press, Cambridge

Roubik DW (2006) Stingless bee nesting biology. Apidologie 37:124–143

Samwald U (2000) Mechanismen der Futterplatzrekrutierung bei Melipona seminigra merillae CCKL (1919) (Hymenoptera; Apidae; Meliponinae). Diploma thesis, University of Vienna, Austria

Sandeman DC, Tautz J, Lindauer M (1996) Transmission of vibration across honeycombs and its detection by bee leg receptors. J Exp Biol 199:2585–2594

Schmidt VM, Zuchi R, Barth FG (2003) A stingless bee marks feeding site in addition to the scent path (*Scaptotrigona* aff. *depilis*). Apidologie 34:237–248

Schmidt VM, Zucchi R, Barth FG (2006) Recruitment in a scent trail laying stingless bee (*Scaptotrigona* aff. *depilis*): changes with reduction but not with increase of the energy gain. Apidologie 37:487–500

Schmidt VM, Hrncir M, Schorkopf DLP, Mateus S, Zucchi R, Barth FG (2008) Food profitability affects intranidal recruitment behaviour in the stingless bee *Nannotrigona testaceicornis*. Apidologie 39:260–272

Schneider P (1975) Versuche zur Erzeugung des Verteidigungstones bei Hummeln. Zool Jahrb allg Zool 79:111–127

Schön A (1911) Bau und Entwicklung des tibialen Chordotonalorgans bei der Honigbiene und bei Ameisen. Zool Jahr Anat 31:439–472

Schorkopf DLP, Jarau S, Francke W, Twele R, Zucchi R, Hrncir M, Schmidt VM, Ayasse M, Barth FG (2007) Spitting out information. *Trigona* bees deposit saliva to signal resource locations. P Roy Soc Lond B 274:895–898

Schorkopf DL, Morawetz L, Bento JM, Zucchi R, Barth FG (2011) Pheromone paths attached to the substrate in meliponine bees: helpful but not obligatory for recruitment success. J Comp Physiol 197:755–764

Schwarz HF (1948) Stingless bees (Meliponidae) of the western hemisphere. B Am Mus Nat Hist 90:1–546

Seeley TD (1992) The tremble dance of the honey bee: message and meanings. Behav Ecol Sociobiol 31:375–383

Simpson J (1964) The mechanism of honey-bee queen piping. Z vergl Physiol 48:277–282

Snodgrass RE (1956) Anatomy of the honey bee. Comstock Publishing Associates, Ithaca

Sommeijer MJ, De Bruijn LLM (1994) Intranidal feeding, trophallaxis and sociality in stingless bees. In: Hunt J, Nalepa C (eds) Nourishment and evolution in insect societies. Westview Press, Oxford, pp 391–418

Srinivasan MV, Zhang S, Altwein M, Tautz J (2000) Honeybee navigation: nature and calibration of the "odometer". Science 287:851–853

Storm J, Kilpinen O (1998) Modelling the subgenual organ of the honeybee, *Apis mellifera*. Biol Cybern 78:175–182

Tautz J (1996) Honeybee waggle dance: recruitment success depends on the dance floor. J Exp Biol 199:1375–1381

Tautz J, Rohrseitz K (1998) What attracts honeybees to a waggle dancer? J Comp Physiol A 183:661–667

Tautz J, Casas J, Sandeman DC (2001) Phase reversal of vibratory signals in honeycomb may assist dancing honeybees to attract their audience. J Exp Biol 204:3737–3746

Vicidomini S (1998) Biology of *Xylocopa* (*Xylocopa*) *violacea* (L., 1758) (Hymenoptera: Apidae): female nest-defence. Annali del Museo Civico di Rovereto 12:85–100

Wille A (1963) Behavioral adaptations of bees for pollen collecting from *Cassia* flowers. Rev Biol Trop 11:205–210

Wille A, Michener CD (1973) The nest architecture of stingless bees with a special reference to those of Costa Rica (Hymenoptera, Apidae). Rev Biol Trop 21(Suppl 1):1–278

# Chapter 19
# The Role of Frequency in Vibrational Communication of Orthoptera

**Nataša Stritih and Andrej Čokl**

**Abstract** The chapter examines characteristics and function of vibratory signals in Orthoptera, which are emitted by different mechanisms. Detection and neural processing of the spectrally diverse signals and the behavioural correlates indicating differential perception of different frequency ranges are discussed. In the light of the knowledge mainly acquired from hearing Ensifera, data from primitively non-hearing cave crickets are highlighted as a comparative system, offering important new insights into the functional organisation and evolution of the vibratory system in Ensifera, and Orthoptera in general. Data from cave crickets, from the behaviour to properties of neuron circuits, stress the importance of perception of low-frequency vibratory signals, which appear to have been underestimated in these insects so far.

## 19.1 Introduction

Orthoptera represent one of the longest established models for studies of insect acoustic and vibratory communication at different levels, from behaviour, ecology, environmental physics and biophysics to neurobiology. The focus of research of their vibratory channel, however, was communication via substrate-borne signals that are produced by stridulation, simultaneously with sound. Here, we revise characteristics of different modes of signalling expressed by these insects, which in addition to stridulation also use drumming or tapping on the substrate (percussion), as well as abdominal and whole-body tremulation, to produce vibration signals

N. Stritih (✉) · A. Čokl
National Institute of Biology, Večna pot 111, 1000 Ljubljana, Slovenia
e-mail: natasa.stritih@nib.si

A. Čokl
e-mail: andrej.cokl@nib.si

R. B. Cocroft et al. (eds.), *Studying Vibrational Communication*,
Animal Signals and Communication 3, DOI: 10.1007/978-3-662-43607-3_19,

during mating, territorial or agonistic interactions (Field 2001a; Virant-Doberlet and Čokl 2004; Benediktov 2009). We aim to show that signals produced in different ways largely differ in spectral characteristics, especially when recorded from the species' natural substrates, and that these signals are detected and processed mostly by different sets of receptors and sensory interneurons. With a note on Caelifera, we focus mainly on signal spectral characteristics, reception, neural processing and the related behaviours from representatives of the suborder Ensifera, the group in which vibratory communication appears to have greatly influenced evolution of audition. The review highlights data from primitively non-hearing cave crickets, from behaviour to different levels of sensory neurobiology. These data are of special significance in understanding both function and evolution of the vibratory system in Ensifera and stress the importance of perception of low-frequency vibratory information, which appears to have been underestimated in these insects so far.

## 19.2 Frequency Characteristics and Function of Signals Produced by Different Mechanisms

### *19.2.1 Vibratory Signals of Caelifera*

Communication via substrate vibration in the orthopteran suborders Ensifera and Caelifera is present apparently in a very different extent. While elaborate sound and/or vibratory signalling is known from representatives of most ensiferan families, in Caelifera, such signalling has been described only for species in three out of eight superfamilies. Among true grasshoppers (Acridoidea), which produce sound via femuro-tegminal stridulation, some Oedipodinae are known to include oscillations and stroking the substrate with the hindlegs in their acoustical mating displays (Loher and Chandrashekaran 1970). Abdominal vibration, hindleg jerking and vibration, and drumming on the substrate have been described as a common mating strategy combined with visual signalling for several soundless neotropical species of Acridoidea that belong to six subfamilies (Riede 1987). Vibratory sexual displays were recently described, together with the recorded signals, for representatives of the long-considered deaf and silent Tetrigoidea (groundhoppers; g. *Tetrix*) and Eumastacoidea (Benediktov 2009; Kočárek 2011). During courtship and copulation, the former produce different types of species- and sex-specific "buzzing" signals, presumably based on thoracic muscle contraction, and in species *Tetrix ceperoi* and *Tetrix bolivari,* signals in addition to those produced by thoracic muscle contraction are emitted by wing vibration. Peak frequencies of the buzzing signals range between 150–500 Hz among the various species, while wing vibration produces somewhat higher frequencies (480–1350 Hz). In *Erianthus versicolor* of the Eumastacoidea, vibratory signalling consists of male abdominal and body vibration, with peak frequencies between 100–250 Hz, and of female abdominal tapping on the substrate (Benediktov 2009; Kočárek 2010). These spectra, however, relate to vibrations of the cellophane paper fixed over the plastic rim, which was part of the recording set-up, and most likely do not entirely reflect the natural signal properties. It is known that the substrate strongly influences frequency characteristics of vibrations, both during emission and transmission, as has been extensively described for stink bugs (see Chap. 10, this volume). Further investigations of the presence of vibratory signals are needed for the soundless Caelifera, which may use the vibratory channel much more often than it appears currently.

## *19.2.2 Vibratory Signals of Ensifera*

### 19.2.2.1 Stridulatory and Drumming Signals

To attract mates from long distances, the terminal taxa of Ensifera, including crickets (Gryllidae), bush crickets (Tettigoniidae) and mole crickets (Gryllotalpidae), produce sound by tegminal stridulation, by which a vibrational signal component is transmitted over the legs to the substrate. In crickets and bush crickets, these stridulatory vibrations improve localisation of the signaller at short range (Latimer and Schatral 1983; Weidemann and Keuper 1987) and probably influence male-male spacing behaviour in bush crickets (Schatral and Kalmring 1985) and mole crickets (Hill and Shadley 1997). In a stridulating cricket *Gryllus bimaculatus,* the vibration spectrum induced in dry soil includes frequency components between 30–500 Hz at low intensity, detectable by conspecifics only a few dm from the emitter (Weidemann and Keuper 1987). Despite strong damping of high frequencies in the ground, Hill and Shadley (1997, 2001) measured ground vibrations 45 cm from the mole cricket *Gryllotalpa major*, which was singing from a burrow with the peak sound frequency of about 2 kHz. Analysis of the ground vibrations found the dominant peak at 100 Hz, prominent harmonics expressed between 100–900 Hz, and a smaller peak at around 2 kHz. Stridulatory vibrations of much broader frequency spectra are induced by bush crickets singing on vegetation. The signals measured for species of the genera *Decticus*, *Ephippiger*, *Tettigonia* and *Isophya* on several types of green plants and on bushes, at distances up to 80 cm from the emitter, included prominent frequency peaks in the range up to 10 kHz that reflected the low-frequency components of the species' airborne sounds (Keuper and Kühne 1983; Keuper et al. 1985). These plant vibrations were characterised by high intensities, with acceleration values between about 0.5 and 1 $m/s^2$ close to the emitter, and by low damping during transmission (20–50 dB/m). Taking into account threshold sensitivity of sensory neuron elements in bush crickets, these values suggested communication ranges of up to 1.5–2 m through plants (Keuper and Kühne 1983).

Some other forms of signal production are expressed by Ensifera that are largely apterous. Non-hearing raspy crickets (Gryllacrididae), Jerusalem crickets

(Stenopelmatidae) and splay-footed crickets (Schizodactylidae) use femuro-abdominal stridulation for defence purposes (Field and Bailey 1997; Weissman 2001), while some of the weta (Anostostomatidae; Deinacridinae), which possess primitive ears, use such signalling also in the social context (Field 2001b). McVean and Field (1996) investigated vibration transmission along the manuka tree for signals induced by defensive stridulation in the tree weta *Hemideina femorata* and for those induced by a falling ball. Both sources induced a resonant oscillation at 0.8–1.5 kHz, which was transmitted with low damping of about 10 dB/m, while an additional peak was induced at 7.5 kHz by stridulation.

Of the above-mentioned non-hearing taxa, raspy crickets and Jerusalem crickets use drumming vibrations for mating purposes. The signals are produced by both sexes, by stroking the hindlegs and abdomen against the substrate in the two groups, respectively (Field and Bailey 1997; Weissman 2001). Field and Bailey (1997) described temporal properties and showed waveforms of signals recorded from sand for two species of the genera *Ametrus* and *Hadrogryllacris*. Weissman (2001) described species-specific temporal patterns of various signal types for a large number of Jerusalem cricket species. Since, however, only the airborne signal component was recorded, frequency characteristics of drumming vibrations for these Ensifera were left to be described.

#### 19.2.2.2 Tremulatory Signals

Vibration of the whole body or the abdomen without touching the substrate, also termed tremulation, is frequently expressed by males of many cricket and bush cricket species during close-range courtship (Alexander and Otte 1967; Gwynne 1977; Bell 1980; Keuper et al. 1985; De Luca and Morris 1998). More complex tremulation signalling that is performed by both sexes, and includes courtship and calling signals transmitted through vegetation, is found among neotropical bush crickets in which auditory signalling was partially or completely reduced (Morris 1980; Morris et al. 1994). In tree wetas, abdominal tremulation is expressed by the males during agonistic battles (Field 2001b). Vibratory courtship and post-copulatory signals of the male cave crickets *Troglophilus neglectus* (Rhaphidophoridae) were recorded recently (Fig. 19.1; Stritih and Čokl 2012). These signals, emitted by abdominal and whole-body tremulation, respectively, provide the first evidence of mechanical signalling for Rhaphidophoridae, which are presumed to be the (most) primitive non-hearing Ensifera (Desutter-Grandcolas 2003; Strauss and Lakes-Harlan 2008a).

Tremulation signals of Ensifera were recorded mostly from plant materials, their natural substrates for mating, which demonstrated the very low-frequency nature of such signals. In bush crickets, the dominant frequency was expressed at around 30 Hz in the signals of *Ephippiger ephippiger*, recorded on a plant about 15 cm from the signaller (Keuper et al. 1985), at around 25 Hz in *Myopophyllum speciosum* on a Hibiscus plant about 30 cm from the male, while the frequency peak was between 8–12 Hz for *Docidocercus gigliotosi* and between 35–45 Hz for

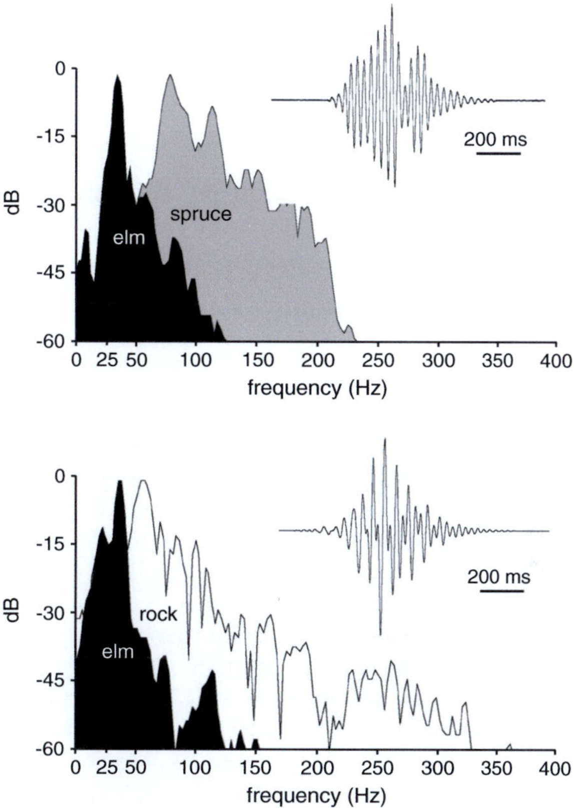


**Fig. 19.1** Spectral characteristics of vibratory signals of the cave cricket *T. neglectus*. Shown are frequency spectra of the velocity component of the signals emitted by males during courtship (*above*) and after copulation (*below*), using abdomen and whole-body tremulation, respectively. The signals were recorded with a laser vibrometer from elm bark, spruce bark and rock (as indicated) a few cm from the male. The display range of 60 dB is normalised to the dominant peak amplitude. At the right are oscillograms of both signal types emitted on elm bark. (see also examples in Stritih and Čokl 2012)

*Copiphora brevirostris* when measured on Heliconia (Morris et al. 1994). In the cave cricket *T. neglectus*, which mates predominantly outside caves, the signals were recorded from bark, moss and rock. Spectral characteristics were similar between courtship and post-copulatory signals, but strongly substrate dependent, so that the dominant frequency ranged around 30 Hz on elm bark, between 80–110 Hz on spruce bark and on moss, and around 40 Hz on rock (Fig. 19.1). In all the signals described above, the spectral energy was concentrated in the range below 200 Hz. These frequencies, and especially dominant peaks of the signals, overlap with the range of wind-induced noise on plants, which predominantly contains components below 100 Hz (Cocroft and Rodriquez 2005; McNett et al. 2010). The possible behavioural and neural mechanisms for an increased efficiency of such signalling in the presence of environmental noise are still to be investigated for the Ensifera, and generally for insects signalling by tremulation.

On artificial substrates, on the other hand, such as metal cages that were used to record tremulations of the bush crickets *Choeroparnops gigliotosi* (Morris et al. 1994) and *Conocephalus nigropleurum* (DeLuca and Morris 1998), prominent frequencies ranged between 100–400 Hz. Similarly, in cave crickets, one of the males was signalling on the terrarium metal net cover with the dominant frequency induced as a resonance at around 150 Hz. These data highlight the importance of recording the signals from the species' natural substrates.

## 19.3 Detection of Vibratory Signals of Different Carrier Frequencies

### *19.3.1 High-Frequency Receptors*

The central role in detection of vibratory signals in Ensifera has been traditionally ascribed to the complex tibial organ (CTO) developed in the proximal tibiae. Most proximal in this chordotonal organ complex is the vibratory subgenual organ (SGO), which is highly developed in Orthopteroidea (Autrum and Schneider 1948; Field and Matheson 1998). In the forelegs of Ensifera, the SGO is distally adjoined by the sound-sensitive tympanal organ in crickets, and the intermediate organ (IO) and the *crista acoustica* (CA), which detect both vibration and sound, in bush crickets (Ball et al. 1989; Rössler et al. 2006). In mid- and hindlegs, as well as in species with no auditory tympana, the homologues of these sensory parts function together with the SGO as substrate vibration receivers (Jeram et al. 1995; Kalmring et al. 1994; Strauss and Lakes-Harlan 2008a, b, 2010).

The frequency range of vibration detected by the SGO was found to be similar across Ensifera. In crickets, responses of the SGO from whole nerve recordings in different legs were tuned to frequencies between 500–1000 Hz with an extremely low threshold (0.002–0.005 $m/s^2$; Dambach 1972). In addition to such broadband responses, tuning to 100–300 Hz with a relatively low threshold (0.01 $m/s^2$) was revealed among individual cricket vibratory receptors; these, however, were only presumed to originate in the SGO (Kühne et al. 1984). In bush crickets, vibratory sensitivity extends to higher frequencies in comparison with crickets. Based on recordings from 113 midleg receptors in *Gampsocleis gratiosa,* three functional types were described with a broad tuning to 200–1000 Hz and minimal thresholds between 0.04 and 0.4 $m/s^2$, three further types tuned to 700–1000 Hz that were the most sensitive (with threshold at 0.005–0.02 $m/s^2$) and four types tuned to 1–4 kHz with an intermediate sensitivity (0.05–0.2 $m/s^2$; Kalmring et al. 1994). These sets of response types, found to be similar in other legs, were attributed to the SGO, the proximal IO and the distal IO with the proximal CA, respectively.

In the TO of cave crickets, the distal part of the IO and the whole CA is missing, while the SGO and the proximal IO are normally developed in all legs (Jeram et al. 1995). Such a situation is exclusive among the Ensifera and is supposed to reflect a primitive condition (see Strauss and Lakes-Harlan 2008a, b, 2010). Accordingly, similar vibratory responses of receptor neurons with best frequency between 200 Hz and 2 kHz were found as in bush crickets, while those tuned to higher frequencies were absent (Fig. 19.2; Čokl et al. 1995). Examples of

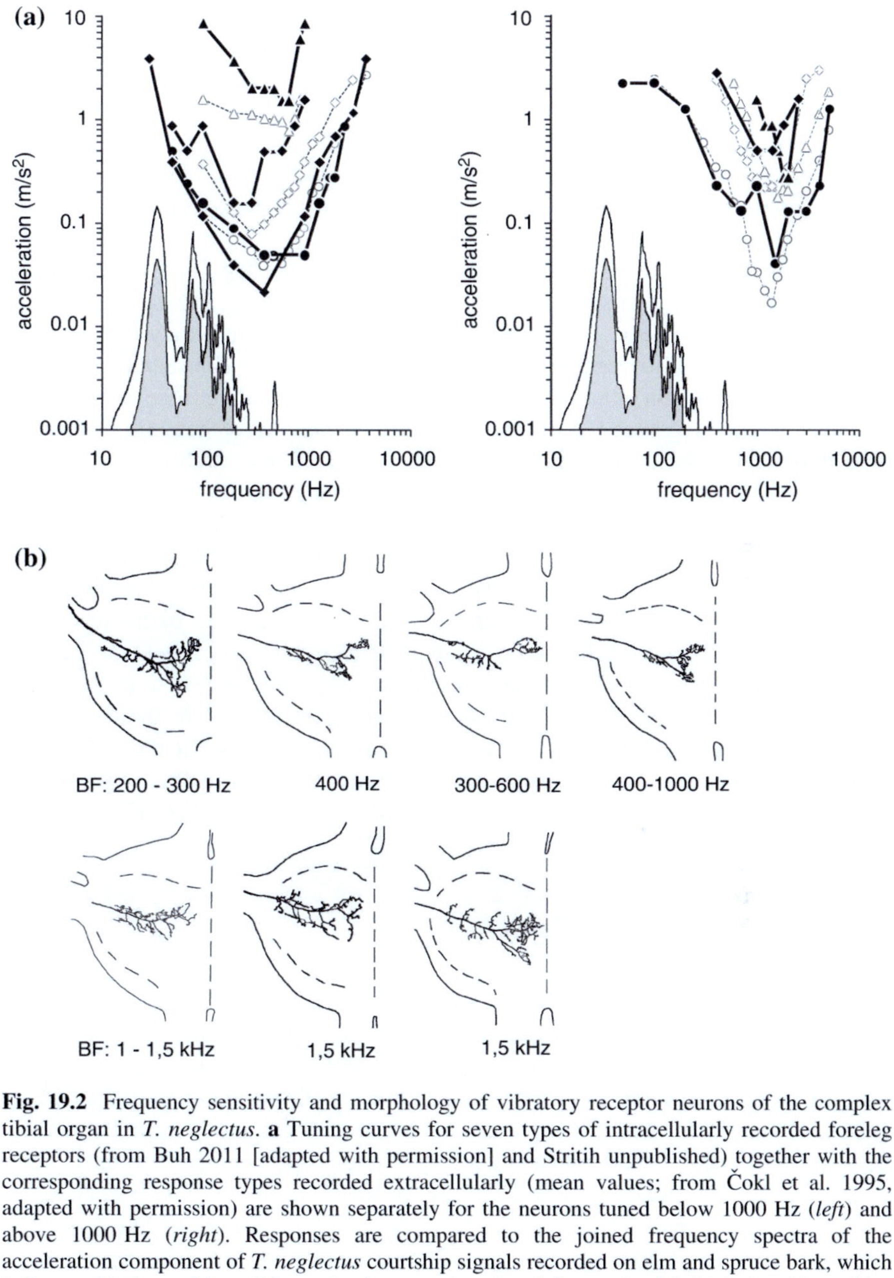


**Fig. 19.2** Frequency sensitivity and morphology of vibratory receptor neurons of the complex tibial organ in *T. neglectus*. **a** Tuning curves for seven types of intracellularly recorded foreleg receptors (from Buh 2011 [adapted with permission] and Stritih unpublished) together with the corresponding response types recorded extracellularly (mean values; from Čokl et al. 1995, adapted with permission) are shown separately for the neurons tuned below 1000 Hz (*left*) and above 1000 Hz (*right*). Responses are compared to the joined frequency spectra of the acceleration component of *T. neglectus* courtship signals recorded on elm and spruce bark, which indicate with the position of the peaks the mean (*grey*) and the maximal (*white*) signal intensities measured. **b** Whole-mount drawings of axonal arborisations in the prothoracic ganglion for different types of receptor neurons, with indicated best frequencies. *Dashed lines* indicate the ganglion midline and the border between the cortex and the sensory neuropile

intracellularly stained receptors in *T. neglectus* were described from the whole CTO response range as possessing a large terminal bifurcation and/or extensive lateral branches of the axon in the neuropile (Fig. 19.2b; Buh 2011, Stritih unpublished). In the weta *Hemideina femorata,* selective backfilling of neurons through small peripheral nerve branches revealed such morphologies for receptors in different parts of the SGO (Nishino and Field 2003). This demonstrates that ascribing peripheral origin of the receptor neurons based solely on their physiology is not entirely supportable.

The function of CA receptors present in atympanate legs and in most non-hearing Ensifera (the crista acoustica homologue (CAH); Strauss and Lakes-Harlan 2008a) is unclear, as with respect to frequency tuning, they appear to be largely redundant. However, in the bush cricket *Polysarcus denticauda,* sensitive responses to vibration were found among fore- and midleg receptors also at 7–8 kHz, suggesting that CA in atympanate legs may serve detection of (very) high-frequency vibration (Kalmring et al. 1996). These additional vibratory receptors may, on the other hand, not serve frequency but intensity range fractionation. Their function needs to be investigated by intracellular studies, including complete morphological identification of the recorded cells (see Stumpner 1996).

We can conclude that the CTO of Ensifera enables detection of vibrations in the broad frequency range above 200 Hz, such as are emitted by stridulation. The extended frequency sensitivity of the organ in bush crickets may reflect the extreme broadband nature of their signals transmitted through plants. In cave crickets, which are unable to stridulate, the lack of an additional sensory part in the organ may reflect the absence of selection for detection of such signals. At the same time, tremulatory signals of cave crickets and similarly of other Ensifera appear to be mostly, if not completely, outside of the CTO detection range (Fig. 19.2a).

### *19.3.2 Low-Frequency Receptors*

A heterogeneous group of mechanoreceptors were recorded in the legs of Orthoptera that respond to vibration frequencies below 200 Hz, but with a relatively low sensitivity. Such responses, typically phase-locked to the sine-wave stimulus, were recorded in bush crickets, crickets and locusts (Kalmring et al. 1978; Kühne 1982a; Kühne et al. 1984) with a tuning to 30–100 Hz and minimal threshold between 0.03 and 0.2 $m/s^2$ following a line of equal displacement ("type 1" after Kühne 1982a), or between 0.1 and 10 $m/s^2$ following a line of equal acceleration ("type 2" after Kühne 1982a). These two response types were attributed to campaniform sensilla (CS) and joint chordotonal organs of the leg, respectively. Simultaneous staining of the recorded receptor axons in the central nervous system later showed that such phase-locked responses could have at least three different origins: campaniform sensilla, fields of hair plates (HP) in the proximity of joints, the distal scoloparium of the femoral chordotonal organ (FeCOd) and possibly the tibio-tarsal chordotonal organ (Mücke 1989; Lakes and Schikorski 1990).

Although the most sensitive among these low-frequency receptors would detect tremulation signals at the intensities characteristic for Ensifera, it is a question to what degree they might be involved in communication. The CS, HP and the FeCOd are known to function primarily as proprioreceptors involved in posture and locomotion control (Field and Pflüger 1989; Kuenzi and Burrows 1995; Zill et al. 1999). Also, their central neuronal projections (Pflüger et al. 1988; Mücke and Lakes-Harlan 1995; Nishino 2000, 2003) terminate outside the medio-ventral association centre (mVAC), which is the primary auditory-vibratory processing neuropile in the ventral nerve cord of Orthoptera (Pflüger et al. 1988; Wohlers and Huber 1985; Stritih and Stumpner 2009; see also Fig. 19.5a). Such receptors may rather be candidates for mediating reflex reactions, such as is the startle response elicited by high-intensity vibration of low frequencies (Friedel 1999; Stritih unpublished, see Sect. 19.5.2).

However, among the functionally diverse scoloparia of the FeCO in Orthoptera, only the distal scoloparium neurons are involved in proprioreception. As demonstrated in locusts, neurons of its proximal scoloparium (FeCOp) do not respond to large-scale tibial movements, but the low-frequency (<300 Hz) vibration of the receptor apodeme (Field and Pflüger 1989). Also, their axons project exclusively into the mVAC, suggesting they may be specialised for detecting vibration (Field and Pflüger 1989; Nishino 2000, 2003). In Ensifera, this scoloparium may contain up to 400 neurons, albeit with tiny axons (Nishino and Sakai 1997), which may be the reason they were not recorded so far individually.

## 19.4 Central Frequency Processing and Corresponding Behaviour

### *19.4.1 Properties of Vibratory Interneurons*

Processing of vibratory information at the central neural level was investigated predominantly in the hearing species of Orthoptera, where vibratory and auditory information converges in large part onto common interneurons (Čokl et al. 1977). Responses from over a hundred tested neurons, ascending from the thoracic region towards the head ganglia in various bush crickets (*P. denticauda*, *Decticus verrucivorus* and *Tettigonia cantans*) and in the locust (*Locusta migratoria*), allowed characterisation of the same physiological types of vibratory neurons across the species (Kalmring and Kühne 1980; Silver et al. 1980; Kühne 1982b; Kalmring et al. 1997). Respecting frequency sensitivity and temporal characteristics of the response, five types of vibration sensitive (V) and five types of bimodal (vibration and sound sensitive; VS) neurons were described. Three types responded to vibration in the narrow range up to 100/200 Hz (V4, V5, VS1), one type was more broadly tuned to 50–200 Hz (V1), while five types expressed broad tuning with highly sensitive responses up to 2 kHz and some response up to the 5 kHz limit of the test (V2, V3, VS2–4). Some other response types were found among the ascending vibratory neurons in the cricket *Gryllus campestris*, either with a sharp tuning to 200 Hz (three types) or 1 kHz (one type) expressed with high sensitivity, and a broad tuning to 400–1000 Hz in less sensitive neurons (three types; Kühne et al. 1984).

It is unclear to what extent these physiological types reflect the actual ratio of coexisting vibratory neurons devoted to processing spectrally diverse signals. With an exception of a prothoracic TN1 neuron (Kühne et al. 1984), morphology of vibration sensitive neurons is unknown for crickets. For bush crickets *Decticus verrucivorus* and *D. albifrons*, scattered examples of vibratory neurons projecting towards the brain were described in different ventral nerve cord segments (Sickmann 1996; Nebeling 2000), including four low-frequency- and seven broadband/high-frequency-tuned neurons. These, however, probably represent only a small fraction of central elements for vibratory processing.

The complexity of the vibratory circuitry at the ventral nerve cord level was recently demonstrated in cave crickets, where twenty-six morpho-functional neuron types were described in only one (prothoracic) neuromere (Stritih 2006, 2009; Stritih and Stumpner 2009). The majority of neurons responded best to vibrations below 400 Hz, including the most highly sensitive units recorded (Fig. 19.3). This contrasts with data from other Ensifera, however, as many local and descending neurons serve such low-frequency processing in cave crickets; elements largely neglected in other studies. The difference might not entirely originate in the functionally modified networks in different species, but also might be due to recording from different neuronal populations.

Among the four local and 22 intersegmental neurons identified in *T. neglectus*, nine putative homologues to sound and/or vibration sensitive neurons of Ensifera were recognised (Stritih 2006; Stritih and Stumpner 2009). The study demonstrated that during evolution, a part of pre-existing vibratory neurons in Ensifera, mostly with pronounced high-frequency inputs in the cave cricket, was exploited for audition. At the same time, vibratory homologues with basically unchanged response properties in different species indicated evolutionary conservation of the low-frequency part of the vibratory system.

Such a low-frequency bias in sensitivity at the central neural level of cave crickets was unexpected, since the CTO responding to high frequencies is generally regarded as the most important vibratory organ. Together with this, the dendritic (input) regions of the majority of highly sensitive, low-frequency-tuned neurons are located in the mVAC (Stritih 2009), strongly indicating that a part of the FeCO, as suggested above, may represent another important source of vibratory information involved in complex behaviour.

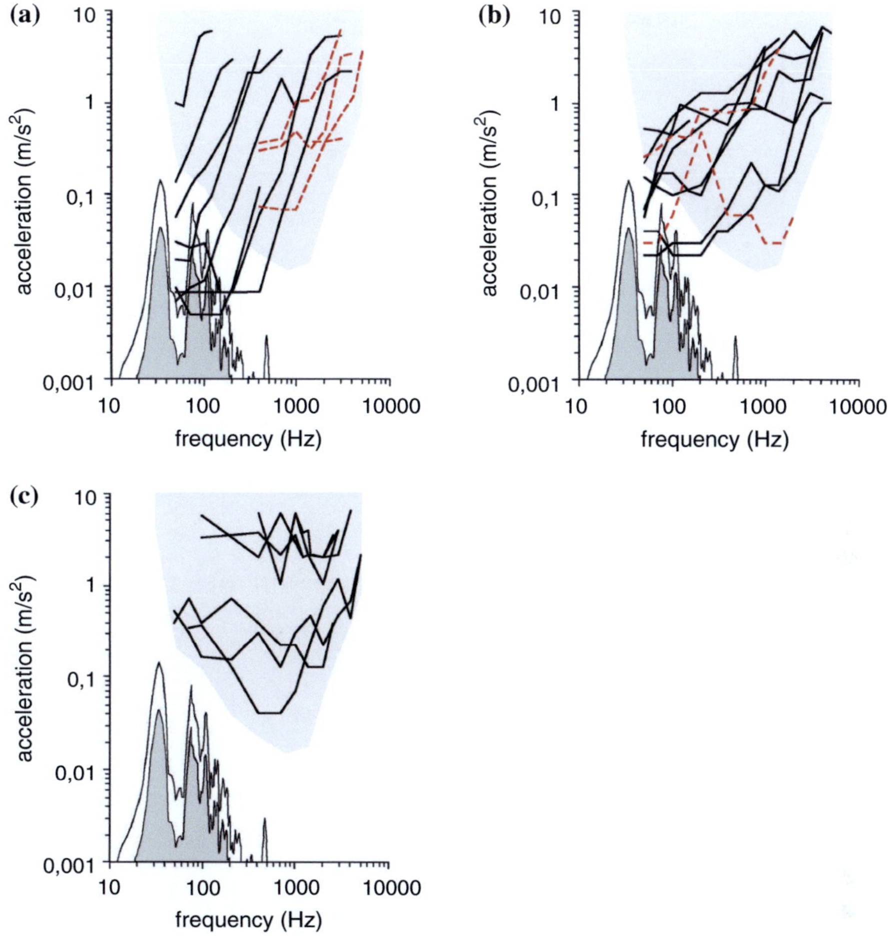


**Fig. 19.3** Summary of the tuning of the prothoracic vibratory interneurons in *T. neglectus*. Shown are lowest values of the superimposed thresholds from different neurons of a certain type, arranged into three classes. *Solid lines* indicate excitation, *dashed lines* indicate inhibition. Frequency spectra of courtship signals at mean and maximal intensities (*lower grey* and *white* areas, respectively; see also text to Fig. 19.2) and the response range of CTO receptors in *T. neglectus* (*upper grey area*) are included. **a** First class is formed of neurons that are sharply tuned to frequencies between 50–400 Hz, including two local neurons, three descending neurons (with a posteriorly projecting axon) and three T-neurons (projecting both anteriorly and posteriorly). In these neurons, thresholds approximately follow equal displacement values (as is also indicated in Fig. 19.4a). **b** Neurons in the second class are broadly tuned to frequencies below 400 Hz, including one local, three descending, three ascending and two T-neurons. Three additional descending neurons and one ascending neuron, tested only in the suprathreshold range, responded best to such low frequencies. **c** Neurons of the third class are most sensitive in the 400–2000 Hz range, including one local, one descending, two ascending and three T-neurons. Note that only the third-class neurons reflect the tuning of the CTO, while responses of the first- and second-class neurons, below the CTO sensitivity, overlap with the intraspecific signals' range. (From Stritih 2009, adapted with permission)

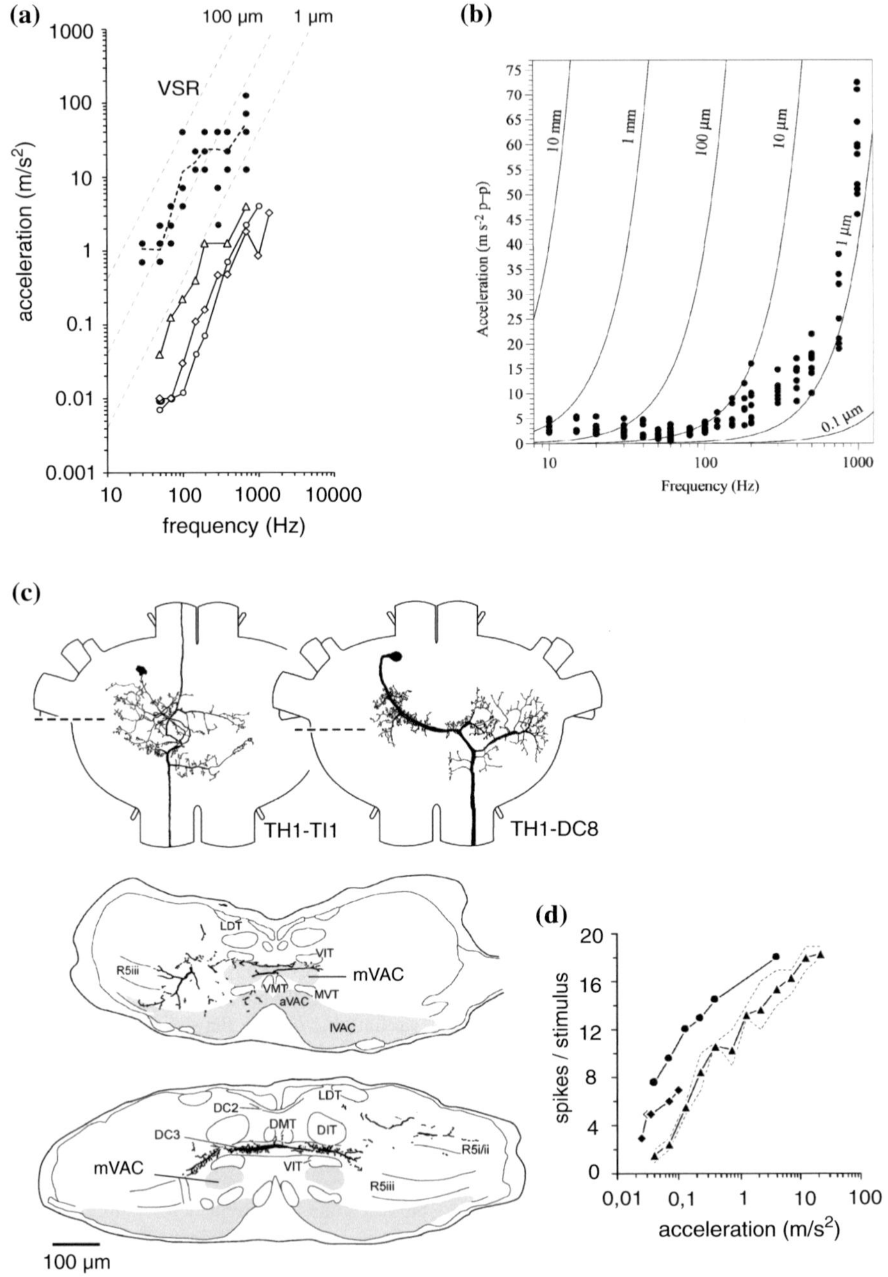
(a)
100 µm
1 µm
VSR
acceleration (m/s²)
frequency (Hz)
(b)
10 mm
1 mm
100 µm
10 µm
1 µm
0.1 µm
Acceleration (m s-2 p–p)
Frequency (Hz)
(c)
TH1-TI1
TH1-DC8
LDT
VIT
R5iii
mVAC
VMT
MVT
aVAC
lVAC
DC2
DMT
DIT
DC3
R5i/ii
100 µm
(d)
spikes / stimulus
acceleration (m/s²)

◀**Fig. 19.4** The vibrational startle response (*VSR*), together with anatomical and functional characteristics of two putatively involved neurons in *T. neglectus*. **a** Threshold of the VSR in ten individuals of *T. neglectus*, vibrated on a loudspeaker membrane (3–9 frequencies between 30 and 700 Hz were randomly tested per animal, 3–7 data points are shown per frequency (*closed circles*) with the mean (*thick dashed line*); responses were detected by the laser vibrometer from the back of the animal; Stritih unpublished), compared to threshold curves of TH1-TI1 neuron from three preparations (*open symbols*). For the comparison with (**b**), the lines of constant displacement are indicated. **b** Threshold of the VSR in the locust *Schistocerca gregaria* (from Friedel 1999, reproduced with permission). **c** Whole-mount drawings of the neurons in the prothoracic ganglion, with the indicated levels of transverse sections that are drawn schematically below. In TH1-TI1, axonal (output) branches terminate in the medio-ventral association centre (mVAC, upper section; for other abbreviations, see Stritih 2009). **d** Intensity-response curves at 50 Hz from three TH1-TI1 preparations. (Data on neurons from Stritih 2009 and Stritih and Stumpner 2009, reproduced with permission)

### *19.4.2 Behavioural Correlates*

Vibratory neuronal inputs at low frequencies may be related both to intraspecific communication by means of tremulation signals (in crickets, bush crickets and cave crickets), as well as to warning the animals of potential danger. In *T. neglectus,* a stationary startle response expressed as a jerky contraction of the legs and the body is elicited by low-frequency vibration at a very high intensity (Fig. 19.4a, Stritih unpublished). The reaction is sharply tuned to 30–50 Hz, reflecting response characteristics of a class of interneurons identified in the species. Such a vibratory startle response was previously demonstrated in the locust *Schistocerca gregaria* (Friedel 1999) with basically the same frequency-intensity sensitivity as in cave crickets (Fig. 19.4b), which directly indicates the existence of at least some functionally conserved circuits for processing low-frequency information across the Orthoptera. Those neurons in the cave cricket with processes outside the mVAC, and at the same time having thick axons for fast signal transmission, are the most likely candidates to mediate such behaviour. The startle may be either triggered directly, by first-order interneurons with axonal processes in the motor neuropiles, or via intercalated centres in the ventral nerve cord or in the brain (Fig. 19.4c). Neurons with a wide response dynamic range could trigger the startle at response rates high above the threshold (Fig. 19.4d), as is the case in the auditory startle pathway of Ensifera (e.g. Nolen and Hoy 1984).

Intensity may represent an important cue for the Ensifera to discriminate spectrally similar signals of different biological significance. In *T. neglectus*, threshold for the startle response was between 0.7 and 1.3 $m/s^2$ at 50 Hz, while peak acceleration values of courtship tremulation signals ranged from 0.004 to 0.14 $m/s^2$ when measured on bark adjacent to the male. Correspondingly, a class of highly sensitive neurons that responded in the signals' range received inhibitory inputs at high frequencies (Fig. 19.5). Such inputs may play a role not only in shaping the neurons' tuning, but also may optimise their sensitivity for lower

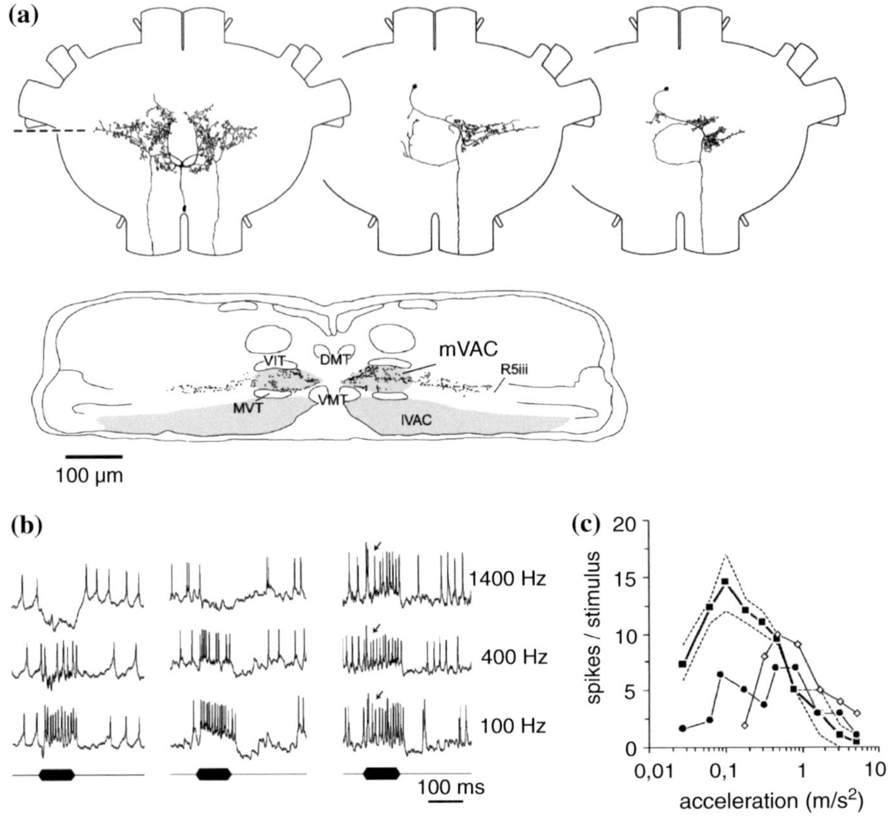


**Fig. 19.5** Anatomy and response properties of descending and local prothoracic neurons of *T. neglectus*, which may be involved in intraspecific communication. Four of the five tuning curves with minimal thresholds between 0.005–0.02 m/s$^2$, as shown in Fig. 19.3a, belong to these neurons. **a** Whole-mount drawings of TH1-DB3, TH1-DC2 and TH1-DC4 (from *left* to *right*) in the prothoracic ganglion. *Dashed line* indicates the level of the transverse section in TH1-DB3 drawn schematically below. As in this example, the dendritic (input) branches of these neurons are located in the medio-ventral association centre (*mVAC*) and the root three of the leg nerve (R5iii; for other abbreviations, see Stritih 2009). **b** Responses of TH1-DB3, TH1-DC2 and TH1-DC4 elicited by vibrations of increasing frequency (100, 400 and 1400 Hz) at constant intensity of 4 m/s$^2$. Vibrated were soma-ipsilateral, soma- contralateral, and both legs (from left to right). *Arrows* point to inhibitory gaps in the spike discharge. **c** Intensity-response curves of TH1-DC4 at different frequencies (*closed circles*, *closed squares* and *open diamonds* for 100, 400 and 1 kHz, respectively). (from Stritih 2009, reproduced with permission)

intensities (Römer 1987; Stumpner 2002; Fig. 19.5c). In addition to the location of dendritic branches in the mVAC, such response characteristics of respective neurons suggest that they may be detecting intraspecific signals. Their descending projections and/or intrasegmental axon terminations suggest either involvement of only local networks in the response to courtship tremulation (which may have an arresting effect on the female; see below) or, alternatively, integration with the inputs from other legs more posteriorly in the ventral nerve cord before transmitting information to the brain.

With some speculation, an analogy from cricket behaviour may suggest the first option to be more likely and that signalling cave crickets may actually be exploiting the network evolved in the context of predator detection (see Greenfield 2002). In a stridulating cricket *Gryllus campestris*, vibration between 50 and 400 Hz caused a "silencing reaction", interruption of singing for several seconds, with a very low threshold between 0.001 and 0.005 $m/s^2$ (Dambach 1989). This reaction, which can be interpreted in the context of predator avoidance, demonstrates existence of highly sensitive, low-frequency-tuned sensory elements, which are unknown in crickets for the range below 200 Hz (Kühne et al. 1985). Such frequency-intensity range is covered by responses of the low-frequency-tuned interneurons in the cave cricket, which together with the inhibitory side bands indicate 400 Hz as a behaviourally relevant high-frequency limit. It is possible that homologous networks perceive such low-intensity signals in crickets and cave crickets, causing sensory inhibition of movements that may be exploited by the males in the context of courtship. In crickets, tremulation signals have not yet been recorded, but they appear to be of very low intensity as well, most likely not exceeding that of stridulatory signals transmitted through the ground (0.005–0.01 $m/s^2$; Weidemann and Keuper 1987).

In bush crickets, tremulation signals of much higher intensities are elicited on plants (about 0.6 $m/s^2$ shown for a signal of *Ephippiger ephippiger*; Keuper et al. 1985). In diurnal species, visual signals may help discriminate the signal context. Elaborate communication by tremulation in some nocturnal species (Morris et al. 1994) suggests use of more specialised networks for signal perception in the low-frequency domain than suggested above.

The function of high-frequency vibratory inputs provided to the central nervous system by the CTO is unclear for non-hearing Ensifera, especially cave crickets. In crickets and bush crickets, (one of) the function(s) of such inputs clearly is detection of high-frequency vibratory signals produced by stridulation. In these species, simultaneous detection of the signals via vibratory and auditory channels improves spatial and temporal resolution of the perceived information (Kalmring et al. 1997). Whether or not signalling by drumming, expressed by some non-hearing Ensifera, also elicits broadband vibration in the substrate (as may be inferred from other insects, such as termites and bugs; Connetable et al. 1999; Žunič et al. 2008) still needs to be investigated. These data would have important implications for discussing evolutionary forces leading to functional elaboration of their CTO. For cave crickets, the role of information in vibration frequencies above 400 Hz is the most speculative. Such inputs may be needed for a differential recognition of signals of various biotic and abiotic origins (see Castellanos and Barbosa 2006). Also, based on the frequency filtering the substrates generally exert on propagated vibrations, the proportion of excitation among the low- and high-frequency-tuned interneurons may enable estimation of distance from the signal source.

**Acknowledgements** We thank the Slovene Ministry for Science and the Slovenian Research Agency for long-term financial support of our research. A part of the work was also covered by the grant Marie Curie Training Sites, Neuronal signals and development (Qualities of life program of the European union, QLGA-1999-51322).

## References

Alexander RD, Otte D (1967) The evolution of genitalia and mating behavior in crickets (Gryllidae) and other Orthoptera. Miscellaneous Publications Museum of Zoology, University of Michigan

Autrum H, Schneider W (1948) Vergleichende Untersuchungen über den Erschütterungssinn der Insekten. Z Vergl Physiol 31:77–88

Ball EE, Oldfield BP, Michel RK (1989) The auditory organ structure, development and function. In: Huber F, Moore TE, Loher W (eds) Cricket behaviour and neurobiology. Cornell University Press, Ithaca, pp 391–421

Bell PD (1980) Multimodal communication by the black-horned tree cricket, *Oecanthus nigricornis* (Walker) (Orthoptera, Gryllidae). Can J Zool 58:1861–1868

Benediktov AA (2009) Vibration communication in orthopteroid insects (Orthoptera) from suborder Caelifera. Moscow Univ Biol Sci Bull 64:126–128

Buh B (2011) Morphological and functional characterization of vibratory receptor neurons in cave crickets of the genus Troglophilus (Orthoptera, Rhaphidophoridae). Graduation thesis, University of Ljubljana

Castellanos I, Barbosa P (2006) Evaluation of predation risk by a caterpillar using substrate-borne vibrations. Anim Behav 72:461–469

Cocroft RG, Rodríguez RL (2005) The behavioral ecology of insect vibrational communication. Bioscience 55:323–334

Connetable S, Robert A, Bouffault F, Bordereau C (1999) Vibratory alarm signals in two sympatric higher termite species: *Pseudacanthotermes spiniger* and *P. militaris* (Termitidae, Macrotermitinae). J Insect Behav 12:329–342

Čokl A, Kalmring K, Wittig H (1977) The responses of auditory ventral-cord neurones of *Locusta migratoria* to vibration stimuli. J Comp Physiol A 120:161–172

Čokl A, Kalmring K, Rössler W (1995) Physiology of atympanate tibial organs in forelegs and midlegs of the cave-living Ensifera, *Troglophilus neglectus* (Rhaphidophoridae, Gryllacridoidea). J Exp Zool 273:376–388

Dambach M (1972) Der Vibrationssinn der Gryllen. I. Schwellenmessungen an Beinen frei beweglicher Tiere. J Comp Physiol A 79:281–304

Dambach M (1989) Vibrational responses. In: Huber F, Moore TE, Loher W (eds) Cricket behaviour and neurobiology. Cornell University Press, Ithaca, pp 179–197

De Luca PA, Morris GK (1998) Courtship communication in meadow katydids: Female preference for large male vibrations. Behaviour 135:777–794

Desutter-Grandcolas L (2003) Phylogeny and the evolution of acoustic communication in extant Ensifera (Insecta, Orthoptera). Zool Scripta 32:525–261

Field LH (2001a) The biology of wetas, king crickets and their allies. CABI Publishing, Oxon

Field LH (2001b) Stridulatory mechanisms and associated behaviour in New Zealand wetas. In: Field LH (ed) The biology of wetas, king crickets and their allies. CABI Publishing, Oxon, pp 271–295

Field LH, Bailey WJ (1997) Sound production in primitive Orthoptera from Western Australia: sounds used in defence and social communication in *Ametrus* sp. and *Hadrogryllacris* sp. (Gryllacrididae: Orthoptera). J Nat History 31:1127–1141

Field LH, Matheson T (1998) Chordotonal organs of insects. Academic Press, London

Field LH, Pfluger H-J (1989) The femoral chordotonal organ: A bifunctional orthopteran (*Locusta migratoria*) sense organ? Comp Biochem Physiol A 93:729–743

Friedel T (1999) The vibrational startle response of the desert locust *Schistocerca gregaria*. J Exp Biol 202:2151–2159

Greenfield MD (2002) Signalers and receivers: Mechanisms and evolution of arthropod communication. Oxford University Press, Oxford

Gwynne DT (1977) Mating behaviour of *Neoconocephalus ensiger* (Orthoptera: Tettigoniidae) with notes on the calling song. Can Ent 109:237–242

Hill PSM, Shadley JR (1997) Substrate vibration as a component of a calling song. Naturwissenschaften 84:460–463

Hill PSM, Shadley JR (2001) Talking back: Sending soil vibration signals to lekking prairie mole cricket males. Am Zool 41:1200–1214

Jeram S, Čokl A, Kalmring K (1995) Structure of atympanate tibial organs in legs of the cave-living Ensifera *Troglophilus neglectus* (Gryllacridoidea, Rhaphidophoridae). J Morph 223:109–118

Kalmring K, Kühne R (1980) The coding of airborne-sound and vibrational signals in bimodal ventral-cord neurons of the grasshopper *Tettigonia cantans*. J Comp Physiol A 139:267–275

Kalmring K, Lewis B, Eichendorf A (1978) The physiological characteristics of primary neurons of the complex tibial organ of *Decticus verrucivorus* L. (Orthoptera, Tettigoniidae). J Comp Physiol A 127:109–121

Kalmring K, Rossler W, Unrast C (1994) Complex tibial organs in the fore-, mid- and hindlegs of the bushcricket *Gampsocleis gratiosa* (Tettigoniidae): comparison of physiology of the organs. J Exp Zool 270:155–161

Kalmring K, Hoffmann E, Jatho M, Sickmann T, Grossbach M (1996) The auditory- vibratory sensory system of the bushcricket *Polysarcus denticauda*. (Phaneropterinae, Tettigoniidae) II. Physiology of receptor cells. J Exp Zool 276:315–329

Kalmring K, Sickmann T, Jatho M, Zhantiev R, Grossbach M (1997) The auditory- vibratory sensory system of the bushcricket *Polysarcus denticauda*. (Phaneropterinae, Tettigoniidae) III. Physiology of the ventral cord neurons ascending to the head ganglia. J Comp Physiol A 279:9–28

Keuper A, Kühne R (1983) The acoustic behaviour of the bushcricket *Tettigonia cantans* II. Transmission of airborne sound and vibration signals in the biotope. Behav Proc 8:125–145

Keuper A, Otto C, Latimer W, Schatral A (1985) Airborne sound and vibration signals of bushcrickets and locusts; their importance for the behaviour in the biotope. In: Kalmring K, Elsner N (eds) Acoustic and vibrational communication in insects. Paul Parey, Berlin Hamburg, pp 135–142

Kočárek P (2010) Substrate-borne vibrations as a component of intraspecific communication in the groundhopper *Tetrix ceperoi*. J Insect Behav 23:348–363

Kočárek P, Holuša J, Grucmanová Š, Musiolek D (2011) Biology of *Tetrix bolivari* (Orthoptera: Tetrigidae). Centr Eur J Biol 6:531–544

Kuenzi F, Burrows M (1995) Central connections of sensory neurones from a hair plate proprioceptor in the thoraco-coxal joint of the locust. J Exp Biol 198:1589–1601

Kühne R (1982a) Neurophysiology of the vibration sense in locusts and bushcrickets: Response characteristics of single receptor units. J Insect Physiol 28:155–163

Kühne R (1982b) Neurophysiology of the vibration sense in locusts and bushcrickets: The responses of ventral-cord neurones. J Insect Physiol 28:615–623

Kühne R, Silver S, Lewis B (1984) Processing of vibratory and acoustic signals by ventral cord neurones in the cricket *Gryllus campestris*. J Insect Physiol 30:575–585

Kühne R, Silver S, Lewis B (1985) Processing of vibratory signals in the central nervous system of the cricket. In: Kalmring K, Elsner N (eds) Acoustic and vibrational communication in insects. Paul Parey, Berlin Hamburg, pp 183–190

Lakes R, Schikorski T (1990) The neuroanatomy of tettigoniids. In: Bailey WJ, Rentz WJ (eds) The Tettigoniidae: Biology, systematics and evolution. Crawford House Press, Bathurst, pp 167–190

Latimer W, Schatral A (1983) The acoustic behaviour of the katydid *Tettigonia cantans* I. Behavioural responses to sound and vibration. Behav Process 8:113–124
Loher W, Chandrashekaran MK (1970) Acoustical and sexual behaviour in the grasshopper *Chimarocephala pacifica pacifica* (Oedipodinae). Ent Exp Appl 13:71–84
McNett G, Luan LH, Cocroft RG (2010) Wind-induced noise alters signaler and receiver behavior in vibrational communication. Behav Ecol Sociobiol 64:2043–2051
McVean A, Field LH (1996) Communication by substratum vibration in the New Zealand tree weta, *Hemideina femorata* (Stenopelmatidae: Orthoptera). J Zool 239:101–122
Morris GK (1980) Calling display and mating behaviour of *Copiphora rhinoceros* Pictet (Orthoptera: Tettigoniidae). Anim Behav 28:42–51
Morris GK, Mason AC, Wall P, Belwood JJ (1994) High ultrasonic and tremulation signals in neotropical katydids (Orthoptera, Tettigoniidae). J Zool 233:129–163
Mücke A (1989) Das Periphere Nervensystem und die Zentralprojektion der Rezeptoren intakter und regenerierten Beine von Schistocerca gregaria und Locusta migratoria. Dissertation, Phillips Universität Marburg
Mücke A, Lakes-Harlan R (1995) Central projections of sensory cells of the midleg of the locust, *Schistocerca gregaria*. Cell Tiss Res 280:391–400
Nebeling B (2000) Morphology and physiology of auditory and vibratory ascending interneurones in bushcrickets. J Exp Zool 286:219–230
Nishino H (2000) Topographic mapping of the axons of the femoral chordotonal organ neurons in the cricket *Gryllus bimaculatus*. J Comp Physiol A 299:145–157
Nishino H (2003) Somatotopic mapping of chordotonal organ neurons in a primitive ensiferan, the New Zealand tree weta *Hemideina femorata*: I. Femoral chordotonal organ. J Comp Neurol 464:312–326
Nishino H, Field LH (2003) Somatotopic mapping of chordotonal organ neurons in a primitive ensiferan, the New Zealand tree weta *Hemideina femorata*: II. Complex tibial organ. J Comp Neurol 464:327–342
Nishino S, Sakai M (1997) Three neural groups in the femoral chordotonal organ of the cricket *Gryllus bimaculatus*: central projections and soma arrangement and displacement during joint flexion. J Exp Biol 200:2583–2595
Nolen TG, Hoy RR (1984) Initiation of behavior by single neurons: the role of behavioral context. Science 226:992–994
Pflüger HJ, Bräunig P, Hustert R (1988) The organization of mechanosensory neuropils in locust thoracic ganglia. Phil Trans R Soc Lond B 321:1–26
Riede K (1987) A comparative study of mating behaviour in some neotropical grasshoppers (Acridoidea). Ethology 76:265–296
Römer H (1987) Representation of the auditory distance within the central neuropile of the bushcricket *Mygalopsis marki*. J Comp Physiol A 161:33–42
Rössler W, Jatho M, Kalmring K (2006) The auditory-vibratory sensory system in bushcrickets. In: Drosopoulos S, Claridge MF (ed) Insect sounds and communication. Physiology, behaviour, ecology and evolution. Taylor & Francis, Boca Raton, pp 35–69
Schatral A, Kalmring K (1985) The role of song for spatial dispersion and agonistic contacts of male bushcrickets. In: Kalmring K, Elsner N (eds) Acoustic and vibrational communication in insects. Paul Parey, Berlin Hamburg, pp 111–116
Sickmann T (1996) Vergleichende funktionelle und anatomische Untersuchung zum Aufbau der Hör- und Vibrationsbahn im thorakalen Bauchmark von Laubheuschrecken. Dissertation, Philipps-Universität Marburg
Silver S, Kalmring K, Kühne R (1980) The responses of central acoustic and vibratory interneurones in bushcrickets and locusts to ultrasonic stimulation. Physiol Entomol 5:427–435
Strauss J, Lakes-Harlan R (2008a) Neuroanatomy and physiology of the complex tibial organ of an atympanate ensiferan, *Ametrus tibialis* (Brunner von Wattenwyl 1888) (Gryllacrididae, Orthoptera) and evolutionary implications. Brain Behav Evol 71:167–180

Strauss J, Lakes-Harlan R (2008b) Neuroanatomy of the complex tibial organ of *Stenopelmatus* (Orthoptera: Ensifera: Stenopelmatidae). J Comp Neurol 511:81–91
Strauss J, Lakes-Harlan R (2010) Neuroanatomy of the complex tibial organ in the splay-footed cricket *Comicus calcaris* Irish 1986 (Orthoptera: Ensifera: Schizodactylidae). J Comp Neurol 518:4567–4580
Stritih N (2006) Response properties, morphology and topographical organisation of the vibratory neurones in the prothoracic ganglion of the cave cricket Troglophilus neglectus Krauss (Orthoptera, Rhaphidophoridae). Dissertation, University of Ljubljana
Stritih N (2009) Anatomy and physiology of a set of low-frequency vibratory interneurons in a nonhearing ensiferan (*Troglophilus neglectus*, Rhaphidophoridae). J Comp Neurol 516:519–532
Stritih N, Čokl A (2012) Mating behaviour and vibratory signalling in non-hearing cave crickets reflect primitive communication of Ensifera. PloS one (7)10:e47646
Stritih N, Stumpner A (2009) Vibratory interneurons of the non-hearing cave cricket indicate evolutionary origin of sound processing elements in Ensifera. Zoology 112:48–68
Stumpner A (1996) Tonotopic organisation of the hearing organ in a bushcricket. Physiological characterisation and complete staining of auditory receptor cells. Naturwissenschaften 83:81–84
Stumpner A (2002) A species-specific frequency filter through specific inhibition, not specific excitation. J Comp Physiol A 188:239–248
Virant-Doberlet M, Čokl A (2004) Vibrational communication in insects. Neotrop Entomol 33:121–134
Weidemann S, Keuper A (1987) Influence of vibratory signals on the phonotaxis of the gryllid *Gryllus bimaculatus* DeGeer (Ensifera: Gryllidae). Oecologia 74:316–318
Weissman DB (2001) Communication and reproductive behaviour in North American Jerusalem crickets (*Stenopelmatus*) (Orthoptera: Stenopelmatidae). In: Field LH (ed) The biology of wetas, king crickets and their allies. CABI Publishing, Oxon, pp 351–375
Wohlers DW, Huber F (1985) Topographical organisation of the auditory pathway within the prothoracic ganglion of the cricket, *Gryllus campestris* L. Cell Tiss Res 239:555–565
Zill S, Ridgel A, DiCaprio R, Frazier S (1999) Load signalling by cockroach trochanteral campaniform sensilla. Brain Res 822:271–275
Žunič A, Čokl A, Virant-Doberlet M, Millar JG (2008) Communication with signals produced by abdominal vibration, tremulation, and percussion in *Podisus maculiventris* (Heteroptera: Pentatomidae). Ann Entomol Soc Am 101:1169–1178

# Chapter 20
# The Tymbal: Evolution of a Complex Vibration-Producing Organ in the Tymbalia (Hemiptera excl. Sternorrhyncha)

**Andreas Wessel, Roland Mühlethaler, Viktor Hartung, Valerija Kuštor and Matija Gogala**

**Abstract** The tymbal is the most complex sound- and vibration-producing organ in arthropods. The tymbal organ was first described from cicadas which use it to produce sound levels of more than 100 dB. Subsequently, it was discovered that leaf- and planthoppers, as well as true bugs and moss bugs, communicate by substrate-borne vibrations, which are also produced by tymbal-like organs. We suggest the name *Tymbalia* for the taxon comprising Cicadomorpha, Fulgoromorpha, and Heteropteroidea (i.e., Hemiptera exclusive of Sternorrhyncha) based on the possession of a tymbal apparatus as an autapomorphic character. While our knowledge of the hoppers' and bugs' "silent songs" is still patchy, vibrational communication is obviously used ubiquitously in the Tymbalia and we hypothesize a common origin for the vibration-producing apparatus more than 300 Mya.

## 20.1 What is a Tymbal?

The "tymbal" sensu stricto is the sound-producing organ of the singing cicadas, the Cicadoidea. Originally, the term was used only for the exoskeletal structure whose deformation (buckling) is the source for airborne sound and, thus, was

A. Wessel (✉) · R. Mühlethaler · V. Hartung
Museum für Naturkunde—Leibniz-Institut für Evolutions- und Biodiversitätsforschung an der Humboldt-Universität zu Berlin, Invalidenstraße 43, 10115 Berlin, Germany
e-mail: andreas.wessel@mfn-berlin.de

A. Wessel
Zoologisches Museum, Universität Hamburg, Martin-Luther-King-Platz 3, 20146 Hamburg, Germany

V. Kuštor
Štefana Kuzmiča 4, 9000 Murska Sobota, Slovenia

M. Gogala
Slovenian Academy of Sciences and Arts, Novi trg 3, 1000 Ljubljana, Slovenia

R. B. Cocroft et al. (eds.), *Studying Vibrational Communication*, Animal Signals and Communication 3, DOI: 10.1007/978-3-662-43607-3_20,

compared to a percussion instrument. At least since the rediscovery of the tymbal by R.-A. F. Réaumur, the vibrating membranes have been described as a drumhead: *timbale* (Réaumur 1740), *tamburo* (Lepori 1869), *Trommelhaut* (Mayer 1877), or simply *drum* (Lloyd Morgan 1886). However, since the first discovery by Casserius (1600/1601), the tymbal organ was recognized as a functional unity of muscles, anchoring and vibrating exoskeletal structures, and resonating cavities.

The specialized and very strongly developed tymbal muscles are the most prominent and defining feature of the organ (Figs. 20.1 and 20.2b). The muscle configuration is very conservative throughout the Cicadoidea, whereas the tymbal plates show a great variety of mostly species-specific forms. These differences in form and structure—e.g., surface geometry, curvature, number and form of ribs, sclerotization, and elasticity—in association with passively resonating parts and cavities are mostly responsible for species-specific differences in airborne vibration (sound) characteristics.

Most startling about the performance of the cicadas tymbal is not only the ability to produce high sound pressures within the body of a small insect, but the transformation of the actions of a pair of muscles working at frequencies between 35 and 100 Hz into sounds with frequencies of up to 25 kHz (Hagiwara and Ogura 1960; Popov et al. 1997). The physics behind this transformation is quite complex and still a matter of research. However, understanding the principles of the process requires at least a distinction between the muscle action frequency, the pulse repetition frequency, and the airborne vibration (sound) frequency in itself. Assuming a simple tymbal plate geometry, a single muscle contraction produces first an inward buckling of the plate as a whole and then with muscle release an outward buckling, resulting in two "clicks" or pulses. However, most tymbal plates are much more sophisticated structures with parallel ribs and/or membraneous parts of different shapes, and it could be shown that in some cases, up to a dozen ribs buckle in succession, with each buckling producing a single sound pulse. (This design even inspired a new musical instrument, the *Tymbalimba* [Smyth and Smith III 2003]). The frequency of the single pulses depends on the eigenfrequency of the ribbed structure. In combination with passively resonating parts like the tympanal membranes, tymbal opercula, abdominal sclerites, or the large air sacs filling the male abdomen very different sounds can result: from clicking sounds, to pure tones and complex tonal sounds with harmonics, to almost white, or rather pink, noise sounds (Hagiwara and Ogura 1960; Young 1972a, b; Young and Josephson 1983).

The term "tymbal" has been adopted for functionally analogous organs in butterflies. S. S. Haldeman described in 1848, the sound production of an arctiid moth "by vibrating the pleura beneath the wings, this part being marked [...] by parallel lines" (p. 435). Laboulène (1864) studied this organ, which is located in the metathorax, more closely in another arctiid species and compared the structure of the vibrating membrane to the cicadas' tymbal plate. The lepidopteran tymbal is, however, moved by the thoracic flight musculature. The fine parallel lines of the vibrating plate act analogously to the ribs of the cicada tymbal plates and are referred to as "microtymbals" (Fenton and Roeder 1974). The patterns of these

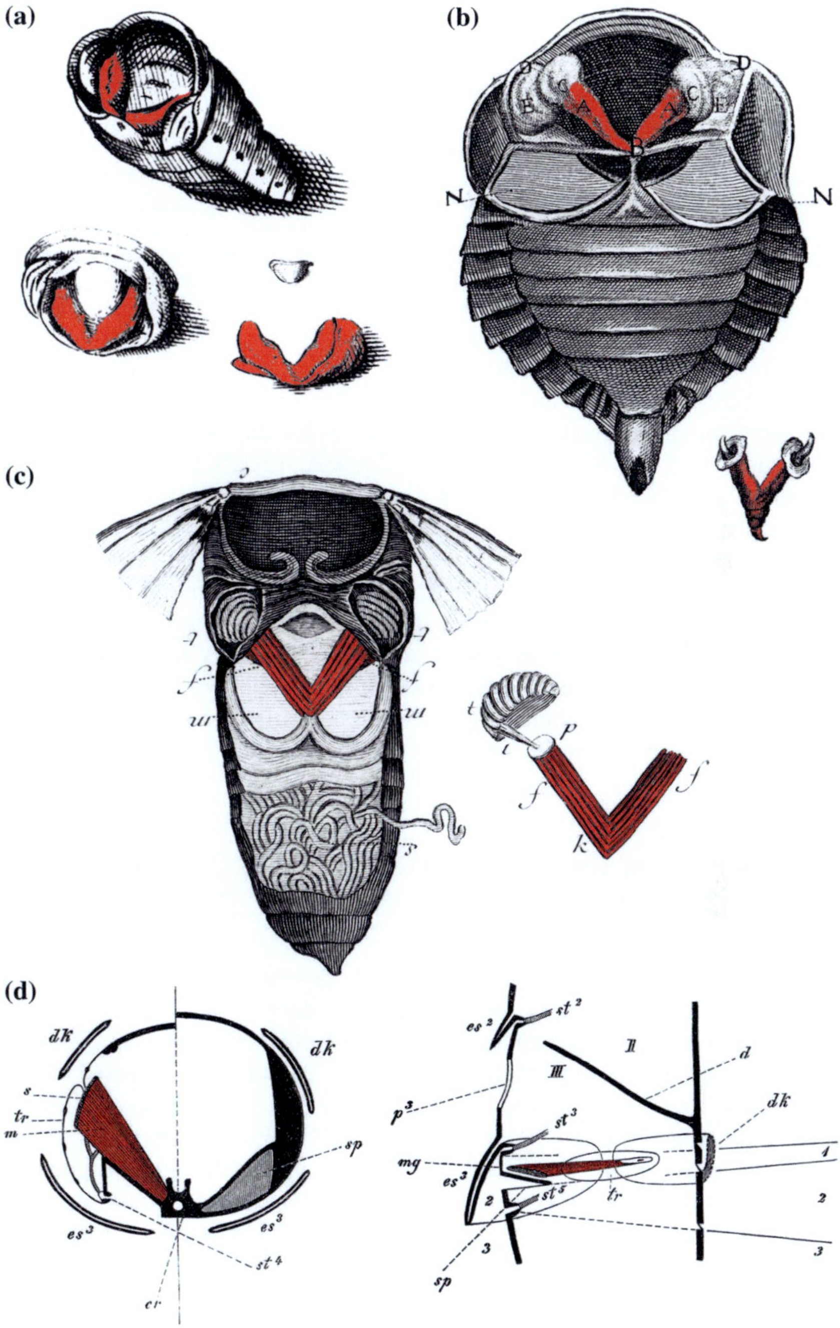
(a)
(b)
D
D
N
N
(c)
f
f
t
p
f
f
k
(d)
dk
dk
s
tr
m
sp
es3
es3
st4
cr
st2
es2
II
III
d
p3
dk
st3
mg
es3
st5
tr
1
2
3
2
3
sp

◀**Fig. 20.1** Discovery of the tymbal organ in the Cicadoidea. (The pictures are modified by highlighting the tymbal muscles [*I a dvm*$_1$] in *red color*.) **a** The first illustrations of the tymbal apparatus by G. Casserius; shown here are the figures "Cicadae V., VII., VIII., and X." of plate XXI in his work *De Vocis Auditusque Organis*… (1600/1601). For a detailed interpretation and the translation of Casserius' original description from New Latin into English see Wessel (2013). **b** Illustrations by Felici (1724), one of the independent discoverers of the cicada's tymbal organ (see Wessel 2013). Shown are Figs. 9 and 10 from plate II depicting the "muscoli moventi le membrane sonore" (p. 76). **c** Illustrations of the tymbal organ by de Réaumur (1740, plate XVII, Fig. 6 [shown upside-down] and 9), depicting "Chaque muscle *f*, est destiné à faire jouer la timbale vers laquelle il se dirige" (p. 198). **d** Illustrations of the tymbal organ by Mayer (1877, Figs. 1 and 2), assigning the "Tonmuskel" (p. 88) correctly to the first abdominal segment

lines are species specific and can be used for taxonomic descriptions and determinations (see, e.g., Forbes and Franclemont 1957 on the "striated bands"). The arctiid moths produce sounds of up to 160 kHz, which they use for defense against bat predators, e.g., for "jamming" their echolocation calls (Fullard et al. 1979; Corcoran et al. 2009; for review see Conner and Corcoran 2012), as well as in mate location and courtship behavior (Conner 1999).

In Lepidoptera, there are at least five completely different types of tymbal-like organs—located at the thorax and the wings, as well as the front and rear part of the abdomen—of independent evolutionary origin (see review in Connor 1999). While it seems very obvious that the tymbals of butterflies and cicadas are not homologous, a recently discovered type of abdominal tymbal in the Nolidae resembles the cicadas' apparatus so closely that it clearly challenges its uniqueness (Skals and Surlykke 1999).

## 20.2 Discovery of the Tymbal Mechanism in Cicadas (Cicadoidea)

Casserius (1600/1601) was the first to correctly describe the tymbal plates as the source of the sound ("Bractealis membranula ſoni auctor," p. 117) and the tymbal muscles as the corresponding effector organs ("Muſculi duo membranas mouentes ad ſonum," p. 117). Moreover, Casserius illustrated the stages of his dissection of the tymbal organ, thus providing the first pictures of this "marvelous work of artful nature" (Casserius 1600/1601, p. 116; Fig. 20.1a). In contrast to later students of the cicadas' sound-producing organ, Casserius did not compare the vibrating plates with a drumhead, though, but rather with the little bells on a children's rattle, apparently referring to a percussion instrument without reverberant properties that produces a "dry sound" (for an English translation and interpretation of Casserius' description, see Wessel 2013). In spite of its extraordinary quality, the description of Casserius was overlooked by his contemporaries and then forgotten for centuries, and the tymbal apparatus was subsequently rediscovered at least four times independently (Wessel 2013, Fig. 20.1b, c). In 1740, R.-A. F. Réaumur published a detailed

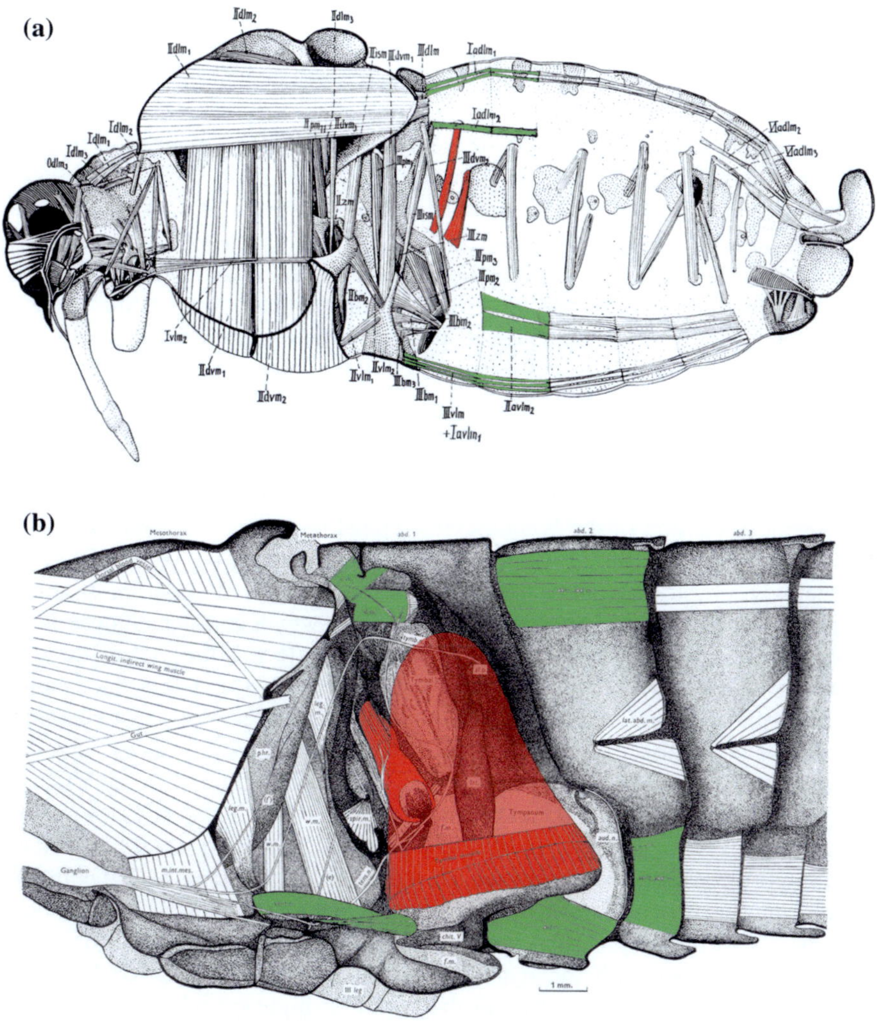


**Fig. 20.2** Configuration of the muscles of the first and second abdominal segments in the Hemiptera. **a** *Left* lateral view of the *right* half of the body of an alate female *black bean aphid* (*Aphis fabae*) showing the musculature, modified from Weber (1928, pl. VIII, Fig. 20). The color scheme used for highlighting the dorsoventral muscles of the first abdominal segment (*I a* $dvm_{1+2}$, *red*) and the longitudinal muscles of the first two abdominal segments (*I a* $dlm_{1+2}$, *II a* $dlm_{1+2}$, *III vlm + I a dvlm*, *II a* $vlm_{1+2}$, *green*) is employed throughout the following figures. The sternorrhynchan muscle configuration shown here is considered to be close to the plesiomorphic state (ground pattern) of the stem line Tymbalia. **b** The configuration of the same set of muscles in the strongly modified abdomen of the singing cicada *Platypleura capitata* (Cicadoidea: Cicadidae: Platypleurinae), modified from Pringle (1954, Fig. 2). The extremely strong developed tymbal muscle *I a* $dvm_1$ is shown partly transparent. (Figure 20.2a is reproduced with kind permission from E. Schweizerbart'sche Verlagsbuchhandlung, www.schweizerbart.de; 20.2b is reproduced with kind permission from The Journal of Experimental Biology, The Company of Biologists, www.jeb.biologists.org)

description of the tymbal structure and an explanation of the sound-producing mechanism (Fig. 20.1c), acclaimed and accepted by most of his fellow naturalists. Not considering the numerous details remaining to be studied, the problem would largely have been settled then if the German polyhistor Landois (1867, 1872, 1874) had not put forward a completely different hypothesis. His explanation of sound production in cicadas by pressing air through specialized spiracles (so-called "Schrill-Stigmen") would not be worth mentioning here had it not caused a lot of attention and confusion [even Darwin (1871) adopted it for a while] and prompted further research (see Wessel 2013). Mayer (1877) finally could study the action of the tymbal muscles in vivo and moreover designated the muscles correctly to the abdominal segments involved (Fig. 20.1d). Nevertheless, the "drum theory" still needed support in the 1880s and was debated in an influential journal like *Nature* (Lloyd Morgan 1886).

The role of the different parts involved in sound production, as well as their anatomical origin, is still a matter of research and comparative studies especially are much needed. Here, we will focus on problems that are important for the subsequent discussion of the homology of tymbal organs in the Hemiptera.

Between 1953 and 1957, J. W. S. Pringle published a series of papers discussing the anatomy, physiology, and evolution of the cicada tymbal organ, most notably "A physiological analysis of cicada song" (1954), which still constitutes the reference point for all studies and interpretations. Pringle was the first to study the physiology of the tymbal muscles itself as well as the influence of associated pairs of muscles—the tensor muscle and dorso-longitudinal muscles of the first abdominal segment—and discussed evolutionary and phylogenetic consequences. The most important problem seemed to be to understand the coordination of the action of the tymbal muscles. Pringle noted that Lloyd Morgan in 1880 already had reported that the muscles (of *Platypleura capensis*) could be set into action by a weak current and that a single electric pulse caused a sequence of muscle contractions. That pointed to a myogenic reaction of the muscles that was also corroborated by Pringle's own experiments with *Platypleura octoguttata* and *P. capitata*. This aspect of muscle physiology, together with the hypothesized metathoracic origin, led Pringle to propose that the tymbal muscles are the modified flight muscles from the third thoracic segment that were relieved of the task of hind-wing movement by wing pair coupling in the Cicadoidea. The only remaining problem was the mechanism of coordination of the two tymbal muscles, which usually work in perfect alternation (cf. Aidley 1969; Young 1972a). Pringle suspected a physical coupling of the muscle actions through their shared anchoring structure, the mighty "chitinous V" at the sternum, or a direct connection of the efferent axons.

At the same time, the experiments of Hagiwara (1953), Wakabayashi and Hagiwara (1953), and Hagiwara and Watanabe (1954) demonstrated a neurogenic activation of tymbal muscles in several Japanese species, with nerve pulse frequencies up to 100 Hz, and subsequent experiments by others, including Pringle, found this in a number of other cicada species as well. These apparently contradictory observations might be explained by the fact that a single muscle can react as both myogenic and neurogenic, depending on the dynamics of the incorporating physical system. For the tymbal muscle, the tymbal plate acts as an antagonist in the intact tymbal organ. An inward buckling of the tymbal plate releases the tymbal muscle immediately from tension while still in an active state; the following outward buckling brings the muscle under tension again and may cause it to contract without the need for a new nerve pulse to trigger this movement. In such a system, muscle contraction frequencies of more than 100 Hz can be facilitated (see Pringle 1965; Penzlin 1991). Hagiwara (1953) could also show (in *Graptosaltria nigrofuscata*) that the tymbal muscles respond to a repetitive stimulus (electrical as well as mechanical) having a frequency of more than 120 Hz with a fusion (tetanus). However, at a frequency of more than 250 Hz, they failed to react to one out of every two or three stimuli, so that a "smooth" tetanus could not be obtained at any frequency of stimulation. A purely neurogenic origin of spike frequencies up to 180 Hz in the muscle itself may also be facilitated by the complex interplay of the motoneurons and the sympathetic nervous system (Voskresenskaya and Svidersky 1960).

A recent study reports that the "superfast sonic muscles" of periodical cicadas also have a unique ultrastructure (Nahirney et al. 2006, see also Stokes and Josephson 2004; Chapman 2005). The muscle filaments are shorter than those in flight muscles and show a unique hexagonal symmetrical arrangement matching hexagonal lattices of the Z-bands. This special design may contribute to high-speed contractions with rapid recovery, long endurance and—especially given the matching symmetry of filaments and Z-band—to supercontraction (Nahirney et al. 2006). This corroborates the non-homology of the tymbal muscles with flight musculature but also points toward the urgent need for comparative studies in the Cicadoidea and related taxa.

Pringle's studies also illustrate the difficulties in assigning the tymbal musculature to the corresponding body segments. First (in 1954), he interpreted the tymbal muscle as of metathoracic origin (and thus being the transformed flight muscles of the hind wings), later (1957)—after the reception of Ossiannilsson's milestone study (1949) on the tymbal organs in leaf- and planthoppers—he recognized it as the dorsoventral muscle of the first abdominal segment (Fig. 20.2b). This problem, however, is not trivial and considerable confusion about the designation of the muscles of the tymbal apparatus exists until today. The reasons for this are as follows: (1) In Hemiptera, generally, the first abdominal segment is well defined in the dorsal part, but is barely recognizable ventrally, being fused with the metasternum. Weber (1928) therefore interpreted the ventral longitudinal muscles of the metathorax and the first abdominal segment as fused, and labeled them accordingly as muscles *III vlm* + *I a vlm* (Fig. 20.2a). (2) In several auchenorrhynchan taxa, a front part of the tergum of the first abdominal segment, divided by an apodeme, is firmly fused with the metapostnotum (see Ossiannilsson 1949; Fig. 20.3). Figure 20.5a illustrates the basal configuration of the muscles belonging to the tymbal organ sensu lato and the labeling according to Weber (1928), which is adopted here.

In most species studied, the pair of tymbal muscles was shown to be working in alternation, i.e., buckling the left and right tymbal, consecutively (see, e.g., Aidley

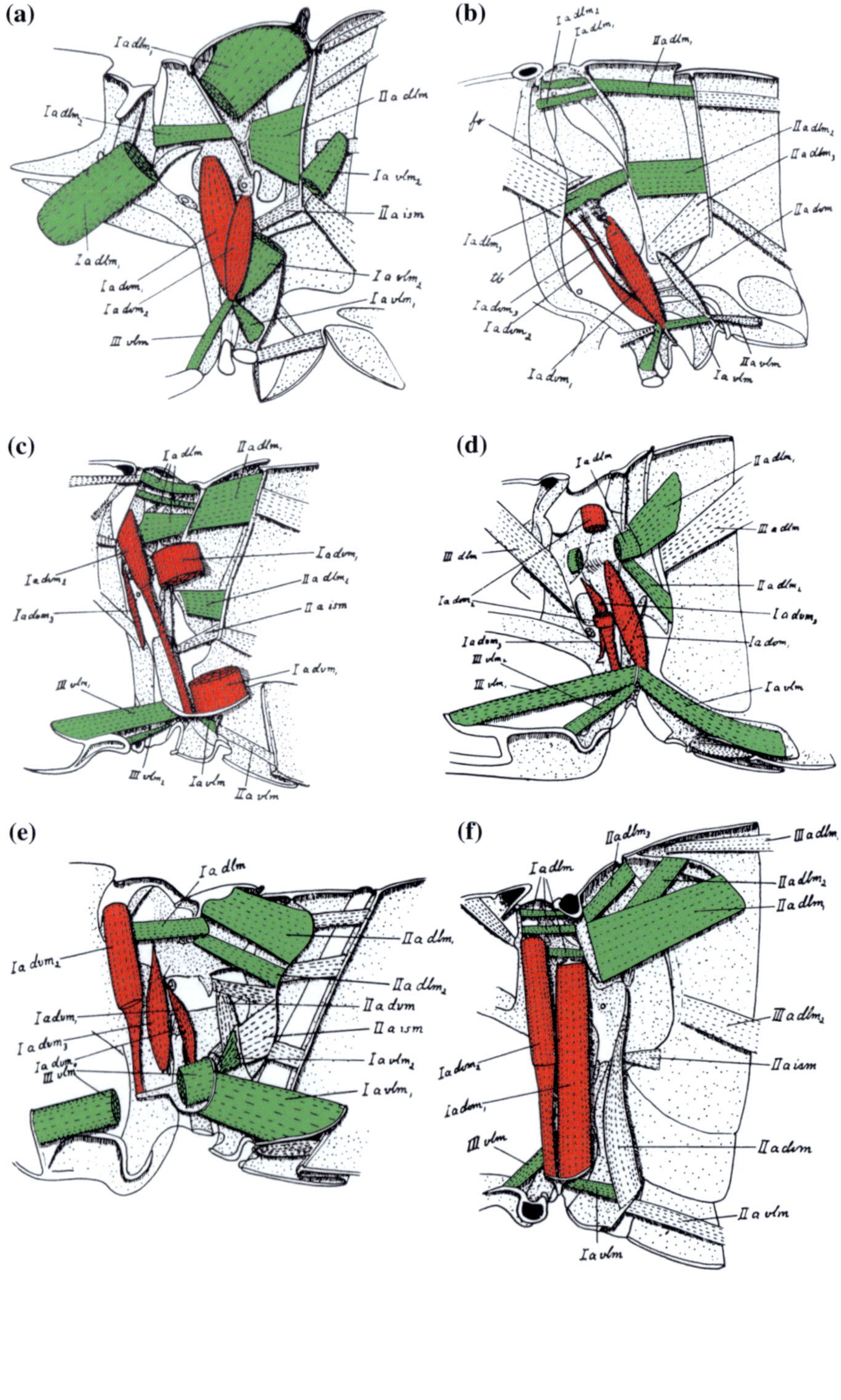
(a)
(b)
(c)
(d)
(e)
(f)
I a dlm
II a dlm
I a vlm
II a vlm
III vlm
I a dvm
II a dvm
II a ism
I a vlm,
II a dlm,
I a dlm,
I a dvm,
III a dlm

◀**Fig. 20.3** The tymbal organ of the leaf- and planthoppers. Configuration of the involved dorsoventral muscles of the first abdominal segment (*red*), and the dorso- and ventro-longitudinal muscles of the first and second abdominal segment (*green*), modified from Ossiannilsson (1949). Note that the muscles labeled *I a vlm* by Ossiannilsson correspond to muscles *II a vlm* throughout the text, following the terminology of Weber (1928, cf. Figure 20.2a). **a** *Dicranotropis hamata* (Delphacidae); **b** *Neophilaenus campestris* (Aphrophoridae); **c** *Evacanthus interruptus* (Cicadellinae); **d** *Cicadella viridis* (Cicadellinae); **e** *Idiocerus lituratus* (Idiocerinae); **f** *Agallia brachyptera* (Agalliinae). See Table A.2 for used species names. (The figures are reproduced with kind permission from Scandinavian Entomology, www.scanentom.se)

1969; Young 1972a). In normal song, there seems to be a fixed cycle of left/right action, causing a sequence of four tymbal plate bucklings (left in/out, right in/out). The phase of the cycles in a song can shift between 0.2 (meaning a time lag between the cycles) and 0.5 (equal time between right and left movements) (Hennig et al. 1994). In disturbance chirps of the same species, the left/right cycle can become irregular, contradicting a direct coupling of the muscles or axons and pointing toward—in conjunction with other evidence—to a central pattern generator governing the tymbal actions (Hennig et al. 1993, 1994).

Carlet (1877) already described an accessory pair of muscles, the tensor muscle, which runs somewhat parallel to the mighty tymbal muscle and inserts at the anterior frame of the tymbal plate. Pringle (1954) could show that the tensor muscle not only modifies the sound by increasing the stiffness and convexity of the tymbal plate but is also necessary for maintaining proper tymbal action during song activity. In some species, the contraction of the tensor muscle seems generally necessary for initiating the sound production (Hennig et al. 1994). The tensor muscle differs significantly from the tymbal muscle in its physiology and acts tonically (Simmons and Young 1978; Hennig et al. 1994). The tensor muscle action affects the sound in several ways that depend on the species-specific characters of the complex sound-producing apparatus. In almost all species studied, it did increase the amplitude of the sound pulses (for a counter-example in the Tibicinidae, see Fonseca et al. 2008), and in *Cystosoma saundersii*, it was also shown to be responsible for keeping the sound sharply tuned to the carrier frequency of the calling song (Simmons and Young 1978).

It has been known since antiquity that the female cicadids do not possess a tymbal organ—the Greek satirist Xenarchus is often quoted with the rather insensitive lines “Happy the cicadas live, since they all have voiceless wives” (cited after Darwin 1871: p. 350). A large motoneuron in the female auditory nerve is considered to be a homologue of the male tymbal muscle motoneuron (Wohlers et al. 1979) and innervates one of the three *tensor tympani* muscles supporting the tympanum in females (Wohlers and Bacon 1980). In experiments, the female “tymbal motoneuron” exhibited an activity pattern very similar to the corresponding male axons with alternating spike sequences in right and left nerves, however, with a different phase relationship and a higher variability (Hennig et al. 1993). The innervated *tensor tympani* is presumably involved in adjusting the auditory threshold to environmental sound and noise levels—how, or if, the "ancient" song motor program is involved in this task is unknown (Hennig et al. 1993).

As already described above, the tymbal plate itself acts as a frequency multiplier (see Michelsen and Nocke 1974), transforming the action of the tymbal muscles into high pitching sounds. A pair of tymbal muscles can produce a cycle of tymbal buckling (left in/out, right in/out) with more than 100 Hz, causing more than 400 buckles per second; ribs on the tymbal plate buckling in succession can amplify this pulse rate tenfold. The structure and arrangement of ribs determine the sound frequency and the harmonic content (Young 1972a; Simmons and Young 1978). The details of the tymbal plate movements vary from species to species and make it difficult to compare the results reported by different researchers, e.g., concerning the role of the tensor muscle in sound modulation (see Fonseca et al. 2008). In *Platypleura kaempferi*, Hagiwara and Ogura (1960) found no obvious correspondence between muscle motor impulses and sound pulses, possibly pointing to a membrane in steady-state vibration (see also Young and Josephson 1983).

The abdomina of male cicadas are completely filled by large paired air sacs of tracheal origin (Pringle 1954). The air-filled abdomen acts as sound radiator and amplifier and is tuned to the fundamental sound frequency of the calling song (Pringle 1954; Fletcher and Hill 1978; Bennet-Clark and Young 1992). However, the "tuned resonator" alone cannot account for the very high levels of acoustic energy radiated by the relatively small animal (MacNally and Young 1980), which are mainly attributed to non-linear characteristics of the signal (Hughes et al. 2009). Singing animals can adjust the resonant properties by stretching and raising the abdomen, thus enlarging the resonant cavity inside as well as the room between thorax and abdomen (Gogala et al. 2004; Boulard 2006). Adjustment of the thorax–abdomen gap may increase the sound volume by tuning the cavity to the eigenfrequency of the tymbal plate, while dilatation of the abdomen might raise the sharpness of tuning and reduce the damping (Pringle 1954; Bennet-Clark and Young 1998).

In summary, the species-specific songs of cicadas depend on the (fixed) physical properties of (1) the tymbal plates, including their mass and the geometry of their ribs, (2) the maximum muscle action frequency, and (3) volume, shape, and mass of (passively) resonating cavities and membranes. The sound can be modified by (1) different muscle contraction rates, (2) a different phasing of tymbal cycles, (3) the action of accessory muscles, and (4) abdomen movements, changing the tension of vibrating parts and the resonating properties of the abdomen (Moore 1962). Neurogenic muscle action depends on a neural central pattern generator, whose plasticity allows for fast evolutionary adaptations in species-specific songs (Fonseca et al. 2008). The heat production by tymbal muscles during singing also considerably influences several parameters of the song, e.g., the pulse rate, structure of pulse groups (interpulse intervals between rib buckling), and amplitude of sound, but not the fundamental sound frequency (Josephson and Young 1979).

The males of the Tettigarctidae, the "hairy cicadas," a relictual taxon with only two extant species (*Tettigarcta crinita* from New South Wales and *T. tomentosa* from Tasmania) were for a long time thought to have lost their ability to produce songs secondarily (Myers 1929; Evans 1941). Both sexes possess striated tymbal plates and tymbal muscles, however, much less developed as in other Cicadoidea—resembling non-cicadoid auchenorrhyncha—and lack the typical tympanal auditory organs (Evans 1941; Pringle 1957). The males (and apparently also the females) of *Tettigarcta crinita* were then found to produce simple, low-intensity, substrate-borne vibrations in courtship, interpreted as a plesiomorphic feature of the only surviving relicts of a Jurassic and Cretaceous radiation of "primitive" cicadas (Claridge et al. 1999). It may be mentioned here that male calling in airborne vibration-producing species induces strong substrate vibrations as well (Gogala et al. 1996).

## 20.3 Discovery of the Tymbal Organ in Plant- and Leafhoppers (Fulgoromorpha and Cicadomorpha excl. Cicadoidea)

In 1889, M. A. Giard described peculiar organs of unknown function in males of the genus *Typhlocyba* (Cicadellidae). Two large apodemes that arise from the ventral side of the second abdominal segment (Giard believed it to be the first segment) and reach backward up to the fourth segment reminded him of the cicadid tymbal apparatus and he hypothesized it to be a homologous sound-producing organ ("Ces organs me paraissent homologues de l'appareil phonateur des Cigales mâles" [Giard 1889, p. 710]). Actual sound production in non-cicadoid Auchenorrhyncha was first described by G. W. Kirkaldy for the delphacid *Perkinsiella saccharicida*, which "has been distinctly, and often, heard to stridulate [...], though specially modified organs have not yet been discovered" (Kirkaldy 1906, p. 285). At the same time, F. A. G. Muir traveled the Fiji Islands and reported "a distinct noise of insects stridulating" from a palm with "the underside of the leaves covered with hundreds of a small red leafhopper" being a member of the Derbidae (Muir in Kirkaldy 1907, p. 7). He observed rapid wing movements as well as abdominal vibrations and identified the rudimental hind wings as the sound-producing stridulation organs. Muir and Kirkaldy described and illustrated the "stridulatory area" at the hind wings for a number of Derbidae and Delphacidae; however, at least for *P. saccharicida,* Muir was "in doubt if this could be the means by which they produce the sound" (Muir in Kirkaldy 1907, p. 8).

More than 40 years after the discovery by Giard—and apparently unaware of his publication—C. J. George (1933) again described the large apodemes of male Typhlocybinae (Cicadellidae) and compared it to the tymbal apparatus of the cicadas. He also described and illustrated the attached muscles and the curved surface of the apodemes and proposed sound production by an internal "timbal plate" buckling: "The contraction of the oblique muscle, attached to the concave [dorsal] side of the plate, will turn the plate convex inside and concave outside. [...] If the plate is thus turned suddenly from concave to convex, and convex to concave, from the nature of the chitin a smart clicking sound ought to result" (George 1933, pp. 55–56). He also observed "that a sharp sound is often produced by the male" of *Empoasca devastans* (George 1933, p. 56).

Ribaut (1936) is the third "discoverer" of the "appareil stridulatoire" in Typhlocybinae (*Eupteryx melissae*). While he was not aware of any sound-producing behavior in leafhoppers, he nevertheless proposed another mechanism of sound production. He correctly described the sternite I as the anterior insertion site of the pair of voluminous "muscles vibrateurs" that posteriorly insert at the dorsal surface of the large apodemes. Ribaut clearly states that in contrast to Giard, he does not consider the apodemes to play a role in actual sound production, and it seems that he interprets the first abdominal sternite as the "membrane vibratoire" (Ribaut 1936, p. 30). Moreover, he labeled a median part of the intersegmental membrane between the abdominal sterna I and II as "tympan" in the accompanying figure, a term not used in the text (Ribaut 1936: Fig. 46).

Several authors also noted that the appearance of the sternal apodemes in Typhlocybinae is species specific and can be used, in addition to the internal male genitalia, for species delimitations, "since they offer distinctive and constant structural characters" (Wheeler 1937, p. 153; see also Paoli 1930, 1933; Poos 1933; Wheeler 1939).

Evans (1946) described a lateral pair of "ridged bosses" on the first abdominal segment—that are stronger developed in males—in two Aetalionidae (p. 42). He states that these structures "suggest comparison with the tymbals of cicadas. [...] The primitive Australian cicada, *Tettigarcta*, has functionless tymbals (see Sect. 20.2) in an identical position (Evans 1941). Whether these structures are sound-producing organs, and whether functional or not, cannot be determined from the dried specimens which have been available for study" (Evans 1946, p. 42).

The doctoral thesis of F. Ossiannilsson, published in 1949 under the title "Insect drummers—a study on the morphology and function of the sound-producing organ of Swedish Homoptera Auchenorrhyncha," was the ultimate breakthrough and still constitutes the fundament for all studies about vibrational communication in non-cicadoid Auchenorrhyncha. Ossiannilsson began his studies on the sound production of Auchenorrhyncha in 1938, and being aware of the importance and novelty of his findings, reported already in 1946, in a "preliminary note," that by "very simple arrangements I have hitherto been able to hear the 'songs' of 74 Swedish species [...] of the families Cixiidae, Araeopidae [Delphacidae], Cercopidae, Ulopidae, Jassidae [Jassinae], and Typhlocybidae [Typhlocybidae]" (p. 83). In a second "preliminary note," Ossiannilsson (1948) described the sound production by females in Aphrophoridae and Cicadellidae and some peculiar sexual dimorphisms. In 1949, he published a detailed paper comprising the description of sound production and the structure of the sound-producing organ in 96, and 80 species, respectively, from 15 (then recognized) families (see Fig. 20.3 and Table A.1).

In addition to the description of vibrational behavior in plant- and leafhoppers, Ossiannilsson intended to test the hypothesis of the homology of their vibration-producing organs with the cicadid tymbal apparatus as suggested by Giard and Evans. He tried to identify a tymbal plate in all investigated species and described the muscles in the first two abdominal segments and their configuration. For 16 species, he provided detailed illustrations of exoskeletal parts, internal apodemes, and muscles and made an attempt to homologize all muscles involved.

In some taxa, e.g., in the Aphrophoridae, Ossiannilsson found at the lateral parts of the first segment an "unpigmented convex surface (tb), which is striated in a characteristic way by about twenty approximately parallel transverse lines [...]. Probably these lines are thickened sclerotic bars. I term this surface the *striated tymbal* by analogy [sic!] with the tymbal in the *Cicadidae*." (1949, p. 21; see Fig. 20.3b). Those "striated tymbals" he could also identify in a membracid (*Centrotus cornutus*), and in some Cicadellidae (Cicadellinae: *Tettigella viridis*, Dorycephalinae: *Eupelix depressa* f. *cuspidata*, Macropsinae: *Oncopsis flavicollis*, *Oncopsis tristis*, *Oncopsis alni*, *Macropsis tiliae*, *Macropsis planicollis*, *Macropsis fuscinervis*, Agalliinae: *Agallia brachyptera*, *Agallia venosa*, Deltocephalinae: *Doratura stylata*, *Doratura homophyla*), whereas he found no equivalents in delphacids, cixiids, and most cicadellids (Cicadellinae: *Euacanthus interruptus*, Ledrinae: *Ledra aurita*, Idiocerinae: *Idiocerus lituratus*, *Idiocerus albicans*, *Idiocerus elegans*, *Idiocerus stigmaticalis*, Megophthalminae: *Paropia scanica*, Deltocephalinae: *Graphocraerus ventralis, Opsius stactogalus, Macrosteles cristatus, Macustus grisescens, Euscelis plebeius*, *Streptanus marginatus*, *Streptanus aemulans*, *Diplocolenus abdominalis, Psammotettix cephalotes, Speudotettix subfusculus, Athysanus argentatus, Limotettix striatulus, Thamnotettix confinis, Solenopyx sulphurellus*, Typhlocybinae: *Empoasca virgator*, *Empoasca smaragdula*, *Empoasca strigilifera*, *Empoasca sordidula*, *Empoasca rufescens*, *Cicadella atropunctata*, Aphrodinae: *Aphrodes bicinctus*, *Aphrodes bifasciatus*, *Aphrodes trifasciatus*, *Aphrodes flavostriatus*). In some species, he instead found: (1) "unstriated" but strongly convex areas in the corresponding place (Aphrodinae: *Aphrodes*); (2) several separated "more strongly sclerotized and pigmented tergites [...] in the region of the first abdominal dorsum" (p. 24) (Cicadellinae: *Euacanthus*, Idiocerinae: *Idiocerus*) where the largest pair of muscles ($I\ a\ dvm_1$) is inserted at a pair of such tergites by a short tendon [as in cicadids] (Fig. 20.3c); or (3) a "tergum of the first abdominal segment furnished with list-shaped strengthenings on the inside" (Ledrinae: *Ledra*, p. 29). He summarized: "In all male forms examined by me the integument in the region of the first abdominal dorsum is partly strongly sclerotized in such a manner that a thin but elastic membrane is present. In certain genera [...] the second abdominal tergum is partly of this nature, too" (p. 104). In some Cercopidae and Cicadellidae with and without a "striated tymbal" Ossiannilsson observed "vigorous vibrations in the lateral parts of the first abdominal tergum, these being made concave and convex in alternation" (p. 111; see also Franz 1978). In a species of Deltocephalinae without striated tymbal, he also observed vibrations with small amplitude, "the membrane not being made concave" (p. 111).

For the 16 species examined anatomically, Ossiannilsson provided detailed descriptions and illustrations of the complicated arrangement of the muscles of the first two abdominal segments and labeled the muscles according to their segmental origin (see Fig. 20.3). However, he was very cautious regarding their homology ("Muscles are not necessarily homologous with those with the same numbers in different superfamilies or higher taxonomic units," p. 11), with the important exception of the dorsoventral muscle $I\ a\ dvm_1$. He states: "This muscle arises on the first abdominal sternum more medially the better it is developed, and is inserted at the hind border of the striated tymbal if such is present or, if there is no striated tymbal, on the corresponding spot somewhat anteriorly of the antecosta of the second abdominal tergum. I believe that this muscle is homologous with the tymbal muscle of the cicadas, too" (p. 104).

Ossiannilsson assumes that the function of the longitudinal muscles is the variation of the tension of the vibrating tergal parts and thus the modification of call characters. He also discusses their more direct involvement in sound production by direct or indirect buckling of the dorsum (e.g., in Deltocephalinae and Delphacidae), but he doubts that those muscles act through high-frequency contractions. Even for Typhlocybinae with their "relatively colossal development of one of the ventral longitudinal muscles in the first abdominal segment" (p. 114), he does not consider this muscle as the primary producer of vibrations. In his functional interpretation, the dorsoventral muscles cause the vibration of a dorsal region, and the mighty ventral longitudinal muscles control the tension of this vibrating part. Ossiannilsson argues that "At the first moment it certainly seems very singular that a muscle that is assumed to have only accessory function should have much larger dimensions than the primarily sound-producing muscles, but after all this is not at all unreasonable. The force necessary to set a thin, elastic membrane in vibration by a muscle inserted directly on the former or on its margins is doubtless insignificant if compared with that required to keep this membrane stretched, especially if the stretching must be effected indirectly" (p. 115). We will remember this argument when discussing the heteropteran tymbal organ below. Ossiannilsson also briefly considers Ribaut's theory of vibrating the first abdominal sternum, but finds some experimental evidence that "the dorsum of the first and second abdominal segments [...] were engaged in high-frequent vibrations of small amplitude" (p. 115).

Simultaneously with Ossiannilsson, but completely neglected by the scientific community, K. Vondráček published in 1949 a comprehensive study on "the sound-producing apparatus in the males of the leafhoppers." Vondráček provides a detailed description of the "sound-producing apparatus [of the male of *Typhlocyba ulmi* (Cicadellidae: Typhlocybinae)] consist[ing] of a chitinous skeleton, formed by the sternite of the first and second abdominal segment, which stretch the connective membrane placed between them, of a resonance equipment amplificating the sound produced, and of muscles that partly set the chitinous skeleton into motion, partly serve to its stabilisation in a definite position" (p. 30). His interpretation of the function of this apparatus—solely based on structure, and not on direct behavioral observations—resembles that of Ribaut (1936): by up-and-down movements of the freely moveable first abdominal sternite, lateral portions of the intersegmental membrane (Ribaut's "tympan") are set in motion. The air-filled cavities in the large apodemes should serve as resonance cavities and amplifiers.

Most notably in Vondráček's study, however, is the description of an "auditory organ" found in the second abdominal segment, being the first and so far only report of an abdominal vibration receiver in non-cicadoid Hemiptera (cf. Lakes-Harlan and Strauß, Chap. 14, this volume). The organ consists of a dorsoventral bundle of scolopidia attached to lateral parts of the second tergum—which Vondráček identifies as "tympanic membranes" or "tympana"—and ventrally attached to a central sclerite of the intersegmental membrane I–II. If we invert the functional interpretation of Vondráček and assume the ventral intersegmental membrane I–II to be the tympanum, it astonishingly resembles—and could be homologized with—the cicadid auditory organ (cf. Vogel 1921, 1923; Young and Hill 1977; Strauß and Lakes-Harlan 2014 Chap. 14, this volume; for a detailed discussion, see Wessel, Mühlethaler and Malenovský, in preperation).

Ossiannilsson reported the occurrence of different types of calls in most species he had studied: (1) "single clackings or beats", (2) "more prolonged drumming or trilling," and (3) "more even tones as from a flute or violin or the human voice" and various combinations of this types (p. 111). For *Doratura stylata* (Deltocephalinae), a species with singing males and females, he reported a female "invitation call," "a call with a sexual significance" (1949, p. 139). However, it remained for the study of Strübing (1958, and Chap. 5, this volume) on the mating behavior in several Delphacidae to conclusively demonstrate the necessity of vibrational communication for successful mating. In 1960, Strübing also described complex songs and highly structured alternating courtship duets in delphacids, accompanied possibly also by some sort of wing-flicking clicks. Neither Ossiannilsson nor Strübing or later researchers offered a satisfactory functional interpretation for the unique and very complicated tymbal organ of the Delphacidae, however.

Ossiannilsson himself considered his work as unfinished: "For the present, however, too much theorizing on the significance of the differences in the various genera would be premature until the studies have been extended even to families not represented in Sweden. [...] I hope to be able to study the biology of the sound-production of the Auchenorrhyncha more closely in the future. The present observations are only a small contribution" (1949, p. 7). These hopes were not fulfilled, and it is astounding how very few studies have added to Ossiannilssons fundamental work.

Some biophysical studies from the group of Strübing (Haße 1974; Mebes 1974; Franz 1978) supported the assumption that energy radiation is accomplished most effectively by substrate vibration. Haße (1974) considered the small insect body of a leafhopper as a point sound source, which does not cause a pressure change but rather a volume displacement, and consequently requires spending a minimum of energy for long-lasting song activity. Mebes (1974) analyzed the vibrational signals of *Euscelis ononidis* and could show that ultrasonic vibrations do not play a role in their communication. Moreover, he discussed the physical limits of vibration production in leafhoppers and concluded that the main energy radiation is in the range of up to 5 kHz.

Smith and Georghiou (1972) used histological sections and scanning electron microscopy for the description of the morphology of the tymbal organ of the beet leafhopper (*Circulifer tenellus*), describing thickly sclerotized, dorso-lateral "tymbal sclerites" surrounded by membraneous areas, "dorsal apodemes" as the attachment structure of the tymbal muscle, and a sternal furca supporting the muscles ventrally. Strübing and Schwarz-Mittelstaedt (1986) finally located the vibrating membrane: "Instead of the paired vibratory plates [i.e., the tymbal plates], there occur dorsally uniform intensely folded soft membranes located at the centre of the first and second abdominal tergum for the species *Euscelis incisus*. They are the attachment for the large V-shaped muscles as well as for further small muscles which altogether participate in the vibration of the membranes" (p. 49).

For delphacids, Mitomi et al. (1984) provided a detailed description of the morphology of the vibration-producing apparatus in the rice brown planthopper, *Nilaparvata lugens*, which coincides with Ossiannilsson's descriptions. Whereas, however, Ossiannilsson assumed a joint (antagonistic) action of the muscles $I\ a\ dlm_1$ and $I\ a\ vlm_2$ in vibrating the whole abdomen, Mitomi et al assume $I\ a\ dlm_1$ to act as the main vibration-producing muscle in males and females, moving in males a large tymbal composed of the first and second abdominal tergites. These two tergites and their lateral membranous parts are completely framed by the metapostnotum and the third abdominal tergite.

In 1985, M. Asche published the only comprehensive comparative study allowing a discussion of evolutionary scenarios for the tymbal organ, treating more than 1,000 species from over three-quarter of the then recognized genera of delphacids. The delphacid ground pattern is a relatively simple song apparatus without sexual dimorphism. As an autapomorphic character of the subfamily Delphacinae, the tymbal organ of the males is much more developed with elaborate apodemes at the metapostnotum (as attachment point for $I\ a\ dlm_1$) and the second abdominal sternite (as attachment for $I\ a\ vlm_2$). In males, the second abdominal tergite is medially developed as a dome-shaped "central plate" whose lateral parts serve as attachment points for the mighty pair of $I\ a\ dlm_1$. Special modifications of the apodemes provide additional characters for phylogenetic inference of delphacine subtaxa. A convergent evolutionary pattern can be observed in Caliscelinae (Asche 1985, pp. 128–129).

## 20.4 Homology of the Tymbal-Like Organs in the "Auchenorrhyncha"

In 1957, Pringle stated for the Auchenorrhyncha: "Within this group there is, in spite of considerable difference in general appearance, such a close similarity in the basic plan [of the sound-producing organs] that it would be unreasonable to suppose that the technique has evolved twice" (p. 154).

In his endeavor to test the hypothesized homology of the vibration-producing organs in small non-cicadoid auchenorrynchans with the tymbal apparatus of the large singing cicadas, Ossiannilsson (1949) initially tried to identify tymbal plate-like structures. The most easily recognizable feature of the cicadid tymbal plates is the (almost always) ribbed surface and thus Ossiannilsson was willing to consider small convex structures with ribs located dorso-laterally at the first abdominal segment (termed by him "striated tymbals") as the homologue of the "real" cicadoid tymbal. However, these "striated tymbals" vary considerably in structure and location in respect to the insertion points of the muscles, and in the majority of species studied he did not find such a "striated tymbal" at all. In some species, he took smooth convex areas or small isolated sclerites (tergites) as tymbal plate homologues, or he proposed larger, membranous parts medially on the first (and sometimes also second) abdominal tergum as vibrating structures.

On the other hand, Ossiannilsson found striking examples of closely related species showing great similarities in the configuration of the muscles involved but with considerable differences in the exoskeletal structures of the first two abdominal segments. Eventually, his conclusion that the vibration-producing organ of the small Auchenorrhyncha is homologous to the cicadid tymbal apparatus was based on the homologization of the muscles involved. In principle, he identified the most medially located dorsoventral muscle of the first abdominal segment ($I\ a\ dvm_1$) as the main song muscle and homologous to the cicadoid tymbal muscle. However, he described for all species a set of accessory dorsoventral as well as dorso- and ventro-longitudinal muscles that assist and modify the actions of the $dvm_1$, and in several cases completely take over its sound-producing function (see Fig. 20.5b). Even when the "tymbal muscle" is not present, the complexity of the tymbal organ makes a homologization possible: "However, whether muscle $I\ a\ dvm_1$ is present or not, the structure of the sound-producing organ is complicated by the presence of several other muscles. As these are as a rule much weaker in mute females, they must play some part in the sound-production" (Ossiannilsson 1949, pp. 112–113).

Ossiannilsson summarized: "In all males thus examined I have found an organ apparently in its essential parts homologous with the tymbal apparatus of the Cicadidae even if it is certainly much modified in some groups. [...] I feel convinced that the possession of a functional tymbal apparatus is general among the Auchenorrhyncha. [...] In some forms the female possesses a functional sound-producing organ of the same kind as that of the male though more weakly developed. [...] In *Doratura* both sexes have a sound-producing organ of the same type and equally well developed. I believe that this is the primary condition" (1949, pp. 138–139).

Later authors mostly followed Ossiannilsson's conclusions and Evans (1957), e.g., even pondered a very early origin of tymbals in the paleozoic. Sweet (1996) suggested the presence of tymbals as a diagnostic character of the Auchenorrhyncha in relation to the Sternorrhyncha ("In the suborder Sternorrhyncha [...] tymbals are absent. [...] In the suborder Auchenorrhyncha [...] tymbals are usually present" [p. 119]). However, he defines "tymbals" exclusively as exoskeletal structures ("The monophyly of the suborder Auchenorrhyncha is probably best indicated by the presence of arched tymbals formed of abdominal segments 1 and 2 in most members of the infraorders and superfamilies of the Auchenorrhyncha" [Sweet 1996, p. 135]). His conclusion, however, that "It seems unlikely that such a unique sound-making mechanism has evolved more than once, and it is more likely that this mechanism for sound production has been lost in those few auchenorrhynchans that apparently lack tymbals, including female cicadids" (Sweet 1996, p. 135) would only be acceptable if the complete, complex tymbal organ was taken into consideration.

In short, the involvement of the same set of muscles (*I a dvm*, *I* + *II a dlm*, *III vlm* + *II a vlm*) in the production of vibrations by distortion of more or less specialized parts of the integument defines the auchenorrhynchan tymbal organ.

## 20.5 Discovery of a Tymbal-Like Organ in True Bugs (Heteroptera) and Moss Bugs (Coleorrhyncha)

Sound production in land-dwelling bugs was first mentioned by J. Ray in his posthumously published *Historia insectorum* (1710). In a reduviid bug (*Reduvius personatus* L., named by Ray "*Musca cimiciformis* 3*a D. Willughby*"), he correctly described stridulation performed by rubbing the tip of the proboscis over a striated furrow at the prosternum ("Sonitum edit Locuſtarum non abſimilem affricando proboſcidem ad ſternum durum, inter priores pedes, ubi proboſcidem fricat parùm canaliculatum. Quieſcens proboſcidem in canali reponit." Ray 1710, p. 56; cf. Handlirsch 1900a). The description of stridulation by a species of Nabidae by Poda von Neuhaus (1761) frequently mentioned in the literature, however, is based on the confusion or rather a mix-up of animals and their ascribed behaviors (for a review see Dolling 1995). Starting with Handlirsch's thorough studies (1900a, b), a host of stridulation mechanisms and behaviors were described during the twentieth century for several heteropteran taxa (for a systematic review see Gogala 2006, and Table A.1).

In April 1953, D. Leston was the first to record complex sound making in pentatominae bugs (*Sciocoris cursitans* and *Stollia fabricii*), but he could not find a stridulation mechanism: "The mating call [...] was at its commencement correlated with up and down movements of the abdomen. The call consists of a high-pitched (compared with the common song) staccato passage with the intervals decreasing and pitch increasing through about a semitone until a steady note is produced with no visible abdominal movement. The call concludes with a mirror image of its commencement and further abdominal movement; the whole lasts about eight to ten seconds. No sound-producing apparatus was detected, but the first and second abdominal terga are depressed laterally to form what is probably a pair of resonating chambers and it is certain that the apparatus involves this part of the abdomen (perhaps including the pair of pre-first sclerites)" (Leston 1954, pp. 49–50).

In 1958, the discovery of a new type of sound production in the acanthosomid bug *Elasmucha grisea* by one of his students was reported by K. H. C. Jordan in two papers (1958a, b): "Meine Schülerin, Helga Slowioczek [...] fand dabei eine besondere Art der Tonerzeugung, die ganz abweichend von der sonst bei Insekten und besonders bei Wanzen vorkommenden Stridulation ist" (Jordan 1958a, p. 393). Jordan and Slowioczek recorded male songs with a basic frequency of 67–83 Hz that would last for up to 45 min and they observed periodical amplitude modulation during single calls. Single observations indicated a function of the songs in courtship as well as in rivalry. During singing, movements of the first and second abdominal tergites, and an up-and-down movement of the whole abdomen were observed. Jordan concluded from their low frequency that the vibrations were not produced by stridulation but rather by a vibrating membrane. He already interpreted the vibration-producing apparatus as a pre-stage of the cicada's tymbal organ without a resonance structure: "Es fehlt dem Lautapparat der wirksame Resonanzboden, wie wir ihn von Zikaden her kennen. Man kann mit gewissen Einschränkungen sagen, der Lautapparat der Pentatomiden ist eine bescheidene Vorstufe zu dem Trommelapparat der Zikaden, die es in höchster Vollendung zeigen"(Jordan 1958a, p. 394).

In the second paper (Jordan 1958b), Jordan and Slowioczek described this new type of vibration production also for some Pentatominae (*Carpocoris pudicus*, *Palomena prasina*, *P. viridissima*, *Dolycoris baccarum*, *Eurydema oleraceum*, *Pentatoma rufipes*, *Aelia acuminata*) and Asopinae (*Picromerus bidens*, *Arma custos*). For some species, they recorded complex courtship behavior with male–female-alternating songs (*C. pudicus*, *P. prasina*), and also the use of strong vibrations as a defense action (*P. prasina*). In most species, they could clearly differentiate between courtship signals and a "common song" in the males. For Pentatomidae as well as for the Acanthosomidae (*E. grisea*), the vibration frequencies were in the range of 55–198 Hz. The songs showed frequency and amplitude modulation and consisted of "simple" as well as "composed sounds" ("zusammengesetzte Laute," Slowioczek in Jordan 1958b, p. 140).

In all studied species, Jordan and Slowioczek observed a back-and-forth movement of the first two abdominal tergites, and whole abdomen movements in synchrony with the vibration pulses produced. For the abdominal vibrations, they assumed a function in "song modulation." The experimental exclusion of wing and spiracle vibrations supports the assumption of vibration production by means of a vibrating membrane. However, a special muscle ("Tonmuskel") as a homologue to the tymbal muscle could not be identified.

This new type (for the true bugs) of vibration-producing apparatus is sometimes referred to as Jordan's organ. However, Jordan himself made it very clear in the text of his two papers that the actual discovery should be accredited to his student H. Slowioczek, thus the name Jordan-Slowioczek's organ would be more appropriate.

Unaware of Jordan's publications, Moore (1961) reported low-frequency sounds (0.25–1.5 kHz, pulse rate around 50 Hz) in a pentatomine bug (the rice stink bug *Oebalus pugnax*) produced "in an unknown manner" (p. 280) and said that "However, the structure of these sounds is very similar to those produced by timbals in certain of the auchenorrhynchous Homoptera" (p. 282). Leston and Pringle (1963b) were actually able to describe seasonal air sacs in *Eysarcoris* (=*Stollia*) males "occupying, during the spring mating period, over half the volume of the abdomen and in more or less immediate contact with the tymbal area. The sacs are analogous to those of Cicadidae" (Leston and Pringle 1963b, p. 798). Whereas Leston and Pringle use (following Jordan's suggestion) the term "tymbal area" for the fused first two abdominal tergites, and even discuss the muscle *I a* $dlm_1$ as "tymbal muscle," they are carefully avoiding an outright homologization: "Both these basically different mechanisms [strigilation and tymbal vibration] are found within the Hemiptera, and each appears to have arisen more than once in evolution. Care is therefore needed in the use of apparent homologies in taxonomic arguments" (Leston and Pringle 1963a, p. 392).

Gogala (1969, 1970) studied the acoustic communication in several species of Cydnidae (*Tritomegas bicolor*, *Sehirus biguttatus*, *Canthophorus melanopterus*, *C. dubius*) and found—in addition to the already known high-frequency stridulation signals—low-frequency signals (100–500 Hz) produced by a tymbal-like mechanism. He identified different sex- and species-specific signals that supported the role of vibrational behavior as an isolating mechanism (see Leston and Pringle 1963a). In 1974, Gogala et al. could then show in a conclusive way (for *T. bicolor* and *C. dubius*) that only the substrate-borne part of the vibratory signals is crucial for communication.

The current model organism for the study of vibrational communication in true bugs is the southern green stink bug, *Nezara viridula* (Pentatominae), an economically important pest, for which the function of vibratory songs during courtship has been studied in detail (Čokl et al. 1972; Čokl 1985; for reviews, see Todd 1989; Čokl and Virant-Doberlet 2003). However, the exact mechanism of vibration production and, thus, the question whether it is produced by a "real" tymbal, homologous to the cicadas' organ, was (and still is) disputed. Pro and contra arguments are mainly based on the properties of the recorded signals. Gogala (1986, p. 47), for example, argued that "Tymbal-signals are usually lower in frequency, may last longer and have in many cases a high content of higher harmonics [compared to stridulatory signals]. [...] the whole abdomen vibrates with the basic frequency of the vibratory signal. The form of the vibratory signal is far from the simple sinusoidal movement, which supports the idea of the click or tymbal mechanism being involved in vibration production."

In a first step toward elucidating the mechanism, Gogala could show that in Cydnidae, the first two tergites are involved in signal production (Gogala 1985b, 2006). However, as outlined above, the core problem for the homologization of the bugs' and the cicadas' tymbal is the configuration and identity of the muscles involved. In his classical study on the "Skeletal motor mechanism of the thorax of the 'stink bug,' *Nezara viridula* L.," N. S. R. Malouf (1932) also described the musculature of the first and (partly) second abdominal segment. Most prominent are the large dorso-longitudinal muscles (TL I and TL II, being *I a* $dlm_1$ and *II a* $dlm_1$), with *II a* $dlm_1$ occupying about two-thirds of the width of the segment and almost fusing at the median line (Fig. 20.4a, d). The first and second segment have one dorsoventral muscle each, *I a dvm* being laterally attached at a conspicuous apodeme (Fig. 20.4a, d). While Malouf (1932, p. 195) found much similarity in the musculature of a cicada (*Tettigia* sp.) studied for comparative purposes, he also described a second dorsoventral muscle: "On a lateral part of the *abdominal* part of the tergum is attached the pleural compressor of the first abdominal segment ([...]LCrI) which attaches ventrally on an inflection of the pleural region of the first abdominal segment. In the cicada there is a smaller accessory lateral compressor, $LCrI^1$. This muscle assigns both the thoracic and the abdominal parts of the tergum hence described to the first abdominal segment."

In 1989, V. Kuštor studied the skeletal parts, muscle configuration, and muscle activity responsible for the production of vibrational signals in *N. viridula* with microanatomical, histological, and electromyographic methods. Kuštor could show that in *N. viridula* an accessory lateral compressor exists as well (named LCr a), which is attached to the same apodeme as the LCr I (see Fig. 20.4d). The most important discovery was another dorsoventral muscle, however, identified as *depressor tymbali* (DrTy) and thus homologized with the cicadas' tymbal muscle. This muscle is relatively small (compared to the dorso-longitudinal muscles) and was only traceable in the histological sections, which also explains its late discovery. The muscle originates dorsally at the antecostal ridge of the tymbal at the lateral edge of muscles *I a* $dlm_1$ and underneath *II a* $dlm_1$ (Fig. 20.4b, c) and runs medially and anteriorly toward the metasternum. It clearly is the homologue of the auchenorrhynchan *I a* $dvm_1$, thus defining the other two lateral compressor muscles as *I a* $dvm_{2+3}$. The electromyogram showed muscles *I a* $dlm_1$, *II a* $dlm_1$, and *I a* $dvm_{2+3}$ to work synchronously and revealed a one-to-one ratio of the number of muscle potentials of any active muscle to the vibration pulses (cycles) produced. The motor activity of *I a* $dvm_1$ could not be studied. Additionally, Amon (1990) was able to show that artificial brain stimulation elicits normal song in *N. viridula*, and the myogram of *I a* $dlm_1$ corresponds to the body vibrations.

However, the exact mechanism of vibration production, i.e., the interaction of muscles and exoskeletal parts, and especially the role of the tergal plate and the tymbal muscle (*I a* $dvm_1$), is still insufficiently understood. Sweet (1996) described the principal phylogenetic pattern emerging as follows: "These tergal apodemes in the Heteroptera, and probably in the Coleorrhyncha, form in part Jordan's organ, a

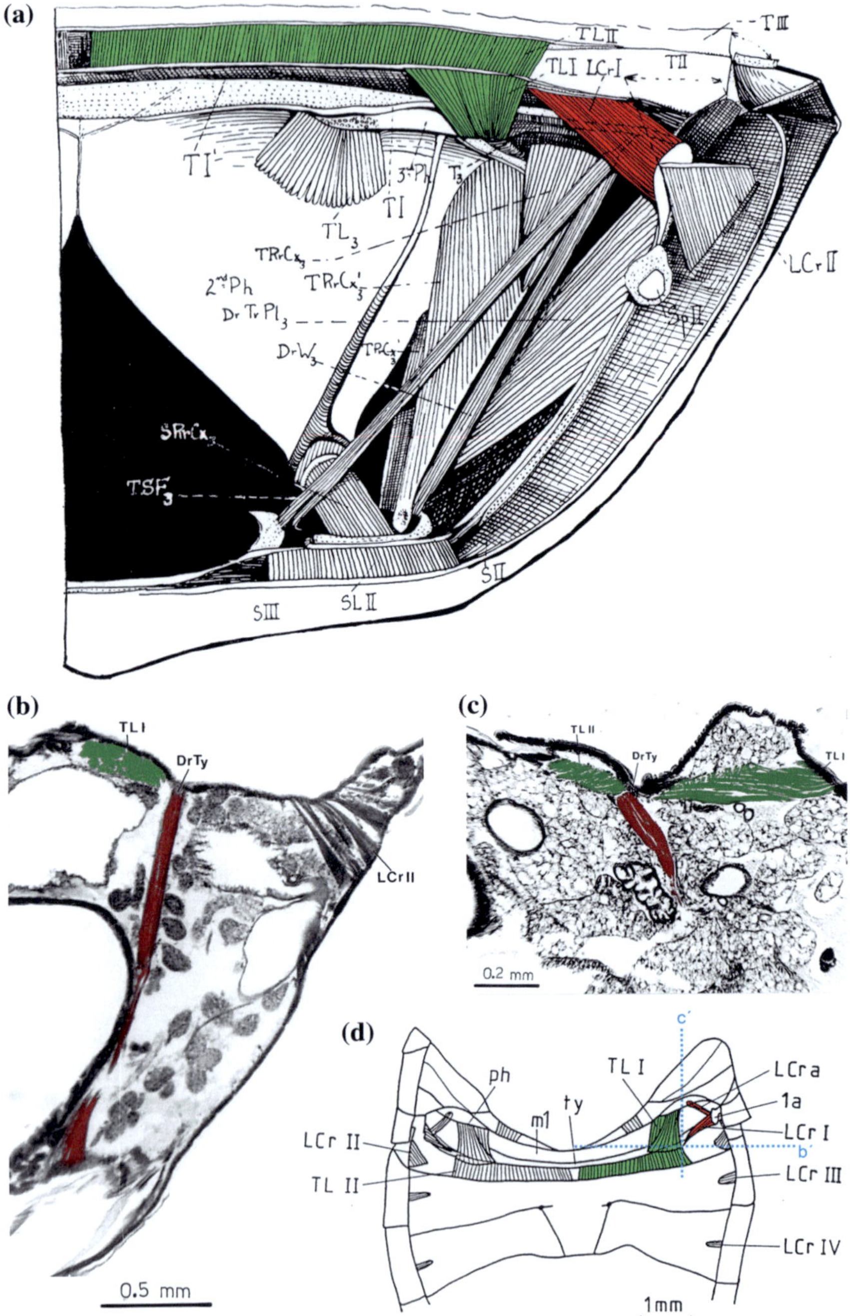
(a)
TIII
TLII
TLI
LCrI
TII
TI'
3rd Ph
TI
TL3
TRrCx3
TRrCx'3
2nd Ph
DrTrPl3
DrW3
SRrCx3
TSF3
LCrII
SpII
SII
SIII
SLII
(b)
TLI
DrTy
LCrII
0.5 mm
(c)
TLII
DrTy
TLI
0.2 mm
(d)
c'
ph
TL I
LCra
1a
ty
m1
LCr I
LCr II
b
TL II
LCr III
LCr IV
1mm

◀**Fig. 20.4** The tymbal organ of the true bug *Nezara viridula* (Heteroptera). **a** Posterior aspect of *right* half of the metathorax, and first and second abdominal segments showing the musculature, modified from Malouf (1932, pl. XVII, Fig. 3). TL I and TL II, = *I a* $dlm_1$ and *II a* $dlm_1$; LCrI = *I a* $dvm_?$. **b** Histological transversal section through the tymbal organ (cf. Figure 20.4d), modified from Kuštor (1989, Fig. 30). **c** Histological sagittal section through the tymbal organ (cf. Fig. 20.4d), modified from Kuštor (1989, Fig. 31). **d** Scheme of the tymbal muscles of the front part of the abdomen, modified from Kuštor (1989, Fig. 21); the *blue dotted lines* indicate the approximate orientation of the sections shown in Fig. 20.4b, c, the point of *intersection* indicates the insertion of muscle *I a* $dvm_1$. (Figure 20.4a is reproduced with permission from The Egyptian Entomological Society, www.ees.eg.net)

tymbal that is vibrated more by dorsal longitudinal muscles than is the tymbal of the Auchenorrhyncha, which is vibrated more by dorsoventral muscles" (p. 120). He added: "From the evidence of the union of terga 1 + 2, tymbal singing will probably prove to be plesiomorphic and widespread throughout the Heteroptera [...]" (Sweet 1996, p. 142). This interpretation, however, is oversimplifying the complex distribution of known (although sometimes not fully understood) vibration-producing mechanisms among the plant- and leafhoppers as described above. In delphacids, it is the (oblique positioned) dorso- and ventro-longitudinal musculature moving a tergal vibrating part, and in typhlocybine leafhoppers (sometimes mighty) ventral longitudinal muscles (*II a vlm*) are attached to the far posterior-stretching apodemes, thus forming a unit of unknown function. As outlined above, Ossiannilsson (1946, p. 115; see Sect. 20.4) discussed that in this latter case, the dorsoventral muscles could still be responsible for vibrating the tymbal plate, with the ventral longitudinal muscles merely providing the tension of the dorsum. In the true bugs, different modifications of the antecostal ridge between the fused terga I + II can be found, suggesting different methods of vibration production, e.g., in cydnid bugs the (in lateral view) V-shaped fusion could act as a clicking tymbal when stretched (see Fig. 20.5b–*d*, and Gogala 1986), whereas a solid apodeme (Fig. 20.5b–*e*) could deform the whole tergal plate (see Leston and Pringle 1963a).

For the sister taxon of the Heteroptera, the Coleorrhyncha or Moss Bugs (comprising a single family, the Peloridiidae with 36 species [Burckhardt 2009; Burckhardt et al. 2011]), Sweet reports in 1996: "It is thus significant that the peloridiids have large apodemes at the posterior margins of terga 1 and 2 that resemble the apodemes found between terga 1 and 2 in heteropterans. Moreover, both taxa have strong apodemes on the antecostal margin of tergum 3. Thus peloridiids, too, may sing by vibrating the anterior terga with dorsal intersegmental muscles, producing sound by tymbals much as do heteropterans (Gogala 1984, 1985a, b, 1986; Cokl 1985; Todd 1989). The male peloridiids I studied have larger, more elaborate apodemes than the females, which would be constant with this hypothesis" (p. 140). Hoch et al. (2006) could confirm this hypothesis by recording vibrational signals in the males of *Hackeriella veitchi*. Following this discovery, Hartung (2007) studied the anatomy of *Hackeriella veitchi* using SEM and μCT and found no developed apodemes at the terga. The sternum II, however, is largely membraneous and deeply excavated and the posterior part of sternum II

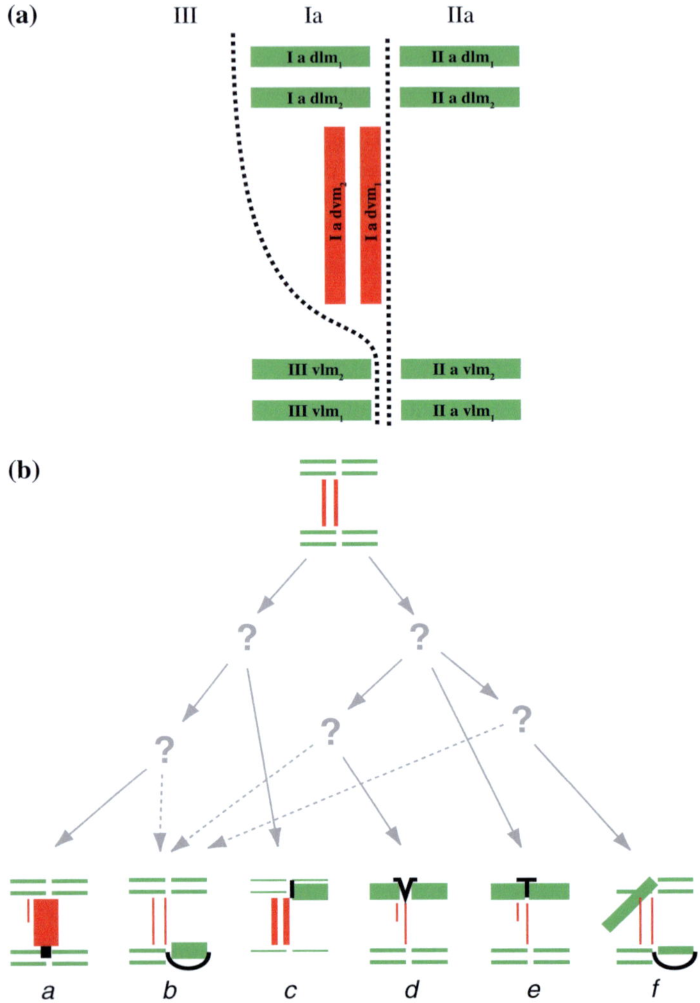


**Fig. 20.5** Configuration of the muscles belonging to the tymbalian tymbal organ. **a** Schematic "ground pattern" configuration showing the affiliation of the muscles to body segments; III, metathoracic segment; Ia and IIa, first and second abdominal segment. **b** Modification of the muscles during the diversification of the tymbal organ in the evolution of the Tymbalia; *a*, in Cicadoidea; *b*, e.g., in Typhlocybinae; *c*, e.g., in Agalliinae; *d* and *e*, e.g., in some Heteroptera; *f*, e.g., in Delphacidae

(the antecosta sternalis III) is conspicuously developed in lateral apodemes, probably as an attachment for muscle(s) *II a vlm*; there is no apparent sexual dimorphism. In two other peloridiids (*Peloridium hammoniorum* and *Xenophyes cascus*), a similar situation was found.

## 20.6 The Tymbal Organ as Autapomorphy of the Taxon Tymbalia or the Tymbal Bugs (Hemiptera excl. Sternorrhyncha)

The studies by Gogala, Kuštor, Sweet, Hoch, et alii reviewed above support the assumption—first expressed by Jordan in 1958—that the vibration-producing organs of the Heteropteroidea and the Auchenorrhyncha are homologous and were "acquired early in the phylogeny of Hemiptera" (Gogala 1988, p. 327). Sweet (1996) was the first to hypothesize explicitly the possession of a tymbal organ as an autapomorphy of a taxon comprising all hemipteran subgroups except for the Sternorrhyncha: "Singing with tymbals may be an important synapomorphy relating the Coleorrhyncha to the Auchenorrhyncha and the Heteroptera. The flattened tergum may therefore be a further synapomorphy of Heteroptera and Coleorrhyncha, reflecting sound production by tymbals using more the longitudinal muscles, whereas the arched terga of Auchenorrhyncha are related to tymbals operated more by dorsoventral muscles, although, as Ossiannilsson (1949) and Pringle (1954) emphasize, both sets of muscles are involved in modulating the sound production" (p. 140, see Figs. 20.5 and 20.6). Following this suggestion, Senter (2008) named the Hemiptera exclusive of Sternorrhyncha the "tymbaled superclade" (p. 264). It should be mentioned that some sternorrhynchan taxa apparently also use vibrational communication in the context of mate finding (Psylloidea: Ossiannilsson 1950, Aleyrodidae: Kanmiya 1996a, b). Stridulation (Tishechkin 2006) and thoracic muscle contractions (Kanmiya 2006), respectively, were proposed as mechanisms of vibration production in these taxa, but there is still need for further research.

Before suggesting a name for the "tymbaled superclade," recent advances in the knowledge about the phylogeny of the Hemiptera need to be discussed briefly (cf. Fig. 20.6). The monophyly of Hemiptera and its five subgroups Sternorrhyncha, Fulgoromorpha, Cicadomorpha, Coleorrhyncha, and Heteroptera are non-controversial (Hennig 1969; Forero 2008; Song et al. 2012; Cui et al. 2013). Furthermore, Hennig (1969) recognized the Auchenorrhyncha and Heteropteroidea as monophyletic, but he only found symplesiomorphic characters for the traditional group "Homoptera." While molecular phylogenetic studies produced conflicting results (see, e.g., von Dohlen and Moran 1995; Song et al. 2012), the morphological evidence is so convincing that the paraphyly of "Homoptera" is generally accepted today (see for review Forero 2008). The monophyly of the Auchenorrhyncha was also doubted early on (see, e.g., Ross 1965), and Hennig himself had to admit that the basis for his conclusion was rather weak (1969, p. 249 ff). Molecular studies yield inconsistent results here as well, and both Fulgoromorpha and Cicadomorpha have been proposed as sister group to the Heteropteroidea (see Forero 2008). However, the monophyly of the taxon, Fulgoromorpha + Cicadomorpha + Heteropteroidea was rarely challenged (for review see Forero 2008; Cui et al. 2013), with the exception of a recent study using mitochondrial genomes that suffers from poor taxon sampling (Song et al. 2012).

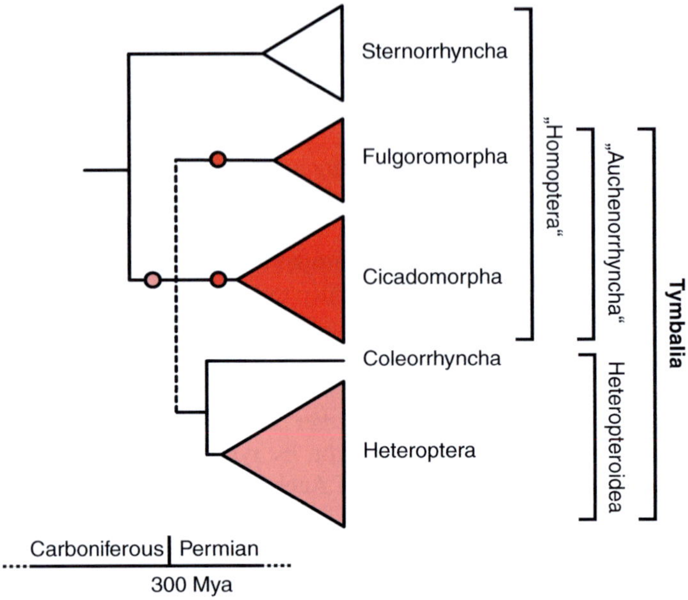


**Fig. 20.6** Phylogeny of the Hemiptera. Hypothetical relationships of the five major subgroups of the Hemiptera as discussed in the text. The *pink dot* on the stem line of the Tymbalia represents the origin of a simple tymbal organ more than 300 Mya; in the auchenorrhynchan taxa, the tymbal organ probably evolved independently into more elaborated forms (*red dots*)

Morphological support comes from some synapomorphies in characters of the wings (including a wing-coupling mechanism), the pronotum, and the inner mouthparts (Emelyanov 1987, see also Kukalová-Peck 1991). The tymbal organ as defined by its function of vibration production performed by a complex of homologous muscles and specialized exoskeletal parts is additional strong evidence for the monophyly of this taxon.

In 1990, J. Zrzavý introduced for "a probable sister group relationship between Auchenorrhyncha and Heteropteroidea" the name "Euhemiptera," which is sometimes used in recent literature (p. 19). However, he used the name only once—and in quotation marks—in a diagram without giving a diagnosis. The name was only mentioned as suitable for "non-sternorrhynchous hemipterans" in a rather informal way (pers. comm., E-mail 9.3.2011). However, the author was obviously not aware of an already available name for this taxon, created by C. F. Fallén more than 160 years ago (Fallén 1829). Fallén divided the Hemiptera into Hymenelytrata (=Sternorrhyncha) and Hemelytrata (the then known "rest" of hemipteran taxa). He gives for the Hemelytrata the diagnosis: "Alæ decussatim impositæ, hemelytris tectæ" (1829, Dispositio Familiarum Hemipterorum synoptica, n. pag.). The name Hemelytrata, however, was never widely accepted or even known, probably not just because of the new and unusual classification but also because the diagnosis is somewhat unclear and impractical. Moreover, the term "hemelytra" today is almost exclusively used for the forewings of the Heteroptera, thus the name Hemelytrata may be misleading. The same is true for the later name "Euhemiptera" as there is no diagnosis available for what may be the "true"(='Eu-') hemipterous characters in this group; additionally, the translation as "true bugs" may here also cause confusion with the Heteroptera. Therefore, we here propose the new name *Tymbalia* for the monophyletic group comprising Fulgoromorpha, Cicadomorpha, Coleorrhyncha, and Heteroptera referring to the possession of a tymbal-like vibration-producing organ as an autapomorphic character.

The oldest fossil insect interpreted as member of the Tymbalia, *Aviorrhyncha magnifica* (Aviorrhynchidae, Nel et al. 2013), is of Moscovian age, living 307–315.2 Mya ago (ICS 2013), and thus implying an origin of vibrational communication by a tymbal apparatus at the latest in the late Carboniferous (see Fig. 20.6).

## 20.7 Again: What is a Tymbalian Tymbal Organ?

If we want to describe in short the "close similarity in the basic plan" (Pringle 1957: p. 154) of the tymbalian tymbal organs, we must refer first and foremost to a homologous set of muscles (*I a dlm* + *II a dlm* + *I a dvm* + *III vlm* + *II a vlm*, see Fig. 20.5), working together in order to produce vibrations for communication purposes. In many taxa, we find that these muscles are combined with more or less specialized integumental parts (the "tymbal plates") that transform the muscle actions into narrow band signals with harmonics, pure tones, or high-pitch pulses. These transformations occur by vibrating membranes or click mechanics or a combination of both. Vibrating or "clicking" sclerites can be found at the lateral and/or dorsal parts of the first two abdominal segments. In some taxa, the signals may even be produced by the vibration/distortion of internal structures (apodemes) in combination with whole abdomen vibrations (see Gogala 2006). Abdominal vibrations (tremulation) may generally facilitate amplitude amplification in small species. In many taxa known to produce vibrational signals, there is as yet no knowledge about the precise mechanism (see Table A.1) and surprises are to be expected.

## 20.8 Some Thoughts on the Evolution of the Tymbalian Tymbal Organ

The assumed homology of the vibration-producing apparatus of the first two abdominal segments in the Tymbalia rises the question what evolutionary scenarios could explain the first steps of development of a basic tymbal organ, its function and the different pathways of its modification.

From a purely morphological point of view, the absence of powerful flight muscles in the metathorax provides a starting point for argumentation. Sweet (1996): "[…] the evolution of tymbals on abdominal segments 1 and 2 was probably promoted by hemipterans having anteromotoral wings. It seems reasonable that the concentration of flight muscles in the mesothorax […] had facilitated the attachment of the tymbal muscles to the metathorax" (p. 143). The problem here is how to explain the process of such a "reattachment" of the tymbal muscles. Pringle speculated early on about a derivation of cicadid tymbal musculature from the flight muscles by a "shift" (see also Sect. 20.3): "This similarity in physiological mechanism is one of the arguments which leads to the conclusion that the tymbal muscle is a modified flight muscle and represents the posterior portion of the vertical indirect wing-muscle complex of the metathorax which has become freed from its connexion with flight through the elaboration of a wing-coupling mechanism and has moved caudad at both ends of its attachment, separating the two functions of flight and sound production which were initially combined in a single mechanism" (Pringle 1954, p. 535). This of course could only work for an airborne sound-producing apparatus, where the flight tone gradually takes over communicative function.

Pringle recognized this problem himself after he became acquainted with Ossiannilsson's work and accepted the ubiquity of the tymbal organ in the Auchenorrhyncha. As a solution, he put forward two different scenarios in 1957, a developmental and a behavioral one: Regarding the musculature he hypothesized, "the development of a dorso-ventral muscle in the first abdominal segment as a case of the serial repetition of a segmental structure extending in this group of insects one segment further back than in the majority, the click mechanism in the tergal region being a parallel development to that of the wing articulation," and continues, "The functional advantage of a method of attracting the sexes to each other might be thought to be sufficient to preserve and encourage such an ontogenetic tendency" (p. 158). Pringle, however, was aware of the problems in understanding the (co-)evolution of a receptor in the second abdominal sternum, and the fact that no parallel case has been ever reported in other insect taxa.

The second scenario arose from contemporary ethological concepts of the evolution of display actions and copulatory movements as appetitive actions necessary to produce (successive) releasers for the final consummatory act. It is worth bringing the argument here in its whole length as it is well suited for stimulating further thought: "If a movement which is originally effective in stimulating the sexual partner only during copulation can evolve into one which produces the correct behavior at a distance, it is likely to be preserved by natural selection. A possible reason for the development of the vibration-producing organ on the first abdominal dorsum and the vibration-receiving organ on the second abdominal sternum may be that these regions came into contact during copulation. If the male mounts the female, then the primitive direction of useful transmission must have been from female to male, but since the evidence suggests that the primitive condition is a similar development in both sexes this may have been enough to start the evolution of the structures in these locations. Only by actual contact could small movements be effective, and small movements of the cuticle must have preceded rapid click movements unless the click mechanism evolved for some other purpose. Once the production and reception of vibration had become established as an intersexual stimulus during copulation, it is not difficult to see how it might have extended to provide a means of attraction between the sexes as a prelude to copulation. At this stage the greater development of the sound-production organ of the male may have started in many species: a development carried to its limit in true cicadas, the females of which are perhaps silent because the great enlargment of the air sacs is inconsistent with the production of large numbers of eggs" (Pringle 1957, p. 158).

The suggestion of a contact signal as starting point for the evolution of the tymbalian tymbal is very tempting, as such mode of communication could play an important role not only in mating, but also in other behavioral and ecological contexts. Furthermore, a function of small amplitude vibrations as contact signals may be widely distributed across tymbalian taxa, but undetected as such a behavior would be difficult to record. An interesting example in this regard, that also nicely illustrates the many blank spots in our systematic knowledge, is a case from the Corixidae.

Jansson (1972) reports—additionally to known stridulatory sounds—for water boatmen (Corixinae: *Cenocorixa blaisdelli*, *C. expleta*, and *Corisella tarsalis*) a "faint sound [that] were always recorded when a male mounted a female, so these may be called mounting signals" (p. 125). Those sounds were of unknown origin, i.e., no mechanism or corresponding movements could be detected, but a sonogram (Jansson 1972: Fig. 7) showed a possibly harmonic frequency structure (which was not discussed by the author). The pulse rate of the "mounting signal" in *C. blaisdelli* "is about 24 pulses per second at 22.8°, this is nearly twice the pulse rate of the first pulse group of the normal male call, or a female call" (Jansson 1971, p. 36). He ends his 1972 paper with the statement: "Mounting signals may be rather common in Corixidae and further investigations are needed to explain the mechanism as well as the function of this signaling" (p. 128); however, this is still an open question.

In addition, this case points to the coexistence of a multitude of different stridulation mechanisms with the tymbal organ in true bugs (see Gogala 2006) and also in some cicadas (see Boulard 2006). In 1963, Leston and Pringle tried to relate the two different mechanisms for the Heteroptera with a fourth evolutionary scenario: "In the Pentatomomorpha there are families in which the movement by the dorso-longitudinal muscles of the 1$^{st}$ abdominal tergum operates a plectrum (the lima) which works on a wing strigil; it is possible that a mechanism of this sort might evolve into a tymbal mechanism by accentuation of the movement of the 1$^{st}$ abdominal tergum and the loss of the wing strigil" (Leston and Pringle 1963a, p. 409). This argument, however, can easily be inverted, as, e.g., done by Sweet: "Stridulatory mechanisms in the Hemiptera are probably secondary to the tymbals to provide higher pitches" (1996, p. 120). Sweet further points out the different advantages of (low-frequency) substrate-borne vibrations and (high-frequency) airborne vibrations that might lead to the evolution of multimodal communication, where the tymbal organ is complemented—and sometimes even directly coupled—with stridulation or other vibration-producing mechanisms.

Generally, it does not do justice to the complexity of tymbal evolution if vibrational communication is just interpreted as a pre-stage to airborne sound production. Vibrational signals are very effective for mate location over relatively large distances compared to the body size of smaller insects (see Traue 1978; Michelsen et al. 1982; Amon and Čokl 1990) and provide a partly more secure communication channel than air vibrations. It is also important to bear in mind that the calling of the large cicadid males induces, in addition to the airborne sound, strong substrate vibrations as well, which as a distinguishable signal can travel more than one meter through rather solid structures (Gogala et al. 1996; see also the near-field courtship "buzzing" described by Alexander and Moore 1958, 1962; Alexander 1968). Moreover, vibrations produced by a tymbal organ may be complemented by other (non-stridulation) methods such as wing clicking (Gogala and Trilar 2003). In some species, female cicadas have "restored" a voice by using such clicking mechanisms to answer their male conspecifics (Trilar and Gogala 2007), and it is long known that males of certain species can be attracted by hand clapping (Lataste 1895, see also Fruhstorfer 1902 for the description of a cicada-clapper ["Cicadenklapper"] of Siamese people, made of a split cane of bamboo, to attract cicadas—to fry them in coconut oil).

It may be finally mentioned that the tymbal-produced signals act in many taxa as an efficient species-specific mate recognition system (Claridge 1985; Drosopoulos and Claridge 2006) and can therefore facilitate fast speciation. Thus, the evolution of the tymbal could account partly for the species richness of the taxon Tymbalia. Tymbal-produced vibrations are even reported to play a pivotal role in the most rapid speciation processes known so far (Wessel et al. 2013), where they serve as an exclusive means of mate recognition in an extreme habitat (Hoch and Wessel 2006).

Nevertheless, there is a host of open questions, as, e.g., the role of nymphal tymbal organs (see Evans 1957), or the coevolution of the tymbal with vibration detection and hearing organs, and we cannot fail to agree on this point with what T. E. Moore said more than 50 years ago: "We have an abundance of hypotheses, and the need for more evidence is all too obvious" (1961, p. 289). There certainly is new hope for morphological studies as a basis for the much needed comparative anatomy. Neglected for a long time, they have been given a boost by the recent development of high-resolution 3D reconstructions using micro- or rather nanoCT that are even capable of elucidating the ultrastructure of neuronal and sensorial structures (see, e.g., Hoch et al. 2014).

**Acknowledgments** We thank Peggy Hill (Tulsa) and Thomas von Rintelen (MfN Berlin) for critically reading and doing language edits of several versions of the manuscript. Hannelore Hoch and Manfred Asche (both MfN Berlin) provided valuable comments and discussions. Igor Malenovský (Brno), Björn Stelbrink, Sina Flügge, Martina Rissberger, and Hans-Ulrich Raake (all MfN Berlin) greatly helped in finding hidden literature and material. AW is grateful to T. von Rintelen for providing working space and resources.

# Appendix A

**Table A.1** Systematic overview of sound and vibrational signals produced in the Tymbalia*

| Taxon | Known signals | v = vibrational s = (sound by) stridulation | Vibration/sound-producing mechanism | t = tymbal organ s = stridulatory organ |
|---|---|---|---|---|
| *Fulgoromorpha* | | | | |
| Acanaloniidae | Wilson, M.R. (unpublished data) | v | | |
| Achilidae | | | | |
| Achilixidae | | | | |
| Caliscelidae | Tishechkin (1998) | v | Ossiannilsson (1946) (no sound recorded) | t |
| Cixiidae | Ossiannilsson (1949), Tishechkin (1997) | v | Ossiannilsson (1949) | t |
| Delphacidae | Strübing (1958), Claridge (1985) | v | Ossiannilsson (1949), Asche (1985) | t |
| Derbidae | Muir in Kirkaldy (1907) | s | | |
| | Tishechkin (2008) | v | | |
| Dictyopharidae | Strübing (1977), Tishechkin (1997) | v | | |
| Eurybrachidae | | | | |
| Flatidae | Moore (1961), Virant-Doberlet and Žežlina (2007) | v | | |
| Fulgoridae | Gogala (unpublished data) | v | | |
| Gengidae | | | | |
| Hypochthonellidae | | | | |
| Issidae | Tishechkin (1998) | v | | |
| Kinnaridae | | | | |
| Lophopidae | Soulier-Perkins et al. (2007) | v | | |
| Meenoplidae | | | | |
| Nogodinidae | | | | |
| Ricaniidae | | | | |
| Tettigometridae | | | | |

(continued)

**Table A.1** (continued)

| Taxon | Known signals | v = vibrational s = (sound by) stridulation | Vibration/sound-producing mechanism | t = tymbal organ s = stridulatory organ |
|---|---|---|---|---|
| Tropiduchidae | Tishechkin (1997) | v | | |
| *Cicadomorpha* | | | | |
| *Cercopoidea* | | | | |
| Aphrophoridae | Moore (1961), Tishechkin (2003), (2011), Ossiannilsson (1946), (1949) | v | Ossiannilsson (1949) | t |
| Cercopidae | Moore (1961), Tischechkin (2003) | v | | |
| Clastopteridae | Moore (1961) | v | | |
| Epipygidae | | | | |
| Machaerotidae | Tishechkin (2003) | v | | |
| *Cicadoidea* | See Sueur (2001) for comprehensive species list | | | |
| Cicadidae | | | | |
| Cicadinae | Since antiquity | s | Casserius (1600/1601), Réaumur (1740), Mayer (1877) | t |
| Platypleurinae | Pringle (1954) | s | Pringle (1954) | t |
| Tibicinidae | Breml (1849) | s | | t |
| | Gogala et al. (1996), Stölting et al. (2002) (substrate born signals) | v | | t |
| | Varley (1939), Jacobi (1907) | s | Jacobi (1907) | s |
| Tettigarctidae | Claridge et al. (1999) | v | Evans (1941), Pringle (1957) | t |
| *Membracoidea* | | | | |
| Cicadellidae | | | | |
| Agalliinae | Ossiannilsson (1946), (1949), Shaw (1976) | v | Ossiannilsson (1949) | t |
| Aphrodinae | Ossiannilsson (1946), (1949), Tishechkin (2000) | v | Ossiannilsson (1949) | t |
| Cicadellinae | Ossiannilsson (1946), (1949), Percy et al. (2008) | v | Ossiannilsson (1949) | t |

(continued)

**Table A.1** (continued)

| Taxon | Known signals | v = vibrational s = (sound by) stridulation | Vibration/sound-producing mechanism | t = tymbal organ s = stridulatory organ |
|---|---|---|---|---|
| Deltocephalinae | Ossiannilsson (1946), (1949), Strübing (1963), Heady et al. (1986), Heady and Nault (1991), Tishechkin (2001) | v | Ossiannilsson (1949), Strübing and Schwarz-Mittelstaedt (1986) | t |
| Dorycephalinae | Ossiannilsson (1946), (1949), Tishechkin (2000), (2005) | v | Ossiannilsson (1949) | t |
| Hecalinae | Tishechkin (2000), (2005) | v | | |
| Iassinae | Ossiannilsson (1949), Tishechkin (2000) | v | Ossiannilsson (1949) | t |
| Idiocerinae | Ossiannilsson (1946), (1949), Tishechkin (2000) | v | Ossiannilsson (1949) | t |
| Ledrinae | Ossiannilsson (1949) | v | Ossiannilsson (1949) | t |
| Macropsinae | Ossiannilsson (1946), (1949), Claridge (1985), Claridge and Nixon (1986), Tishechkin (2002) | v | Ossiannilsson (1949) | t |
| Megophthalminae | Ossiannilsson (1949), Tishechkin (2003) | v | Ossiannilsson (1949) | t |
| Penthimiinae | Tishechkin (2000) | v | | |
| Stegelytrinae | | | | |
| Typhlocybinae | Ossiannilsson (1946), (1949), Shaw et al. (1974), Saxena and Kumar (1984), Tishechkin (2001) | v | Giard (1889), Ossiannilsson (1949) | t |
| Xestocephalinae | | | | |
| Membracidae | | | | |
| Centronodinae | | | | |
| Centrotinae | Cocroft and McNett (2006), Tishechkin (2003) | v | Ossiannilsson (1949) (no sound recorded) | t |
| Darninae | Cocroft and De Luca (2006) | v | | |
| Endoiastinae | | | | |
| Heteronotinae | Cocroft and De Luca (2006) | v | | |
| Membracinae | Cocroft and De Luca (2006) | v | | |

(continued)

**Table A.1** (continued)

| Taxon | Known signals | v = vibrational s = (sound by) stridulation | Vibration/sound-producing mechanism | t = tymbal organ s = stridulatory organ |
|---|---|---|---|---|
| Nicomiinae | Cocroft and De Luca (2006) | v | | |
| Oxyrhanchinae | Cocroft and De Luca (2006) | v | | |
| Smiliinae | Moore (1961), Hunt (1993), Strübing (1999), Tishechkin (2003), Cocroft and De Luca (2006) | v | | |
| Stegaspidinae | Cocroft and De Luca (2006) | v | | |
| Aetalionidae | | | | |
| Aetalioninae | Cocroft and De Luca (2006) | v | Evans (1946) | t |
| Biturritinae | Cocroft and De Luca (2006) | v | | |
| Melizoderidae | | | | |
| Ulopidae | Ossiannilsson (1946), (1949), Tishechkin (2003) | v | Ossiannilsson (1949) | t |
| Myerslopiidae | | | | |
| *Coleorrhyncha* | | | | |
| Peloridiidae | Hoch et al. (2006) | v | Hartung (this Chap. and unpublished data) | t |
| *Heteroptera* | | | | |
| *Cimicomorpha* | | | | |
| Anthocoridae | No reports according to Gogala (2006) | | | |
| Cimicidae | No reports according to Gogala (2006) | | | |
| Joppeicidae | | | | |
| Lasiochilidae | | | | |
| Lyctocoridae | | | | |
| Medocostidae | | | | |
| Microphysidae | | | | |
| Miridae | Gogala (1984) | s | Schuh and Slater (1974), Akingbohungbe (1979) | s |

(continued)

**Table A.1** (continued)

| Taxon | Known signals | v = vibrational s = (sound by) stridulation | Vibration/sound-producing mechanism | t = tymbal organ s = stridulatory organ |
|---|---|---|---|---|
| Miridae | Strong et al. (1970), Groot et al. (1998) | v | | |
| Nabidae | No reports after Gogala (2006) | | Leston (1957) (description but no sound recorded) | s |
| Pachynomidae | | | | |
| Plokiophilidae | | | | |
| Polyctenidae | | | | |
| Thaumastocoridae | | | | |
| Tingidae | Gogala (1984), (1985b) | v | | |
| Reduviidae | Ray (1710) | s | Ray (1710) | s |
| | Gogala and Čokl (1983) | v | Gogala (1984), Lazzari et al. (2006) | t |
| Velocipedidae | | | | |
| *Dipsocoromorpha* | | | | |
| Ceratocombidae | | | | |
| Dipsocoridae | | | | |
| Hypsipterygidae | | | | |
| Schizopteridae | | | Leston (1957) (no sound recorded); China (1946) | s |
| Stemmocryptidae | | | | |
| *Enicocephalomorpha* | | | | |
| Enicocephalidae | | | | |
| Aenictopecheidae | | | | |
| *Gerromorpha* | | | | |
| Gerridae | Koga and Hayashi (1991), Wilcox (1995) | v | Wilcox (1995) | t |
| Hebridae | | | | |

(continued)

**Table A.1** (continued)

| Taxon | Known signals | v = vibrational s = (sound by) stridulation | Vibration/sound-producing mechanism | t = tymbal organ s = stridulatory organ |
|---|---|---|---|---|
| Hermatobatidae | | | | |
| Hydrometridae | | | | |
| Macroveliidae | | | | |
| Mesoveliidae | | | | |
| Paraphrynoveliidae | | | | |
| Veliidae | Hungerford (1929) | s | Hungerford (1929) | s |
| | Wilcox (1995) | v | Wilcox (1995) | t |
| *Leptopodomorpha* | | | | |
| Aepophilidae | | | | |
| Leptopodidae | | | | |
| Omaniidae | | | | |
| Saldidae | Leston (1957) | s | Leston (1957), Drake and Hottes (1951) | s |
| *Nepomorpha* | | | | |
| Aphelocheiridae | | | | |
| Belostomatidae | Wilcox (1972), (1995) | v | Wilcox (1995) | t |
| Corixidae | Ball (1845), Mitis (1936), Moore (1961), Finke (1968), Aiken (1982), Jansson (1972) | s | Mitis (1936), Moore (1961) | s |
| | Jansson (1971), (1972) | (v) | Jansson (1971), (1972) | (t) |
| Gelastocoridae | | | | |
| Helotrephidae | | | | |
| Naucoridae | | | | |
| Nepidae | Torre-Bueno (1903) | s | Torre-Bueno (1903) | s |
| Notonectidae | Hungerford (1924) | s | Wilcox (1975) | s |
| Ochteridae | | | | |

(continued)

**Table A.1** (continued)

| Taxon | Known signals | v = vibrational s = (sound by) stridulation | Vibration/sound-producing mechanism | t = tymbal organ s = stridulatory organ |
|---|---|---|---|---|
| Pleidae | | | | |
| Potamocoridae | | | | |
| *Pentatomorpha* | | | | |
| Acanthosomatidae | Jordan (1958a, b), Gogala (unpublished data) | v | Jordan (1958a, b) | t |
| Alydidae | Schaefer and Pupedis (1981) | s | | |
| | Numata et al. (1989), Gogala (1990) | s + v | Numata et al. (1989), Gogala (1990) | s + t |
| Aradidae | No records according to Gogala (2006) | | Usinger (1954), Usinger and Matsuda (1959) | s |
| Berytidae | | | | |
| Colobathristidae | Stys (1966) | s | Schuh and Slater (1974) | s |
| Canopidae | | | | |
| Coreidae | Stys (1961) | s | | |
| | Gogala (1984), (1990) | v | | |
| Corimelaenidae | | | | |
| Cydnidae | Dupuis (1953), Haskell (1957), Moore (1961), Gogala (1969) | s | | |
| | Gogala (1984) | s + v | Gogala (2006) | s + t |
| Dinidoridae | | | | |
| Hyocephalidae | | | | |
| Idiostolidae | | | | |
| Largidae | Lattin (1958) | s | | |
| Lestoniidae | | | | |
| Lygaeidae | Leston (1957) | s | Leston (1957), Ashlock and Lattin (1963) | s |

(continued)

**Table A.1** (continued)

| Taxon | Known signals | v = vibrational s = (sound by) stridulation | Vibration/sound-producing mechanism | t = tymbal organ s = stridulatory organ |
|---|---|---|---|---|
| Lygaeidae | Thorpe and Harrington (1981), Gogala (1984) | v | Gogala (1984) | t |
| Malcidae | | | | |
| Megarididae | | | | |
| Parastrachiidae | No reports (probably similar to Cydnidae: Gogala 2006) | | | |
| Pentatomidae | Leston (1954) | s | Leston (1954) | s |
| | Čokl (1985), Čokl et al. (1972), Čokl and Virant-Doberlet (2003) | v | Čokl and Virant-Doberlet (2003) | t |
| Phloeidae | | | | |
| Piesmatidae | Haskell (1961) | s | | s |
| | Leston (1957) | s + v | Leston (1957) | s + t |
| Plataspididae | Gogala (1990) | v | | |
| Pyrrhocoridae | | | Lattin (1958) | s |
| Rhopalidae | Gogala (1990) | v | Gogala (1990) | t |
| | Zych et al. (2012) | s | Zych et al. (2012) | s |
| Saileriolidae | | | | |
| Scutelleridae | Leston (1954) | s | Au (1969), Schaefer (1980) | s |
| | Gogala (1984) | s + v | | |
| Stenocephalidae | | | | |
| Termitaphididae | | | | |
| Tessaratomidae | Leston (1957), Puranik et al. (1981), Gogala (1984) | s | Puranik et al. (1981) | s |
| Thaumastellidae | Štys (1964), Gogala (1984) | s | | |
| Urostylididae | | | Grazia et al. (2008) | s |

*Only the first publications on song recordings and additional important reviews are listed. (An empty field means that, we found no record in the literature, negative records are cited)

**Table A.2** Species names used in the text

| Species names as used by the cited authors and in the text | Valid taxon names |
|---|---|
| *Aelia acuminata* | *Aelia acuminata* (Linnaeus, 1758) |
| *Agallia brachyptera* | *Agallia brachyptera* (Boheman, 1847) |
| *Agallia venosa* | *Anaceratagallia venosa* (Fourcroy, 1785) |
| *Aphis fabae* | *Aphis fabae* Scopoli, 1763 |
| *Aphrodes bicinctus* | *Aphrodes bicinctus* (Schrank, 1776) |
| *Aphrodes bifasciatus* | *Planaphrodes bifasciatus* (Linnaeus, 1758) |
| *Aphrodes flavostriatus* | *Anoscopus flavostriatus* (Donovan, 1799) |
| *Aphrodes trifasciatus* | *Planaphrodes trifasciatus* (Fourcroy, 1785) |
| *Arma custos* | *Arma custos* (Fabricius, 1794) |
| *Athysanus argentatus* | *Athysanus argentarius* Metcalf, 1955 |
| *Aviorrhyncha magnifica* | *Aviorrhyncha magnifica* Nel et al., 2013 |
| *Canthophorus dubius* | *Canthophorus dubius* (Scopoli, 1763) |
| *Canthophorus melanopterus* | *Canthophorus melanopterus* (Herrich-Schäffer, 1835) |
| *Carpocoris pudicus* | *Carpocoris pudicus* (Poda, 1761) |
| *Cenocorixa* | *Cenocorixa* Hungerford, 1948 |
| *Cenocorixa blaisdelli* | *Cenocorixa blaisdelli* (Hungerford, 1930) |
| *Cenocorixa expleta* | *Cenocorixa expleta* (Uhler, 1895) |
| *Centrotus cornutus* | *Centrotus cornutus* (Linnaeus, 1758) |
| *Cicadella atropunctata* | *Eupteryx atropunctata* (Goeze, 1778) |
| *Circulifer tenellus* | *Circulifer tenellus* (Baker, 1896) |
| *Corisella tarsalis* | *Corisella tarsalis* (Fieber, 1851) |
| *Cystosoma saundersii* | *Cystosoma saundersii* Westwood, 1842 |
| *Dicranotropis hamata* | *Dicranotropis hamata* (Boheman, 1847) |
| *Diplocolenus abdominalis* | *Verdanus abdominalis* (Fabricius, 1803) |
| *Dolycoris baccarum* | *Dolycoris baccarum* (Linnaeus, 1758) |
| *Doratura homophyla* | *Doratura homophyla* (Flor, 1861) |
| *Doratura stylata* | *Doratura stylata* (Boheman, 1847) |
| *Elasmucha grisea* | *Elasmucha grisea* (Linnaeus, 1758) |
| *Empoasca devastans* | *Amrasca biguttula* (Ishida, 1912) |
| *Empoasca rufescens* | *Kybos rufescens* Melichar, 1896 |
| *Empoasca smaragdula* | *Kybos smaragdula* (Fallén, 1806) |
| *Empoasca sordidula* | *Kybos sordidulus* (Ossiannilsson, 1941) |
| *Empoasca strigilifera* | *Kybos strigilifer* (Ossiannilsson, 1941) |
| *Empoasca virgator* | *Kybos virgator* (Ribaut, 1933) |
| *Euacanthus interruptus* | *Evacanthus interruptus* (Linnaeus, 1758) |
| *Eupelix depressa f. cuspidata* | *Eupelix cuspidata* (Fabricius,1775) |
| *Eupteryx melissae* | *Eupteryx melissae* Curtis, 1837 |
| *Eurydema oleraceum* | *Eurydema oleraceum* (Linnaeus, 1758) |
| *Euscelis incisus* | *Euscelis incisus* (Kirschbaum, 1858) |
| *Euscelis ononidis* | *Euscelis ononidis* Remane, 1967 |
| *Euscelis plebeius* | *Euscelis incisus* (Kirschbaum, 1858) |
| *Eysarcoris (= Stollia)* | *Eysarcoris* Hahn, 1834 |
| *Graphocraerus ventralis* | *Graphocraerus ventralis* (Fallén, 1806) |
| *Graptosaltria nigrofuscata* | *Graptosaltria nigrofuscata* (Motschulsky, 1866) |

(continued)

**Table A.2** (continued)

| Species names as used by the cited authors and in the text | Valid taxon names |
|---|---|
| *Hackeriella veitchi* | *Hackeriella veitchi* (Hacker, 1932) |
| *Idiocerus albicans* | *Populicerus albicans* (Kirschbaum, 1868) |
| *Idiocerus elegans* | *Metidiocerus elegans* (Flor, 1861) |
| *Idiocerus lituratus* | *Idiocerus lituratus* (Fallén, 1806) |
| *Idiocerus stigmaticalis* | *Idiocerus stigmaticalis* Lewis, 1834 |
| *Ledra aurita* | *Ledra aurita* (Linnaeus, 1758) |
| *Limotettix striatulus* | *Ophiola decumana* (Kontkanen, 1949) |
| *Macropsis fuscinervis* | *Macropsis fuscinervis* (Boheman, 1845) |
| *Macropsis planicollis* | *Macropsis cerea* (Germar, 1837) |
| *Macropsis tiliae* | *Pediopsis tiliae* (Germar, 1831) |
| *Macrosteles cristatus* | *Macrosteles cristatus* (Ribaut, 1927) |
| *Macustus grisescens* | *Macustus grisescens* (Zetterstedt, 1828) |
| *Neophilaenus campestris* | *Neophilaenus campestris* (Fallén, 1805) |
| *Nezara viridula* | *Nezara viridula* (Linnaeus, 1758) |
| *Nilaparvata lugens* | *Nilaparvata lugens* (Stål, 1845) |
| *Oebalus pugnax* | *Oebalus pugnax* (Fabricius, 1775) |
| *Oncopsis alni* | *Oncopsis alni* (Schrank, 1801) |
| *Oncopsis flavicollis* | *Oncopsis flavicollis* (Linnaeus, 1761) |
| *Oncopsis tristis* | *Oncopsis tristis* (Zetterstedt, 1840) |
| *Opsius stactogalus* | *Opsius stactogalus* Fieber, 1866 |
| *Palomena prasina* | *Palomena prasina* (Linnaeus, 1761) |
| *Palomena viridissima* | *Palomena viridissima* (Poda, 1761) |
| *Paropia scanica* | *Megophthalmus scanicus* (Fallén, 1806) |
| *Peloridium hammoniorum* | *Peloridium hammoniorum* Breddin, 1897 |
| *Pentatoma rufipes* | *Pentatoma rufipes* (Linnaeus, 1758) |
| *Perkinsiella saccharicida* | *Perkinsiella saccharicida* Kirkaldy, 1903 |
| *Picromerus bidens* | *Picromerus bidens* (Linnaeus, 1758) |
| *Platypleura capensis* | *Platypleura capensis* (Linnaeus, 1764) |
| *Platypleura capitata* | *Platypleura capitata* (Oliver, 1790) |
| *Platypleura kaempferi* | *Platypleura kaempferi* (Fabricius, 1794) |
| *Platypleura octoguttata* | *Platypleura octoguttata* (Fabricius, 1798) |
| *Psammotettix cephalotes* | *Psammotettix cephalotes* (Herrich-Schäffer, 1834) |
| *Reduvius personatus* | *Reduvius personatus* (Linnaeus, 1758) |
| *Sciocoris cursitans* | *Sciocoris cursitans* (Fabricius, 1794) |
| *Sehirus biguttatus* | *Adomerus biguttatus* (Linnaeus, 1758) |
| *Solenopyx sulphurellus* | *Elymana sulphurella* (Zetterstedt, 1828) |
| *Speudotettix subfusculus* | *Speudotettix subfusculus* (Fallén, 1806) |
| *Stollia fabricii* | *Eysarcoris venustissimus* (Schrank, 1776) |
| *Streptanus aemulans* | *Streptanus aemulans* (Kirschbaum, 1868) |
| *Streptanus marginatus* | *Streptanus marginatus* (Kirschbaum, 1858) |
| *Tettigarcta crinita* | *Tettigarcta crinita* Distant, 1883 |
| *Tettigarcta tomentosa* | *Tettigarcta tomentosa* White, 1845 |
| *Tettigella viridis* | *Cicadella viridis* (Linnaeus, 1758) |
| *Tettigia sp.* | *unidentified* Cicadoidea |

(continued)

**Table A.2** (continued)

| Species names as used by the cited authors and in the text | Valid taxon names |
|---|---|
| *Thamnotettix confinis* | *Thamnotettix confinis* Zetterstedt, 1828 |
| *Tritomegas bicolor* | *Tritomegas bicolor* (Linnaeus, 1758) |
| *Typhlocyba* | *Unidentified* Typhlocybinae |
| *Typhlocyba ulmi* | *Ribautiana ulmi* (Linnaeus, 1758) |
| *Xenophyes cascus* | *Xenophyes cascus* Bergroth, 1924 |

## References


Aidley DJ (1969) Sound production in a Brazilian cicada. J Exp Biol 51:325–337

Aiken RG (1982) Sound production and mating in a waterboatman, *Palmacorixa nana* (Heteroptera: Corixidae). Anim Behav 30:54–61

Akingbohungbe AE (1979) A new genus and four new species of Hyaliodinae (Heteroptera: Miridae) from Africa with comments on the status of the subfamily. Rev Zool Afr 93:500–522

Alexander RD (1968) Arthropods. In: Sebeok TA (ed) Animal communication: techniques of study and results of research. Indiana UP, Bloomington, London, pp 167–216

Alexander RD, Moore TE (1958) Studies on the acoustical behavior of seventeen-year cicadas (Homoptera: Cicadidae: *Magicicada*). Ohio J Sci 58:107–127

Alexander RD, Moore TE (1962) The evolutionary relationships of 17-year and 13-year cicadas, and three new species (Homoptera, Cicadidae, *Magicicada*). Misc Publ Mus Zool Univ Michigan 121:1–59

Amon T (1990) Electrical brain stimulation elicits singing in the bug *Nezara viridula*. Naturwissenschaften 77:291–292

Amon T, Čokl A (1990) Transmission of the vibratory song of the bug *Nezara viridula* (Pentatomidae, Heteroptera) on the *Hedera helix* plant. Scopolia Suppl 1:133–141

Asche M (1985) Zur Phylogenie der Delphacidae Leach, 1815 (Homoptera, Cicadina, Fulgoromorpha). Marburg Entomol Publ II(1):1–910

Ashlock PD, Lattin JD (1963) Stridulatory mechanisms in the Lygaeidae, with a new American genus of Orsillinae (Hemiptera: Heteroptera). Ann Entomol Soc Am 56:693–703

Au EC (1969) The taxonomic value of the metathoracic wing in the Scutelleridae (Hemiptera: Heteroptera). MA Thesis, Oregon State University

Ball R (1845) On the noises produced by one of the Notonectidae. Rep Brit Assoc 15:64–65

Bennet-Clark HC, Young D (1992) A model of the mechanism of sound production in cicadas. J Exp Biol 173:123–153

Bennet-Clark HC, Young D (1998) Sound radiation by the bladder cicada *Cystosoma saundersii*. J Exp Biol 201:701–715

Boulard M (2006) Acoustic signals, diversity and behavior of cicadas (Cicadidae, Hemiptera). In: Drosopoulos S, Claridge MF (eds) Insect sounds and communication: physiology, behaviour, ecology and evolution. Taylor & Francis, New York, pp 331–349

Breml H (1849) Übersicht der schweizerischen Rhynchoten. Mitt nat Ges Zürich 34:325–339

Burckhardt D (2009) Taxonomy and phylogeny of the Gondwanan moss bugs or Peloridiidae (Hemiptera, Coleorrhyncha). Deut Entomol Z 56:173–235

Burckhardt D, Bochud E, Damgaard J, Gibbs GW, Hartung V, Larivière M-C, Wyniger D, Zürcher I (2011) A review of the moss bug genus *Xenophyes* (Hemiptera: Coleorrhyncha: Peloridiidae) from New Zealand: systematics and biogeography. Zootaxa 2923:1–26

Carlet MG (1877) Mémoire sur l'appareil musical de la cigale. Ann Sci Nat 6th Ser Zool 5:1–36

Casserius I (1600/1601) De vocis auditusque organis historia anatomica singulari fide methodo ac industria concinnata tractatibus duobus explicata ac variis iconibus aere excusis illustrata. 2 vols. Vittorio Baldini, Ferrariae (Ferrara)

Chapman GB (2005) A light and transmission electron microscope study of some cells and tissues associated with the tymbal muscle of a periodical cicada (Homoptera, Cicadidae). Invertebr Biol 124:321–331

China WE (1946) New Cryptostemmatidae from Trinidad. Proc Ent Soc London B 15:148

Claridge MF (1985) Acoustic signals in the Homoptera: behavior, taxonomy, and evolution. Annu Rev Entomol 30:297–317

Claridge MF, Nixon GA (1986) *Oncopsis flavicollis* (L.) associated with tree birches (*Betula*): a complex of biological species or a host plant utilization polymorphism? Biol J Linn Soc 27:381–397

Claridge MF, Morgan JC, Moulds MS (1999) Substrate-transmitted acoustic signals of the primitive cicada, *Tettigarcta crinita* Distant (Hemiptera Cicadoidea, Tettigarctidae). J Nat Hist 33:1831–1834

Cocroft RB, DeLuca PA (2006) The influence of size on variation in the plant-borne vibrational signals of insects. In: Drosopoulos S, Claridge MF (eds) Insect sounds and communication: physiology, behaviour, ecology and evolution. Taylor & Francis, New York, pp 99–110

Cocroft RB, McNett GD (2006) Vibratory communication in treehoppers (Hemiptera: Membracidae). In: Drosopoulos S, Claridge MF (eds) Insect sounds and communication: physiology, behaviour, ecology and evolution. Taylor & Francis, New York, pp 305–317

Čokl A (1985) Problems of sound communication in a land bug species *Nezara viridula* L. (Heteroptera, Pentatomidae). In: Kalmring K, Elsner (eds). Acoustic and vibrational communication in insects. Paul Parey, Berlin, pp 163–168

Čokl A, Virant-Doberlet M (2003) Communication with substrate-borne signals in small plant-dwelling insects. Annu Rev Entomol 48:29–50

Čokl A, Gogala M, Jez M (1972) The analysis of the acoustic signals of the bug *Nezara viridula*. Biol Vestnik (Ljubljana) 20:47–53

Conner WE (1999) 'Un chant d'appel amoureux': Acoustic communication in moths. J Exp Biol 202:1711–1723

Conner WE, Corcoran AJ (2012) Sound strategies: the 65-million-year-old battle between bats and insects. Annu Rev Entomol 57:21–39

Corcoran AJ, Barber JR, Conner WE (2009) Tiger moth jams bat sonar. Science 325:325–327

Cui Y, Xie Q, Hua J, Dang K, Zhou J, Liu X, Wang G, Yu X, Bu W (2013) Phylogenomics of Hemiptera (Insecta: Paraneoptera) based on mitochondrial genomes. Syst Entomol 38:233–245

Darwin C (1871) The descent of man, and selection in relation to sex, 1st edn. John Murray, London

Dolling WR (1995) The identity of 'Cicada longicornis Poda' (Hem., Nabidae). Entomologist's Monthly Magazine 131:113

Drake CJ, Hottes FC (1951) Stridulatory organs in Saldidae. Great Basin Nat 11:43–46

Drosopoulos S, Claridge MF (2006) Insect sounds and communication: physiology, behaviour, ecology, and evolution. CRC (Taylor & Francis), Boca Raton

Dupuis C (1953) Notes, remarques et observations diverses sur les Hémiptères Note VI—Appareil stridulatoire et stridulation des Cydnidae et Tessaratomidae (Heteroptera, Pentatomidae). Cahiers des Naturalistes, Bulletin des N.P. n.s. 8:25–27

Emelyanov AF (1987) Филогения цикадовых (Homoptera, Cicadina) по сравнительно-морфологическим данным [The phylogeny of Cicadina based on comparative morphological data.] Тр ВЭО [Trudy Vsesoyuznogo Entomologicheskogo Obshchestva] 69:19–109

Evans JW (1941) The morphology of *Tettigarcta tomentosa* White (Homoptera, Cicadidae). Pap P Roy Soc Tasmania 1940:36–49

Evans JW (1946) A natural classification of leaf-hoppers (Homoptera, Jassoidea). Part 2: Aetalionidae, Hylicidae, Eurymelidae. T Roy Ent Soc Lond 97:39–54

Evans JW (1957) Some aspects of the morphology and inter-relationships of extinct and recent Homoptera. T Roy Ent Soc London 109:275–294
Fallén CF (1829) Hemiptera Sveciæ. Sectio prior, Hemelytrata. Cimicides eorumque familiæ affines. Berling, Lund
Felici G (1724) Lettera del Sig. Dott. Giovambatista Felici, Fiorentino, intorno al canto delle cicale, scritta l'anno 1717, al Sig. Dott. Pascasio Giannetti, Lettore ordinario di medicina, nell' universita di Pisa. Giornale De' Letterati d'Italia (Venezia) XXXVI:56–77
Fenton MB, Roeder KD (1974) The microtymbals of some Arctiidae. J Lepid Soc 28:205–211
Finke C (1968) Lautäußerungen und Verhalten von *Sigara striata* und *Callicorixa praeusta* (Corixidae Leach., Hydrocorisae Latr.). J Comp Physiol 58:398–422
Fletcher NH, Hill KG (1978) Acoustics of sound production and of hearing in the Bladder cicada *Cystosoma saundersii* (Westwood). J Exp Biol 72:43–55
Fonseca PJ, Serrão EA, Pina-Martins F, Silva P, Mira S, Quartau JA, Paulo OS, Cancela L (2008) The evolution of cicada songs contrasted with the relationships inferred from mitochondrial DNA (Insecta, Hemiptera). Bioacoustics 18:17–34
Forbes WTM, Franclemont JG (1957) The striated band (Lepidoptera, chiefly Arctiidae). Lep News 11:147–150
Forero D (2008) The systematics of the Hemiptera. Rev Colomb Entomol 34:1–21
Fullard JH, Fenton MB, Simmons JA (1979) Jamming bat echolocation: the clicks of arctiid moths. Can J Zool 57:647–649
Franz PU (1978) Zum Lauterzeugungsmechanismus der Zikaden (Homoptera-Cicadina)—ein funktionsmorphologischer Vergleich. Verh Dtsch Zool Ges 1978:158
Fruhstorfer H (1902) [Siamese cicada-clapper.] Berl Entomol Z, Sber 47:(28)
George CJ (1933) A suspected sound producing organ in *Empoasca devastans*. J Univ Bombay 1:54–57
Giard MA (1889) Sur la castration parasitaire des *Typhlocyba* par une larve d'Hyménoptère (*Aphelopus melaleucus* Dalm.) et par une larve de Diptère (*Atelenevra spuria* Meig.). C R Acad Sci (Paris) 109:708–710
Gogala M (1969) Die akustische Kommunikation bei der Wanze *Tritomegas bicolor* (L.) (Heteroptera, Cydnidae). Z Vergl Physiol 63:379–391
Gogala M (1970) Artspezifität der Lautäußerungen bei Erdwanzen (Heteroptera, Cydnidae). Z Vergl Physiol 70:20–28
Gogala M (1984) Vibration producing structures and songs of terrestrial Heteroptera as systematic character. Biol Vestnik (Ljubljana) 32:19–36
Gogala M (1985a) Vibrational communication in insects (Biophysical and behavioural aspects). In: Kalmring K, Elsner (eds) Acoustic and vibrational communication in insects. Paul Parey, Berlin, Hamburg, pp 117–126
Gogala M (1985b) Vibrational songs of land bugs and their production. In: Kalmring K, Elsner (eds) Acoustic and vibrational communication in insects. Paul Parey, Berlin, Hamburg, pp 143–150
Gogala M (1986) Structures producing vibrational signals in Heteroptera. Proceedings 2nd international congress Rhynchota Balkan, Microlimni, Greece, pp 47–48
Gogala M (1988) Vibrational structures and signals in land bugs and Auchenorrhyncha: differences and similarities. In: Vidano C, Arzone A (eds) 6th Auchenorrhyncha meeting Turin, Italy. Cosiglio Nazionale delle Ricerche, pp 327
Gogala M (1990) Distribution of low frequency vibrational songs in local Heteroptera. Scopolia Suppl 1:125–132
Gogala M (2006) Vibratory signals produced by Heteroptera—Pentatomorpha and Cimicomorpha. In: Drosopoulos S, Claridge MF (eds) Insect sounds and communication: physiology, behaviour, ecology and evolution. Taylor & Francis, New York, pp 275–295
Gogala M, Čokl A (1983) The acoustic behavior of the bug *Phymata crassipes* (F.) (Heteroptera). Rev Can Biol Exptl 42:249–256
Gogala M, Trilar T (2003) Video analysis of wing clicking in cicadas of the genera *Cicadatra* and *Pagiphora* (Homoptera: Auchenorrhyncha: Cicadoidea). Acta Entomol Sloven 11:5–15

Gogala M, Trilar T, Kozina U, Duffels H (2004) Frequency modulated song of the cicada *Maua albigutta* (Walker 1856) (Auchenorrhyncha: Cicadoidea) from South East Asia. Scopolia 54:1–16

Gogala M, Čokl A, Drašlar K, Blaževič A (1974) Substrate-borne sound communication in Cydnidae. J Comp Physiol 94:25–31

Gogala M, Popov AV, Ribarič D (1996) Bioacoustics of singing cicadas of the Western Palaearctic: *Cicadetta tibialis* (Panzer) (Cicadoidea: Tibicinidae). Acta Entomol Sloven 4:45–62

Grazia J, Schuh RT, Wheeler WC (2008) Phylogenetic relationships of family groups in Pentatomoidea based on morphology and DNA sequences (Insecta: Heteroptera). Cladistics 24:1–45

Groot AT, van der Wal E, Schuurman A, Visser JH, Blommers LHM, Van Beek TA (1998) Copulation behaviour of *Lygocoris pabulinus* under laboratory conditions. Entomol Exp Appl 88:219–228

Hagiwara S (1953) Neuro-muscular transmission in insects. Jpn J Physiol 3:284–296

Hagiwara S, Ogura K (1960) Analysis of songs of Japanese cicadas. J Insect Physiol 5:259–263

Hagiwara S, Watanabe A (1954) Action potential of insect muscle examined with intracellular electrode. Jpn J Physiol 4:65–78

Haldeman SS (1848) A new organ of sound in Lepidoptera. Am J Sci Arts V:435

Handlirsch A (1900a) Zur Kenntnis der Stridulationsorgane bei den Rhynchoten. Ann K-K Naturhist Hofmus Wien 15:127–141

Handlirsch A (1900b) Neue Beiträge zur Kenntnis der Stridulationsorgane bei den Rhynchoten. Verh K-K Zool-Bot Ges Wien 50:555–560

Hartung V (2007) Investigations of internal anatomy of *Hackeriella veitchi* (Hemiptera: Coleorrhyncha: Peloridiidae) pertaining to some newly discovered behavioural traits. Diploma thesis, Humboldt-Universität zu Berlin

Haße A (1974) Lauterzeugungsmechanismus und Lautrezeption von Kleinzikaden. Naturwissenschaften 61:81–82

Haskell PT (1957) Stridulation and its analysis in certain Geocorisae (Hemiptera Heteroptera). P Zool Soc Lond 129:351–358

Haskell PT (1961) Insect sounds. HF and G Witherby Ltd, London, p 189

Heady SE, Nault LR (1991) Acoustic signals of *Graminella nigrifrons* (Homoptera: Cicadellidae). Great Lakes Entomol 24:9–16

Heady SE, Nault LR, Shambaugh GF, Fairchild L (1986) Acoustic and mating behavior of *Dalbulus* leafhoppers (Homoptera: Cicadellidae). Ann Entomol Soc Am 79:727–736

Hennig RM, Weber T, Huber E, Kleindienst H-U (1993) A new function for an old structure: the “timbal muscle” in cicada females. Naturwissenschaften 80:324–326

Hennig RM, Weber T, Moore TE, Huber F, Kleindienst H-U, Popov AV (1994) Function of the tensor muscle in the cicada *Tibicen linnei*. J Exp Biol 187:33–44

Hennig W (1969) Die Stammesgeschichte der Insekten. W. Kramer, Frankfurt/M

Hoch H, Wessel A (2006) Communication by substrate-borne vibrations in cave planthoppers. In: Drosopoulos S, Claridge MF (eds) Insect sounds and communication: physiology, behaviour, ecology and evolution. Taylor & Francis, New York, pp 187–197

Hoch H, Deckert J, Wessel A (2006) Vibrational signalling in a Gondwanan relict insect (Hemiptera: Coleorrhyncha: Peloridiidae). Biol Lett 2:222–224

Hoch H, Wessel A, Asche M, Baum D, Beckmann F, Bräunig P, Ehrig K, Mühlethaler R, Riesemeier H, Staude A, Stelbrink B, Wachmann E, Weintraub P, Wipfler B, Wolff C, Zilch M (2014) Non-sexual abdominal appendages in adult insects challenge a 300 million years old bauplan. Curr Biol 24:R16–R17

Hughes DR, Nuttall AH, Katz RA, Carter GC (2009) Nonlinear acoustics in cicada mating calls enhance sound propagation. J Acoust Soc Am 125:958–967

Hungerford HB (1924) Stridulation of *Buenoa limnocastoris* Hungerford and systematic notes on the *Buenoa* of the Douglas Lake region of Michigan, with the description of a new form. Ann Entomol Soc Am 17:223–227

Hungerford HB (1929) A new *Velia* from Peru. Hemiptera Veliidae. Entomologisk Tidskrift 50:146–147

Hunt RE (1993) Role of vibrational signals in mating behavior of *Spissistilus festinus* (Homoptera: Membracidae). Ann Entomol Soc Am 86:356–361

ICS—International Commission on Stratigraphy (2013) International Chronostratigraphic Chart v2013/01. http://www.stratigraphy.org/ICSchart/ChronostratChart2013-01.pdf

Jacobi A (1907) Ein Schrillapparat bei Singcicaden. Zool Anz 32:67–71

Jansson A (1971) Stridulation and its significance in the waterbug genus *Cenocorixa*. PhD thesis, Univ British Columbia, Vancouver

Jansson A (1972) Mechanisms of sound production and morphology of the stridulatory apparatus in the genus *Cenocorixa* (Hemiptera, Corixidae). Ann Zool Fennici 9:120–129

Jordan KHC (1958a) Die Biologie von *Elasmucha grisea* L. (Heteroptera: Acanthosomidae). Beitr Entomol 8:385–397

Jordan KHC (1958b) Lautäußerungen bei den Hemipteren-Familien der Cydnidae, Pentatomidae und Acanthosomidae. Zool Anz 161:130–144

Josephson RK, Young D (1979) Body temperature and singing in the Bladder cicada, *Cystosoma saundersii*. J Exp Biol 80:69–81

Kanmiya K (2006) Mating behavior and vibratory signals in whiteflies (Hemiptera: Aleyrodidae). In: Drosopoulos S, Claridge MF (eds) Insect sounds and communication: physiology, behaviour, ecology and evolution. Taylor & Francis, New York, pp 365–379

Kanmiya K (1996a) Discovery of male acoustic sounds in the Greenhouse whitefly, *Trialeurodes vaporariourum* (Westwood) (Homoptera: Aleyrodidae). Appl Entomol Zool 31:255–262

Kanmiya K (1996b) Discovery of male acoustic sounds in whitefly genera, *Bemisia*, *Trialeurodes* and *Pealius* (Homoptera: Aleyrodidae). Proc XX Int Congr Entomol 1996:475

Kirkaldy GW (1906) Leaf-hoppers and their natural enemies. Report of work of the experimental station of the Hawaiian Sugar Planters' Association I:271–479

Kirkaldy GW (1907) Leaf-hoppers—supplement (Hemiptera). Report of work of the experimental station of the Hawaiian Sugar Planters' Association III:1–186

Koga T, Hayashi K (1991) Territorial behavior of both sexes in the water strider *Metrocoris histrio* (Hemiptera: Gerridae) during the mating season. J Insect Behav 6:65–77

Kukalová-Peck J (1991) Fossil history and the evolution of hexapod structures. In: Nauman ID, Carne PB, Lawrence JF, Nielsen ES, Spradbery JP, Taylor RW, Whitten MJ, Littlejohn MJ (eds) The insects of Australia: a textbook for students and research workers, 2nd edn. Cornell Univ Press, Ithaca

Kuštor V (1989) Aktivnost mišic pri vibracijskem organu stenice *Nezara viridula* (L.). [Activity of vibratory organ muscles in the bug *Nezara viridula* (L.)] MsD thesis, University of Ljubljana

Laboulbène A (1864) Sur l'organe musical de la *Chelonia pudica*. Ann Soc Ent Fr 4:689–704

Landois H (1867) Die Ton- und Stimmapparate der Insecten in anatomisch-physiologischer und akustischer Beziehung. Z Wiss Zool 17:104–184

Landois H (1872) Ueber ein dem sogenannten Tonapparat der Cikaden analoges Organ bei den hiesigen Gryllen. Z Wiss Zool 22:348354

Landois H (1874) Thierstimmen. Herder'sche Verlagsbuchhandlung, Freiburg i. Br

Lataste F (1895) Un procédé pour capturer les Cigales. Feuil Jeu Nat 25:157–158

Lattin JD (1958) A stridulatory mechanism in *Arhaphe cicindeloides* Walker (Hemiptera: Heteroptera: Pyrrhocoridae). Pan-Pac Entomol 34:217–219

Lazzari CR, Manrique G, Schilman PE (2006) Vibratory communication in Triatominae (Heteroptera). In: Drosopoulos S, Claridge MF (eds) Insect sounds and communication: physiology, behaviour, ecology and evolution. Taylor & Francis, New York, pp 297–304

Lepori C (1869) Nuove ricerche anatomiche e fisiologiche sopra l'organo sonoro delle cigale. Boll Soc Entomol Ital 1:221–235

Leston D (1954) Strigils and stridulation in Pentatomoidea (Hem.): Some new data and a review. Ent Month Mag 90:49–56

Leston D (1957) The stridulatory mechanisms in terrestrial species of Hemiptera Heteroptera. P Zool Soc Lond 128:369–386
Leston D, Pringle JWS (1963a) Acoustical behaviour of Hemiptera. In: Busnel R-G (ed) Acoustic behaviour of animals. Elsevier, Amsterdam, pp 391–411
Leston D, Pringle JWS (1963b) Addendum to Chapter 14 (Acoustical behaviour of Hemiptera). In: Busnel R-G (ed) Acoustic behaviour of animals. Elsevier, Amsterdam, pp 798–799
Lloyd Morgan C (1880) Note on the "Singerjie" (*Platypleura capensis*). Trans South Afr Phil Soc 1:161–164
Lloyd Morgan C (1886) On the sound-producing apparatus of the cicada. Nature 33:368–369
MacNally R, Young D (1980) Song energetics of the bladder cicada, *Cystosoma saundersii*. J Exp Biol 90:185–196
Malouf NSR (1932) The skeletal motor mechanism of the thorax of the "stink bug", *Nezara viridula* L. Bull Soc Roy Entomol D'Égypte 16:161–203
Mayer P (1877) Der Tonapparat der Zikaden. Z Wiss Zool (Leipzig) 28:79–92
Mebes H-D (1974) Zur Biophysik der Schallerzeugung bei Kleinzikaden—The biophysics of sound production of leafhoppers. Forma et Functio 7:95–118
Michelsen A, Nocke H (1974) Biophysical aspects of sound communication in insects. Adv Insect Physiol 10:247–296
Michelsen A, Fink F, Gogala M, Traue D (1982) Plants as transmission channels for insect vibrational songs. Behav Ecol Sociobiol 11:269–281
von Mitis H (1936) Zur Biologie der Corixiden. Stridulation. Z Morphol Ökol Tiere 30:479–495
Mitomi M, Ichikawa T, Okamoto H (1984) Morphology of the vibration-producing organ in adult rice brown planthopper, *Nilaparvata lugens* (Stål) (Homoptera: Delphacidae). Appl Ent Zool 19:407–417
Moore TE (1961) Audiospectrographic analysis of sounds of Hemiptera and Homoptera. Ann Entomol Soc Am 54:273–291
Moore TE (1962) Acoustical behavior of the cicada *Fidicina pronoe* (Walker) (Homoptera: Cicadidae). Ohio J Sci 62:113–119
Myers JG (1929) Insect singers. G Routledge & Sons, London
Nahirney PC, Forbes JG, Morris HD, Chock SC, Wang K (2006) What the buzz was all about: superfast song muscles rattle the tymbals of male periodical cicadas. FASEB J 20:2017–2026
Nel A, Roques P, Nel P, Prokin AA, Bourgoin T, Prokop J, Szwedo J, Azar D, Desutter-Grandcolas L, Wappler T, Garrouste R, Coty D, Huang D, Engel MS, Kirejtshuk AG (2013) The earliest known holometabolous insects. Nature 503:257–261
Numata H, Kon M, Fujii H, Hidaka T (1989) Sound production in the bean bug *Riptortus clavatus* Thunberg (Heteroptera: Alydidae). Appl Entomol Zool 24:169–173
Ossiannilsson F (1946) On the sound-production and the sound-producing organ in Swedish Homoptera Auchenorrhyncha (A Preliminary Note). Opusc Entomol XI:82–84
Ossiannilsson F (1948) On the sound-production of the females of certain auchenorrhynchous Homoptera. Ent Tidskr (Stockholm) 69:125–126
Ossiannilsson F (1949) Insect drummers. A study on the morphology and function of the sound-producing organ of Swedish Homoptera Auchenorrhyncha with notes on their sound-production. Opusc Entomol Suppl 10:1–145
Ossiannilsson F (1950) Sound-production in psyllids (Hem. Hom.). Opusc Entomol 15:202
Paoli G (1930) Caratteri diagnostici delle *Empoasca* e descrizione di nuove specie. Atti Soc Tosc Sci Nat Proc Verb 39:64–75
Paoli G (1933) Specie nuove di *Empoasca* (Hemiptera-omoptera), e appunti di corologia. Mem Soc Entomol Ital 11:109–122
Penzlin H (1991) Lehrbuch der Tierphysiologie, 5th edn. Gustav Fischer Verlag, Jena
Percy DM, Boyd EA, Hoddle MS (2008) Observations of acoustic signaling in three sharpshooters: *Homalodisca vitripennis*, *Homalodisca liturata*, and *Graphocephala atropunctata* (Hemiptera: Cicadellidae). Ann Entomol Soc Am 101:253–259
Poos FW (1933) Four new species of *Empoasca* (Homoptera: Cicadellidae). Proc Ent Soc Wash 35:174–179

Poda N (1761) Insecta Musei Græcensis, quæ in ordines, genera et species juxta systema naturæ Caroli Linnæi digessit. Græcii: Widmanstad

Popov AV, Beganović A, Gogala M (1997) Bioacoustics of singing cicadas of the Western Palaearctic: *Tettigetta brullei* (Fieber, 1876) (Cicadoidea: Tibicinidae). Acta Entomol Sloven 5:89–101

Pringle JWS (1953) Physiology of song in cicadas. Nature 172:248–249

Pringle JWS (1954) A physiological analysis of cicada song. J Exp Biol 31:525–560

Pringle JWS (1957) The structure and evolution of the organs of sound-production in cicadas. Proc Linn Soc London 167:144–159

Pringle JWS (1965) Locomotion: flight. In: Rockstein M (ed) The physiology of insects, vol II. Academic Press, New York, pp 283–329

Puranik PG, Ahmed A, Siddiqui MA (1981) The mechanism of sound production in the pentatomid bug *Tessaratoma javanica* Thunberg. Proc Ind Acad Sci 90:173–186

Ray J (1710) Historia insectorum. Opus posthumum Jussu Regiæ Societatis Londinensis Editum. Cui subjungitur appendix de scarabæis Britannicis, autore M. Lister S. R. S. ex Mss. Musæi Ashmolæani. A. & J. Churchill, Londini

Réaumur R-AF de (1740) Mémoires pour servir à l'Histoire des Insectes. Tom. V. Suite de l'Histoire des Mouches à deux aîles, et l'Histoire de plusieurs Mouches à quatre aîles, sçavoir, des Mouches à Scies, des Cigales, et des Abeilles. Imprimerie royale, Paris

Ribaut H (1936) Homoptères Auchénorhynques. I (Typhlocybidae). Faune de France, vol 31. P. Lechevalier, Paris

Ross HH (1965) A textbook of entomology, 3rd edn. Wiley, New York

Saxena KN, Kumar H (1984) Acoustic communication in the sexual behavior of the leafhopper *Amrasca devastans*. Physiol Entomol 9:77–86

Schaefer C (1980) The sound-producing structures of some primitive Pentatomoidea (Hemiptera: Heteroptera). J New York Entomol Soc 88:230–235

Schaefer CW, Pupedis RJ (1981) A stridulatory device in certain Alydinae (Hemiptera: Heteroptera: Alydidae. J Kans Entomol Soc 54:143–152

Schuh RT, Slater JA (1974) True bugs of the world (Hemiptera: Heteroptera). Cornell University Press, Ithaca and London

Senter P (2008) Voices of the past: a review of Paleozoic and Mesozoic animal sounds. Hist Biol 20:255–287

Shaw KC (1976) Sounds and associated behavior of *Agallia constricta* and *Agalliopsis novella* (Homoptera: Auchenorrhyncha: Cicadellidae). J Kans Entomol Soc 49:1–17

Shaw KC, Vargo A, Carlson OV (1974) Sounds and associated behaviour of some species of *Empoasca* (Homoptera: Cicadellidae). J Kans Entomol Soc 47:284–307

Simmons P, Young D (1978) The tymbal mechanism and song pattern of the Bladder cicada, *Cystosoma saundersii*. J Exp Biol 76:27–45

Skals N, Surlykke A (1999) Sound production by abdominal tymbal organs in two moth species: the green silver-line and the scarce silver-line (Noctuoidea: Nolidae: Chloephorinae). J Exp Biol 202:2937–2949

Smith JW, Georghiou GP (1972) Morphology of the tymbal organ of the beet leafhopper, *Circulifer tenellus*. Ann Entomol Soc Am 65:221–226

Smyth T, Smith JO III (2003) A musical controller inspired by the cicada's efficient buckling mechanism. J New Music Res 32:361–368

Song N, Liang A-P, Bu C-P (2012) A molecular phylogeny of Hemiptera inferred from mitochondrial genome sequences. PLoS One 7:e48778

Soulier-Perkins A, Sueur J, Hoch A (2007) Substrate-borne acoustic production: first record for an Australian Lophopid (Hemiptera, Lophopidae) and test of its historical use within Hemiptera. Aust J Entomol 46:129–132

Stokes DR, Josephson RK (2004) Power and control muscles of cicada song: structural and contractile heterogeneity. J Comp Physiol A 190:279–290

Stölting H, Moore TE, Lakes-Harlan R (2002) Substrate vibrations during acoustic signalling in the cicada *Okanagana rimosa*. J Insect Sci 2:1–7

Strauß J, Lakes-Harlan R (2014) Evolutionary and phylogenetic origins of tympanal hearing organs in insects. In: Hedwig B (ed) Insect hearing and acoustic communication. Springer, Berlin, pp 5–26

Strong FE, Sheldahl JA, Hughes PR, Hussein EMK (1970) Reproductive biology of *Lygus hesperus* Knight. Hilgardia 40:105–143

Strübing H (1958) Lautäusserungen—der entscheidende Faktor für das Zusammenfinden der Geschlechter bei Kleinzikaden (Homoptera, Auchenorrhyncha): Vorläufige Mitteilung. Zool Beitr NF 4: 15–21 [for English translation see Chap. 5, this volume]

Strübing H (1960) Paarungsverhalten und Lautäusserung von Kleinzikaden, demonstriert an Beispielen aus der Familie der Delphacidae (Homoptera, Auchenorrhyncha). Proc Int Congr Entomol Wien 11:12–14

Strübing H (1963) Lautäusserungen von *Euscelis*-Bastarden (Homoptera, Auchenorrhyncha). Verh Dtsch Zool Ges Stuttgart 57:268–281

Strübing H (1977) Lauterzeugung ode Substratvibration als Kommunikationsmittel bei Kleinzikaden? Diskutiert am Beispiel von *Dictyophara europaea* (Homoptera, Cicadina: Fulgoroidea). Zool Beitr NF 23:323–332

Strübing H (1999) Anmerkungen zu den Vibrationssignalen von *Stictocephala bisonia* KOPP et YONKE und *Dictyophara europaea* (L.) (Hemiptera: Auchenorrhyncha: Cicadomorpha et Fulgoromorpha: Membracidae et Dictyopharidae). Reichenbachia 33:63–69

Strübing H, Schwarz-Mittelstaedt G (1986) The vibratory membranes in the genus *Euscelis*. In: Proceedigngs of 2nd International Congress Rhynchota Balkan, Microlimni, pp 49–52

Štys P (1961) The stridulatory mechanism in *Centrocoris spiniger* (F.) and some other Coreidae (Heteroptera). Acta Entomol Mus Nat Prag 34:427–431

Štys P (1964) Thaumastellidae—a new family of pentatomoid Heteroptera. Acta Soc Entomol Cechoslov 61:238–253

Štys P (1966) Revision of the genus *Dayakiella* Horv. and notes on its systematic position (Heteroptera, Colobathristidae). Acta Bohemoslov 63:27–39

Sweet MH (1996) Comparative external morphology of the pregenital abdomen of the Hemiptera. In: Schaefer CW (ed) Studies on Hemipteran phylogeny. Thomas Say Publications in entomology: proceedings. Entomological Society of America, Lanham/MD, pp 119–158

Sueur J (2001) Audiospectrographical analysis of cicada sound production: a catalogue (Hemiptera, Cicadidae). Deut Entomol Z 48:33–51

Thorpe KV, Harrington BJ (1981) Sound production and courtship behavior in the seed bug *Ligyrocoris diffusus*. Ann Entomol Soc Am 74:369–373

Tishechkin DY (1997) Calling signals emitted by males of Cixiidae (Homoptera: Cicadinea) as compared with acoustic signals of some other Fulgoroidea (Homoptera, Cicadinea, Fulgoroidea). Entomol Rev 77:1070–1077

Tishechkin DY (1998) Acoustic signals of Issidae (Homoptera, Cicadinea, Fulgoroidea) compared with signals of some other Fulgoroidea and notes on taxonomic status of the subfamily Caliscelinae. Entomol Rev 78:884–892

Tishechkin DY (2000) Vibrational communication in Aphrodinae leafhoppers (Deltocephalinae auct., Homoptera: Cicadellidae) and related groups with notes on classification of higher taxa. Russ Entomol J 9:1–66

Tishechkin DY (2001) Vibrational communication in Cicadellinae sensu lato and Typhlocybinae leafhoppers (Homoptera: Cicadellidae) with notes on classification of higher taxa. Russ Entomol J 9:283–314

Tishechkin DY (2002) Review of the Species of the Genus *Macropsis* Lewis, 1834 (Homoptera, Cicadellidae, Macropsinae) from European Russia and Adjacent Territories. Russ Entomol J 11:123–184

Tishechkin DY (2003) Vibrational communication in Cercopoidea and Fulgoroidea (Homoptera: Cicadina) with notes on classification of higher taxa. Russ Entomol J 12:129–181

Tishechkin DY (2005) The description of vibrational signals of *Anoterostemma ivanoffi* Homoptera, Cicadellidae with notes on taxonomic position of the tribe Anoterostemmatini. Entomol Rev 85:1039–1044

Tishechkin DY (2006) Vibratory communication in Psylloidea (Hemiptera). In: Drosopoulos S, Claridge MF (eds) Insect sounds and communication: physiology, behaviour, ecology and evolution. Taylor & Francis, New York, pp 357–363

Tishechkin DY (2008) On the similarity of temporal pattern of vibrational calling signals in different species of Fulgoroidea (Homoptera: Auchenorrhyncha). Russ Entomol J 17:349–357

Tishechkin DY (2011) Calling signals in sympatric species of the far-eastern *Aphrophora* (Homoptera: Auchenorrhnyncha: Aphrophoridae): regularities of communication channel segregation. Russ Entomol J 20:31–38

Todd JW (1989) Ecology and behavior of *Nezara viridula*. Ann Rev Entomol 34:273–292

de la Torre-Bueno JR (1903) A day's collecting in February. Can Entomol 35:123–125

Traue D (1978) Zur Biophysik der Schallabstrahlung bei Kleinzikaden am Beispiel von *Euscelis incisus* Kb. (Homoptera-Cicadina: Jassidae). Zool Beitr NF 24:155–164

Trilar T, Gogala M (2007) The song structure of *Cicadetta podolica* (Eichwald, 1830) (Hemiptera: Cicadidae). Acta Entomol Sloven 15:5–20

Usinger RL (1954) A new genus of Aradidae from the Belgian Congo, with notes on stridulatory mechanisms in the family. Ann Mus Congo Tervuren (Zool) 1:540–543

Usinger RL, Matsuda R (1959) Classification of the Aradidae (Hemiptera, Heteroptera). British Museum (N.H.), London

Varley GC (1939) Unusual methods of stridulation in a cicada (*Clidophleps distanti* (Van D.)) and a grasshopper (*Oedaleonotus fuscipes* Scud.) in California. Proc Roy Entomol Soc Lond A 14:97–100

Virant-Doberlet M, Žežlina I (2007) Vibrational communication of *Metcalfa pruinosa* (Hemiptera: Fulgoroidea: Flatidae). Ann Entomol Soc Am 100:73–82

Vogel R (1921) Bericht über ein Gehörorgan bei Singzikaden. Naturwissenschaften 9:427–431

Vogel R (1923) Über ein tympanales Sinnesorgan, das mutmassliche Hörorgan der Singzikaden. Z Anat Entwickl-gesch, Abt 1(67):190–231

Vondráček K (1949) Příspěvek k poznání zvukového ústrojí u samcu křísu / Contribution to the knowledge of the sound-producing apparatus in the males of the leafhoppers (Homoptera-Auchenorrhyncha). Acta Acad Sci Nat Moravo-siles (Brno) 21(8):1–36

von Dohlen CD, Moran NA (1995) Molecular phylogeny of the Homoptera: a paraphyletic taxon. J Mol Evol 41:211–223

Voskresenskaya AK, Svidersky VL (1960) The role of the central and sympathetic nervous system in the function of the tymbal muscles of cicadas. J Insect Physiol 6:26–35

Wakabayashi T, Hagiwara S (1953) Mechanical and electrical events in the main sound muscle of cicada. Jpn J Physiol 3:249–253

Weber H (1928) Skelett, Muskulatur und Darm der schwarzen Blattlaus *Aphis fabae* Scop., mit besonderer Berücksichtigung der Funktion der Mundwerkzeuge und des Darms. Zoologica (Stuttgart) 76:1–120

Wessel A (2013) Casserius and the secret of the cicada's voice. Deut Entomol Z 60:139–150

Wessel A, Hoch H, Asche M, von Rintelen T, Stelbrink B, Heck V, Stone FD, Howarth FG (2013) Rapid species radiation initiated by founder effects in Hawaiian cave planthoppers. Proc Nat Acad Sci USA 110:9391–9396

Wheeler NH (1937) Trap-light studies on leafhoppers belonging to the genus *Empoasca* (Homoptera: Cicadellidae), with the description of two new species. Proc Entomol Soc Wash 39:141–156

Wheeler NH (1939) The North American empoascan leafhoppers of the *radiata* group (Homoptera: Cicadellidae). J Wash Acad Sci 29:292–300

Wilcox RS (1972) Communication by surface waves. Mating behavior of a water strider (Gerridae). J Comp Physiol 80:255–266

Wilcox RS (1975) Sound-producing mechanisms of *Buenoamacrotibialis* Hungerford (Hemiptera: Notonectidae). Int J Insect Morphol Embryol 4:169–182

Wilcox RS (1995) Ripple communication in aquatic and semiaquatic insects. Ecoscience 2:109–115

Wohlers D, Bacon J (1980) Sexual dimorphism of motoneurons: Timbal muscle innervation in male periodical cicadas and homologous structures in females. Cell Tissue Res 209:371–382

Wohlers DW, Williams JLD, Huber F, Moore TE (1979) Central projections of fibers in the auditory and tensor nerves of cicadas (Homoptera: Cicadidae). Cell Tissue Res 203:35–51

Young D (1972a) Neuromuscular mechanism of sound production in Australian cicadas. J Comp Physiol 79:343–362

Young D (1972b) Analysis of songs of some Australian cicadas (Homoptera: Cicadidae). J Aus Entomol Soc 11:237–243

Young D, Hill KG (1977) Structure and function of the auditory system of the cicada, *Cystosoma saundersii*. J Comp Physiol 117:23–45

Young D, Josephson RK (1983) Pure-tone songs in cicadas with special reference to the genus *Magicicada*. J Comp Physiol 152:197–207

Zrzavý J (1990) Evolution of Hemiptera: an attempt at synthetic approach. Proc Int Symp Scale Insect Stud VI (Kraków) II:19–22

Zych AF, Mankin RW, Gillooly JF, Foreman E (2012) Stridulation by *Jadera haematoloma* (Hemiptera: Rhopalidae): Production mechanism and associated behaviors. Ann Entomol Soc Am 105:118–127

# Alphabetical Taxa Index




R. B. Cocroft et al. (eds.), *Studying Vibrational Communication*, Animal Signals and Communication 3, DOI: 10.1007/978-3-662-43607-3,

# Systematic Family and Species Index




R. B. Cocroft et al. (eds.), *Studying Vibrational Communication*, Animal Signals and Communication 3, DOI: 10.1007/978-3-662-43607-3,